High Performance Computing in Science and Engineering '18

Wolfgang E. Nagel · Dietmar H. Kröner ·
Michael M. Resch
Editors

High Performance Computing in Science and Engineering '18

Transactions of the High Performance
Computing Center, Stuttgart (HLRS) 2018

 Springer

Editors
Wolfgang E. Nagel
Zentrum für Informationsdienste und
Hochleistungsrechnen (ZIH)
Technische Universität Dresden
Dresden, Germany

Dietmar H. Kröner
Abteilung für Angewandte Mathematik
Universität Freiburg
Freiburg, Germany

Michael M. Resch
Höchstleistungsrechenzentrum Stuttgart
(HLRS)
Universität Stuttgart
Stuttgart, Germany

Front cover figure: Stress distribution in a section of a dual-phase steel after quenching predicted by a multiphase-field simulation. The martensitic transformation leads both to high stresses in the former austenitic islands and to a considerable stress input into the surrounding ferritic matrix. Details can be found in "Multiphase-Field Modeling and Simulation of Martensitic Phase Transformation in Heterogeneous Materials" by E. Schoof, C. Herrmann, D. Schneider, J. Hötzer, and B. Nestler, Institute of Applied Materials, Karlsruhe Institute of Technology (KIT), Germany and Institute of Digital Materials Science, Karlsruhe University of Applied Sciences, Germany, on page 475 ff.

ISBN 978-3-030-13324-5   ISBN 978-3-030-13325-2   (eBook)
https://doi.org/10.1007/978-3-030-13325-2

Library of Congress Control Number: 2019932797

Mathematics Subject Classification (2010): 65Cxx, 65C99, 68U20, 90-08, 97M50

This Springer imprint is published by the registered company Springer Nature Switzerland AG
The registered company address is: Gewerbestrasse 11, 6330 Cham, Switzerland

Contents

Part II Molecules, Interfaces, and Solids

Part III Reactive Flows

Part IV Computational Fluid Dynamics

Part I
Physics

In this section, six physics projects are described, which achieved important scientific results in 2017/18 by using Hazel Hen at the HLRS and ForHLR II of the Steinbuch Center.

Fascinating new results are being presented in the following pages on astrophysical systems (simulations of galaxy formation, neutron star and binary merger evolutions, and of photoionization and radiative charge transfer), soft matter systems (interaction between blood proteins and nanoparticles), many body quantum systems (simulations of trapped ultracold quantum systems) and high energy physics systems (simulations of the QCD phase diagram on the lattice).

The studies of the astrophysical systems have focused on galaxy formation, neutron star and binary merger evolutions, and on photoionization and radiative charge transfer of several particles.

D. Nelson, A. Pillepich, V. Springel, R. Pakmor, L. Hernquist, R. Weinberger, S. Genel, M. Vogelsberger, F. Marinacci, P. Torrey, J. Naiman from Garching (D.N., V.S.,), Heidelberg (A.P., V.S., R.P., R.W.), Cambridge USA (L.H., J.N., M.V., F.M., P.T.), and New York (S.G.), present in their project *GCS-DWAR* results from the TNG50 run, the third and final volume of the cosmological, magnetohydrodynamical simulations in the IllustrisTNG project (AREPO code), with over 20 billion resolution elements and a spatial resolution of 100 parsecs. The authors conclude that the results of the TNG50 run has been successful for the analysis of many observational data, e.g., the star formation histories and abundances of dwarf galaxies in the first few billion years, the detailed mapping of our Galaxy's stellar system, the imaging and spectroscopic characterisation of the galaxy populations residing in nearby galaxy clusters like Virgo, Fornax, and Coma, the X-ray maps of the outskirts of galaxy clusters, integral-field spectroscopy campaigns of nearby galaxies, the characterisation of circumgalactic and IGM gas, and magnetic field measurements in halos and in the disks of galaxies.

L. Rezolla, F. Guercilena, and E. Most from the University of Frankfurt investigated in their project *HypeBBHs* interesting phenomena of neutron stars and black holes, as well of mergers, by numerically solving the Einstein equations for space-time evolution and the equations of hydrodynamics and magnetohydrodynamics, using a variety of codes, e.g. the open-source McLachlan code, self-developed codes (WhiskeyTHC, WhiskeyMHD, WhiskeyResitive), and the FIL code, an extension of the open-source code IllinoisGRMHD. The focus of the project is on the general behaviour of black hole systems in vacuum, the study of rigidly and differentially rotating isolated neutron stars, the evolution of neutron stars and accretion disks systems, the evolution and properties of magnetized neutron stars in the contexts of both ideal and resistive magnetohydrodynamics, and on the study of ejecta in binary neutron star systems and the resulting implications for nucleosynthesis. A new approach to relativistic hydrodynamics has been developed, the accuracy of numerical waveforms from binary mergers has been investigated, as well as the nucleosynthesis in the ejecta of binary neutron star mergers.

B. M. McLaughlin, C. P. Ballance, M. S. Pindzola, P. C. Stancil, J.F. Babb, S. Schippers and A. Müller from the Universities of Belfast (B.M.M., C.P.B.), Auburn (M.S.P.), Georgia (P.C.S.), Cambridge USA (J.F.B.), and Giessen (S.S., A.M.) investigated in their project *PAMOP2* atomic, molecular and optical collisions on petaflop machines, relevant for astrophysical applications, for magnetically-confined fusion and plasma modeling, and as theoretical support for experiments and satellite observations. The Schrödinger and Dirac equations have been solved with the R-matrix or R-matrix with pseudo-states approach, and the time dependent close-coupling method has been used. Various systems and phenomena have been investigated, in detail the photoionisation of calcium ions and of zinc ions, the electron impact excitation of molybdenum, the electron impact double ionization of the H_2 molecule, the X-ray emission in collisions of O^{8+} ions with H and He atoms, the radiative charge transfer of He^+ ions with Ar atoms, and the radiative association of Si and O atoms and of C and S atoms.

Studies of the soft matter systems have focused on the interaction of blood proteins with nanoparticles.

T. Schäfer, C. Muhl, M. Barz, F. Schmid and G. Settanni from the University of Mainz investigated in their project *Flexadfg* interaction of blood proteins with nanoparticles by Molecular Dynamics simulations, using the program NAMD and CHARMM force fields, a cell-list algorithm, and a smooth particle mesh Ewald (PME) method, and in selected cases a "Hamiltonian replica exchange with solute tempering" (REST2) sampling technique. One of the project goals was to identify the features that make a polymer amenable for coating nanoparticles for drug delivery, with particular focus on salt dependent behaviour of poly(ethylene glycol) (PEG). The simulations helped to explain how PEG chains shrink in the presence of potassium ions and they are being used to reveal how PEG and PSAR show similar stealth effects. In addition, interactions of blood proteins with silica has been explored.

In the last granting period, quantum mechanical properties of high energy physics systems have been investigated as well as the quantum many body dynamics of trapped bosonic systems.

S. Borsanyi, Z. Fodor, J. Günther, S. D. Katz, A. Pasztor, I. P. Vazquez, C. Ratti, and K. K. Szabó from Wuppertal (S.B., Z.F., A.P., K.K.S.), Budapest (Z.F., S.D.K., J.G., S.D.K.), the FZ Jülich (Z.F., K.K.S.), Regensburg (J.G., S.D.K.) and Houston (I.P.V., C.R.) studied in their project *GCS-POSR* the QCD phase diagram from the lattice. In the project, several diagonal and non-diagonal fluctuations of conserved charges have been computed in a system of 2+1+1 quark flavors with physical masses, on a lattice with size $48^3 \times 12$. From high statistics lattice QCD simulations several μ_B-derivatives of observables have been computed, which account for the confinement-deconfinement transition between the hadron gas and quark gluon plasma phases. Higher order fluctuations at $\mu_B = 0$ were obtained as derivatives of the lower order ones. From these correlations and fluctuations ratios of net-baryon number cumulants have been constructed as functions of temperature and chemical potential, satisfying important experimental conditions. In addition, in the project fluctuations were computed from a simple model, assuming the absence of a nearby critical end point in the phase diagram, and good agreement with the simulation results was found. Based on their findings, the authors conclude that the experimental finding of a critical end-point below $\mu_B / T < 2$ is unlikely.

O.E. Alon, R. Beinke, C. Bruder, L.S. Cederbaum, S. Klaiman, A.U.J. Lode, K. Sakman, M. Theisen, M.C. Tsatsos, S.E. Weiner, and A.I. Streltsov from the Universities of Haifa (O.E.A.), Heidelberg (R.B., L.S.C., S.K., M.T., A.I.S.), Basel (C.B., A.U.J.L.), Wien (K.S.), Sao Paulo (M.C.T.), and Berkeley (A.S.) studied in their project *MCTDHB* trapped ultracold atomic systems by their method termed multi-configurational time-dependent Hartree for bosons (MCTDHB and MCTDH-X software packages). The principal investigators investigated entropies and correlations of ultracold bosons in a lattice, crystallization of bosons with dipole-dipole interactions and its detection in single-shot images, studied the management of correlations in ultracold gases, explored the pulverising of a Bose-Einstein-Condensate (BEC), the dynamical pulsation of ultracold droplet crystals by laser light, two-component bosons interacting with photons in a cavity, analysed the impact of the range of the interaction on the quantum dynamics of a bosonic Josephson junction, computed the exact many-body wavefunction and properties of trapped bosons in the infinite-particle limit, analyzed the angular-momentum conservation in a BEC, investigated the variance of an anisotropic BEC and enhanced many-body effects in the excitation spectrum of a weakly-interacting rotating BEC.

Peter Nielaba
Fachbereich Physik, Universität Konstanz
78457 Konstanz, Germany
e-mail: peter.nielaba@uni-konstanz.de

The TNG50 Simulation of the IllustrisTNG Project: Bridging the Gap Between Large Cosmological Volumes and Resolved Galaxies

Dylan Nelson, Annalisa Pillepich, Volker Springel, Rüdiger Pakmor, Lars Hernquist, Rainer Weinberger, Shy Genel, Mark Vogelsberger, Federico Marinacci , Paul Torrey and Jill Naiman

Abstract Cosmological hydrodynamical simulations of galaxy formation are a powerful theoretical tool, and enable us to directly calculate the observable signatures resulting from the complex process of cosmic structure formation. Here we present early results from the ongoing TNG50 run, an unprecedented 'next generation' cosmological, magnetohydrodynamical simulation—the third and final volume of the IllustrisTNG project, with over 20 billion resolution elements and capturing spatial scales of ~100 parsecs. It incorporates the comprehensive TNG model for galaxy formation physics, and we here describe the simulation scope, novel achievements, and early investigations on resolved galactic and halo structural properties.

D. Nelson (✉) · V. Springel
Max-Planck-Institut für Astrophysik, Karl-Schwarzschild-Str. 1, 85741 Garching, Germany
e-mail: dnelson@mpa-garching.mpg.de

V. Springel
e-mail: volker.springel@h-its.org

A. Pillepich
Max-Planck-Institut für Astronomie, Königstuhl 17, 69117 Heidelberg, Germany
e-mail: pillepich@mpia-hd.mpg.de

V. Springel · R. Pakmor · R. Weinberger
Heidelberg Institute for Theoretical Studies, Schloss-Wolfsbrunnenweg 35,
69118 Heidelberg, Germany
e-mail: ruediger.pakmor@h-its.org

R. Weinberger
e-mail: rainer.weinberger@h-its.org

V. Springel
Zentrum für Astronomie der Universität Heidelberg, ARI, Mönchhofstr. 12,
69120 Heidelberg, Germany

© Springer Nature Switzerland AG 2019
W. E. Nagel et al. (eds.), *High Performance Computing in Science and Engineering '18*, https://doi.org/10.1007/978-3-030-13325-2_1

1 Introduction

Observed galaxies today range in mass from a few thousand to a few trillion times the mass of the sun, range in physical size from a fraction to tens of kilo-parsecs, and span a wide variety of morphologies, from ultra-diffuse extended disks to ultra-compact spheroidal nuggets. Galaxies can exist in isolation, or as members of rich clusters—one out of thousands. Throughout space their distribution traces a filamentary cosmic web of matter which has arisen from 13.8 billion years of cosmic evolution, starting from the nearly homogeneous distribution of matter in the early universe. According to the current paradigm this large-scale structure is driven by the dominant presence of cold dark matter (CDM), which results in a hierarchical growth governing the formation and transformation of galaxies.

The mathematical equations governing the dominant physical processes are non-linear, complex, and highly coupled. As a result, computational methods are an absolute requirement for their solution, although the inherently multi-scale, multi-physics nature of the problem poses significant challenges. Numerical simulations of the formation and evolution of galaxies in the full cosmological context of our evolving universe have reached impressive levels of sophistication, achieving along the way a number of notable successes as well as failures. Early, pioneering beginnings with dark-matter only simulations over the past thirty years [1–3] paved the way for keystone projects of the past decade [4, 5]. These DM-only simulations do not, however, provide direct predictions for the *observable* properties of galaxies, and additional modeling is required to make any firm connection to observational data. To do so, it has recently become possible to simultaneously model, in addition to gravity, the hydrodynamics of cosmic gas (i.e., CFD of relatively diffuse plasmas) together with the principal baryonic physics of relevance. More accurate fluid codes are coupled to increasingly sophisticated physical models, which are then used to

L. Hernquist · J. Naiman
Harvard-Smithsonian Center for Astrophysics, 60 Garden Street, Cambridge, MA 02138, USA
e-mail: lars@cfa.harvard.edu

S. Genel
Center for Computational Astrophysics, 162 Fifth Avenue, Flatiron, New York 10010, USA
e-mail: shygenelastro@gmail.com

M. Vogelsberger · F. Marinacci · P. Torrey
Department of Physics, MIT Kavli Institute for Astrophysics and Space Research, Massachusetts Institute of Technology, Cambridge, MA 02139, USA
e-mail: mvogelsb@mit.edu

F. Marinacci
e-mail: fmarinac@mit.edu

execute simulations of ever increasing scope. Theoretically, these calculations yield a self-consistent and uniquely predictive realization of a given cosmological model.

Large volumes allow statistically robust comparisons of entire populations of galaxies—we no longer are relegated to simulating single objects. Cosmological hydrodynamical simulations have emerged as powerfully predictive theoretical models, and this recent development has been made clear with the Illustris [6–9] and EAGLE [10, 11] simulations. Together with other cosmological simulations [12, 13] these projects have finally succeeding in demonstrating that hydrodynamical simulations of structure formation, run at kilo-parsec spatial resolutions, can reasonably reproduce a number of fundamental scaling relations and properties of observed galaxy populations. This zeroth order agreement has buttressed many theoretical investigations and predictions. At the same time, however, it has revealed many telling shortcomings in the current generation of models.

IllustrisTNG is a 'next generation' series of large, cosmological, gravo-magneto-hydrodynamical simulations incorporating a comprehensive model for galaxy formation physics. Conducted over the past two years on the Hazel Hen machine at the High Performance Computing Center Stuttgart (HLRS) and supported by two Gauss Centre for Supercomputing allocations (GCS-ILLU in 2014, and GCS-DWAR in 2016) it includes three main runs: TNG50, TNG100, and TNG300. The latter two have been completed and recently presented [14–18]. TNG50 is the most computational demanding of the three simulations by far: with an anticipated cost of ∼100 M core hours the simulation is still ongoing, and we present herein early results from this project.

2 IllustrisTNG: Physical and Numerical Developments

The IllustrisTNG simulation project[1] expands upon the original Illustris simulation in two principal aspects. First, it alleviates the major problems—that is, model deficiencies [19]—with respect to strongly constraining low redshift observations. Second, it significantly expands upon the original scope, with simulations of larger volumes, at higher resolution, and with new physics.

2.1 The TNG Galaxy Formation Model

In terms of physical modeling, we retain the same basic approach as in Illustris, acknowledging that physics below a given spatial scale, of order a hundred to a few hundred parsecs, cannot be resolved and must be treated by approximate, sub-resolution models. This includes, most importantly, the process of star formation, the detailed action of individual supernova events, the formation and growth of

---

[1]http://www.tng-project.org.

supermassive blackholes, and the near-field coupling of blackhole feedback energy
to the surroundings. The updated TNG model for galaxy formation is described in
[20, 21], and we employ it unchanged in all three flagship TNG simulations.[2]

The physical framework includes models of the most important physical processes
for the formation and evolution of galaxies; (i) gas radiative microphysics, including
primordial (H/He) and metal-line cooling and heating with an evolving ultraviolet/
x-ray background field, (ii) star formation in dense interstellar medium (ISM) gas, (iii)
the evolution of stellar populations and chemical enrichment, tracking supernovae
Ia, II, and AGB stars, and individual species: H, He, C, N, O, Ne, Mg, Si, and Fe,
(iv) galactic-scale outflows launched by stellar feedback, (v) the formation, binary
mergers, and gas accretion by supermassive blackholes, (vi) blackhole feedback,
operating in a thermal mode at high accretion rates and a kinetic 'wind' mode at low
accretion rates. Aspects (iv) and (vi) have been substantially revised in TNG, and we
describe them each in more detail.

2.1.1 Galactic Winds

First, galactic-scale outflows generated by stellar feedback are modeled using a
kinetic wind approach [22], where the energy available from Type II (core-collapse)
supernovae is used to stochastically kick star-forming gas cells away from galaxies. Winds particles are hydrodynamically decoupled from surrounding gas until
they exit the dense, star-forming environment, which avoids small-scale interactions
which cannot be modeled at the available resolution. The velocity of this wind-phase
gas at injection is $v_w \propto \sigma_{DM}$ where σ_{DM} is the local one-dimensional dark matter
velocity dispersion, with a minimum launch speed of $v_{w,min} = 350$ km/s enforced.
The mass loading of the winds, $\eta = \dot{M}_w / \dot{M}_{SFR}$ is then given by $\eta = 2(1 - \tau_w)e_w/v_w^2$
where $\tau_w = 0.1$ is a fraction of energy chosen to be thermal, and e_w is a metallicity
dependent factor which modifies the canonical 10^{51} erg available per event. As a
result, the total energy available for these outflows scales with the star formation rate
as $\Delta E = e_w \dot{M}_{SFR} \simeq 10^{41} - 10^{42}$ erg/s $\dot{M}_{SFR}/(M_\odot/\text{yr})$, depending on the local gas
metallicity [full details in 21].

2.1.2 Black Hole Feedback

The second major change in TNG is blackhole feedback. Blackholes are formed
in DM halos above a mass threshold, and are then able to accrete nearby gas at
the Bondi rate, limited to the Eddington accretion rate. This accretion rate then
determines their feedback mode. If the Eddington ratio exceeds a value of $\chi =$
$\min[0.002(M_{BH}/10^8 M_\odot)^2, 0.1]$ we inject thermal energy (continuously) into the

---

[2]We also keep fixed all the relevant physical and numerical parameters when varying the numerical
resolution, an approach not always adopted in computational galaxy formation. This allows us to
establish a well-posed numerical problem towards which we can demonstrate convergence.

surrounding gas. The energy injection rate is $\Delta E_{\text{high}} = \varepsilon_{\text{f,high}} \varepsilon_{\text{r}} \dot{M}_{\text{BH}} c^2$ where $\varepsilon_{\text{f,high}}$ is a coupling efficiency parameter in this 'high accretion state', and ε_{r} is the usual radiative accretion efficiency. Together, $\varepsilon_{\text{f,high}} \varepsilon_{\text{r}} = 0.02$, i.e. the net efficiency in this mode is about one percent. Below this accretion rate threshold, we instead inject kinetic energy (time-pulsed), in the form of an oriented, high-velocity wind which has a direction which changes between each event. Here the energy injection rate is $\Delta E_{\text{low}} = \varepsilon_{\text{f,low}} \dot{M}_{\text{BH}} c^2$ with $\varepsilon_{\text{f,low}} \leq 0.2$, the efficiency modulated as a function of environmental density [full details in 20].

This new, kinetic wind feedback from blackholes successfully terminates star formation in massive galaxies, transforming them from 'blue' disk-like systems similar to our own Milky Way, into 'red and dead' elliptical or spheroidal quenched systems [16, 23]. This bimodality of observed galaxy properties, which is seen in their colors, star forming rates, and structural properties for example, is a critical challenge for galaxy formation models [24].

2.1.3 Model Confirmations Across Observables

Other encouraging investigations and positive comparisons of TNG galaxy properties with observational data include, for example, their stellar contents, spatial clustering, and magnetic properties [14, 15, 17], as well as galaxy sizes, metal contents, ionized oxygen abundances, blackhole properties, and dark matter contents [23, 25–29].[3] These successes of the first two TNG simulations (TNG100 and TNG300), motivated and supported an even more ambitious project.

2.2 New Physics and Numerical Improvements

The IllustrisTNG simulation suite is founded upon the AREPO code [30]. AREPO solves the coupled equations of self-gravity and ideal, continuum magnetohydrodynamics [MHD; 31, 32], the former computed with the Tree-PM approach and the latter employing a Godunov/finite-volume method with a spatial discretization based on an unstructured, moving, Voronoi tessellation. The scheme is quasi-Lagrangian, second order in both space and time, and uses individual particle time-stepping.

The principle new physics that sets apart IllustrisTNG from previous numerical campaigns is the addition of magnetic fields. We have successfully implemented a method for the solution of the ideal MHD equations, using an 8-wave Powell cleaning scheme to maintain the zero divergence constraint [31, 32]. We have demonstrated the fidelity of this approach on both idealized test cases as well as galaxy simulations [33]. Since its original implementation, we have also made improvements to the MHD solver, through an additional timestep criterion which limits the magnitude of the source term applied for divergence cleaning.

---

[3] See http://www.tng-project.org/results/ for an ongoing publication list.

The AREPO code has been designed to efficiently execute large, parallel astrophysical simulations—the TNG50 simulation reviewed here is run on 16320 cores. At these scales there are several challenges, particularly given the highly coupled, high dynamic range of the galaxy formation problem. The TNG50 simulation itself captures a spatial dynamic range of $\sim 10^7$; the time hierarchy similarly requires evolution on timescales which differ by ten thousand or more. The previous MUSCL-Hancock time integration scheme has been replaced with a approach following Heun's method, a second-order Runge-Kutta variant which exhibits significantly better convergence properties [34]. The method for obtaining linear gradients of primitive fluid quantities has also been replaced with an iterative least-squares method, improving the convergence properties of the code as well as the conservation of angular momentum.

Finally, the long-range gravity FFT calculation now uses a new, column-based MPI-parallel FFT which improves scaling at high core numbers. Furthermore, the gravity calculation has been rewritten using a decomposition based on a recursive splitting of the N-body Hamiltonian into short- and long- timescale particle systems, such that the forces required to integrate the rapidly evolving particles can be computed with improved parallel efficiency, and in a mathematically robust way without any untreated coupling terms.

2.3 Project Scope

TNG50 is the third and final volume of the IllustrisTNG project. This simulation occupies a unique combination of large volume and high resolution—its details and numerical properties are given in Table 1 in comparison to the TNG series as a whole, while Fig. 1 gives a visual comparison of the volumes. Our 50 Mpc box is sampled by 2160^3 dark matter particles (with masses of $4 \times 10^5\,M_\odot$) and 2160^3 initial gas cells (with masses of $8 \times 10^4\,M_\odot$). The total number of resolution elements is therefore slightly over 20 billion. The *average* spatial resolution of star-forming interstellar medium (ISM) gas is ~ 90 (~ 140) parsecs at $z = 1$ ($z = 6$). TNG50 has 2.5 times better spatial resolution, and 15 times better mass resolution, than TNG100 (or equivalently, original Illustris). This resolution approaches or exceeds that of modern 'zoom' simulations of individual galaxies [35, 36], while the volume contains $\sim 20,000$ resolved galaxies with $M_\star > 10^7\,M_\odot$ (at $z = 1$).

Table 1 Details of the TNG50 simulation in comparison to its two larger volume counterparts

Run name	Volume	N_{GAS}	N_{DM}	N_{TR}	m_{baryon}	m_{DM}	$\varepsilon_{\mathrm{gas,min}}$	$\varepsilon_{\mathrm{DM,stars}}$
	(Mpc^3)	–	–	–	(M$_\odot$)	(M$_\odot$)	(pc)	(pc)
TNG50	**51.7^3**	**2160^3**	**2160^3**	**2160^3**	**8.5 × 10^4**	**4.5 × 10^5**	**74**	**288**
TNG100	110.7^3	1820^3	1820^3	2 × 1820^3	1.4 × 10^6	7.5 × 10^6	185	740
TNG300	302.6^3	2500^3	2500^3	2500^3	1.1 × 10^7	5.9 × 10^7	370	1480

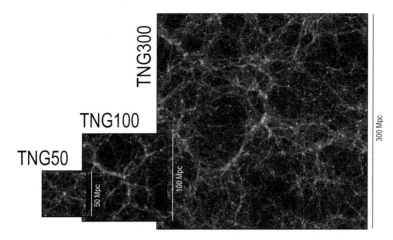

Fig. 1 Comparison of the TNG50 simulation to its larger volume counterparts. Due to the significantly higher resolution, the TNG50 run is by far the most computationally challenging of the three, and offers a unique configuration for a cosmological simulation

At the time of writing, the TNG50 simulation has been evolved from the initial conditions of the Universe to $z \sim 0.6$ (8 billion years after), and we are currently completing the run to reach the current epoch of cosmic evolution ($z = 0$).

TNG50 contains roughly 100 Milky Way mass-analogs, enabling detailed comparisons to our own galaxy at $z = 0$. It also hosts one massive galaxy cluster with a total mass $\sim 10^{14} \, M_\odot$, a Virgo-like analog, and dozens of group sized halos at $\sim 10^{13} \, M_\odot$. All of these massive objects are simulated at higher numerical resolution than in any previously published study, enabling studies not only of the gaseous halos and central galaxies, but also of the large populations of their satellite galaxies.

3   The TNG50 Simulation: Early Results

We proceed to show selected early results from TNG50, focusing on the internal structural properties of the gaseous components of galaxies and their halos, topics which are explored in a number of papers currently in preparation. We can now analyze a fully representative, simulated galaxy population spanning $10^7 < M_\star/M_\odot < 10^{11.5}$ across intermediate to high redshifts, $1 < z < 8$. The high resolution of TNG50 is specifically exploited to investigate scales, regimes, and scientific questions not addressable using the other two TNG simulations.

This coverage in redshift range and galaxy stellar mass enables us to make quantitative predictions for signatures observable with the James Webb Space Telescope (JWST), now anticipated to launch in 2019, as well as recent ground-based IFU instruments such as MUSE, SINFONI, and KCWI. The key science drivers of TNG50 focus not only on the present day ($z = 0$), but also at earlier epochs, from cosmic noon ($z \sim 2$) through reionization ($z \sim 6$).

3.1 Resolving Galactic Structure

The internal structures of the progenitors of present-day galaxies are observed at $z \sim 2$ from ground based telescopes using adaptive-optics techniques, which enable kiloparsec scale structure by tracing bright emission lines from hydrogen (such as Hα) or metals (i.e. OIII), as in [37]. Using TNG50 we can provide model predictions for projected radial profiles and resolved 2D maps of gas density, star formation rate (Hα), gas-phase metallicity, and gas line-of-sight velocity dispersion.

In Fig. 2 we show nine examples of star-formation surface density of $z = 1$ massive galaxies. Quantitative measurements of the morphology of the star-forming gas, as will be traceable at high-redshift through rest-frame UV and optical nebular emission lines (e.g. MgII, FeII, OII, and OIII), discriminate between galaxy evolutionary stages. In general, star formation peaks towards the center of galaxies as it follows the local gas density according to the Schmidt-Kennicutt relation. Small scale structure at sub-kiloparsec scale is easily resolved by TNG50, revealing rich morphological features including flocculent spiral arms of extended, low-density gas as well as clumpy sub-structures. We see directly how outflows generated from the nuclear regions of disks leave signatures in the gas left behind, evidenced in the central depressions, which result from the direct expulsion of gas due to the action of the kinetic supermassive blackhole feedback.

Despite their strong ordered rotation, galactic disks at $z = 1$ are highly turbulent gaseous reservoirs. These settle with time: the peak velocity dispersion at $M_\star = 10^{10.5} \, M_\odot$ drops from >400 km/s at $z = 6$ to ~100 km/s at $z = 1$. These trends, as a function of galaxy mass and cosmic time, can be directly tested against upcoming observational programs, providing a strong constraint on our galaxy formation model. At the same time, by constructing synthetic observations of, for example, the Hα emission from hydrogen atoms excited by nearby young stars, we can assess observational biases and aid in the robust interpretation of observational results which do not have access to the intrinsic properties of the galaxies.

3.2 Heavy Metal Distribution and Metallicity Gradients

Metallicity gradients—that is, radial variations in the enrichment of galaxies by heavy elements—have been observed in local $z = 0$ systems as part of the SDSS MANGA survey [38], but not yet at higher redshifts. Within the framework of inside-out disk

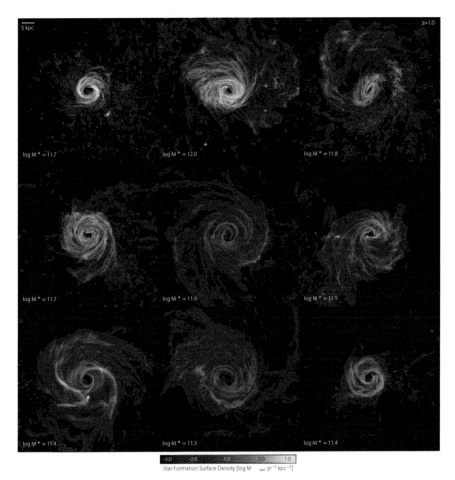

Fig. 2 Nine examples of massive galaxies at redshift one. Each is rotated so that it is viewed face-on, emphasizing the spiral structure of disk-like galaxies. Here we show projections of the star formation rate surface density, in units of $M_\odot \, yr^{-1} \, kpc^{-2}$. Star formation activity increases towards the centers, following the gas density, while fine-grained structure on hundred parsec scales is able to develop as a result of the high gas spatial resolution in TNG50

growth, metallicities in central regions of massive galaxies versus their outskirts reflects an equilibrium between ongoing stellar nucleosynthesis, supernova feedback and gas inflows. That is, they encode the relation between chemical enrichment from stars and ejective outflows due to feedback on local scales.

In Fig. 3 we show average metallicity profiles, stacking many galaxies at $z \sim 0.7$ together at each mass (different colored lines), as a function of radius. Metallicities monotonically increase with mass, while the extended gradients of the stacked profiles gradually flatten. At some point a critical threshold is reached at $M_\star \simeq 10^{11}$ $M_\odot$, above which the central profiles become constant with distance within < 10 kpc.

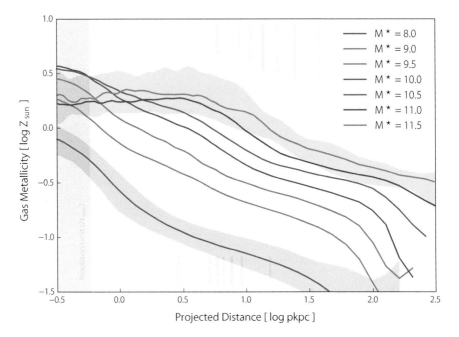

Fig. 3 Median radial profiles of gas-phase metallicity—that is, the relative enrichment of gas by elements heavier than helium as a function of distance from the galaxy center. Shown in 2D projection, as would be observed, where each colored line represents a stack of all galaxies with approximately the stellar mass M_* as indicated in the legend. Each stack contains between a few, and a few thousand, galaxies. Metallicities generally increase with mass, while the gradients encode a wealth of information on the interplay between gas-poor inflows and gas-rich outflows

This complete flattening of metallicity gradients is due to post-outflow, metal-poor inflows, and corresponds to an inversion of the central gradient of star formation rate. Whether these gas inflows arise from cooling out of the gaseous halo atmosphere, primordial cosmological accretion, or radial flux through the disk due to angular momentum loss processes remains to be investigated.

At higher redshifts (not shown), we also see that galaxies as small as $10^8 \, M_\odot$ reach solar metallicity in their centers already at $z = 6$, although this value drops quickly to $Z_\odot/10$ at $\sim 10 \, \text{kpc}$ as a result of a nearly time-invariant metallicity gradient for galaxies of a given mass, which is place already by $z = 6$ and supported by a pre-enriched cosmological inflow. This early distribution of metals into low-density environments occurs as a result of high velocity outflows driven by the blackhole winds, leading to large-distance pollution of the circumgalactic and intergalactic medium (IGM) by metal-enriched gas at high redshift.

3.3 Gas-Dynamical Processes in the Intracluster Medium (ICM)

The thermodynamical state of the intracluster medium of galaxy clusters is a sensitive probe of galaxy formation physics, plasma physics, and even cosmology. Observations at X-ray and radio wavelengths have revealed a rich level of detail, including sub-cluster components, ram pressure stripped tails of gas, and sharp jumps in the temperature and density of the gas resulting from bow shocks [39].

In Fig. 4 we show four different views of a single massive galaxy cluster in TNG50. Stars (upper left) trace individual galaxies, while the temperature distribution (upper right) peaking at over 100 million Kelvin in the halo center might lead one to conclude a rather homogeneous and uniform intra-cluster medium (ICM). In reality, this gaseous reservoir is highly dynamic and constantly being stirred up by the influence of baryonic feedback processes from the central galaxy and its many luminous satellites, as seen by the distribution and structure of hydrodynamical shocks (lower left) and leading to significant chemical enrichment of the cluster volume (lower right). We can quantify the energy dissipation rates and Mach numbers of shock fronts, which are typically mildly supersonic. The cooling properties of the hot halo gas are modulated not only by feedback from the central blackhole, but also by supernovae-driven winds launched by infalling satellite galaxies. The ICM is substantially enriched by iron produced by both supernovae type Ia and type II, and substantially so out to a significant fraction of the virial radius. This metal-rich gas has a turbulent velocity structure down to the smallest resolvable scales, which can influence the physical interpretation of high-resolution spectroscopic x-ray line observations [40].

The resolution of TNG50 also allows us to study cluster-member luminous satellites down to $\sim 10^7 \, M_\odot$, which will provide a detailed picture of the temporal evolution of such an ensemble of heirarchically accumulating galaxies.

3.4 Characteristic Sizes of Galaxies and Halos

The relationship between the size of a galaxy and the size (or mass) of its dark matter halo is fundamentally related to the theory of hierarchical assembly and collapse [41], while observations of galaxy sizes in different tracers—star-forming gas, optical stellar light, or molecular line emission, as examples—provide direct observational probes [42]. As TNG50 evolves all of the relevant baryonic components self-consistently together, we can provide direct predictions for the relationships between different 'sizes' of galaxies and halos, as shown in Fig. 5 at $z \simeq 0.7$. Here we compare, across five order of magnitude in stellar mass, the halo virial radius (green), the radius enclosing half the gaseous mass (blue), the radius enclosing half the stellar mass (orange), the optical 'effective radius' R_e (red), the half light radius of star-forming gas (purple), and the extent of neutral hydrogen (HI; brown).

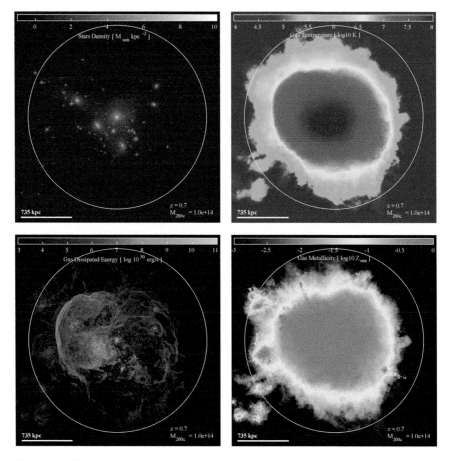

Fig. 4 Four different views of one of the most massive galaxy clusters in the TNG50 simulation at $z \simeq 0.7$. In each case, the circle demarcates the virial radius ($r_{200,\mathrm{crit}}$). We show: stellar mass (upper left), gas temperature (upper right), energy dissipation rate due to hydrodynamical shocks (lower left), and gas metallicity (lower right). Complex galaxy interactions are encoded in the small-scale features of the intracluster medium (ICM) gas, particularly in hydrodynamical shock structures

After the dark matter halo itself, the gas is the most extended component. This is followed by the neutral hydrogen component, which is more concentrated than the gas overall as it becomes ionized in low-density halo outskirts due to photoionization from the ultraviolet/x-ray background radiation field. Surprisingly, the predicted Hα emission is generally as concentrated as the optical light from the stellar component itself, although this breaks down at small (large) masses, where it is smaller (larger), respectively. At $M_\star \simeq 10^{10}\,M_\odot$, the gas of a galactic halo extends to roughly an order of magnitude larger distances than its stars, but in quantitative detail this depends on the observational tracer used for both components. TNG50 provides details not only on the extent, but also the shape (e.g. axis ratios) and morphology (e.g. asymmetry)

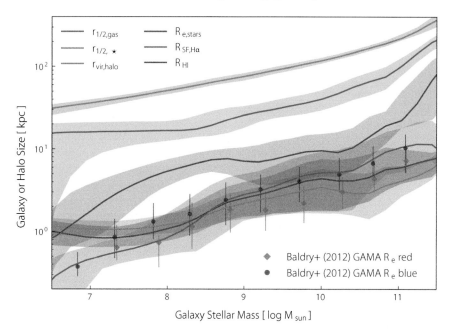

Fig. 5 Characteristic sizes of galaxies and galaxy halos as a function of stellar mass, in the TNG50 simulation at $z \simeq 0.7$. Increasing massive galaxies are always larger, and here we compare: the halo virial radius (green), the radii enclosing half the gaseous mass (blue) and half the stellar mass (orange), the optical 'effective radius' R_e in a putative observation with the F115W filter aboard the James Web Space Telescope (JWST; red), the Hα half light radius of star-forming gas (purple), and the neutral hydrogen (HI) size (brown). Observations from [43] at $z = 0$ are shown for reference

of each component. Although already well constrained at low redshift, these aspects of the galaxy population are largely unknown at earlier epochs and are a predictive regime for the current generation of cosmological simulations.

4　Conclusions

Despite the numerous theoretical achievements of recent large volume simulations such as Illustris, Eagle, or even TNG100 and TNG300, their limited mass and spatial resolution complicate any study of the structural details of galaxies less massive than a few times 10^9 M$_\odot$. In contrast, projects focused on 'zoom' simulations of one or a handful of galaxies are constrained by their small sample sizes and inability to construct cosmologically representative volumes for statistical population analyses. For the most massive galaxy clusters, simulations with sufficient resolution to simultaneously model the co-evolving population of satellite galaxies have been prohibited by the large computational requirements as well as the complexity of the physical mechanisms which operate to shape the circumgalactic and intracluster gas.

TNG50 is already proving to be an instrumental theoretical tool for the comparison—via mock observations of the simulated data—with existing and upcoming observational datasets. These include, for example, the star formation histories and abundances of dwarf galaxies in the first few billion years of the history of the Universe (with HST, JWST); the detailed mapping of our Galaxy's stellar system (e.g. with Gaia-ESO, Rave, APOGEE, GALAH, SEGUE, DES); the imaging and spectroscopic characterization of the galaxy populations residing in nearby galaxy clusters like Virgo, Fornax, and Coma (SAURON, ATLAS-3D, NGVS); the X-ray maps of the outskirts of galaxy clusters with the SUZAKU telescope; integral-field spectroscopy campaigns of nearby galaxies (MANGA, CALIFA); the characterization of circumgalactic and IGM gas (e.g. COS-Halo); and magnetic field measurements in halos (LOFAR, SKA) and in the disks of galaxies (ASKAP). In each case, TNG50 will be able to provide a foundation for theoretical interpretation.

The IllustrisTNG project aims to redefine the state-of-the-art of cosmological hydrodynamical simulations of galaxy formation. Even beyond our immediate scientific investigations, the TNG50 simulation will be a unique platform to pursue as of yet unimagined future projects, as we are now able to treat cosmological simulations as almost open ended laboratories for studying galaxy formation physics.

Acknowledgements The authors acknowledge the Gauss Centre for Supercomputing (GCS) for providing computing time for the GCS Large-Scale Projects GCS-ILLU (2014) and GCS-DWAR (2016) on the GCS share of the supercomputer Hazel Hen at the High Performance Computing Center Stuttgart (HLRS). DN acknowledges additional simulations and analysis carried out on supercomputers at the Max Planck Computing and Data Facility (MPCDF, formerly known as RZG). VS, RW, and RP acknowledge support through the European Research Council under ERC-StG grant EXAGAL-308037 and would like to thank the Klaus Tschira Foundation.

References

1. C.S. Frenk, S.D.M. White, M. Davis, ApJ **271**, 417 (1983). https://doi.org/10.1086/161209
2. M. Davis, G. Efstathiou, C.S. Frenk, S.D.M. White, ApJ **292**, 371 (1985). https://doi.org/10.1086/163168
3. J.F. Navarro, C.S. Frenk, S.D.M. White, ApJ **490**, 493 (1997). https://doi.org/10.1086/304888
4. A.E. Evrard, T.J. MacFarland, H.M.P. Couchman, J.M. Colberg, N. Yoshida, S.D.M. White, A. Jenkins, C.S. Frenk, F.R. Pearce, J.A. Peacock, P.A. Thomas, ApJ **573**, 7 (2002). https://doi.org/10.1086/340551
5. V. Springel, S.D.M. White, A. Jenkins, C.S. Frenk, N. Yoshida, L. Gao, J. Navarro, R. Thacker, D. Croton, J. Helly, J.A. Peacock, S. Cole, P. Thomas, H. Couchman, A. Evrard, J. Colberg, F. Pearce, Nature **435**, 629 (2005). https://doi.org/10.1038/nature03597
6. M. Vogelsberger, S. Genel, V. Springel, P. Torrey, D. Sijacki, D. Xu, G. Snyder, D. Nelson, L. Hernquist, MNRAS **444**, 1518 (2014). https://doi.org/10.1093/MNRAS/stu1536
7. M. Vogelsberger, S. Genel, V. Springel, P. Torrey, D. Sijacki, D. Xu, G. Snyder, S. Bird, D. Nelson, L. Hernquist, Nature **509**, 177 (2014). https://doi.org/10.1038/nature13316
8. S. Genel, M. Vogelsberger, V. Springel, D. Sijacki, D. Nelson, G. Snyder, V. Rodriguez-Gomez, P. Torrey, L. Hernquist, MNRAS **445**, 175 (2014). https://doi.org/10.1093/mnras/stu1654

9. D. Sijacki, M. Vogelsberger, S. Genel, V. Springel, P. Torrey, G.F. Snyder, D. Nelson, L. Hernquist, MNRAS **452**, 575 (2015). https://doi.org/10.1093/mnras/stv1340

10. J. Schaye, R.A. Crain, R.G. Bower, M. Furlong, M. Schaller, T. Theuns, C. Dalla Vecchia, C.S. Frenk, I.G. McCarthy, J.C. Helly, A. Jenkins, Y.M. Rosas-Guevara, S.D.M. White, M. Baes, C.M. Booth, P. Camps, J.F. Navarro, Y. Qu, A. Rahmati, T. Sawala, P.A. Thomas, J. Trayford, MNRAS **446**, 521 (2015). https://doi.org/10.1093/mnras/stu2058

11. R.A. Crain, J. Schaye, R.G. Bower, M. Furlong, M. Schaller, T. Theuns, C. Dalla Vecchia, C.S. Frenk, I.G. McCarthy, J.C. Helly, A. Jenkins, Y.M. Rosas-Guevara, S.D.M. White, J.W. Trayford, MNRAS **450**, 1937 (2015). https://doi.org/10.1093/mnras/stv725

12. Y. Dubois, S. Peirani, C. Pichon, J. Devriendt, R. Gavazzi, C. Welker, M. Volonteri, MNRAS **463**, 3948 (2016). https://doi.org/10.1093/mnras/stw2265

13. K. Dolag, E. Komatsu, R. Sunyaev, MNRAS **463**, 1797 (2016). https://doi.org/10.1093/mnras/stw2035

14. A. Pillepich, D. Nelson, L. Hernquist, V. Springel, R. Pakmor, P. Torrey, R. Weinberger, S. Genel, J.P. Naiman, F. Marinacci, M. Vogelsberger, MNRAS **475**, 648 (2018). https://doi.org/10.1093/mnras/stx3112

15. V. Springel, R. Pakmor, A. Pillepich, R. Weinberger, D. Nelson, L. Hernquist, M. Vogelsberger, S. Genel, P. Torrey, F. Marinacci, J. Naiman, MNRAS **475**, 676 (2018). https://doi.org/10.1093/mnras/stx3304

16. D. Nelson, A. Pillepich, V. Springel, R. Weinberger, L. Hernquist, R. Pakmor, S. Genel, P. Torrey, M. Vogelsberger, G. Kauffmann, F. Marinacci, J. Naiman, MNRAS **475**, 624 (2018). https://doi.org/10.1093/mnras/stx3040

17. F. Marinacci, M. Vogelsberger, R. Pakmor, P. Torrey, V. Springel, L. Hernquist, D. Nelson, R. Weinberger, A. Pillepich, J. Naiman, S. Genel, ArXiv e-prints (2017)

18. J.P. Naiman, A. Pillepich, V. Springel, E. Ramirez-Ruiz, P. Torrey, M. Vogelsberger, R. Pakmor, D. Nelson, F. Marinacci, L. Hernquist, R. Weinberger, S. Genel, ArXiv e-prints (2017)

19. D. Nelson, A. Pillepich, S. Genel, M. Vogelsberger, V. Springel, P. Torrey, V. Rodriguez-Gomez, D. Sijacki, G.F. Snyder, B. Griffen, F. Marinacci, L. Blecha, L. Sales, D. Xu, L. Hernquist, Astron. Comput. **13**, 12 (2015). https://doi.org/10.1016/j.ascom.2015.09.003

20. R. Weinberger, V. Springel, L. Hernquist, A. Pillepich, F. Marinacci, R. Pakmor, D. Nelson, S. Genel, M. Vogelsberger, J. Naiman, P. Torrey, MNRAS **465**, 3291 (2017). https://doi.org/10.1093/mnras/stw2944

21. A. Pillepich, V. Springel, D. Nelson, S. Genel, J. Naiman, R. Pakmor, L. Hernquist, P. Torrey, M. Vogelsberger, R. Weinberger, F. Marinacci, MNRAS **473**, 4077 (2018). https://doi.org/10.1093/mnras/stx2656

22. V. Springel, L. Hernquist, MNRAS **339**, 289 (2003). https://doi.org/10.1046/j.1365-8711.2003.06206.x

23. R. Weinberger, V. Springel, R. Pakmor, D. Nelson, S. Genel, A. Pillepich, M. Vogelsberger, F. Marinacci, J. Naiman, P. Torrey, L. Hernquist, ArXiv e-prints (2018)

24. S. Cole, C.G. Lacey, C.M. Baugh, C.S. Frenk, MNRAS **319**, 168 (2000). https://doi.org/10.1046/j.1365-8711.2000.03879.x

25. D. Nelson, G. Kauffmann, A. Pillepich, S. Genel, V. Springel, R. Pakmor, L. Hernquist, R. Weinberger, P. Torrey, M. Vogelsberger, F. Marinacci, MNRAS (2018). https://doi.org/10.1093/mnras/sty656

26. P. Torrey, M. Vogelsberger, F. Marinacci, R. Pakmor, V. Springel, D. Nelson, J. Naiman, A. Pillepich, S. Genel, R. Weinberger, L. Hernquist, ArXiv e-prints (2017)

27. S. Genel, D. Nelson, A. Pillepich, V. Springel, R. Pakmor, R. Weinberger, L. Hernquist, J. Naiman, M. Vogelsberger, F. Marinacci, P. Torrey, MNRAS **474**, 3976 (2018). https://doi.org/10.1093/mnras/stx3078

28. M. Vogelsberger, F. Marinacci, P. Torrey, S. Genel, V. Springel, R. Weinberger, R. Pakmor, L. Hernquist, J. Naiman, A. Pillepich, D. Nelson, MNRAS **474**, 2073 (2018). https://doi.org/10.1093/mnras/stx2955

29. M.R. Lovell, A. Pillepich, S. Genel, D. Nelson, V. Springel, R. Pakmor, F. Marinacci, R. Weinberger, P. Torrey, M. Vogelsberger, L. Hernquist, ArXiv e-prints (2018)

30. V. Springel, MNRAS **401**, 791 (2010). https://doi.org/10.1111/j.1365-2966.2009.15715.x
31. R. Pakmor, A. Bauer, V. Springel, MNRAS **418**, 1392 (2011). https://doi.org/10.1111/j.1365-2966.2011.19591.x
32. R. Pakmor, V. Springel, MNRAS **432**, 176 (2013). https://doi.org/10.1093/mnras/stt428
33. R. Pakmor, F. Marinacci, V. Springel, ApJL **783**, L20 (2014). https://doi.org/10.1088/2041-8205/783/1/L20
34. R. Pakmor, V. Springel, A. Bauer, P. Mocz, D.J. Munoz, S.T. Ohlmann, K. Schaal, C. Zhu, MNRAS **455**, 1134 (2016). https://doi.org/10.1093/mnras/stv2380
35. R.J.J. Grand, F.A. Gómez, F. Marinacci, R. Pakmor, V. Springel, D.J.R. Campbell, C.S. Frenk, A. Jenkins, S.D.M. White, ArXiv e-prints (2016)
36. P.F. Hopkins, A. Wetzel, D. Keres, C.A. Faucher-Giguere, E. Quataert, M. Boylan-Kolchin, N. Murray, C.C. Hayward, S. Garrison-Kimmel, C. Hummels, R. Feldmann, P. Torrey, X. Ma, D. Angles-Alcazar, K.Y. Su, M. Orr, D. Schmitz, I. Escala, R. Sanderson, M.Y. Grudic, Z. Hafen, J.H. Kim, A. Fitts, J.S. Bullock, C. Wheeler, T.K. Chan, O.D. Elbert, D. Narananan, ArXiv e-prints (2017)
37. N.M. Förster Schreiber, A. Renzini, C. Mancini, R. Genzel, N. Bouché, G. Cresci, E.K.S. Hicks, S.J. Lilly, Y. Peng, A. Burkert, C.M. Carollo, A. Cimatti, E. Daddi, R.I. Davies, S. Genel, J.D. Kurk, P. Lang, D. Lutz, V. Mainieri, H.J. McCracken, M. Mignoli, T. Naab, P. Oesch, L. Pozzetti, M. Scodeggio, K. Shapiro Griffin, A.E. Shapley, A. Sternberg, S. Tacchella, L.J. Tacconi, S. Wuyts, G. Zamorani, ArXiv e-prints (2018)
38. F. Belfiore, R. Maiolino, C. Tremonti, S.F. Sánchez, K. Bundy, M. Bershady, K. Westfall, L. Lin, N. Drory, M. Boquien, D. Thomas, J. Brinkmann, MNRAS **469**, 151 (2017). https://doi.org/10.1093/mnras/stx789
39. H.R. Russell, J.S. Sanders, A.C. Fabian, S.A. Baum, M. Donahue, A.C. Edge, B.R. McNamara, C.P. O'Dea, MNRAS **406**, 1721 (2010). https://doi.org/10.1111/j.1365-2966.2010.16822.x
40. Hitomi Collaboration, Nature **535**, 117 (2016). https://doi.org/10.1038/nature18627
41. H.J. Mo, S. Mao, S.D.M. White, MNRAS **295**, 319 (1998). https://doi.org/10.1046/j.1365-8711.1998.01227.x
42. R.S. Somerville, P. Behroozi, V. Pandya, A. Dekel, S.M. Faber, A. Fontana, A.M. Koekemoer, D.C. Koo, P.G. Pérez-González, J.R. Primack, P. Santini, E.N. Taylor, A. van der Wel, MNRAS **473**, 2714 (2018). https://doi.org/10.1093/mnras/stx2040
43. I.K. Baldry, S.P. Driver, J. Loveday, E.N. Taylor, L.S. Kelvin, J. Liske, P. Norberg, A.S.G. Robotham, S. Brough, A.M. Hopkins, S.P. Bamford, J.A. Peacock, J. Bland-Hawthorn, C.J. Conselice, S.M. Croom, D.H. Jones, H.R. Parkinson, C.C. Popescu, M. Prescott, R.G. Sharp, R.J. Tuffs, MNRAS **421**, 621 (2012). https://doi.org/10.1111/j.1365-2966.2012.20340.x

44065 HypeBBH Yearly Report—High Performance Computing Services at HLRS

Federico Guercilena, Elias Most and Luciano Rezzolla

1 Introduction

The allocation "44065 HypeBBHs" on the system CRAY XC40 (HAZEL HEN) at HLRS has been awarded to our research group in February 2015 for one year, then extended in February 2016 until December 2016, becoming one of the main computing resources available to us. A request for a further extension until December 2017 and new resources allocation has been submitted, and we have been granted access to the machine with full approval of our application.

The scientific research carried on in the group relates to a number of broad topics in the field of general relativistic astrophysics. These topics, which fall in general under the label of "physics of compact objects", include:

- the general behaviour of black hole systems in vacuum
- the study of rigidly and differentially rotating isolated neutron stars
- the evolution of neutron stars and accretion disks systems
- the evolution and properties of magnetized neutron stars in the contexts of both ideal and resistive magnetohydrodynamics
- the study of ejecta in binary neutron star systems and the resulting implications for nucleosynthesis.

Due to the sheer complexity of the physical models involved in this field of research, no realistic analytical solutions are available. The main tools of our research are therefore large scale numerical simulations of the physical system of interest. As such another important line of research we follow is the development of better numerical methods to improve the accuracy and/or performance of our results. This state of affairs in turn makes computing resources such as the ones provided by HLRS a very important asset to us.

F. Guercilena · E. Most · L. Rezzolla (✉)
Institute for Theoretical Physics (ITP), Goethe University, Max-von-Laue-Str. 1, 60438 Frankfurt, Germany
e-mail: rezzolla@th.physik.uni-frankfurt.de

© Springer Nature Switzerland AG 2019
W. E. Nagel et al. (eds.), *High Performance Computing in Science and Engineering '18*, https://doi.org/10.1007/978-3-030-13325-2_2

2 Numerical Methods

Our simulations are performed using several codes developed by members of the group and others publicly available as part of the `Einstein Toolkit` suite [1–3], of which some of the members of our group are developers and/or contributors.

We rely on the Einstein Toolkit as the main infrastructure for all our simulations, in particular to handle the set up of the numerical grid and mesh refinement, the time evolution of fields (in a method-of-lines fashion using Runge-Kutta integrators), the output, checkpointing, memory access and management.

We discretize the Einstein equations for the spacetime evolution using a standard finite-difference (e.g., [4]) method where all the derivatives, with the exception of the terms associated with the advection along the shift vector, for which we use a stencil upwinded by one grid point, are computed with a centered stencil. Typically all these terms are computed with an eighth-order accurate scheme in vacuum and fourth-order in the presence of matter. To ensure the nonlinear stability of the scheme we add a ninth-order (fifth-order respectively) Kreiss-Oliger [5] style artificial dissipation.

Specifically, we use the publicly available finite-difference `McLachlan` code [6], which uses the so-called BSSN [7–9] or CCZ4 [10] formulations of Einstein's equations solved by high order finite-differencing. It contains state-of-the art gauge conditions and supports OpenMP-parallelization and explicit instruction vectorisation. `McLachlan` is freely available under the GPL. It makes use of the `Kranc` framework [11, 12] to automatically generate optimized, robust C++ code from Mathematica formulas describing the partial differential equations. More recently we have developed a new framework to automatically generate lengthy tensor expressions using modern compile-time evaluated C++ 14 expression templates. Using this framework we have not only reimplemented the BSSN, CCZ4 and Z4c [13] equations, but also managed to a get a significant speed-up compared to the previous `McLachlan` implementation.

To discretize the equations of hydrodynamics and magnetohydrodynamics and couple them to the evolution of the spacetime, we employ the `WhiskyTHC`, `WhiskyMHD` and `WhiskyResistive` codes. These are proprietary codes mainly developed in our group. They provide finite-difference and finite-volume based discretizations of the Euler and Maxwell-Euler equations along with state-of-the-art algorithms (such as high-order reconstruction methods, Riemann solvers and flux limiters) to handle the difficulties arising in the solutions of such equations.

The `WhiskyTHC` [14–16] code in particular also includes algorithms which allow the inclusion of realistic microphysical processes in the hydrodynamical simulations, such as realistic equations of state and neutrino leakage schemes.

Additionally we make use of the recently developed `FIL` code that is a high-order extension of the publically available `IllinoisGRMHD` code [17]. More specifically, it evolves the staggered vector potential using the upwind constraint transport method to preserve the divergence constraint on the magnetic field. It also uses a fourth-order accurate WENO flux update and includes support for currently used temperature dependent equations of state and neutrino leakage and can be coupled to an M0 neutrino absorption scheme.

These evolution codes were developed in the context of the `Cactus` framework [18, 19], a framework providing MPI parallelization, data management and an API to bind together different modules written in different computer languages, which is distributed under GPL 2.0.

All the codes are internally OpenMP-parallelized, leading to a very efficient hybrid OpenMP/MPI parallelization model. The use of mesh-refinement techniques is of fundamental importance in our simulations. For this reason, we use the `Carpet` driver [20] that implements a vertex- or cell-centered adaptive-mesh-refinement scheme adopting nested and moving grids. This enables the code to follow the moving compact objects (black holes or neutron stars) with fine-resolution grids, while covering with coarser grids a much larger volume needed for the accurate treatment of the gravitational waves.

3 Job Statistics and Resources Usage

The typical job size for our projects is largely dependent on the type of physical system we are modelling, the microphysical processes involved and of course the level of accuracy needed, which sets constraints on numerical parameters such as the resolution, domain size and order of the method. Nonetheless we can provide some estimates. A typical job can require the use of 50 nodes (1200 cores) for a period of 24 ours at low resolution and twice as much (100 nodes/2400 cores) for several days at mid-high resolution. In the hybrid MPI/OpenMP parallelization scheme that we employ, this results in a total of 50–100 MPI processes (2 per node), each of them spawning 12 threads each. We normally do not exploit hyperthreading and attempt to bind each thread to a separate physical core.

A grand total of 15,808,872 core-hours was awarded to us, and we have exhausted it, with a usage of 99.97% as of 06/04/2018. At present 14 researchers within our group have access to computing resources at HLRS, but of these 4 none are actually active, as we have currently taken to completion all projects on this platform.

4 Completed Projects and Publications

4.1 Entropy-Limited Hydrodynamics: A Novel Approach to Relativistic Hydrodynamics

We developed a new approach for the computation of numerical fluxes arising in the discretization of hyperbolic equations in conservation form, which we call ELH: entropy-limited hydrodynamics. ELH is based on the hybridisation of an unfiltered high-order scheme with the first-order Lax-Friedrichs method. The activation of the low-order part of the scheme is driven by a measure of the locally generated entropy inspired by the artificial-viscosity method proposed by Guermond et al. [21]. We present ELH in the context of high-order finite-differencing methods and of the equations of general-relativistic hydrodynamics. We study the performance of ELH in a series of classical astrophysical tests in general relativity involving isolated, rotating and nonrotating neutron stars, and including a case of gravitational collapse to black hole. We present a detailed comparison of ELH with the fifth-order mono-tonicity preserving method MP5 [22], one of the most common high-order schemes currently employed in numerical-relativity simulations. We find that ELH achieves comparable and, in many of the cases studied here, better accuracy than more tradi-tional methods at a fraction of the computational cost (up to $\sim$50% speedup). Given its accuracy and its simplicity of implementation, ELH is a promising framework for the development of new special- and general-relativistic hydrodynamics codes well adapted for massively parallel supercomputers. The research conducted on this subject on the HazelHen platform and summarized here has led has been published in [23].

The most commonly used numerical methods in the context of general-relativistic hydrodynamics are collectively known as high-resolution shock-capturing (HRSC) techniques [24]. HRSC methods are very effective in dealing with shocks and sup-pressing spurious oscillations, and have been employed with varying degree of suc-cess in astrophysical simulations. Many of such schemes, however, potentially suffer from a few shortcomings: (i) they are complex to derive and implement, or to extend and modify; (ii) they often depend on many coefficients that require some degree of optimisation (e.g., the WENO methods); (iii) they can lead to load imbalance in parallel implementations as a result of their complexity.

We propose a different approach that is able to address some of these shortcomings. The underlying concept is relatively straightforward: the hydrodynamical fluxes are computed using an unfiltered, high-order stencil, to which a contribution from a low-order, stable numerical flux (the Lax-Friedrichs flux) is added in order to ensure stability. To determine which gridpoints are in need of the low-order contribution, we employ a "troubled-cell" indicator, which not only marks region of the computational domain requiring the limiter, but also determines the relative weights of the high and low-order fluxes.

To drive the activation of the Lax-Friedrichs method, a criterion to flag generi-cally problematic points of the computational domain is needed. Such a criterion is offered by the entropy viscosity function described by [21, 25, 26], in which the local production of entropy is used to identify shocks. Since entropy is produced only in the presence of shocks, this results in a stable method able to recover high-order in regions of smooth flow.

We subjected the ELH scheme to a variety of tests, ranging from one dimensional, special-relativistic tests, to more realistic, three-dimensional, general relativistic sim-ulations of isolated neutron stars. The former allowed us to verify that ELH scheme is convergent (to the nominal order of convergence expected, see Fig. 1), as well as to check for stability in standard hydrodynamics tests such as shock tubes, see Fig. 2.

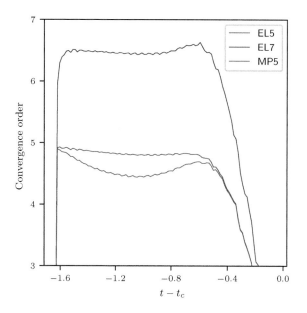

Fig. 1 Convergence order computed on the smooth nonlinear-wave test as a function of time to caustic formation, for two variants (5th and 7th order) of the ELH scheme, plus the MP5 scheme

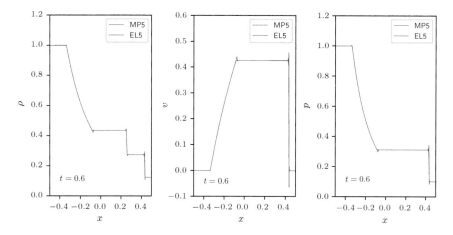

Fig. 2 Profiles of the rest-mass density (left), velocity (center) and pressure (right) for the special-relativistic Sod test at $t = 0.6$. The solution is computed on a grid of 800 points. The EL5 scheme correctly captures the features of the solution despite oscillations at the discontinuities

$$\log(\rho/\rho_{0,\max})$$

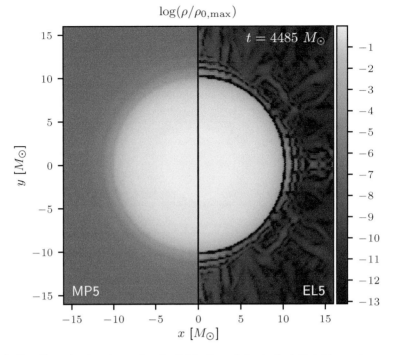

Fig. 3 Two-dimensional rest-mass density distribution relative to the initial data maximum value on the equatorial (x, y) plane at time $t = 4485\, M_\odot$ for the dynamical TOV test. The matter tails are even more extended in MP5 case compared with the Cowling test, EL5 instead preserves its behaviour at the stellar surface and exterior

The neutron star tests results are well summarized by the plots in Figs. 3, 4 and 5. These refer to the evolution of an isolated, stable, unperturbed (apart from perturbations induced by truncation errors) neutron star in a dynamically evolved spacetime. We closely compare the results obtained with the ELH and MP5 methods.

As it can be seen from Fig. 3, both schemes accurately capture the solution in the stellar interior, but significant differences arise at the surface and in the exterior. The MP5 scheme shows a rather diffusive behaviour, with a smooth transition to the external "vacuum" (i.e., to a region close to the rest-mass density floor) and extended low-density tails. The ELH scheme, on the other hand, produces a sharper edge. Oscillations in the solution can be seen just outside of the star, resulting in shell-like structures around the surface, which are particularly noticeable in the coordinate axes directions. The stellar exterior is much closer to the vacuum with the ELH scheme and, in contrast to MP5, it also displays small-scale dynamics at very low densities.

A more quantitative comparison of the two schemes can be seen in the conservation of the rest mass (left panel of Fig. 4). The ELH scheme is able to conserve the initial value to an accuracy roughly two orders of magnitude better than the MP5 scheme.

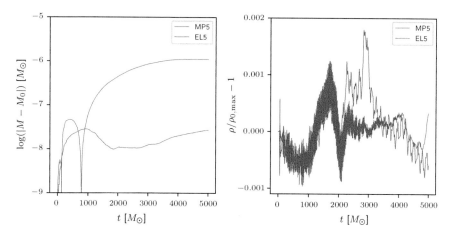

Fig. 4 Deviation of the total rest mass (left panel) and central rest-mass density (right panel) from the initial values for the dynamical TOV test

Fig. 5 PSD of the central rest-mass density evolution and physical eigenfrequencies of the stellar model for the dynamical TOV test. Note the good agreement between the first eigenfrequencies and the peaks in the data

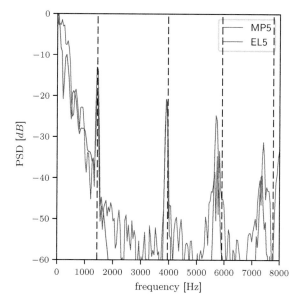

Furthermore, it is interesting to notice how the behaviour of the ELH scheme is much more smooth and predictable; MP5 by contrast displays both spurious losses and gains of mass, which lead to the zero crossings clearly visible in the figure. This is due to interpolation errors arising during the restriction and prolongation operations between different refinement levels. These errors are most likely more severe with MP5 due to the presence of long tails of low density matter in the stellar exterior. In contrast, the ELH scheme is less affected since the exterior of the star (especially away from the coordinate axes) is nearly vacuum.

The evolution of the central rest-mass densitsy shown in the right panel of Fig. 4, with both schemes varying no more than 0.2% from the initial value, but with MP5 displaying oscillations at much higher frequency. The significance of this data can be appreciated by computing the frequency content of the signal, as in Fig. 5. Peaks are clearly visible above the noise, of which the lowest-frequency ones correspond to the fundamental oscillation mode of the star and its first overtone (as computed from perturbation theory), while the following ones are higher overtones and are progressively more offset from the corresponding perturbative eigenfrequencies. The peaks in the ELH data appear to be more clearly identifiable and less broad than in the MP5 case. Above $\sim$8000 Hz, and not shown in Fig. 5, the MP5 scheme shows significant high-frequency components, clearly visible in the first part of the corresponding curve in Fig. 4, but with a rather disordered spectrum. These same frequencies are instead greatly suppressed in the ELH scheme. Overall, also this test highlights that the ELH scheme captures quite well the physical behaviour of the system as expected from perturbative methods and is free from some of the artefacts which appear instead in the evolution with the MP5 scheme.

It has to be stressed that the ELH scheme offers a definite advantage in performance, even though the precise magnitude of it depends on external factors (such as the MPI communication overhead): all the tests we performed showed a $\sim$50% speedup with respect to MP5. This was achieved despite our implementation not being particularly optimized. This open interesting prospects for the exploitation of this scheme on modern computing architectures, such the many-core architecture or GPUs.

A part of the computational resources awarded to our group has been devoted to an extension of this project, involving the application of the ELH method to the simulation of the dynamics of binary neutron stars, with the goal of extracting the gravitational-wave signal from this physical scenario and compare the accuracy of our method, code and implementation to different ones. This investigation has highlighted that the ELH method requires a certain amount of tuning to produce sufficiently accurate waveforms and the exploration of optimal setup for this type of problem is still under investigation.

4.2 Assessing the Accuracy of Numerical Waveforms

The advent of gravitational-wave astronomy initially started with the detection of GW15... and recently culminated into the detection of the first binary neutron star merger GW170817 [27] that together with the coincident optical counterparts has led to a significants improvements of our understanding of neutron stars and nuclear matter at high densities. In particular constraints on the maximum mass and radius of neutron stars [28–34] have been set using information from the gravitational-wave observation and kilonova modelling. In particular, using results of previous work partially carried on Hazel Hen [35] we have been able to constrain the maximum mass of non-rotating neutron stars $2.01^{+0.04}_{-0.04} \leq M_{\text{TOV}}/M_\odot \lesssim 2.16^{+0.17}_{-0.15}$, which presents a significant step forward in our understanding of the equation of state of nuclear matter.

One of the most important information extracted from the gravitational wave is the tidal deformability $\tilde{\Lambda}$ which depends both on the equation of state and the mass and radius of the neutron star. In order to extract this quantity from the observed gravitational waveform a delicate matching to template banks from numerical simulations and analytical models is necessary.

Since these models also need input from numerical simulations on the last to orbits before merger, over which tidal effects become significant, understanding the accuracy and limitations of gravitational-wave modelling by large numerical simulation is now more crucial than ever. In order to systematically study the error involved in numerical waveform modelling and in particular to assess the accuracy of numerous codes used by various groups, a code comparison project was started that includes eight groups from Europe and the US that typically supply numerical waveforms for gravitational-wave matching.

As part of this effort we have performed various simulations at different resolutions and with different codes and space-time formulations. In particular we have used the newly developed high-order GRMHD code FIL together with the space-time solver Antelope, which is based on modern C++14 compile-time evaluated expression templates. Using these codes and our WhiskyTHC code, we have assessed the error of using different numerical methods in solving the general-relativistic hydrodynamics equations and of using different formulations for the space-time evolution, e.g., Z4c [13] and CCZ4 [10]. While we do not find any differences between the two space-time formulations we find that it is important to use the right combination of numerical methods if higher than second order convergence is to be achieved. In the two panels of Fig. 6 we show the gravitational-wave strain and phase evolution of a 2.7 $M_\odot$ binary simulated using the SLy equation of state. Most importantly

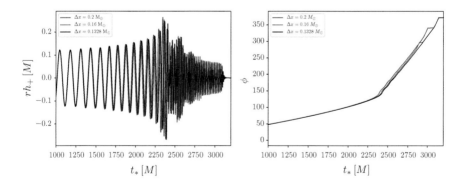

Fig. 6 *Left panel*: Gravitational-wave strain component h_+^{22} normalized to the gravitational mass of the binary. The simulations were performed at various resolutions as part of a multi-group code comparison. *Right panel*: Gravitational-wave phase ϕ normalized to the gravitational mass of the binary. The simulations were performed at various resolutions as part of a multi-group code comparison

slightly higher than second order convergence is achieved in the inspiral phase up to the merger. This allows us to give a robust error on our numerical waveforms, which also agree well with waveforms found by other groups. As the code comparison will soon be finished we hope to report on the final results in the near future.

4.3 Nucleosynthesis in the Ejecta of Binary Neutron Stars Mergers

The recent detection of gravitational waves from binary neutron stars [27] as well as the simultaneous detection of an electromagnetic counterpart has led to many advancement in the study of such systems, such as shedding light on the long-standing puzzle of the origin of short gamma-ray bursts (SGRBs) [36–40]. An electromagnetic counterpart from a merger that has recently received significant attention is that of a kilonova [41–56], i.e., an infrared/optical signal powered by the decay of a variety of heavy elements. These elements can be formed after a BNS merger due to the onset of rapid neutron-capture process (r-process). A kilonovae has been indeed already observed in GW 170817.

In this work, we have conducted a number of high-resolution numerical-relativity simulations of BNS mergers to investigate the effects of the neutron-star initial masses, mass ratios and most importantly the microphysical equation of state (EOS) on the resulting r-process nucleosynthesis. We consider three fully temperature-dependent EOSs spanning a wide range of stiffness (namely: (i) DD2 [57]; (ii) LS220 [58] with compressibility parameter $K = 220\,\text{MeV}$; SFHO [59]). For each EOS, we consider three equal-mass initial setups covering a realistic range of initial BNS masses. Additionally, we consider for each EOS one unequal-mass case.

To follow the evolution of the fluid, we use a combination of techniques, namely outflow detectors and passively advected fluid tracers. The properties and use of the latter in general-relativistic simulations have been discussed in [60]. We then post-process the data using a complete nuclear-reaction network [61, 62] to obtain the final r-process abundances. We also compute the associated kilonova light curves using the model outlined in [42].

Although some variation in the properties of the ejected mass (i.e., typical values of the electron fraction, entropy or velocity) are observed and appear to correlate with the choice of EOS or neutron-star mass for a given BNS model, these differences have minimal influence on the final r-process nucleosynthesis yields. Given the kilonova light curves associated to our simulations, we find that the prospects for their direct observation are rather limited; however, in view of the approximations made in our current analysis, this may be not a conclusive statement. Finally, we have uncovered an interesting geometrical structure in the angular distribution of the ejecta which could have important implications on the properties of the kilonova signal.

The work that we have conducted and summarize in this section has been published in [63]. Among the several results reported there, three deserve special mention. First, we have shown that the r-process nucleosynthesis from BNS mergers is very robust and essentially universal in that it depends only very weakly on the properties of the binary system, such as the EOS, the total mass or the mass (see Fig. 7). In all cases considered, in fact, and modulo small differences, the yields of our nuclear-reaction networks are in very good agreement with the solar abundances for mass numbers $A \gtrsim 120$. Our study strengthens the evidence that BNS mergers are the site of production of the r-process elements in the galaxy.

Second, we have employed two different approaches to measure the amount of matter ejected dynamically and found that it is $\lesssim 10^{-3} M_\odot$, see the plot in Fig. 8, which is smaller than what usually assumed. There are a number of factors that need to be taken into account when deriving these estimates, namely: the EOS, the neutrino treatment, the criterion for unboundedness, the resolution, the numerical methods used. Although these systematic factors can lead to differences as large as one order of magnitude even for the same initial data, we find it unlikely that the mass

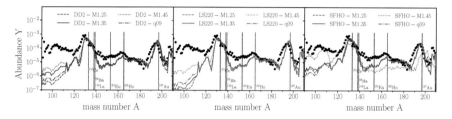

Fig. 7 Relative heavy-elements abundances for all the 12 BNS models as a function of mass number A. The abundances are normalized so that the total mass fraction is unity, while the different panels and lines refer to the various EOSs, masses and mass ratios, respectively. The black filled circles report the observed solar abundances, while the vertical lines mark a few representative r-process elements: ^{133}Cs, ^{138}Ba, ^{139}La, ^{153}Eu, ^{165}Ho, ^{197}Au

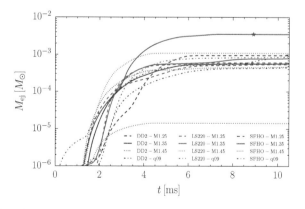

Fig. 8 Evolution of the dynamically ejected unbound mass $M_{\rm ej}$ as measured through a detector at radius 300 km when using the geodesic criterion and for the various binaries considered. The star denotes the time of black-hole formation for model SFHO-M1.35. Binaries LS220-M1.45 and SFHO-M1.35 collapse shortly after merger and are not visible in the plot

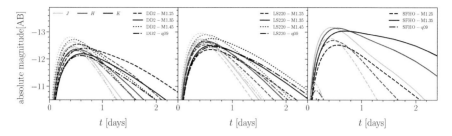

Fig. 9 Synthetic light curves in the infrared 2MASS J, H and K-bands for all of the binaries considered

ejected dynamically can ever reach the values sometimes assumed in the literature of 10^{-2}–$10^{-1}\ M_\odot$. Clearly, a more detailed and comparative study is necessary to better constrain the uncertainties behind the amount of mass lost by these systems.

Third, using a simplified and gray-opacity model we have assessed the observability of the infrared transients associated to the decay of r-process elements (plotted in Fig. 9), i.e., of the kilonova emission. We have found that all of our binaries show a very similar behaviour, peaking around $\sim 1/2$ day in the H-band and rapidly decreasing in luminosity after one day, reaching a maximum magnitude of -13. These rather low luminosities are most probably the direct consequence of the small amounts of ejected matter, thus making the prospects for detecting kilonovae rather limited. More sophisticated calculations with improved neutrino treatments will be needed to settle this issue conclusively.

As a final remark we note that even though the r-process abundance pattern does not give us simple clues to the original BNS parameters, e.g., it does not allow to disentangle various EOS and mass configurations, there are distinguishing features in the ensuing kilonova signal relatable through the difference in ejecta properties obtained in our simulations. In particular, we have found that softer EOSs tend to result in a higher average electron fractions, which implies differences in the type of kilonova produced (blue vs. red kilonovae). Additionally, we have found that this difference in electron fraction is highly angular dependent with higher electron fractions around the polar regions and lower along the orbital plane. Even though there is significantly less material ejected along the poles versus the plane, our simulations show that the simplified models of kilonova modelling, such as that of a homogeneously expanding group of material, need to be adjusted to account for this anisotropic emission. We reserve the investigation of this issue to future studies, where an improved neutrino treatment will also be implemented.

5   Conclusion

The relativistic astrophysics group in Frankfurt has a number of active projects benefiting from the 44065 HypeBBHs allocation on the HazelHen cluster. Despite a number of difficulties and setbacks both in the scientific part of said projects and their computational implementation, progress has been made which has already led to interesting results as well as publications. We plan on continuing their development on the HLRS HazelHen cluster.

References

1. Einstein Toolkit: Open Software for Relativistic Astrophysics. http://einsteintoolkit.org
2. F. Löffler, J. Faber, E. Bentivegna, T. Bode, P. Diener, R. Haas, I. Hinder, B.C. Mundim, C.D. Ott, E. Schnetter, G. Allen, M. Campanelli, P. Laguna, The Einstein Toolkit: a community computational infrastructure for relativistic astrophysics. Class. Quantum Grav. **29**(11), 115001 (2012)
3. M. Zilhão, F. Löffler, An introduction to the Einstein toolkit. Int. J. Mod. Phys. A **28**, 40014 (2013)
4. R.J. Leveque, *Numerical Methods for Conservation Laws* (Birkhauser Verlag, Basel, 1992)
5. H.O. Kreiss, J. Oliger, *Methods for the Approximate Solution of Time Dependent Problems* (GARP publication series No. 10, Geneva, 1973)
6. McLachlan, A Public BSSN Code. https://www.cct.lsu.edu/~eschnett/McLachlan/
7. T.W. Baumgarte, S.L. Shapiro, Numerical integration of Einstein's field equations. Phys. Rev. D **59**(2), 024007 (1999)
8. J. David, Brown, covariant formulations of Baumgarte, Shapiro, Shibata, and Nakamura and the standard gauge. Phys. Rev. D **79**(10), 104029 (2009)
9. M. Shibata, T. Nakamura, Evolution of three-dimensional gravitational waves: Harmonic slicing case. Phys. Rev. D **52**, 5428–5444 (1995)
10. D. Alic, C. Bona-Casas, C. Bona, L. Rezzolla, C. Palenzuela, Conformal and covariant formulation of the Z4 system with constraint-violation damping. Phys. Rev. D **85**(6), 064040 (2012)
11. Kranc: Automated Code Generation
12. S. Husa, I. Hinder, C. Lechner, Kranc: a mathematica application to generate numerical codes for tensorial evolution equations. Comput. Phys. Comm. **174**, 983–1004 (2006)
13. D. Hilditch, S. Bernuzzi, M. Thierfelder, Z. Cao, W. Tichy, B. Brügmann, Compact binary evolutions with the Z4c formulation. Phys. Rev. D **88**(8), 084057 (2013)
14. D. Radice, L. Rezzolla, THC: a new high-order finite-difference high-resolution shock-capturing code for special-relativistic hydrodynamics. Astron. Astrophys. **547**, A26 (2012)
15. D. Radice, L. Rezzolla, F. Galeazzi, Beyond second-order convergence in simulations of binary neutron stars in full general-relativity. Mon. Not. R. Astron. Soc. L. **437**, L46–L50 (2014)
16. D. Radice, L. Rezzolla, F. Galeazzi, High-order fully general-relativistic hydrodynamics: new approaches and tests. Class. Quantum Grav. **31**(7), 075012 (2014)
17. Z.B. Etienne, V. Paschalidis, R. Haas, P. Mösta, S.L. Shapiro, IllinoisGRMHD: an open-source, user-friendly GRMHD code for dynamical spacetimes. Class. Quantum Grav. **32**(17), 175009 (2015)
18. G. Allen, W. Benger, T. Goodale, H. Hege, G. Lanfermann, A. Merzky, T. Radke, E. Seidel, J. Shalf, The cactus code: a problem solving environment for the grid, in *Proceedings of Ninth IEEE International Symposium on High Performance Distributed Computing, HPDC-9, August 1-4 2000, Pittsburgh* (IEEE Press, 2000), pp. 253–260. http://www.cactuscode.org/Articles/Cactus_Allen00f.pre.pdf
19. http://www.cactuscode.org
20. E. Schnetter, CARPET: A Mesh Refinement driver for CACTUS
21. J.L. Guermond, R. Pasquetti, B. Popov, Entropy viscosity method for nonlinear conservation laws. J. Comput. Phys. **230**(11), 4248–4267 (2011)
22. A. Suresh, H.T. Huynh, Accurate monotonicity-preserving schemes with runge-kutta time stepping. J. Comput. Phys. **136**(1), 83–99 (1997)

23. F. Guercilena, D. Radice, Luciano Rezzolla, Entropy-limited hydrodynamics: a novel approach to relativistic hydrodynamics. Comput. Astrophys. Cosmol. **4**(1), 3 (2017)
24. L. Rezzolla, O. Zanotti, *Relativistic Hydrodynamics* (Oxford University Press, Oxford, UK, 2013)
25. J.L. Guermond, R. Pasquetti, Entropy-based nonlinear viscosity for Fourier approximations of conservation laws. C. R. Acad. Sci. Paris **346**(346), 801–806 (2008)
26. V. Zingan, J.L. Guermond, J. Morel, B. Popov, Implementation of the entropy viscosity method with the discontinuous Galerkin method. Comput. Methods Appl. Mech. Eng. **253**, 479–490 (2013)
27. B.P. Abbott et al., Gw170817: observation of gravitational waves from a binary neutron star inspiral. Phys. Rev. Lett. **119**, 161101 (2017)
28. E. Annala, T. Gorda, A. Kurkela, A. Vuorinen, Gravitational-Wave constraints on the Neutron-Star-Matter equation of state. Phys. Rev. Lett. **120**(17), 172703 (2018)
29. A. Bauswein, O. Just, H.-T. Janka, N. Stergioulas, Neutron-star radius constraints from GW170817 and future detections. Astrophys. J. Lett. **850**, L34 (2017)
30. B. Margalit, B.D. Metzger, Constraining the maximum mass of Neutron Stars from multi-messenger observations of GW170817. Astrophys. J. Lett. **850**, L19 (2017)
31. D. Radice, A. Perego, F. Zappa, S. Bernuzzi, GW170817: joint constraint on the Neutron Star equation of state from multimessenger observations. Astrophys. J. Lett. **852**, L29 (2018)
32. L. Rezzolla, E.R. Most, L.R. Weih, Using gravitational-wave observations and Quasi-Universal relations to constrain the maximum mass of Neutron Stars. Astrophys. J. Lett. **852**, L25 (2018)
33. M. Ruiz, S.L. Shapiro, A. Tsokaros, GW170817, general relativistic magnetohydrodynamic simulations, and the neutron star maximum mass. Phys. Rev. D **97**(2), 021501 (2018)
34. M. Shibata, S. Fujibayashi, K. Hotokezaka, K. Kiuchi, K. Kyutoku, Y. Sekiguchi, M. Tanaka, Modeling GW170817 based on numerical relativity and its implications. Phys. Rev. D **96**(12), 123012 (2017)
35. M. Hanauske, K. Takami, L. Bovard, L. Rezzolla, J.A. Font, F. Galeazzi, H. Stöcker, Rotational properties of hypermassive neutron stars from binary mergers. Phys. Rev. D **96**(4), 043004 (2017)
36. I. Bartos, P. Brady, Szabolcs Marka, How gravitational-wave observations can shape the Gamma-ray Burst Paradigm. Class. Quantum Grav. **30**, 123001 (2013)
37. E. Berger, Short-Duration Gamma-Ray Bursts. Ann. Rev. Astron. Astrophys. **52**, 43–105 (2014)
38. D. Eichler, M. Livio, T. Piran, D.N. Schramm, Nucleosynthesis, neutrino bursts and gamma-rays from coalescing neutron stars. Nature **340**, 126–128 (1989)
39. R. Narayan, B. Paczynski, T. Piran, Gamma-ray bursts as the death throes of massive binary stars. Astrophys. J. Lett. **395**, L83–L86 (1992)
40. L. Rezzolla, B. Giacomazzo, L. Baiotti, J. Granot, C. Kouveliotou, M.A. Aloy, The missing link: merging Neutron Stars naturally produce jet-like structures and can power Short Gamma-ray Bursts. Astrophys. J. Lett. **732**, L6 (2011)
41. J. Barnes, D. Kasen, M.-R. Wu, G. Martínez-Pinedo, Radioactivity and thermalization in the ejecta of compact object mergers and their impact on kilonova light curves. Astrophys. J. **829**, 110 (2016)
42. D. Grossman, O. Korobkin, S. Rosswog, T. Piran, The long-term evolution of neutron star merger remnants—II. Radioactively powered transients. Mon. Not. R. Astron. Soc. **439**, 757–770 (2014)
43. O. Just, A. Bauswein, R.A. Pulpillo, S. Goriely, H.-T. Janka, Comprehensive nucleosynthesis analysis for ejecta of compact binary mergers. Mon. Not. R. Astron. Soc. **448**, 541–567 (2015)
44. O. Just, M. Obergaulinger, H.-T. Janka, A. Bauswein, N. Schwarz, Neutron-star merger ejecta as obstacles to neutrino-powered Jets of Gamma-Ray Bursts. Astrophys. J. Lett. **816**, L30 (2016)
45. L.-X. Li, B. Paczyński, Transient events from Neutron Star Mergers. Astrophys. J. **507**, L59–L62 (1998)
46. B.D. Metzger, C. Zivancev, Pair fireball precursors of neutron star mergers. Mon. Not. R. Astron. Soc. **461**, 4435–4440 (2016)

47. A. Perego, S. Rosswog, R.M. Cabezón, O. Korobkin, R. Käppeli, A. Arcones, M. Liebendörfer, Neutrino-driven winds from neutron star merger remnants. Mon. Not. R. Astron. Soc. **443**, 3134–3156 (2014)
48. T. Piran, E. Nakar, S. Rosswog, The electromagnetic signals of compact binary mergers. Mon. Not. R. Astron. Soc. **430**, 2121–2136 (2013)
49. D. Radice, F. Galeazzi, J. Lippuner, L.F. Roberts, C.D. Ott, L. Rezzolla, Dynamical Mass Ejection from Binary Neutron Star Mergers. Mon. Not. R. Astron. Soc. **460**, 3255–3271 (2016)
50. S. Rosswog, U. Feindt, O. Korobkin, M.-R. Wu, J. Sollerman, A. Goobar, G. Martinez-Pinedo, Detectability of compact binary merger macronovae. Class. Quantum Grav. **34**(10), 104001 (2017)
51. S. Rosswog, T. Piran, E. Nakar, The multimessenger picture of compact object encounters: binary mergers versus dynamical collisions. Mon. Not. R. Astron. Soc. **430**, 2585–2604 (2013)
52. Y. Sekiguchi, K. Kiuchi, K. Kyutoku, M. Shibata, Dynamical mass ejection from binary neutron star mergers: radiation-hydrodynamics study in general relativity. Phys. Rev. D **91**(6), 064059 (2015)
53. Y. Sekiguchi, K. Kiuchi, K. Kyutoku, M. Shibata, K. Taniguchi, Dynamical mass ejection from the merger of asymmetric binary neutron stars: radiation-hydrodynamics study in general relativity. Phys. Rev. D **93**(12), 124046 (2016)
54. M. Tanaka, Kilonova/Macronova Emission from compact binary mergers. Adv. Astron. **2016**, 634197 (2016)
55. S. Wanajo, Y. Sekiguchi, N. Nishimura, K. Kiuchi, K. Kyutoku, M. Shibata, Production of all the r-process Nuclides in the dynamical ejecta of Neutron Star Mergers. Astrophys. J. **789**, L39 (2014)
56. R.T. Wollaeger, O. Korobkin, C.J. Fontes, S.K. Rosswog, W.P. Even, C.L. Fryer, J. Sollerman, A.L. Hungerford, D.R. van Rossum, A.B. Wollaber, Impact of ejecta morphology and composition on the electromagnetic signatures of neutron star mergers. Mon. Not. R. Astron. Soc. **478**, 3298–3334 (2018)
57. S. Typel, G. Röpke, T. Klähn, D. Blaschke, H.H. Wolter, Composition and thermodynamics of nuclear matter with light clusters. Phys. Rev. C **81**(1), 015803 (2010)
58. J.M. Lattimer and F.D. Swesty, A generalized equation of state for hot, dense matter. Nucl. Phys. A **535**, 331–376 (1991)
59. A.W. Steiner, M. Hempel, T. Fischer, Core-collapse supernova equations of state based on Neutron Star observations. Astrophys. J. **774**, 17 (2013)
60. L. Bovard, L. Rezzolla, On the use of tracer particles in simulations of binary neutron stars. Class. Quantum Grav. **34**(21), 215005 (2017)
61. O. Korobkin, S. Rosswog, A. Arcones, C. Winteler, On the astrophysical robustness of the neutron star merger r-process. Mon. Not. R. Astron. Soc. **426**, 1940–1949 (2012)
62. C. Winteler, R. Käppeli, A. Perego, A. Arcones, N. Vasset, N. Nishimura, M. Liebendörfer, F.-K. Thielemann, Magnetorotationally driven supernovae as the origin of early Galaxy r-process elements? Astrophys. J. Lett. **750**, L22 (2012)
63. L. Bovard, D. Martin, F. Guercilena, A. Arcones, L. Rezzolla, O. Korobkin, On r-process nucleosynthesis from matter ejected in binary neutron star mergers. Phys. Rev. D **96**, 124005 (2017)

PAMOP2: Towards Exascale Computations Supporting Experiments and Astrophysics

B. M. McLaughlin, C. P. Ballance, M. S. Pindzola, P. C. Stancil, J. F. Babb, S. Schippers and A. Müller

Abstract Our prime computational effort is to support current and future measurements of atomic photoionization cross-sections at various synchrotron radiation facilities, and ion-atom collision experiments, together with plasma, fusion and astrophysical applications. In our work we solve the Schrödinger or Dirac equation using the R-matrix or R-matrix with pseudo-states approach from first principles. Finally, we present cross-sections and rates for radiative charge transfer, radiative association, and photodissociation collision processes between atoms and ions of interest for several astrophysical applications.

B. M. McLaughlin (✉) · C. P. Ballance
School of Mathematics & Physics, Molecular and Optical Physics (CTAMOP),
Centre for Theoretical Atomic, Queen's University Belfast, Belfast BT7 1NN, UK
e-mail: bmclaughlin899@btinternet.com

C. P. Ballance
e-mail: c.ballance@qub.ac.uk

M. S. Pindzola
206 Allison Laboratory, Department of Physics, Auburn University, Auburn AL 36849, USA
e-mail: pindzola@physics.auburn.edu

P. C. Stancil
Department of Physics and Astronomy and the Center for Simulational Physics,
University of Georgia, Athens GA 30602-2451, USA
e-mail: stancil@physast.uga.edu

J. F. Babb
Institute for Theoretical Atomic and Molecular Physics, Harvard Smithsonian Center
for Astrophysics, 60 Garden Street, Cambridge, MA 02138, USA
e-mail: jbabb@cfa.harvard.edu

S. Schippers
I. Physikalisches Institut, Justus-Liebig-Universität Giessen, 35392 Giessen, Germany
e-mail: Stefan.Schippers@physik.uni-giessen.de

A. Müller
Institut für Atom- und Molekülphysik, Justus-Liebig-Universität Giessen, 35392 Giessen,
Germany
e-mail: Alfred.Mueller@iamp.physik.uni-giessen.de

© Springer Nature Switzerland AG 2019
W. E. Nagel et al. (eds.), *High Performance Computing in Science
and Engineering '18*, https://doi.org/10.1007/978-3-030-13325-2_3

1 Introduction

Our research efforts continue to focus on the development of computational methods to solve the Schrödinger and Dirac equations for atomic and molecular collision processes. Access to leadership-class computers such as the Cray XC40 at HLRS, for example, enable us to benchmark theoretical solutions against experiments at synchrotron radiation facilities such as the Advanced Light Source (ALS), ASTRID II, BESSY II, SOLEIL and PETRA III and to provide atomic and molecular data for ongoing research in ion-atom collisions, plasma and fusion science. Our codes are positioned to take account of the trend towards exascale computing.

In order to have direct comparisons with experiment, and to achieve spectroscopic accuracy, semi-relativistic, fully relativistic R-matrix, or R-matrix with pseudo-states (RMPS) computations, an exponentially increasing number of target-coupled states is required and such computations could not be even attempted without access to high performance computing (HPC) resources such as those available at computational centers in Europe (HLRS) and the USA (NERSC, NICS or ORNL).

The motivation for our work is multi-fold; (a) Astrophysical applications [1–4], (b) Magnetically-confined fusion and plasma modeling, (c) Fundamental studies and (d) Support of experiment and satellite observations. For photoabsorption by heavy atomic systems [5, 6], little atomic data exists and our work provides results for new frontiers on the applications of the R-matrix semi-relativistic (BREIT- PAULI) or fully relativistic Dirac Atomic R-matrix (DARC) parallel suite of codes. Our highly efficient R-matrix codes have evolved over the past decade and have matured to a stage now that large-scale collision calculations can be carried out in a timely manner for electron or photon impact of heavy systems where inclusion of relativistic effects is essential. These codes are widely applicable for the theoretical interpretation of current experiments being performed at leading synchrotron radiation facilities. Examples of our results are presented below in order to illustrate the predictive nature of the methods employed compared to experiment.

2 Valence Shell Studies

For comparison with the measurements made at the ALS, state-of-the-art theoretical methods using highly correlated wavefunctions were applied that include relativistic effects. An efficient parallel version of the DARC [7–9] suite of codes continues to be developed and applied to address electron and photon interactions with atomic systems, providing for hundreds of levels and thousands of scattering channels. These codes are presently running on a variety of parallel high performance computing architectures world wide [10–14]. The input wavefunctions for the DARC codes are determined by the GRASP0 code [15–17].

GRASP0 [15, 16] is used to construct a bound orbital basis set for the residual ion of the system of interest, with the Dirac–Coulomb Hamiltonian H_D, via the relation,

$$H_D = \sum_i -ic\alpha\nabla_i + (\beta-1)c^2 - \frac{Z}{r_i} + \sum_{i<j} \frac{1}{|r_j - r_i|}, \tag{1}$$

where the electrons are labellred by i and j and the summation is taken over all electrons of the system. The matrices α and β are directly related to the Pauli spin matrices, c is the speed of light and the atomic number is Z. The relativistic orbitals are described with a large component, $\mathscr{P}_{n\ell}$ and small component $\mathscr{Q}_{n\ell}$. The residual ion target wavefunctions are appropriately defined on a radial grid for input into the relativistic R-matrix -codes (DARC) [9].

The total cross-section σ for photoionization (in Megabarns, $1\text{Mb} = 10^{-18}$ cm^2) by unpolarized light is obtained by integrating over all electron-ejection angles $\hat{k}$ and averaging over photon polarization [17] to give

$$\sigma = \frac{8\pi^2\alpha a_0^2\omega}{3(2J_i + 1)} \sum_{l,j,J} |\langle \Psi_f^- |M_1|\Psi_i\rangle|^2, \tag{2}$$

where M_1 represents the dipole length operator and the equations are simplified by use of the Wigner-Eckart theorem [17]. The cross-section σ can be cast in both length (L) and velocity gauges (V) through the dipole moment operator M_1, where for the velocity gauge $\omega = 2\pi\nu$, with ν the frequency of the photon is replaced by ω^{-1}, and a_0 is the Bohr radius, α is the fine structure constant, and $g_i = (2J_i + 1)$ is the statistical weighting of the initial state, with Ψ_i, Ψ_f^- being the initial and final state scattering wavefunctions.

As the scale of excitation and photoionisation calculations grows exponentially in complexity, one of the bottlenecks has been the memory management for tasks that run concurrently but require orders of magnitude difference in resources between tasks. To this end, we introduced FORTRAN modules throughout the code that allow the calculation, in terms of integers and real numbers to be specified at the accuracy required by the user whether it is REAL*8, REAL*16, or REAL*32, ...etc.

It is difficult to know in advance the dimensions of the largest matrices involved in our work. Hence, we have implemented a scheme in which each process keeps track of these dimensions during runtime and if a dimension limit is exceeded, or is going to be exceeded, the code internally resizes matrices and transfers data through *call move_alloc()* commands. The efficiency of smaller tasks ensures that their memory footprint is small, or if the task is completed, creating the necessary temporary space required to transfer information from one large matrix to another.

2.1 Photoionization of Atomic Calcium Ions: Ca$^+$

DARC calculations on photoionization of heavy ions carried out for Se$^+$ [5], Xe$^+$ [6], Cl$^+$ [18, 19], Co$^+$ [20], Ar$^+$ [23–25], Fe$^+$ [26], Xe^{7+} [27], Kr$^+$ [28], Se^{2+} [29], ions showed suitable agreement with high resolution ALS measurements. Large-scale DARC photoionization cross-section calculations on neutral sulfur compared to photolysis experiments, carried out in Berlin [30], and measurements performed at SOLEIL for 2p-removal in Si^+ ions by photons [31] both showed suitable agreement. In addition to this, our theoretical work on W^+ [32–34], W^{2+}, W^{3+} [35], and W^{4+} [36] ions compared with measurements made at the ALS indicated good agreement. All of these cross-section calculations using either the DARC or BREIT- PAULI R-matrix codes [9] showed satisfactory agreement with the measurements made at the Advanced Light Source, BESSY II and SOLEIL radiation facilities.

Relative cross sections for the valence shell photoionisation (PI) of ^2S ground level and ^2D metastable Ca^+ ions were measured with high energy resolution by using the ion-photon merged-beams technique at the Advanced Light Source. Overview measurements were performed with a full width at half maximum bandpass of $\Delta E = 17$ meV, covering the energy range 20–56 eV. Details of the PI spectrum were investigated at energy resolutions reaching the level of $\Delta E = 3.3$ meV. The photon energy scale was calibrated with an uncertainty of ± 5 meV. By comparison with previous absolute measurements the present experimental high-resolution data were normalised to an absolute cross-section scale and the fraction of metastable Ca^+ ions that were present in the parent ion beam was determined to be 18±4%. Large-scale R-matrix calculations using the Dirac Coulomb approximation and employing 594 levels in the close-coupling expansion were performed for the $Ca^+(3s^23p^64s\ ^2S_{1/2})$ and $Ca^+(3s^23p^63d\ ^2D_{3/2,5/2})$ levels. The experimental data are compared with the results of from our large-scale DARC calculations in Fig. 1. The structures in the mea-

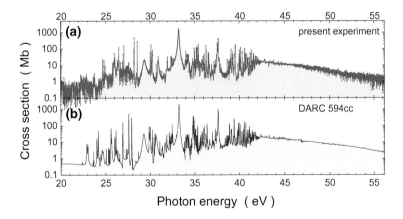

Fig. 1 Overview of the present experimental (panel a) and theoretical (panel b) results obtained for PI of Ca^+ ions in the energy range 20–56 eV. For the comparison with the present 594-level DARC calculations the specific experimental conditions had to be accounted for. The experimental data were taken at a photon-energy resolution of 17 meV. Theoretical results were convoluted with a Gaussian having a profile of 17 meV FWHM, assuming that the spectral distribution of the photon beam can be approximated by a Gaussian function. The ion beam used in the experiment consisted of a fraction of 82% of Ca^+ ions in the $^2S_{1/2}$ ground level and 18% of Ca^+ ions in the ^2D metastable term with fine-structure levels $^2D_{3/2}$ and $^2D_{5/2}$ [21]

sured and calculated spectra are similar and resonances from theory and experiment can be associated, although resonance energies do not match perfectly. An additional criterion for the identification of specific autoionizing levels is provided by the comparison of theoretical and experimental natural linewidths of resonances. Asymmetries in resonance cross sections have been observed in the experiments indicating interference of specific PI channels. The extraction of individual Fano asymmetry parameters, natural linewidths and lifetimes from the measurements is hampered by the blending of resonance features associated with experimentally unresolvable fine-structure contributions.

When compared with the present measurements, large-scale DARC calculations show satisfactory agreement over the photon energy region investigated (see Fig. 1) and reproduce the main features in the experimental spectrum. Some minor discrepancies are found between the theoretical results obtained from the 594-level DARC collision model and the measurements. These are attributed to the limitations of electron-correlation incorporated in the present 594-level collision model. Larger-scale calculations, beyond the scope of the present investigation, would likely be required to resolve this. Further details can be found in the recent work by Müller and co-workers [21].

2.2 Photoionization of Atomic Zinc Ions: Zn^+

About half of the heavy elements ($Z \geq 30$) in the Universe were formed during the Asymptotic Giant Branch (AGB) phase through slow neutron-capture (n-capture) nucleosynthesis (the s-process). In the inter-shell between the H- and He-burning shells, Fe-peak nuclei experience neutron captures interlaced with β-decays, which transform them into heavier elements. For example, in spite of its cosmic rarity, atomic selenium has been detected in the spectra of stars and of astrophysical nebulae. The chemical composition of these objects illuminate details of stellar nucleosynthesis and the chemical evolution of galaxies. To interpret such astrophysical spectra requires three types of fundamental parameters: (i) the energy of internal states (or transition frequencies), (ii) transition probabilities (or Einstein A-coefficients), and (iii) collisional excitation rate coefficients (or collision strengths in the case of electron impact). While these parameters have been studied over the past century for many atomic and molecular species, the necessary information is far from complete or is of insufficient reliability. Extensive databases do exist for the first two items (e.g., NIST: National Institute of Standards and Technology), referred to as spectroscopic data; the situation for the last item is far less satisfactory.

However, atomic data, such as PI cross sections, are unknown for the vast majority of *trans*-Fe, neutron-capture element ions. We note that the n-capture elements, Cd, Ge, and Rb, have recently been detected in planetary nebulae. Measurements for the photoionization (PI) cross-section on the *trans-Fe* element Se, in various ionized stages provided critically needed data to astrophysical modelers, and was

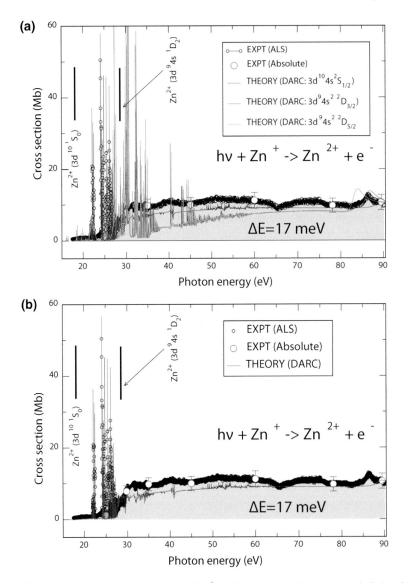

Fig. 2 Single photoionization cross section of Zn^+: ALS experimental data (open circles) and the 355-level Dirac R-matrix theoretical results. The ALS spectrum consists of Zn^{2+} ion yield spectra normalized to the absolute cross section measurements (magenta open circles). The thresholds for $Zn^{2+}(3d^{10}\,^1S_0)$ at 17.964 eV and $Zn^{2+}(3d^9 4s\,^1D_2)$ at 28.317 eV, respectively are indicated by vertical lines. The correction factor f_c has not been applied to the ALS data illustrated in the figures. **a** ALS measurements and the DARC PI calculations for the ground $(3d^{10}4s\,^2S_{1/2})$ (blue line) and metastable $(3d^9 4s^2\,^2D_{3/2})$ (red line) and $3d^9 4s^2\,^2D_{5/2}$ (green line) contributions. **b** ALS measurements and the DARC PI calculations with an appropriate mixture of the ground and metastable states

benchmarked against results from the R-matrix DARC parallel codes with excellent agreement being achieved.

Photoionization cross sections producing Zn^{2+} from Zn^+ ions were measured in the photon energy range of 17.5–90 eV with a photon energy resolution of 17 meV FWHM. In Fig. 2 , we see the good agreement between experiment and theory. Below 30 eV the spectrum is dominated by excitation autoionizing resonance states. The experimental results are compared with large-scale photoionization cross-section calculations performed using the relativistic R-matrix DARC parallel codes. Comparisons are made with previous experimental studies, resonance states are identified and contributions from metastable states of Zn^+ determined. All the sharp resonance features present in the spectrum (for the energy region 20–28 eV) have been analyzed and identified. These resonance features we attribute to Rydberg transitions originating from the ground state Zn^+ into excited autoionizing states of the Zn^+ ion that ionize to the different $3d^94s$ $^3D_{3,2,1}$ and $3d^94s$ $^1D_2$ threshold states of Zn^{2+}. We note that the DARC calculations, for resonance energies and peak heights, favour the present ALS measurements, as opposed to the previous measurements from the Daresbury radiation facility [37].

Interpretation of the experimental data was made possible using photoionization cross section results from large-scale DARC calculations that included the interaction between closed and open channels. We traced the contributions from different initial states in the ion beam and determined the metastable fraction present in the experiment to be small, in the region of 2%. Further details can be found in the recent work of Hinojosa and co-workers [38].

3   Electron Impact Excitation of Molybdenum: Mo I

A recent PISCES- B Mod experiment [39] has revealed up to a factor of five discrepancy between measurement and the two existing theoretical models [40, 41] providing important diagnostics for Mo I. We address this by employing relativistic atomic structure and R-matrix scattering calculations to improve upon the available models for future applications and benchmark results against a recent Compact Toroidal Hybrid experiment [42].

The atomic structure of Mo I was determined using the GRASP0 code. The electron-impact excitation of Mo I is investigated using the relativistic R-matrix DARC parallel suite of codes. Electron-impact excitation cross sections are presented and compared to the few available theoretical cross sections. Throughout, our emphasis is on improving the results for the $z^5P^o_{1,2,3} \rightarrow a^5S_2$, $z^7P^o_{2,3,4} \rightarrow a^7S_3$, and $y^7P^o_{2,3,4} \rightarrow a^7S_3$ electric dipole transitions of particular relevance for diagnostic work. Our model includes the following 12 non-relativistic configurations: $4d^55s, 5p, 6s$; $4d^45s^2, 5s5p, 5p^2, 6s^2$; $4d^6$; $4d^35s6s^2$; $4p^54d^65s$; $4p^54d^7$; and finally $4p^44d^75s$. This generated 2,298 fine structure levels; however, we chose only the first 800 of these to be included in the close coupling expansion, providing

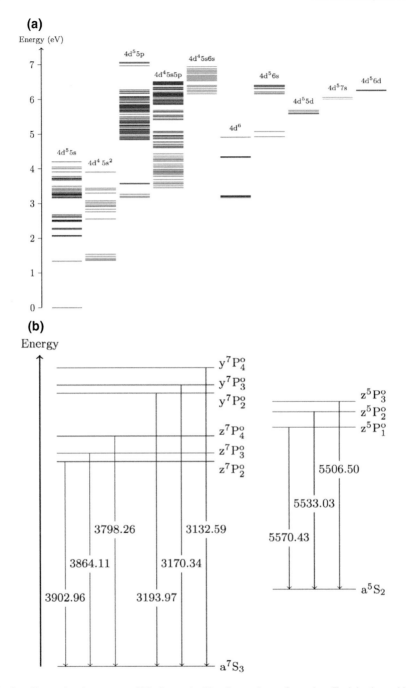

Fig. 3 **a** Energy level spectrum of Mo I organised by electronic configuration. Each horizontal line designates a specific fine structure level listed in the NIST database [22]. **b** Fine structure transitions are illustrated (with wavelengths in Å) of experimental interest

us with a comprehensive target description up to 10 eV. The significance of such an energy range is given by Badnell and co-workers [40] who report on the fractional abundance of both Mo I and Mo II, showing that the abundance of Mo I is dominant below energies of approximately 10 eV, while Mo II is dominant at higher energies. In this work we have carried out extensive atomic structure and electron-impact excitation calculations for neutral molybdenum. We have highlighted the incomplete nature of the two existing models and improved upon these, helping to rectify the current discrepancies between theory and experiment. In Fig. 3a, b, we illustrate a sample of the Mo I energy levels included in our collision model and the experimental wavelengths of interest.

The atomic structure was performed using the relativistic multi-configurational Dirac-Fock method and results are compared to available experimental data, with a particular focus on the transitions of spectroscopic interest. The atomic structure was then carried through to an 800 state relativistic R-matrix scattering calculation where excitation cross sections are compared to the results of two available theoretical models. Comparisons show large differences between the new and previous cross sections, highlighting the deficiencies contained within the previous models.

In Fig. 4 we present excitation cross sections (in units of Mb) from the 800 state, fully relativistic R-matrix scattering calculation. These are compared with the results of the two previous calculations in LS coupling, for which we must statistically

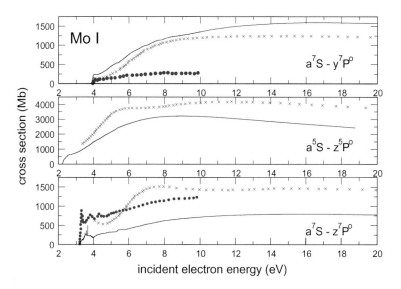

Fig. 4 Comparison of R-matrix calculations for the excitation cross section in (Mb) for selected dipole transitions in Mo I. The black curve is the 800-level DARC, R-matrix result [43], statistically avergaed over the fine structure levels. The red crosses are the 6-level R-matrix work of Badnell and co-workers [40], and the blue circles are the 67-level R-matrix results of Bartschat and co-workers [41], both performed in LS coupling

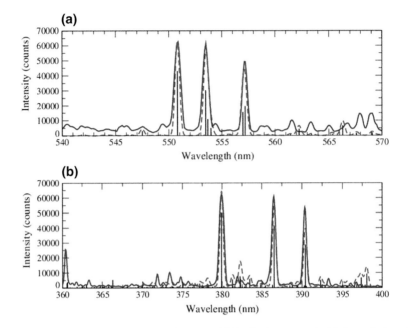

Fig. 5 Measured spectrum from the CTH plasma (solid blue line), compared with theoretical results. The solid black sticks show the PECFs coefficients for the Mo I transitions, while the dashed red curve shows a theoretical spectrum based upon Gaussian convolved PECF data. A FWHM for the Gaussian convolution of 0.15 nm was used, based upon the instrument resolution, and the PECFS are shown for an electron temperature of 6 eV and an electron density of 1×10^{12} cm^{-3}

average our fine structure results. We focus on the $a^7S \rightarrow z^7P^o$, $a^7S \rightarrow y^7P^o$, and $a^5S \rightarrow z^5P^o$ electric dipole transitions of interest to diagnostic work.

In Fig. 5 we show a plot of the photon emissivity coefficients (PECFs) (in units of number of photons cm^3 s^{-1}) calculated using our structure and collision data, covering the lines in the 379–391 nm range and the 550–558 nm range and for the electron temperature and density values expected in CTH in the region of the Mo probe. We compare with a spectrum obtained from the CTH experiment [42], as described in the previous section. It is clear that there is very good agreement with the strongest $z^7P^o_{2,3,4} \rightarrow a^7S_3$ and $z^5P^o_{1,2,3} \rightarrow a^5S_2$ transitions. This high level of agreement gives credence to the accuracy of the present atomic structure data. It must be noted that additional peaks are seen in the experimental spectrum due to the presence of impurities within the plasma. A full comparison of the CTH spectrum and the new theoretical data will be the subject of future work and will include investigations of the other Mo I lines that were observed, the sensitivity of the lines to plasma electron temperature and density, and the role of metastable states. Further details may be found in the recent work of Smyth and co-workers [43].

4   Time Dependent Close-Coupling

4.1   *Electron-Impact Double Ionization of the H_2 Molecule*

There are three main electron-impact ionization processes for diatomic hydrogen. The first is single ionization leaving the molecular ion H_2^+. The second is single ionization leaving the atom H and the atomic ion H^+. The third is double ionization leaving two atomic ions H^+. Experiments over a wide incident electron energy range have been made to obtain total cross sections for each process, for example electron-beam target gas measurements.

A time-dependent close-coupling (TDCC) method [44] in spherical polar coordinates was developed to calculate the electron-impact double ionization of the H_2 molecule. The full wavefunction was represented by an expansion in products of 6D radial-angular numerical functions and 3 analytic rotational functions. For an incident energy of 100 eV, the total cross section was calculated for the electron-impact double ionization of H_2 and compared with experiment.

Our result is a total double ionization cross section of 489 kb, which is somewhat above the experimental value of 291 kb at 150 eV incident energy for the production of two atomic ions H^+. This measured value only accounts for H^+ atomic ions emitted at a relative angle of 90° due to limitations of the experimental apparatus at that time. Therefore, it is likely to underestimate the total double ionization cross section. We also note that, unfortunately, no measurements are available below incident energies of 150 eV. The measured double ionization cross section may increase as the incident energy decreases. The TDCC- 6D calculations at larger incident energies require further significant computing resources. It is clear though that new measurements of the cross sections for full breakup of H_2 are desirable. Further details can be found in the recent work of Pindzola and co-workers [45].

4.2   *X-Ray Emission in Collisions of O^{8+} ions with H and He Atoms*

Recent interest in line ratios for x-ray emission based on ion-atom charge exchange collisions is based in part on the discovery of highly charged ions in the solar wind. The relative velocity between the interacting ion and neutral particles directly impacts both the principal quantum state and and the angular momentum state into which an electron is captured from the neutral into the ion. These capture states and radiative selection rules then dictate the radiative emission resulting from electron cascades in the excited ion. For interaction velocities that are typical of solar wind ions interacting with astrophysical neutral species, it is difficult to gauge which theoretical models are applicable for different interaction pairs and velocities. It is therefore important to assess the outcomes of different theoretical models with respect to each other and to experimental results, when available, for modeling and analysis of plasma

Fig. 6 Cross sections as a function of energy for O^{8+} + H collisions. Solid (red): time-dependent lattice (TDL), dashed (blue): molecular orbital close-coupling (MOCC), dotted-dashed (green): atomic orbital close-coupling (AOCC), double-dotted-dashed (orange): advanced adiabatic (AA), and diamonds (black): experiment [46]

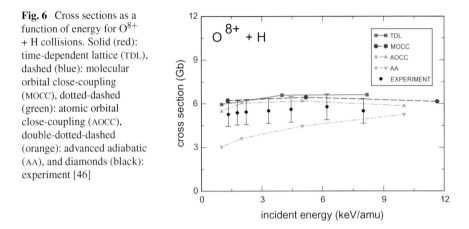

environments that may contain a diversity of ion and neutral species interacting over a range of velocities.

A time-dependent lattice method was used to calculate state selective charge transfer cross sections in O^{8+} collisions with H and He atoms at incident energies of 1.00, 4.00, and 8.18 keV/amu. Figure 6 shows our cross section results for O^{8+} ions in collision with H atom compared to other theoretical models and with experiment. Using standard radiative transitions, line ratios, $L(\beta/\alpha)$, $L(\gamma/\alpha)$, and $L(\delta/\alpha)$, were calculated using the time-dependent lattice capture cross sections for both H and He. Figure 7, illustrates our line ratio results for O^{8+} ions in collision with H atoms. In addition, the line ratios for He were also found to be in good agreement with direct experimental measurements at 3.11 keV/amu [46].

In the future, we plan to use the time-dependent lattice method to calculate state selective charge transfer cross sections for many highly charged ion collisions with light atoms of astrophysical interest. The $n\ell$ subshell cross sections will then be used to calculate line ratios for x-ray emission to compare with observations. We are currently working on Mg^{12+} collisions with H at 1.00 keV/amu and above.

5   Radiative Processes in Collisions of Atoms and Ions

The molecular ion $^{36}ArH^+$ (argonium) was identified through its rotational transitions in the sub-millimeter wavelengths towards the Crab Nebula using SPIRE onboard the *Herschel Space Observatory*, with earlier and later observations revealing $^{36}ArH^+$ and $^{38}ArH^+$ in the diffuse interstellar medium using *Herschel*/HIFI. With the ALMA Observatory, the red-shifted spectra of ArH^+ was detected in an external galaxy. Chemical modeling indicates that the presence of ArH^+ may serve as a tracer of atomic gas in the very diffuse interstellar medium (ISM) and as an indicator of cosmic ray ionization rates.

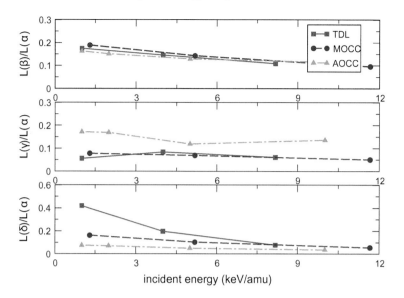

Fig. 7 Line ratios as a function of energy for O^{8+} + H collisions. Solid (red): time-dependent lattice (TDL), dashed (blue): molecular orbital close-coupling (MOCC), and dotted-dashed (green): atomic orbital close-coupling (AOCC) [46]

Ar created in the supernova is predominately ^{36}Ar and it was the isotope differences of ^{36}ArH$^+$ absorption spectra compared to ^{38}ArH$^+$ absorption spectra that provided the crucial clues leading to the initial identification in the Crab nebula. Recently, Priestly and co-workers [47] modeled ArH$^+$ formation in the filaments of the Crab nebula, considering the specific high-energy synchrotron radiation found there, in addition to the cosmic ray flux considered earlier for applications to the ISM.

A detailed exploration of model parameters led Priestly and co-workers [47] to the conclusion that in order to reproduce observed abundances, the relevant cosmic ray ionization rate is 10^7 times larger than the standard interstellar value, though sensitivity to other parameters, such as to the rate for dissociative recombination of ArH$^+$, was noted.

5.1 Radiative Charge Transfer of He$^+$ Ions with Ar Atoms

In models of the Crab Nebula, Ar$^+$ ions are produced through cosmic ray, UV and X-ray ionizations and by the reaction of H_2^+ with Ar. The ArH$^+$ is formed by the ion-atom interchange reaction

$$Ar^+ + H_2 \rightarrow ArH^+ + H \tag{3}$$

and destroyed by reaction with H_2. Sources of Ar^+ that enter the chemical models were outline by Schilke and co-workers [48] and Priestly and co-workers [47] and include photoionization of Ar and photodissociation of ArH^+. As He II ions are present in the Crab nebula, it is fitting to investigate whether an additional source of Ar^+ might arise from charge transfer in Ar and He^+ collisions. Previous studies [49–51] indicate that the direct charge transfer process

$$Ar(^1S) + He^+(^2S) \rightarrow Ar^+(3p^5\ ^2P^o) + He(2s^2\ ^1S) \tag{4}$$

occurs through mechanisms involving couplings to excited molecular states that are inaccessible at thermal energies. We conclude that charge transfer at thermal energies will proceed by the radiative charge transfer (RCT) mechanism

$$Ar + He^+ \rightarrow Ar^+ + He + h\nu, \tag{5}$$

where $h\nu$ is the photon energy. Experimentally, the rate coefficient for the loss of helium ions in argon gas was found by Johnsen and co-workers [52] to be no more than 10^{-13} cm^3/s at 295 K, while Jones and co-workers [53] found that the rate coefficient is greater than 2×10^{-14} cm^3/s at 300 K; both consistent with the early estimate of less than 10^{-13} cm^3/s at 300 K [54]. Thus, (5) is likely to be relatively slow, but we are unaware of any detailed calculations for the cross sections and rate coefficients of (5).

Potential energy curves (PECs) and transition dipole moments (TDMs) as a function of internuclear distance R were calculated for a range of values of internuclear distance R between 3 and 60 a_0, using the latest version of the MOLPRO quantum chemistry package, running on parallel architectures [55]. We used a state-averaged-multi-configuration-self-consistent-field (SA- MCSCF) approach, followed by multi-reference configuration interaction (MRCI) calculations [56], together with the Davidson correction (MRCI+Q) [57]. Figure 8 illustrates the $HeAr^+$ potential energy curves used in our work.

The radiative loss cross sections were calculated using the optical potential approach, which accounts for radiative loss in collisions of He^+ with Ar. The theory is described in [58–62] and references therein. In carrying out the calculations of (5), the probability of approach in the $B\,^2\Sigma^+$ state is unity. The reduced mass of the colliding system corresponds to ^{4}He and ^{36}Ar.

The calculated cross sections for radiative loss (radiative charge transfer) are shown in Fig. 9. The $X\,^2\Sigma^+$ to $B\,^2\Sigma^+$ transition is significantly larger than the $B\,^2\Sigma^+$ to $A\,^2\Pi$ transition. The radiative loss cross sections include the possibility of radiative association, however, because the well depth of the $X\,^2\Sigma^+$ state is only 262 cm^{-1} and the well depth of the $A\,^2\Pi$ state is even less, and because the transition dipole moments are diminished at the equilibrium distances characteristic of these states (about 5 a_0), the relative contributions of radiative association to radiative loss are very small.

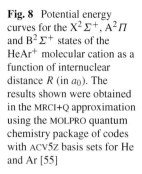

Fig. 8 Potential energy curves for the $X^2\Sigma^+$, $A^2\Pi$ and $B^2\Sigma^+$ states of the HeAr$^+$ molecular cation as a function of internuclear distance R (in a_0). The results shown were obtained in the MRCI+Q approximation using the MOLPRO quantum chemistry package of codes with ACV5Z basis sets for He and Ar [55]

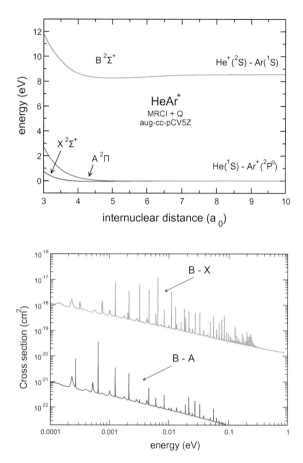

Fig. 9 Cross sections in cm^2 for radiative loss in collisions of He$^+$ with Ar. Upper curve for $B^2\Sigma^+$ to $X^2\Sigma^+$ transitions, lower curve for $B^2\Sigma^+$ to $A^2\Pi$ transitions [55]

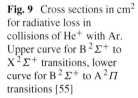

The corresponding radiative charge transfer reaction rates calculated from the data in Fig. 9 were added to the chemical network relevant to the ArH$^+$ molecular ion, as described in [63] in the PDRLight version of the Meudon PDR code [64].[1] The model was run both for diffuse galactic cloud conditions (standard interstellar radiation field, $n_H = 50$ cm^{-3}, $T = 50$ K, low visual extinction, cosmic ionization rate $\zeta = 10^{-16}$ s^{-1}) and for conditions corresponding to the Crab Nebula environment as reported in [47] ($n_H = 2000$ cm^{-3}, $\zeta = 10^{-9}$ s^{-1}). Using the value 10^{-14} cm^3/s for the rate coefficient of the radiative charge transfer reaction (5), no significant modifications of the chemical equilibrium of the ArH$^+$ molecular ion were found for either case. Further details can be found in the recent work of Babb and McLaughlin [55].

---

[1] Available at http://ism.obspm.fr.

5.2   Radiative Association of Silcon and Oxygen Atoms

Radiative association is the process whereby the formation of a molecular complex is created when two systems collide, stabilizing the complex by the emission of a photon. The formation of dust in environments such as the inner winds of AGB stars, the ejecta of novae, and the ejecta of supernovae depends on the production of molecules, either as precursors of dust grains or as competitors to dust production.

The formation of molecules like CO and SiO in such environments can be modeled using large networks of chemical reactions, which depend on the knowledge of rate constants for individual chemical reactions relating to the molecule species. Such models have been applied in explaining the observations of CO and SiO in the ejecta of Supernova 1987A, the inner winds of AGB stars, and the outer envelopes of carbon-rich and oxygen-rich stars. In fact, the observation of SiO in carbon-rich stars suggests a shock-induced chemistry which dissociates CO freeing-up atomic oxygen to form SiO and other oxides.

While in typical interstellar medium (ISM) environments, SiO is formed primarily by the exchange reaction

$$Si + OH \rightarrow SiO + H, \tag{6}$$

the radiative association (RA) process may become competitive, particularly in hydrogen-deficient gas.

The RA process for SiO formation is believed to be a key step in the subsequent formation of silicates and dust [67]. Therefore, it is of considerable interest to provide a comprehensive set of reliable rate constants for all contributions to the RA processes

$$Si + O \rightarrow SiO + h\nu, \tag{7}$$

and

$$Si \cdots O \rightarrow SiO + h\nu, \tag{8}$$

where $h\nu$ represents an emitted photon. Here, the $Si \cdots O$ metastable state in (8) may be formed as an intermediate step in process (7) or by an independent process such as inverse predissociation.

The molecular electronic structure calculations used in our work were determined from a multi-reference configuration-interaction approximation with the Davidson correction (MRCI+Q) and an AV6Z basis [56, 57]. We used the MOLPRO quantum chemistry package running on parallel architectures to determine all the molecular data used in our dynamical calculations. The molecular orbitals were determined using the state-averaged-complete-active-space-self-consistent field (SA- CASSCF) approximation. Structure calculations were reported previously in the literature on this molecule using the internally-contracted multi-reference configuration-interaction (IC- MRCI) approximation with an AV5Z basis set with the molecular orbitals (MOs) obtained from dynamically-weighted (DW) CASSCF calculations. Figure 10 shows the comparison with the previous work of Bauschlicher [66] for the potential energy curves and

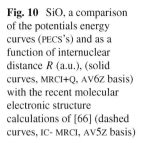

Fig. 10 SiO, a comparison of the potentials energy curves (PECS's) and as a function of internuclear distance R (a.u.), (solid curves, MRCI+Q, AV6Z basis) with the recent molecular electronic structure calculations of [66] (dashed curves, IC- MRCI, AV5Z basis)

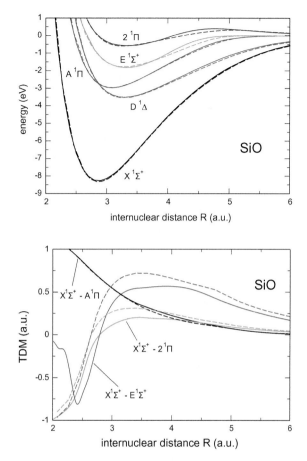

Fig. 11 SiO, comparison of the transition dipole moments (TDM's), coupling the ground $X^1\Sigma^+$, to the excited states as a function of internuclear distance R (a.u.), (solid curves, MRCI+Q, AV6Z basis) with the recent molecular electronic structure calculations of [66] (dashed curves, IC- MRCI, AV5Z basis) [65]

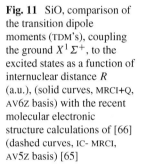

Fig. 11 illustrates the comparison for the transitions dipole moments coupling the ground state as a function of the internuclear distance [65].

Dynamical calculations were performed for SiO formation by RA via approach on the $E^1\Sigma^+$, $2^1\Pi$, $3^1\Pi$, and $D^1\Delta$ molecular states [65]. Figures 12 and 13 illustrates the results for the rates for SiO formation by RA via these initial states. In addition to computing the direct RA contribution for each intermediate electronic state, we also calculate tunneling and radiative lifetimes for all of the quasi-bound vibrational states in order to provide estimates of the indirect resonant contribution for different kinetic conditions.

Having considered all possible RA contributions, it is clear that formation of SiO is dominated by the $A^1\Pi$, $E^1\Sigma^+$, and $2^1\Pi$ states at low temperatures, and by the $3^1\Pi$ state at high temperatures, as illustrated from the results presented in Figs. 12 and 13. Further details can be found in the recent work of Cairne and co-workers [65].

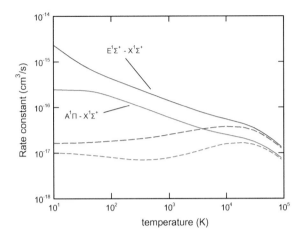

Fig. 12 Radiative association rate constants for the $E^1\Sigma^+ \to X^1\Sigma^+$, and $A^1\Pi \to X^1\Sigma^+$ transitions in SiO. Solid curves correspond to LTE and dashed curves to NLTE [65]

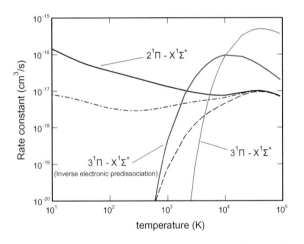

Fig. 13 Radiative association rate constants for $2^1\Pi \to X^1\Sigma^+$ and $3^1\Pi \to X^1\Sigma^+$ transitions in SiO. Solid curves correspond to LTE and broken curves to NLTE. The dashed broken curve, which uses only the tunneling widths for the $2^1\Pi$ potential, shows a threshold due to a barrier in the potential. The dot-dashed curve includes tunneling widths from the $A^1\Pi$ potential, which removes the threshold for the $2^1\Pi$ rate constant. The red curve corresponds to approach on the $3^1\Pi$ state. The blue curve corresponds to a radiation-less transition from the $A^1\Pi$ state to an intermediate bound level of the $3^1\Pi$ state prior to radiative decay to the $X^1\Sigma^+$ state [65]

5.3 Photodissociation of Carbon and Sulfur Atoms: CS

CS is a molecule of great astrophysical interest and is isoelectronic with the SiO molecule. The CS molecule is also one of the most abundant sulfur-bearing compounds in interstellar clouds and is found in a variety of astrophysical objects includ-

Fig. 14 CS, potentials energy curves (PEC's) as a function of internuclear distance R (a.u.), [70] (solid lines, $^1\Sigma^+$ states, dashed lines, $^1\Pi$ states, at the MRCI+Q level of approximation using AV6Z basis sets for both atoms) [70]

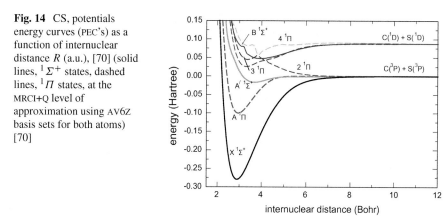

ing star-forming regions, protostellar envelopes, dense interstellar clouds, carbon-rich stars, oxygen-rich stars, planetary nebulae, and comets.

Photodissociation is an important mechanism for the destruction of molecular environments with an intense radiation field, so accurate photodissociation rates are necessary to estimate the abundance of CS. Heays and co-workers [68] presented photodissociation cross sections and photorates for CS using previous estimates [69] applying measured wavelengths for transitions to the $B\,^1\Sigma^+$ (or $3\,^1\Sigma^+$) states from the ground state and vertical excitation energies of higher states. However, comprehensive photodissociation cross sections are required to compute photorates in many environments.

Similar to our work on SiO, the CS molecular electronic structure calculations used in our calculations were determined from a multi-reference configuration-interaction approximation together with the Davidson correction (MRCI+Q) and an AV6Z basis [56, 57]. The potential energy curves used in our work are illustrated in Fig. 14 and the transition dipole moments coupling the ground state to the excited states are shown in Fig. 15 as a function of internuclear separation [70].

Fig. 15 CS, transition dipole moments (TDM's), coupling the ground $X^1\Sigma^+$ state to the excited states, as a function of internuclear sdistance R (a.u.), [70] (solid lines, $^1\Sigma^+$ states, dashed lines, $^1\Pi$ states, at the MRCI+Q level of approximation using AV6Z basis sets for both atoms) [70]

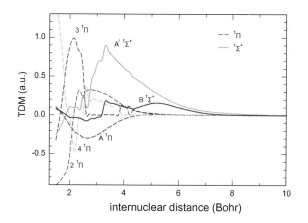

Fig. 16 CS LTE cross
sections at 3,000 K for each
of the six considered
photodissociation transitions
[70]

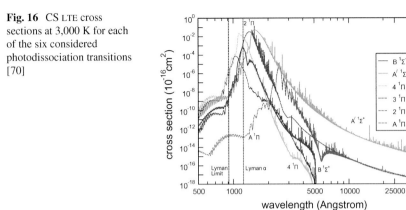

In the present study we have calculated photodissociation cross sections for the CS molecule for several electronic transitions from a wide range of initial rovibrational levels. Photodissociation cross sections for transitions from the $X^1\Sigma^+$ electronic ground state to the $A\,^1\Pi$, $A'\,^1\Sigma^+(2\,^1\Sigma^+)$, $2\,^1\Pi$, $3\,^1\Pi$, $B^1\Sigma^+(3\,^1\Sigma^+)$, and $4\,^1\Pi$ electronic states are studied here. Calculations have been performed for transitions from 14,908 initial bound rovibrational levels v'', J'' of the X state. We also explore predissociation out of the $3\,^1\Pi$, $B^1\Sigma^+$, and $4\,^1\Pi$ excited electronic states.

The present cross section calculations were performed using quantum-mechanical techniques. Applications of the cross sections to environments appropriate for local thermodynamic equilibrium (LTE) conditions are included, where a Boltzmann distribution of initial rovibrational levels is assumed. Photodissociation rates are computed for the standard interstellar radiation field (ISRF) and for black-body radiation fields at a wide range of temperatures. LTE cross sections have been computed for each transition using the state-resolved cross sections from 1,000 K to 10,000 K in 1,000 K intervals. A comparison of LTE cross sections for each transition as a function of photon wavelength (Angstroms) at 3,000 K is displayed in Fig. 16. The $A'\,^1\Sigma^+ \leftarrow X^1\Sigma^+$ transition is the dominant transition at longer wavelengths, while the $4\,^1\Pi \leftarrow X^1\Sigma^+$ transition dominates for short wavelengths. The $A'\,^1\Sigma^+ \leftarrow X^1\Sigma^+$ transition is dominant for the majority of wavelengths. Finally, we consider a situation where a gas containing CS is in LTE at a certain temperature and is immersed in a radiation field generated by a black-body at the same temperature (i.e., equal gas kinetic and radiation temperatures). The photodissociation rates of CS in such a situation are computed using the LTE cross sections; a plot of these rates against temperature is shown in Fig. 17.

Accurate cross sections for the photodissociation of the CS molecule have been computed for transitions to several excited electronic states using new *ab initio* potentials and transition dipole moment functions determine at the MRCI+Q level [56, 57] using the MOLPRO quantum chemistry suite of codes, running on parallel architectures.

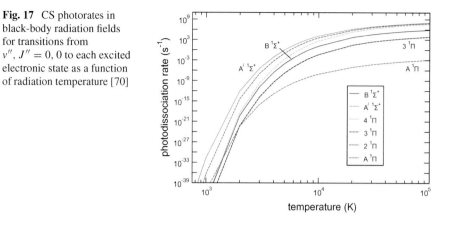

Fig. 17 CS photorates in black-body radiation fields for transitions from v'', $J'' = 0, 0$ to each excited electronic state as a function of radiation temperature [70]

The state-resolved cross sections have been computed for nearly all rotational transitions from vibrational levels $v'' = 0$ through $v'' = 84$ of the $X\,^1\Sigma^+$ ground electronic state of CS. Predissociation is found to be significantly smaller than direct photodissociation for CS. Additionally, LTE cross sections have been computed for temperatures ranging from 1,000 to 10,000 K. The computed cross sections are applicable to the photodissociation of CS in a variety of UV-irradiated interstellar environments including diffuse and translucent clouds, circumstellar disks, and protoplanetary disks. Photodissociation rates in the interstellar medium and in regions with a black-body radiation field have been computed as well. To facilitate the calculation of local photo-rates for particular astrophysical environments, all photodissociation cross section data can be obtained from the UGA Molecular Opacity Project Web site.[2] Further details can be found in the recent work of Patillo and co-workers [70].

6 Summary

Access to leadership architectures is essential to our research work (like the Cray XC40 at HLRS, the Cray XC40 at NERSC, the Cray XK7 or Summit architectures at ORNL) providing essential resources to our computational effort in atomic, molecular and optical collision processes, without which we simply could not do these calculations. The theoretical results obtained from access to and use of such high performance resources illustrates vividly the predictive nature of the methods employed and the symbiotic and synergistic relationship between theory, experiment and astrophysical applications.

---

[2] Available at, http://www.physast.uga.edu/ugamop/.

Acknowledgements A Müller acknowledges support by Deutsche Forschungsgemeinschaft under project number Mu-1068/20. B. M. McLaughlin acknowledges support from the US National Science Foundation through a grant to ITAMP at the Harvard-Smithsonian Center for Astrophysics, a visiting professorship from the University of Georgia at Athens, and a visiting research fellowship (VRF) from Queen's University Belfast. ITAMP is supported in part by a grant from the NSF to the Smithsonian Astrophysical Observatory and Harvard University. M. S. Pindzola acknowledges support by NSF and NASA grants through Auburn University. P. C. Stancil acknowledges support from NASA grants through University of Georgia at Athens. This research used computational resources at the National Energy Research Scientific Computing Center (NERSC) in Berkeley, CA, USA, and at the High Performance Computing Center Stuttgart (HLRS) of the University of Stuttgart, Stuttgart, Germany. The Oak Ridge Leadership Computing Facility at the Oak Ridge National Laboratory, provided additional computational resources, which is supported by the Office of Science of the U.S. Department of Energy under Contract No. DE-AC05-00OR22725. The Advanced Light Source is supported by the Director, Office of Science, Office of Basic Energy Sciences, of the US Department of Energy under Contract No. DE-AC02-05CH11231.

References

1. B.M. McLaughlin, J.-M. Bizau, D. Cubaynes, S. Guilbaud, S. Douix, M.M. Al Shorman, M.O.A. El Ghazaly, I. Sahko, M.F. Gharaibeh, K-Shell photoionization of O^{4+} and O^{5+} ions: experiment and theory. Mon. Not. Roy. Astro. Soc. (MNRAS) **465** 4690 (2017)
2. B.M. McLaughlin, Inner-shell Photoionization, Fluorescence and Auger Yields, in *Spectroscopic Challenges of Photoionized Plasma*, ed. by G. Ferland, D.W. Savin. Astronomical Society of the Pacific, ASP Conference Series, vol. 247, San Francisco (2001), p. 87
3. T.R. Kallman, Challenges of plasma modelling: current status and future plans. Space Sci. Rev. **157**, 177 (2010)
4. B.M. McLaughlin, C.P. Ballance, Photoionization, fluorescence and inner-shell processes, in *McGraw-Hill Yearbook of Science and Technology*, ed. by McGraw-Hill (Mc Graw Hill, New York, 2013), p. 281
5. B.M. McLaughlin, C.P. Ballance, Photoionization cross section calculations for the halogen-like ions Kr^+ and Xe^+. J. Phys. B: At. Mol. Opt. Phys. **45**, 085701 (2012)
6. B.M. McLaughlin, C.P. Ballance, Photoionization cross sections for the trans-iron element Se^+ from 18 to 31 eV. J. Phys. B: At. Mol. Opt. Phys. **45**, 095202 (2012)
7. C.P. Ballance, D.C. Griffin, Relativistic radiatively damped R-matrix calculation of the electron-impact excitation of W^{46+}. J. Phys. B: At. Mol. Opt. Phys. **39**, 3617 (2006)
8. P.H. Norrington, I.P. Grant, Low-energy electron scattering by Fe XXIII and Fe VII using the Dirac R-matrix method. J. Phys. B: At. Mol. Opt. Phys. **20**, 4869 (1987)
9. R-matrix DARC and BREIT- PAULI codes (2016). http://connorb.freeshell.org
10. B.M. McLaughlin, C.P. Ballance, Petascale computations for large-scale atomic and molecular collisions, in *Sustained Simulated Performance 2014*, ed. by M.M. Resch, Y. Kovalenko, E. Fotch, W. Bez, H. Kobaysahi (Springer, New York, 2014), ch 15
11. B.M. McLaughlin, C.P. Ballance, M.S. Pindzola, A. Müller, PAMOP: petascale atomic, molecular and optical collisions, in *High Performance Computing in Science and Engineering'14*, ed. by W.E. Nagel, D.H. Kröner, M.M. Resch (Springer, New York, 2015), ch 4
12. B.M. McLaughlin, C.P. Ballance, M.S. Pindzola, S. Schippers, A. Müller, PAMOP: petascale computations in suport of experiments, in *High Performance Computing in Science and Engineering'15*, ed. by W.E. Nagel, D.H. Kröner, M.M. Resch (Springer, New York, 2016), ch 4
13. B.M. McLaughlin, C.P. Ballance, M.S. Pindzola, P.C. Stancil, S. Schippers, A. Müller, PAMOP project: computations in suport of experiments and astrophysical applications, in *High Performance Computing in Science and Engineering'16*, ed. by W.E. Nagel, D.H. Kröner, M.M. Resch (Springer, New York, 2017), ch 4

14. B.M. McLaughlin, C.P. Ballance, M.S. Pindzola, P.C. Stancil, J.F. Babb, S. Schippers, A. Müller, PAMOP: Large-Scale computations in suport of experiments and astrophysical applications, in *High Performance Computing in Science and Engineering'17* ed. by W.E. Nagel, D.H. Kröner, M.M. Resch (Springer, New York, 2018), ch 4
15. K.G. Dyall, I.P. Grant, C.T. Johnson, E.P. Plummer, GRASP: a general-purpose relativistic atomic structure program. Comput. Phys. Commun. **55**, 425 (1989)
16. I.P. Grant, *Quantum Theory of Atoms and Molecules: Theory and Computation.* (Springer, New York, USA, 2007)
17. P.G. Burke, *R-Matrix Theory of Atomic Collisions: Application to Atomic, Molecular and Optical Processes.* (Springer, New York, USA, 2011)
18. E.M. Hernández, A.M. Juárez, A.L.D. Kilcoyne, A. Aguilar, L. Hernández, A. Antillón, D. Macaluso, A. Morales-Mori, O. González-Magaña, D. Hanstorp, A.M. Covington, V. Davis, D. Calabrese, G. Hinojosa, Absolute measurements of chlorine Cl^+ cation single photoionization cross section. JQSRT **151**, 217 (2015)
19. B.M. McLaughlin, Photoionisation of Cl^+ from the $3s^2 3p^4\ ^3P_{2,1,0}$ and the $3s^2 3p^4\ ^1D_2,{}^1S_0$ states in the energy range 19–28 eV. Mon. Not. Roy. Astro. Soc. (MNRAS) **464**, 1990 (2017)
20. N.B. Tyndall, C.A. Ramsbottom, C.P. Ballance, A. Hibbert, Photoionization of Co^{2+} using the Dirac R—matrix method. Mon. Not. Roy. Astro. Soc. (MNRAS) **462**, 3350 (2016)
21. A. Müller, S. Schippers, R.A. Phaneuf, A.M. Covington, A., Hinojosa, G., Bozek, J. Sant'Anna, M.M. Sant'Anna, A.S. Schlachter, C. Cisneros, B.M. McLaughlin, Photoionization of Ca^+ in the valence energy region 20–56 eV: experiment and theory. J. Phys. B: At. Mol. Opt. Phys. **50**, 205001 (2017)
22. A.E. Kramida, Y. Ralchenko, J. Reader, NIST ASD team, in *NIST Atomic Spectra Database (version 5.3),* (National Institute of Standards and Technology, Gaithersburg, MD, USA, 2015)
23. A.M. Covington, A. Aguilar, I.R. Covington, G. Hinojosa, C.A. Shirley, R.A. Phaneuf, I. Álvarez, C. Cisneros, I. Dominguez-Lopez, M.M. Sant'Anna, A.S. Schlachter, C.P. Ballance, B.M. McLaughlin, Valence-shell photoionization of chlorinelike Ar^+ ions. Phys. Rev. A **84**, 013413 (2011)
24. C. Blancard et al., L-shell photoionization of Ar^+ to Ar^{3+} ions. Phys. Rev. A **85**, 043408 (2012)
25. N.B. Tyndall, C.A. Ramsbottom, C.P. Ballance, A. Hibbert, Valence and L-shell photoionization of Cl-like argon using R-matrix techniques. Mon. Not. Roy. Astro. Soc. (MNRAS) **456**, 366 (2016)
26. V. Fivet, M.A. Bautista, C.P. Ballance, Fine-structure photoionization cross sections of Fe II. J. Phys. B: At. Mol. Opt. Phys. **45**, 035201 (2012)
27. A. Müller, S. Schippers, D. Esteves-Macaluso, M. Habibi, A. Aguilar, A.L.D. Kilcoyne, R.A. Phaneuf, C.P. Ballance, B.M. McLaughlin, High resolution valence shell photoionization of Ag-like (Xe^{7+}) Xenon ions: experiment and theory. J. Phys. B: At. Mol. Opt. Phys. **47**, 215202 (2014)
28. G. Hinojosa, A.M. Covington, G.A. Alna'Washi, M. Lu, R.A. Phaneuf, M.M. Sant'Anna, C. Cisneros, I. Álvarez, A. Aguilar, A.L.D. Kilcoyne, A.S. Schlachter, C.P. Ballance, B.M. McLaughlin, Valence-shell single photoionization of Kr^+ ions: experiment and theory. Phys. Rev. A **86**, 063402 (2012)
29. D.A. Macaluso, A. Aguilar, A.L.D. Kilcoyne, E.C. Red, R.C. Bilodeau, R.A. Phaneuf, N.C. Sterling, B.M. McLaughlin, Absolute single-photoionization cross sections of Se^{2+}: experiment and theory. Phys. Rev. A **92**, 063424 (2015)
30. M. Barthel, R. Flesch, E. Rühl, B.M. McLaughlin, Photoionization of the $3s^2 3p^4\ ^3P$ and the $3s^2 3p^4\ ^1D,{}^1S$ states of sulfur: experiment and theory. Phys. Rev. A **91**, 013406 (2015)
31. E.T. Kennedy, J.-P. Mosnier, P. Van Kampen, D. Cubaynes, S. Guilbaud, C. Blancard, B.M. McLaughlin, J.-M. Bizau, Photoionization cross sections of the aluminumlike Si^+ ion in the region of the $2p$ threshold (94–137 eV). Phys. Rev. A **90**, 063409 (2014)
32. A. Müller, Precision studies of deep-inner-shell photoabsorption by atomic ions. Phys. Scr. **90**, 054004 (2015)
33. A. Müller, S. Schippers, J. Hellhund, A.L.D. Kilcoyne, R.A. Phaneuf, C.P. Ballance, B.M. McLaughlin, Single and multiple photoionization of W^{q+} tungsten ions in charged states $q = 1, 2, 5$: experiment and theory. J. Phys. Conf. Ser. **488**, 022032 (2014)

34. A. Müller, S. Schippers, J. Hellhund, K. Holosto, A.L.D. Kilcoyne, R.A. Phaneuf, C.P. Ballance, B.M. McLaughlin, Single-photon single ionization of W^+ ions: experiment and theory. J. Phys. B: At. Mol. Opt. Phys. **48**, 2352033 (2015)

35. B.M. McLaughlin, C.P. Ballance, S. Schippers, J. Hellhund, A.L.D. Kilcoyne, R.A. Phaneuf, A. Müller, Photoionization of tungsten ions: experiment and theory for W^{2+} and W^{3+}. J. Phys. B: At. Mol. Opt. Phys. **49**, 065201 (2016)

36. A. Müller, S. Schippers, J. Hellhund, A.L.D. Kilcoyne, R.A. Phaneuf, B.M. McLaughlin, Photoionization of tungsten ions: experiment and theory for W^{4+}. J. Phys. B: At. Mol. Opt. Phys. **50**, 085007 (2017)

37. B. Peart, I.C. Lyon, K. Dolder, Measurements of absolute photoionisation cross sections of Ga^+ and Zn^+ *ions*. J. Phys. B: At. Mol. Phys. **21** 225701 (1987)

38. G. Hinojosa, V.T. Davis, A.M. Covington, J.S. Thompson, A.L.D. Kilcoyne, A. Antillón, E.M. Hernández, D. Calabrese, A. Morales-Mori, A.M. Juárez, O. Windelius, B.M. McLaughlin, Single-Photoionization of the Zn II ion in the photon energy range 17.5–90 eV: experiment and theory. Mon. Not. Roy. Astro. Soc. (MNRAS) **470**, 4048 (2017)

39. D. Nishijima, R.P. Doerner, D.G. Whyte, M.J. Baldwin, T. Schwarz-Selinger, Measurements of Mo I S/XB values. J. Phys. B: At. Mol. Opt. Phys. **43**, 225701 (2010)

40. N.R. Badnell, T.W. Gorczyca, M.S. Pindzola, H.P. Summers, Excitation and ionization of neutral Cr and Mo, and the application to impurity flux. J. Phys. B: At. Mol. Opt. Phys. **29**, 3683 (1996)

41. K. Bartschat, A. Dasgupta, J. Giuliani, Electron-impact excitation of molybdenum from the $(4d^5 5s)a^7 S$ ground state. J. Phys. B: At. Mol. Opt. Phys. **35**, 2899 (2002)

42. G.J. Hartwell, S.F. Knowlton, J.D. Hanson, D.A. Ennis, D.A. Maurer, Design, construction, and operation of the compact toroidal hybrid. Fusion Sci. Technol. **72**, 76 (2017)

43. R.T. Smyth, C.A. Johnson, D.A. Ennis, S.D. Loch, C.A. Ramsbottom, C.P. Ballance, Relativistic R-matrix calculations for the electron-impact excitation of neutral molybdenum. Phys. Rev. A **96**, 042713 (2017)

44. M.S. Pindzola et al., The time-dependent close-coupling method for atomic and molecular collision processes. J. Phys. B: At. Mol. Opt. Phys. **40**, R39 (2007)

45. M.S. Pindzola, J.P. Colgan, B.M. McLaughlin, Electron-impact double ionization of the H_2 molecule. J. Phys. B: At. Mol. Opt. Phys. **51**, 035206 (2018)

46. M.S. Pindzola, M. Folge, P.C. Stancil, Line ratios for x-ray emission in O^{8+} collisions with H and He atoms. J. Phys. B: At. Mol. Opt. Phys. **51**, 065204 (2018)

47. F.D. Priestly, M.J. Barlow, S. Viti, Modelling the ArH^+ emission from the Crab nebula. Mon. Not. Roy. Astro. Soc. (MNRAS) **472**, 4444 (2017)

48. P. Schilke, D.A. Neufeld, H.S.P. Müller, C. Comito, E.A. Bergin, D.C. Lis, M. Gerin, J.H. Black, M. Wolfire, N. Indriolo, J.C. Pearson, K.M. Menten, B. Winkel, Á. Sánchez-Monge, T. Möller, B. Godard, E. Falgarone, Ubiquitous argonium (ArH^+) in the diffuse interstellar medium: a molecular tracer of almost purely atomic gas. Astron. Astrophys. **566**, A29 (2014)

49. Smith, F.T., Fleischmann, H.H., Young, R.A. (1970) Collision Spectroscopy. III. Scattering in Low-Energy Charge-Transfer Collisions of He^+ and Ar. Phys. Rev. A **2**, 379 (1970)

50. R.C. Isler, Inelastic collisions of He^+ and Ar. Phys. Rev. A **10**, 117 (1974)

51. R. Albat, B. Wisram, A four-state calculation of charge exchange in He^+—Ar collisions using a quasimolecule model. J. Phys. B: At. Mol. Phys. **10**, 81 (1977)

52. R. Johnsen, M.T. Leu, M. Biondi, Studies of nonresonant charge transfer between atomic ions and atoms. Phys. Rev. A **8**, 1808 (1973)

53. J.D.C. Jones, D.G. Lister, N.D. Twiddy, Charge-transfer reaction rate coefficients for He^+ and Ne^+ with Ar at 300 K. J. Phys. B: At. Mol. Phys. **12**, 2723 (1979)

54. F.C. Fehsenfeld, A.L. Schmeltekopf, P.D. Goldan, H.I. Schiff, E.E. Ferguson, Thermal Energy Ion–Neutral Reaction Rates. I. Some Reactions of Helium Ions. J. Chem. Phys. **44**, 4087 (1966)

55. J.F. Babb, B.M. McLaughlin, Radiative charge transfer between He^+ and Ar. Astrophys. J. **860**, 151 (2018)

56. T. Helgaker, P. Jorgesen, J. Oslen, *Molecular electronic-structure theory* (Wiley, New York, 2000)
57. S. Langhoff, E.R. Davidson, Configuration interaction calculations on the nitrogen molecule. Int. J. Quantum Chem. **8**, 61 (1974)
58. B. Zygelman, A. Dalgarno, M. Kimura, N.F. Lane, Radiative and nonradiative charge transfer in $He^+ + H$ collisions at low energy. Phys. Rev. A **40**, 2340 (1989)
59. B.M. McLaughlin, H.L.D. Lamb, I.C. Lane, J.F. McCann, Ultracold radiative charge transfer in hybrid Yb ion—Rb atom traps. J. Phys. B: At. Mol. Opt. Phys. **47**, 145201 (2014)
60. J.F. Babb, B.M. McLaughlin, Radiative charge transfer in collisions of C with He^+. J. Phys. B: At. Mol. Opt. Phys. **50**, 044003 (2017)
61. P.C. Stancil, B. Zygelman, Radiative charge transfer in collisions of Li with H^+. Astrophys. J. **472**, 102 (1996)
62. X.J. Liu, Y.Z. Qu, B.J. Xiao, C.H. Liu, Y. Zhou, J.G. Wang, R.J. Buenker, Radiative charge transfer and radiative association in $He^+ + Ne$ collisions. Phys. Rev. A **81**, 022717 (2010)
63. D.A. Neufeld, M.G. Wolfire, The chemistry of interstellar argonium and other probes of the molecular fraction in diffuse clouds. Astrophys. J. **826**, 183 (2016)
64. J. Le Bourlot, F. Le Petit, C. Pinto, E. Roueff, F. Roy, Surface chemistry in the interstellar medium. Astron. Astrophys. **541**, A76 (2012)
65. M. Cairne, R.C. Forrey, J.F. Babb, P.C. Stancil, B.M. McLaughlin, Rate constants for the formation of SiO via radiative association. Mon. Not. Roy. Astro. Soc. (MNRAS) **471** 2481 (2017)
66. C.W. Bauschlicher Jr., Low-lying electronic states of SiO. Chem. Phys. Letts. **658**, 76 (2016)
67. S. Marassi, R. Schneider, M. Limongi, A. Chieffi, M. Bocchio, S. Bianchi, The metal and dust yields of the first massive stars. Mon. Not. Roy. Astro. Soc (MNRAS) **454**, 4250 (2015)
68. A.N. Heays, A.D. Bosman, E.F. van Dishoeck, Photodissociation and photoionisation of atoms and molecules of astrophysical interest. Astron. Astrophys. **602**, A105 (2017)
69. van E.F. Dishoeck, in *Rate Coefficients in Astrochemistry*, ed. by T.J. Millar, D.A. Williams. vol. 49 (Kluwer Academic Publishers, Dordrecht, Boston, MA, 1988)
70. R.J. Patillo, R. Cieszewsk, P.C. Stancil, R.C. Forrey, J.F. Babb, J.F. McCann, B.M. McLaughlin, Photodissociation of CS from excited rovibrational levels. Astrophys. J. **858**, 10 (2018)

Interactions Between Blood Proteins and Nanoparticles Investigated Using Molecular Dynamics Simulations

Timo Schafer, Christian Muhl, Matthias Barz, Friederike Schmid
and Giovanni Settanni

Abstract In the development of new therapeutic agents based on nanoparticles it is of fundamental importance understanding how these substances interact with the underlying biological milieu. Our research is focussed on simulating *in silico* these interactions using accurate atomistic models, and gather from these information general pictures and simplified models of the underlying phenomena. Here we report results about the interactions of blood proteins with promising hydrophilic polymers used for the coating of therapeutic nanoparticles, about the salt dependent behavior of one of these polymers (poly-(ethylene glycol)) and about the interactions of blood proteins with silica, one of the most used materials for the production of nanoparticles.

1 Introduction

Huge steps in the production of nanosized materials with specific functionalities have opened the way to a vast variety of applications, one of which is their use as drug delivery systems [1]. A wide spectrum of nanoparticles are being tested for their capabilities to load different kinds of cargos, remain soluble in the biological milieu (e.g. blood, mucosa, etc.), evade secretion and immune response, target specific tissues, and specifically release their cargo. The efficiency of most of these processes depends on how the nanoparticle surface interacts with the biological medium in which it is introduced. These interactions determine the composition of the layer of biological material (protein corona) that forms around nanoparticles as they come in contact with an organism. It is precisely the protein corona that has been shown to determine the fate of the nanoparticle in the host organism, in terms for example of circulation time, cell uptake or immunogenicity [2, 3].

Understanding protein-nanoparticles interactions, thus, represents a fundamental step for designing effective nano materials for drug delivery purposes. A molecular

T. Schafer · F. Schmid · G. Settanni (✉)
Institut für Physik, Johannes Gutenberg University, Mainz, Germany
e-mail: settanni@uni-mainz.de

C. Muhl · M. Barz
Institut für Organische Chemie, Johannes Gutenberg University, Mainz, Germany

© Springer Nature Switzerland AG 2019
W. E. Nagel et al. (eds.), *High Performance Computing in Science
and Engineering '18*, https://doi.org/10.1007/978-3-030-13325-2_4

level description of these interactions would enable a fine grained tuning of the nanoparticle to enhance the properties we are interested in and suppress those that have a negative impact on its functionality. A direct experimental characterization of the molecular interactions involved is made difficult by the small length scales and the complexity of the systems, which is often not easily and completely controllable. An *in silico* approach, on the other hand, allows for a reduction of the complexity of the problem by focussing only on selected aspects and could, in principle, provide a high resolution picture of the involved mechanisms.

The main tool of our research is classical atomistic molecular dynamics (MD), where the position and velocities of each atom of the simulated system are evolved according to Newton's equations of motion [4] and an empirical force field, that is classical approximations of the underlying quantum mechanical interactions between the atoms [5]. During the last 40 years increased availability of computer resources and improvements in the accuracy of the force fields have made possible the atomistic simulation of biological systems (proteins, nucleic acids, lipids, sugars, etc.) of the size of tens to hundreds nanometers on time scales of hundreds to thousand nanoseconds (even milliseconds on specially designed architectures) [6].

Here, we address with the help of MD the interface between biological materials and nanoparticle surfaces. We do that by exploiting further recent improvements in force field development, namely the availability of force field parameters describing materials like silica [7] or polymers [8–10] and compatible with those available for biological matter, like proteins, lipids and sugars.

The manuscript is structured in the following way: in the general method section we describe the basic algorithms that are used throughout our research. Then, results are separately reported for the three research directions: 1. Characterization of blood proteins interactions with polymers used for coating nanoparticles. 2. Properties of poly(ethylene glycol) in saline solutions. 3. Characterization of blood protein interactions with solid nanoparticle materials. In each of these subsections we will also report the methodologies and simulation set-up that are specific to them.

2 Methods

Molecular dynamics simulations are carried out using the program NAMD [11], which has been specifically compiled to exploit the Cray XC40 architecture of Hazelhen. Unless specified differently, simulations are carried out using the CHARMM force field [12], with a base time step of 1fs. Direct space non-bonded interactions are cut-off at 1.2 nm with a switch function smoothing them out from 1.0 to 1.2 nm. The neighbor list cutoff is 1.4 nm. A specially designed cell-list algorithm is used to speed up neighbor search [13]. Long range electrostatic interactions are treated using the smooth particle mesh Ewald (PME) method [14]. A multiple time step scheme is implemented [15], so that non-bonded interactions are computed every 2 steps and long range electrostatics every 4. Water molecules are simulated explicitly using the TIP3P model [16]. A Langevin piston [17, 18] is used to control the temperature and the pressure of the system at 300 K and 1 atm, respectively. Unless specified differently, all the simulations are carried out at physiological ion concentration [NaCl] = 150 mM by replacing few water molecules with sodium and chlorine ions during the simulation setup. This setup insures good scaling on Hazelhen up to about 100 nodes for typical systems of the size of hundreds thousand atoms which we normally simulate. The job length of our typical jobs on Hazelhen generally does not exceed 3 h, so that we do not need to write intermediate restart files, an operation that, as a matter of experience on Hazelhen, may lead to crashes in case the disk write is delayed. In this way we generally manage to collect trajectories from 1 to 8 ns per hour per 100 nodes on Hazelhen. Multiple runs with different initial conditions are used to improve the statistical accuracy of the results.

Beside standard classical MD simulations, we also adopted in selected cases an enhanced sampling technique, namely Hamiltonian replica exchange with solute tempering (REST2) [19, 20]. According to this technique multiple replicas of the system are run at the same time, where the interactions between a selected part of the system and the rest are rescaled differently in each replica. The replicas are then allowed to swap the scaling factor at regular intervals during the simulations, according to a Metropolis-like criterion. It can be demonstrated that this is equivalent to selectively increase the temperature of the selected part of the system [19], thus helping the crossing of high free energy barriers like in standard replica exchange/parallel tempering techniques, but with the advantage that the number of replicas necessary to allow for good replica swap probability and a large temperature range across the replicas depends on the number of degrees of freedom of the selected part of the system and not the total number of degrees of freedom, which make standard replica exchange/parallel tempering inconvenient for large atomistic systems with explicit solvent.

3 Protein-Polymers Interactions

The aim of this part of the project is to identify the features that make a polymer amenable for coating nanoparticles for drug delivery. The objective is investigated by simulating at atomistic resolution the behavior of several blood proteins in contact with polymers in an aqueous environment using molecular dynamics and cross validating the outcomes with experimental results from collaboration partners. We have been considering several different polymers which are being investigated by our experimental collaborators, in particular poly(ethylene glycol) (PEG), poly-sarcosine (PSAR) and PHPMA (the case of poly-phosphonates has been addressed indirectly, as it will be explained later). The proteins that we have simulated include human serum albumin (HSA), transferrin (TFA), complement Cq1 subcomponent subunit C, bovine serum albumin (BSA) and apolipoprotein A1 (APOA1). Beside standard classical atomistic molecular dynamics we have also investigated the use of Hamiltonian replica exchange with solute tempering [19, 20], which can help to speed up sampling. Details of the simulation techniques and set up are discussed more extensively below.

In the case of PEG, the simulations revealed that the PEG density around each amino acid depends mainly on its type, with negatively charged residues showing the lowest densities and non-polar residues showing the highest. In other words effective attractive interactions between PEG and non-polar residues were observed in the simulations, while the effective interactions between PEG and negatively charged residues were mainly repulsive. From those observations we have derived a model that describes the interactions of proteins with densely PEGylated nanoparticles using only the amino acid composition of the protein surface. We then applied the model to a large set of blood proteins for which the three-dimensional structure has been determined and verified a good correlation between the expected PEG density around the protein as derived from the model and the adsorption free energy of the proteins on PEGylated nanoparticles measured using mass spectroscopy experiments [21]. After noting that the adsorption free energies of the blood proteins on the PEGylated nanoparticles are highly correlated to those measured on nanoparticles densely grafted with poly-phosphonates, we showed that exactly the same model obtained for protein interactions with the PEGylated nanoparticle can be applied to the poly-phosphonated ones [21]. Further analysis of the simulations also allowed to measure the differential binding coefficients of several proteins in PEG water mixtures [22].

On the basis of the results obtained for PEG and poly-phosphonate we have extended the calculations to PSAR. PSAR is a poly-peptoid (the monomer is similar to alanine but with the residue C_β atom bound to the nitrogen atom rather than to the C_α atom), which, like PEG, can help reduce unspecific interactions with proteins, but, unlike PEG, it can be metabolized by the organism [23]. These facts make PSAR a very promising substitute for PEG. As done for PEG, we measured the affinity of the protein surface for PSAR using atomistic molecular dynamics simulations of HSA in PSAR/water mixtures. Then, we extracted the amino acid/PSAR interaction parameter as done for PEG. The results show that the affinity of the various amino acids for PSAR are similar to those measured for PEG (Fig. 1), which is in agreement with the similarity in the behavior of the two polymers in contact with proteins. We are now in the process of analyzing these simulations in relation to the mass spectroscopy data of PSAR-coated nanoparticles made available by our collaboration partners. The work on PSAR is made in collaboration with the group of Dr. Barz in the chemistry department of our university. Along the same line of research we have also carried out simulations of proteins in PHPMA/water mixture and we are now in the process of analyzing the results. Similar to the protocol used with PEG [21] the simulation setup for PSAR-protein simulations and PHPMA-protein simulations, consisted in an initial preparation and equilibration of the polymer/water mixture, followed by insertion of the protein (HSA in this case) and subsequent equilibration. In addition to the CHARMM force field [5], a specific force field for PSAR has been adopted

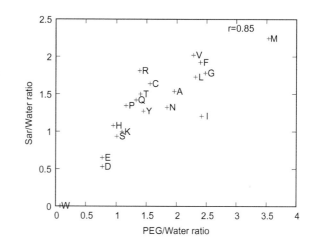

Fig. 1 The PSAR/water ratio measured along the simulations of HSA around each amino acid type is plotted versus the PEG/water ratio for the same protein. The high correlation coefficient between the two data sets highlights the similarity in the behavior of the two polymers in contact with blood proteins

[9], while the one for PHPMA has been generated using CHARMM generalized force field [10]. This led to cubic simulation box sizes of about 10 nm side length and 100 k atoms. For the production phase, five independent runs were carried out each for 200 ns. The typical job on Hazelhen consisted of a 125-node run lasting approximately 40 min and collecting cumulatively 5 ns of trajectory. The trajectories are then analyzed using VMD [24] and WORDOM [25].

The standard simulations of PEG-protein interactions showed very long auto-correlation times for the protein-PEG pair interaction energy [21], leading to long convergence times. To try and shorten these times we set up several REST2 calculations where we selected parts of the simulation box to undergo parallel tempering. In a first setup we simulated HSA in PEG/water mixture and selected only the PEG molecules for the tempering. We run 50 replicas in a temperature range from 300 to 650 K for 10 ns. These simulations showed good replica swapping probabilities, however the replicas at high temperature displayed a phase separation between PEG and water due to the reduction of the interaction parameters between the two subsystem. This resulted in limited diffusion of replicas in the temperature space (Fig. 2). We then set up a new system where we included the protein in the tempered set of atoms, with the idea of enhancing PEG-protein interactions at high temperature. The protein was kept folded also at high temperature by means of artificial springs among the CA atoms. In this case we reduced the number of replicas to 25 and, because of the increased number of tempered degrees of freedom we also reduced the temperature range to 300–373 K to keep the replica swap probabilities similar to the previous case. We are now in the process of analyzing these results.

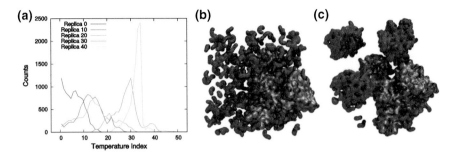

Fig. 2 Results from early REST2 simulations: **a** distribution of the temperatures visited by few selected replicas. Although the distributions are relatively broad in most cases, they are not uniform across the whole range of temperatures, indicating limited temperature diffusion of the replicas. **b** and **c** Typical snapshots from replicas with PEG tempered at 300 and 650 K, where the protein is represented as a colored surface and PEG molecules as red surfaces (water and ions have been omitted for clarity). PEG is properly dissolved in the simulation box at 300 K while it forms a compact phase at high temperature due to the reduced interaction with the surrounding water. This phenomenon explains the reduced temperature diffusion of replicas.

4   Properties of PEG in Saline Solutions

Although PEG has widespread use in bio- and medicinal chemistry, its properties are not yet fully understood. In collaboration with the lab of Prof. Sönnichsen, in the Physical Chemistry Institute of our university, we have studied the elastic properties of PEG in saline solutions. Experiments using plasmon rulers performed in our partners lab showed an increase of the stiffness of PEG single polymer chains and reduced end-to-end distance in the presence of small concentrations of potassium ions in the water solution. Our atomistic simulations of PEG 36-mer and 18-mer showed that potassium induced massive formation of crown-like structures where about 7 adjacent PEG monomers surrounded a single potassium ion (Fig. 3). Control simulations showed that this did not occur when potassium was replaced by sodium, and this fact was subsequently validated experimentally. Thus, the increased stiffness and reduced end-to-end distance of PEG chains could be attributed to the formation of the crown-like structures in the presence of potassium ions in solution [26]. In this case the simulations consisted in REST2 runs of a single PEG chain immersed in different solvent boxes. We used 2 chain lengths of 18 and 36 monomers and cubic boxes of 7.2 nm side length. Three different solvents were used to fill the boxes, pure water, saline water with [NaCl] = 0.15 M and saline water with [KCl] = 0.15M. The salt ions were introduced by replacing few water molecules to reach the expected concentration. Each REST2 run consisted of 10 replicas spanning temperatures from 300 to 650 K for the polymer chain, while the rest of the system was kept to 300 K. Each replica was run for 100 ns with swap attempts every 1ps, leading to good swap probabilities for both chains and a uniform distribution of the replicas over the 10 temperatures. The stretches of trajectory from replicas running at 300 K were used

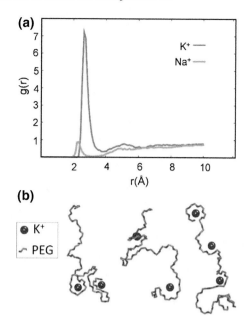

Fig. 3 **a** The radial distribution functions of potassium (*blue line*) and sodium (*green line*) cations as a function of distance to the oxygen atoms of the PEG chain. The strong peak at about 2.5 Å observed for potassium but not sodium suggests a much stronger coordination to PEG. **b** Coordination of potassium ion in three representative snapshots from the simulations shows the crown pattern around the potassium cations. The picture is adapted with permission from reference [26]. Copyright (2017) American Chemical Society

for analysis. VMD [24] and WORDOM [25] were used to analyze the simulations. In this case the typical Hazelhen job consisted of a 80-node run of the 10 replicas in parallel for 1 ns, which took approximately 45 min in real time.

5   Characterization of Blood Protein Interactions with Solid Nanoparticle Materials

Silica nanoparticles represent a vast field of research. The large variety of techniques that have been developed to produce them address a wide range of design requirements. For this reason, understanding how blood proteins interact with silica surfaces is of crucial importance. Along this line of research, we have simulated at the atomistic level the early adsorption stages of fibrinogen and HSA on silica surfaces. The crystal structure of human Fg (PDB ID: 3GHG) and HSA (PDB ID: 1AO6) [27, 28], were used as initial structure for the simulations. The model of the silica slab is based on the Interface forcefield [7]. The initial model consists of a Q3 surface with a silanol group surface density of 4.7 nm^{-2} and an ionization degree of 14%, fit to the properties of silica nanoparticle surfaces at pH 7. Based on this surface model, a $30 \times 17 \times 2$ nm^3 and a $13 \times 14 \times 2$ nm^3 pseudo-infinite slab were generated for Fg and HSA respectively. In addition to the methods already listed above, in this case, all covalent bonds involving hydrogen were fixed in their length to allow a time-steps of 2 fs. The dimensions of the system in the plane of the slab were fixed. Additionally, the bottom layer of silica atoms in the slab was restrained using a set of quadratic potentials to keep the slab from drifting. While building the simulation box, The protein was rotated to generate different starting conformations with a distance of 0.8 nm from the surface. For the production simulation, the colvars module [29] of NAMD was used to keep the proteins from moving too far away from the surface and increase the frequency of adsorption attempts. For each starting configuration several independent simulations were performed with different initial velocities. The length of the trajectories reaches 100 ns for Fg and 200 ns for HSA. Typical jobs on Hazelhen for Fg involved 100 nodes for 70 min and collected 1.2 ns of trajectory. For HSA jobs involved 120 nodes for 30 min and prodcuced 1.2 ns of trajectory.

In all simulations, we observe the formation of contacts between protein and silica (a contact occurs when a protein heavy atom reaches within 0.5 nm of the average position of the highest layer of silica atoms on the slab [30]). The top layer of silica atoms is highlighted in Fig. 4. The average number of contacts observed over time is shown in Fig. 5a. For Fg the number of contacts increases with some fluctuations and a small plateau towards the end. For HSA a plateau is reached much earlier, with around 5 contacts from 30 to 200 ns. To understand the determinants of contact formation, amino acids have been grouped according to their characteristics. Figure 5b shows the fraction of contacts formed by different regions of Fg. At the earliest times, all contacts are formed via the glycosilation sites, as these are the most exposed. They can easily "anchor" Fg to the silica surface, while the rest of

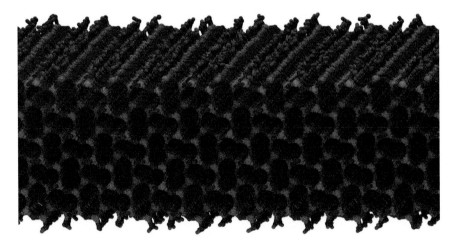

Fig. 4 Silica slab structure. Silica slab model shown with ionized silanol group oxygen atoms highlighted in purple and top silica atom layer highlighted in bright yellow

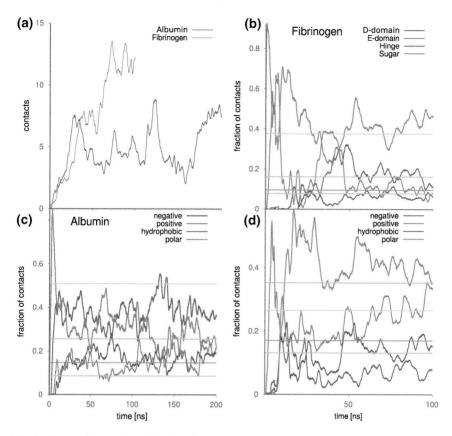

Fig. 5 Contact formation. a Number of contacts between protein and silica over time, averaged over all trajectories. **b** Fraction of contacts formed by different regions of Fg. Straight lines are expected fractions according to exposed surface area **c** Fraction of contacts formed by different types of amino acids of Alb. **d** Fraction of contacts formed by different types of amino acids of Fg

the protein slowly approaches. After the glycosilation sites, the E-domain makes up a majority of contacts. Overall, the E-domain formed contacts significantly more often than expected from its fraction of exposed surface, while the D-domain is underrepresented. Figure 5c, d shows that for both Fg and Alb, positive amino acids take part in contact formation more frequently than expected. Polar and especially negative amino acids are underrepresented. This is likely due to the high negative charge of ionized silanol groups on the silica surface. Notably, positive amino acids are a main part of the initial contact formation (in the case of Fg, the glycosilation sites contribute as well). It should be noted that the local environment of the amino acids matters, as negative amino acids may form contacts with the surface not because it is favorable, but because they are close to positive amino acids. One amino acids that stands out for both proteins is lysine. The high amount of contacts formed by positively charged amino acids is almost entirely due to lysine (Fig. 6). Another

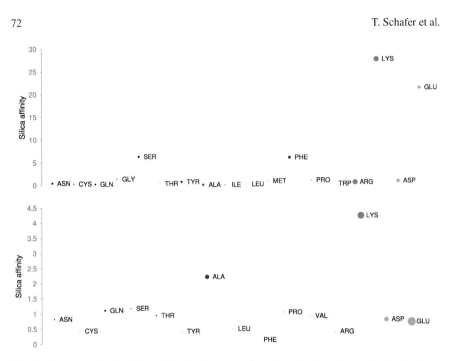

Fig. 6 Amino acids affinities. a Affinity for contact formation for Fg amino acids. Ordered and colored by electrostatics. Point size represent absolute number of contacts formed. **b** Affinity for contact formation for Alb amino acids. Same presentation as in (**a**)

noteworthy amino acids is glutamic acid which plays a significant role in contact formation for Fg despite its negative charge.

The mode of adsorption observed for silica resembles the one already observed for mica [30]. The analysis will be further extended to identify possible preferential adsorption orientations which may play an important role in the determination of adsorption-related functional changes in blood proteins. As a further step along the same line of research, we intend to extend this kind of investigations to polystyrene, a material which has also been used extensively in nanoparticle research.

6   Conclusions

Here, we have shown how MD simulations, which efficiently exploit the Cray XC40 architecture of Hazelhen, can be used to asses protein-nanoparticle interfaces in a way that is complementary to experiments. The atomistic resolution attainable with this technique provides important insights and has helped explaining experimental observations. The simulations have helped to explain how PEG chains shrink in the presence of potassium ions and they are being used to reveal how PEG and PSAR show similar stealth effects. They are being further used to describe protein interaction with solid surfaces like silica and, in the future, polystyrene, with the final aim to get a more complete view of how biological matter may interact with nanoparticles and possibly predicting their efficacy.

Acknowledgements TS gratefully acknowledges financial support from the Graduate School Materials Science in Mainz. GS gratefully acknowledges financial support from the Max-Planck Graduate Center with the University of Mainz. We gratefully acknowledge support with computing time from the HPC facility Hazelhen at the High performance computing center Stuttgart and the HPC facility Mogon at the university of Mainz. This work was supported by the German Science Foundation within SFB 1066 project Q1.

References

1. K. Cho, X. Wang, S. Nie, Z.G. Chen, D.M. Shin, Clin. Cancer Res. **14**(5), 1310 (2008). https://doi.org/10.1158/1078-0432.CCR-07-1441. http://clincancerres.aacrjournals.org/content/14/5/1310.abstract
2. M.P. Monopoli, C. Aberg, A. Salvati, K.A. Dawson, Nat. Nanotechnol. **7**(12), 779 (2012). https://doi.org/10.1038/nnano.2012.207. URL http://dx.doi.org/10.1038/nnano.2012.207
3. A. Lesniak, F. Fenaroli, M.P. Monopoli, C. Aberg, K.A. Dawson, A. Salvati, ACS Nano **6**(7), 5845 (2012). https://doi.org/10.1021/nn300223w. URL http://dx.doi.org/10.1021/nn300223w
4. D. Frenkel, B. Smit, *Understanding Molecular Simulations, Computational Science*, vol. 1, 2nd edn. (Academic Press, 2002)
5. A.D. MacKerell, D. Bashford, Bellott, R.L. Dunbrack, J.D. Evanseck, M.J. Field, S. Fischer, J. Gao, H. Guo, S. Ha, D. Joseph-McCarthy, L. Kuchnir, K. Kuczera, F.T.K. Lau, C. Mattos, S. Michnick, T. Ngo, D.T. Nguyen, B. Prodhom, W.E. Reiher, B. Roux, M. Schlenkrich, J.C. Smith, R. Stote, J. Straub, M. Watanabe, J. Wiórkiewicz-Kuczera, D. Yin, M. Karplus, J. Phys. Chem. B **102**(18), 3586 (1998). https://doi.org/10.1021/jp973084f. URL http://pubs.acs.org/doi/abs/10.1021/jp973084f
6. K. Lindorff-Larsen, S. Piana, R.O. Dror, D.E. Shaw, Science **334**(6055), 517 (2011). https://doi.org/10.1126/science.1208351. http://www.sciencemag.org/cgi/content/abstract/sci;334/6055/517
7. H. Heinz, T.J. Lin, R.K. Mishra, F.S. Emami, Langmuir **29**(6), 1754 (2013). https://doi.org/10.1021/la3038846. URL http://dx.doi.org/10.1021/la3038846. PMID: 23276161
8. H. Lee, R.M. Venable, A.D. Mackerell, R.W. Pastor, Biophys. J. **95**(4), 1590 (2008). https://doi.org/10.1529/biophysj.108.133025
9. D.T. Mirijanian, R.V. Mannige, R.N. Zuckermann, S. Whitelam, J. Comput. Chem. **35**(5), 360 (2014). https://doi.org/10.1002/jcc.23478
10. K. Vanommeslaeghe, E. Hatcher, C. Acharya, S. Kundu, S. Zhong, J. Shim, E. Darian, O. Guvench, P. Lopes, I. Vorobyov, A.D. Mackerell, J. Comput. Chem. **31**(4), 671 (2010). https://doi.org/10.1002/jcc.21367. URL http://dx.doi.org/10.1002/jcc.21367
11. J.C. Phillips, R. Braun, W. Wang, J. Gumbart, E. Villa, C. Chipot, R.D. Skeel, L. Kale, K. Schulten, J. Comput. Chem. **26**, 1781 (2005)
12. A.D. Mackerell, M. Feig, C.L. Brooks, J. Comput. Chem. **25**(11), 1400 (2004). https://doi.org/10.1002/jcc.20065
13. M.P. Allen, D.J. Tildesley, *Computer Simulation of Liquids* (Clarendon Press, 1987)
14. U. Essmann, L. Perera, M.L. Berkowitz, T. Darden, H. Lee, L.G. Pedersen, J. Chem. Phys. **103**(19), 8577 (1995). https://doi.org/10.1063/1.470117. http://scitation.aip.org/content/aip/journal/jcp/103/19/10.1063/1.470117
15. R.D. Skeel, J.J. Biesiadecki, Ann. Numer. Math. **1**, 191 (1994)
16. W.L. Jorgensen, J. Chandrasekhar, J.D. Madura, R.W. Impey, M.L. Klein, J. Chem. Phys. **79**(2), 926 (1983). https://doi.org/10.1063/1.445869

17. G.J. Martyna, D.J. Tobias, M.L. Klein, J. Chem. Phys. **101**(5), 4177 (1994). https://doi.org/10.1063/1.467468. http://link.aip.org/link/?JCP/101/4177/1
18. S.E. Feller, Y. Zhang, R.W. Pastor, B.R. Brooks, J. Chem. Phys. **103**(11), 4613 (1995). https://doi.org/10.1063/1.470648. http://link.aip.org/link/?JCP/103/4613/1
19. L. Wang, R.A. Friesner, B.J. Berne, J. Phys. Chem. B **115**(30), 9431 (2011). https://doi.org/10.1021/jp204407d
20. S. Jo, W. Jiang, Comput. Phys. Commun. **197**, 304 (2015). https://doi.org/10.1016/j.cpc.2015.08.030
21. G. Settanni, J. Zhou, T. Suo, S. Schöttler, K. Landfester, F. Schmid, V. Mailänder, Nanoscale **9**(6), 2138 (2017). https://doi.org/10.1039/C6NR07022A. http://pubs.rsc.org/en/content/articlelanding/2017/nr/c6nr07022a
22. G. Settanni, J. Zhou, F. Schmid, J. Phys. Conf. Ser. **921**(1), 012002 (2017)
23. D. Zhang, S.H. Lahasky, L. Guo, C.U. Lee, M. Lavan, Macromolecules **45**(15), 5833 (2012). https://doi.org/10.1021/ma202319g. URL https://doi.org/10.1021/ma202319g
24. W. Humphrey, A. Dalke, K. Schulten, J. Mol. Gr. **14**, 33 (1996)
25. M. Seeber, M. Cecchini, F. Rao, G. Settanni, A. Caflisch, Bioinformatics **23**(19), 2625 (2007)
26. L. Tüting, W. Ye, G. Settanni, F. Schmid, B. Wolf, R. Ahijado-Guzmann, C. Sönnichsen, J. Phys. Chem. C **121**, 22396 (2017)
27. J. Kollman, L. Pandi, M. Sawaya, M. Riley, R. Doolittle, Biochemistry **48**(18), 3877 (2009). https://doi.org/10.1021/bi802205g
28. S. Sugio, A. Kashima, S. Mochizuki, M. Noda, K. Kobayashi, Protein Eng. **12**(6), 439 (1999)
29. G. Fiorin, M.L. Klein, J. Hénin, Mol. Phys. **111**(22–23), 3345 (2013)
30. S. Köhler, F. Schmid, G. Settanni, Langmuir **31**(48), 13180 (2015). https://doi.org/10.1021/acs.langmuir.5b03371. PMID: 26569042

The QCD Phase Diagram from the Lattice

Higher Order Fluctuations and Correlations of Conserved Charges from Lattice QCD
Status Report: POSR 44090

S. Borsanyi, Z. Fodor, J. Günther, S. D. Katz,
A. Pasztor, I. P. Vazquez, C. Ratti and K. K. Szabó

Abstract We calculate several diagonal and non-diagonal fluctuations of conserved charges in a system of $2 + 1 + 1$ quark flavors with physical masses, on a lattice with size $48^3 \times 12$. Higher order fluctuations at $\mu_B = 0$ are obtained as derivatives of the lower order ones, simulated at imaginary chemical potential. From these correlations and fluctuations we construct ratios of net-baryon number cumulants as functions of temperature and chemical potential, which satisfy the experimental conditions of strangeness neutrality and proton/baryon ratio. Our results qualitatively explain the behavior of the measured cumulant ratios by the STAR collaboration. We explain the obtained simulation results with a simple model, and find consistent behaviour with a scenario with no nearby critical end point in the QCD phase diagram.

S. Borsanyi (✉) · Z. Fodor · A. Pasztor · K. K. Szabó
Department of Physics, Wuppertal University, Gaussstr. 20, 42119 Wuppertal, Germany
e-mail: borsanyi@uni-wuppertal.de

Z. Fodor · S. D. Katz
Institute for Theoretical Physics, Eötvös University, Pázmány P. sétány 1/A,
Budapest 1117, Hungary

Z. Fodor · K. K. Szabó
Jülich Supercomputing Centre, Forschungszentrum Jülich, 52425 Jülich, Germany

J. Günther · S. D. Katz
Department of Physics, University of Regensburg, 93053 Regensburg, Germany

J. Günther · S. D. Katz
MTA-ELTE Lendület Lattice Gauge Theory Research Group, Pázmány P. sétány 1/A,
Budapest 1117, Hungary

I. P. Vazquez · C. Ratti
Department of Physics, University of Houston, Houston, TX 77204, USA

© Springer Nature Switzerland AG 2019
W. E. Nagel et al. (eds.), *High Performance Computing in Science
and Engineering '18*, https://doi.org/10.1007/978-3-030-13325-2_5

1 Introduction

One of the most challenging goals in the study of Quantum Chromodynamics (QCD) is a precise mapping of the phase diagram of strongly interacting matter. First principle, lattice QCD simulations predict that the transition from hadrons to de confined quarks and gluons is a smooth crossover [1–6], taking place in the temperature range $T \simeq 145 - 165$ MeV. Lattice simulations cannot presently be performed at finite density due to the sign problem, thus leading to the fact that the QCD phase diagram is still vastly unexplored when the asymmetry between matter and antimatter becomes large.

With the advent of the second Beam Energy Scan (BES-II) at the Relativistic Heavy Ion Collider (RHIC), scheduled for 2019–2020, there is a renewed interest in the heavy ion community towards the phases of QCD at moderate-to-large densities. A rich theoretical effort is being developed in support of the experimental program; several observables are being calculated, in order to constrain the existence and location of the QCD critical point and to observe it experimentally.

Fluctuations of conserved charges (electric charge Q, baryon number B and strangeness S) are among the most relevant observables for the finite-density program for several reasons. One possible way to extend lattice results to finite density is to perform Taylor expansions of the thermodynamic observables around chemical potential $\mu_B = 0$ [7–11]. Fluctuations of conserved charges are directly related to the Taylor expansion coefficients of such observables, thus, they are needed to extend first principle approaches to the regions of the phase diagram relevant to RHIC. An other popular method to extend observables to finite density is the analytical continuation from imaginary chemical potentials [12–16]. The agreement between the analytical continuation and Taylor expansion was shown for the transition temperature by Bonati et al. in Ref. [17].

Fluctuations can also be measured directly, and a comparison between theoretical and experimental results allows to extract the chemical freeze-out temperature T_f and chemical potential μ_{Bf} as functions of the collision energy [18–22]. Such fluctuations have been recently calculated and extrapolated using the Taylor method in Ref. [23]. Finally, higher order fluctuations of conserved charges are proportional to powers of the correlation length and are expected to diverge at the critical point, thus providing an important signature for its experimental detection [9, 24, 25].

In this paper, we calculate several diagonal and non-diagonal fluctuations of conserved charges up to sixth-order and give estimates for higher orders, in the temperature range 135 MeV $\leq T \leq$ 220 MeV, for a system of $2 + 1 + 1$ dynamical quarks with physical masses and lattice size $48^3 \times 12$. We simulate the lower-order fluctuations at imaginary chemical potential and extract the higher order fluctuations as derivatives of the lower order ones at $\mu_B = 0$. This method has been successfully used in the past and proved to lead to a more precise determination of the higher order fluctuations, compared to their direct calculation [26, 27]. The direct method (see e.g. [7]) requires the evaluation of several terms and is affected by a signal-to-noise ratio which is decreasing as a power law of the spatial volume V, with an exponent that grows with the order of the susceptibility.

2 Fluctuations and Imaginary Chemical Potentials

The chemical potentials are implemented on a flavor-by-flavor basis, their relation to the phenomenological baryon (B), electric charge (Q) and strangeness (S) chemical potentials are given by

$$\mu_u = \frac{1}{3}\mu_B + \frac{2}{3}\mu_Q$$
$$\mu_d = \frac{1}{3}\mu_B - \frac{1}{3}\mu_Q$$
$$\mu_s = \frac{1}{3}\mu_B - \frac{1}{3}\mu_Q - \mu_S. \tag{1}$$

The observables we are looking at are the derivatives of the free energy with respect to the chemical potentials. Since the free energy is proportional to the pressure, we can write:

$$\chi^{B,Q,S}_{i,j,k} = \frac{\partial^{i+j+k}(p/T^4)}{(\partial\hat{\mu}_B)^i (\partial\hat{\mu}_Q)^j (\partial\hat{\mu}_S)^k}, \tag{2}$$

with

$$\hat{\mu}_i = \frac{\mu_i}{T}. \tag{3}$$

These are the generalized fluctuations we calculated around $\mu = 0$ in our previous work [28].

At $\mu_B = i\pi T$ there is a first order phase transition at all temperatures above the Roberge-Weiss critical end point T_{RW} [29]. When μ_B crosses $i\pi T$ in the imaginary direction, the imaginary baryon density is discontinuous. This behaviour is illustrated in Fig. 1, where the imaginary baryon density as a function of the imaginary chemical potential is shown. At low temperature the Hadron Resonance Gas model predicts $\langle B \rangle \sim \sinh(\mu_B/T)$, thus for imaginary values we expect a sine function below T_c: $\mathrm{Im}\langle B \rangle \sim \sin(\mathrm{Im}\mu_B/T)$. At temperatures slightly above T_c, we observe that further Fourier components appear in addition to $\sin(\mathrm{Im}\mu_B/T)$ with alternating coefficients, these are consistent with a repulsive interaction between baryons [30]. At very high temperatures, on the other hand, $\langle B \rangle$ is a polynomial of μ_B since the diagrams contributing to its $\sim\mu_B^5$ and higher order components are suppressed by asymptotic freedom [31, 32]. The Stefan-Boltzmann limit is non-vanishing only for two Taylor coefficients of $\mathrm{Im}\langle B \rangle$, giving $\mathrm{Im}\langle B \rangle|_{\mu_B/T=i\pi-\epsilon} = 8\pi/27$. At finite temperatures above T_{RW} this expectation value is smaller but positive. It implies a first order transition at $\mu_B = i\pi T$.

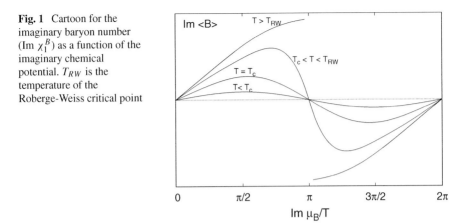

Fig. 1 Cartoon for the imaginary baryon number (Im χ_1^B) as a function of the imaginary chemical potential. T_{RW} is the temperature of the Roberge-Weiss critical point

3 Lattice Setup

In this work we calculate high order fluctuations by studying the imaginary chemical potential dependence of various generalized quark number susceptibilities.

We use a tree-level Symanzik improved gauge action, with four times stout smeared ($\rho = 0.125$) staggered fermions. We simulate $2 + 1 + 1$ dynamical quarks, where the light flavors are tuned in a way to reproduce the physical pion and kaon masses and we set $\frac{m_c}{m_s} = 11.85$ [33]. For the zero-temperature runs that we used for the determination of the bare masses and the coupling, the volumes satisfy $Lm_\pi > 4$. The scale is determined via f_π. More details on the scale setting and lattice setup can be found in [28].

Our lattice ensembles are generated at eighteen temperatures in the temperature range 135...220 MeV. We simulate at eight different values of imaginary μ_B given as: $\mu_B^{(j)} = iT \frac{j\pi}{8}$ for $j \in \{0, 1, 2, 3, 4, 5, 6, 7\}$. In this work the analysis is done purely on a $48^3 \times 12$ lattice, we leave the continuum extrapolation for future work.

In terms of quark chemical potentials we generate ensembles with $\mu_u = \mu_d = \mu_s = \mu_B/3$. In each simulation point we calculate all derivatives in Eq. (2) up to fourth order. Thanks to our scan in Im $\hat{\mu}_B$, we can calculate additional μ_B derivatives. Reference [27] uses various "trajectories" in the $\mu_B - \mu_Q - \mu_S$ space, allowing the numerical determination of higher e.g. μ_Q and μ_S derivatives. We find relatively good signal for the μ_Q and μ_S derivatives by directly evaluating Eq. (2) within one simulation. We recently summarized the details of the direct calculation in Ref. [28].

4 Results for the High Order Fluctuations

In the actual simulations we determine χ_{ijk}^{BQS} as a function of temperature and imaginary chemical potential

$$\chi_1^B(\hat{\mu}_B) = 2c_2\hat{\mu}_B + 4c_4\hat{\mu}_B^3 + 6c_6\hat{\mu}_B^5 + \frac{4!}{7!}c_4\epsilon_1\hat{\mu}_B^7 + \frac{4!}{9!}c_4\epsilon_2\hat{\mu}_B^9 \tag{4}$$

$$\chi_2^B(\hat{\mu}_B) = 2c_2 + 12c_4\hat{\mu}_B^2 + 30c_6\hat{\mu}_B^4 + \frac{4!}{6!}c_4\epsilon_1\hat{\mu}_B^6 + \frac{4!}{8!}c_4\epsilon_2\hat{\mu}_B^8 \tag{5}$$

$$\chi_3^B(\hat{\mu}_B) = 24c_4\hat{\mu}_B + 120c_6\hat{\mu}_B^3 + \frac{4!}{5!}c_4\epsilon_1\hat{\mu}_B^5 + \frac{4!}{7!}c_4\epsilon_2\hat{\mu}_B^7 \tag{6}$$

$$\chi_4^B(\hat{\mu}_B) = 24c_4 + 360c_6\hat{\mu}_B^2 + c_4\epsilon_1\hat{\mu}_B^4 + \frac{4!}{6!}c_4\epsilon_2\hat{\mu}_B^6. \tag{7}$$

We perform a correlated fit for the four measured observables, thus obtaining the values of c_2, c_4 and c_6 for each temperature, and the corresponding χ_2^B, χ_4^B and χ_6^B. We repeat the fit for 1000 random draws for ϵ_1 and ϵ_2. The result is weighted using the Akaike Information Criterion [34]. Through these weights we get a posterior distribution from the prior distribution. Our final estimate for χ_8^B represents this posterior distribution. We do not show the posterior for χ_{10}^B, which is mostly noise.

These results are shown in Fig. 2, together with an estimate of χ_8^B, related to χ_4^B by the prior condition.

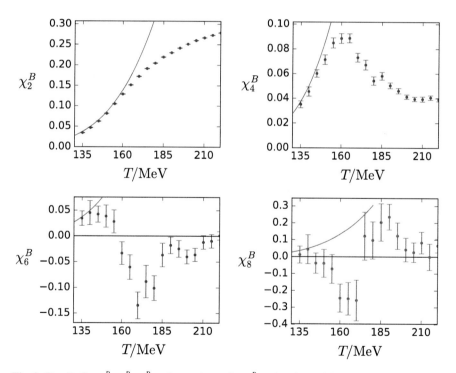

Fig. 2 Results for χ_2^B, χ_4^B, χ_6^B and an estimate for χ_8^B as functions of the temperature, obtained from the single-temperature analysis. We plot χ_8^B in green to point out that its determination is guided by a prior. The red curve in each panel corresponds to the Hadron Resonance Gas (HRG) model result

For cross-correlators let's take the example of the channel set by the observables χ_{ijk}^{BQS} with $j = 1, k = 0, i = 0, 1, 2$. Our ansatz for cross-correlators is analogous to Eqs. (4)–(7):

$$\chi_{01}^{BS}(\hat{\mu}_B) = \chi_{11}^{BS}\hat{\mu}_B + \frac{1}{3!}\chi_{31}^{BS}\hat{\mu}_B^3 + \frac{1}{5!}\chi_{51}^{BS}\hat{\mu}_B^5 + \frac{1}{7!}\chi_{71}^{BS}\hat{\mu}_B^7 + \frac{1}{9!}\chi_{91}^{BS}\hat{\mu}_B^9 \qquad (8)$$

We truncated the expression at tenth order. The priors assume $|\chi_{71}^{BS}| \lesssim |\chi_{31}^{BS}|$ and $|\chi_{91}^{BS}| \lesssim |\chi_{31}^{BS}|$, as it is certainly true at high temperature and within the HRG model. The prior distribution is wider than 1, we used the same mean and variance as in the channel with no μ_S derivative.

When we use Eq. (8) we take χ_1^S, χ_{11}^{BS}, χ_{21}^{BS} and χ_{31}^{BS} as correlated quartets for each imaginary chemical potential and determine the three free coefficients of Eq. (8). This fitting procedure is repeated 1000 times with random $\chi_{71}^{BS}/\chi_{31}^{BS}$ and $\chi_{91}^{BS}/\chi_{31}^{BS}$ coefficients. Again, using the Akaike weights we constrain the prior distribution. The resulting estimate for χ_{71}^{BS} along with the fit coefficients are shown in Fig. 3. The posterior for χ_{91}^{BS} is not only noisy, but it is probably heavily contaminated by the higher orders that we did not account for.

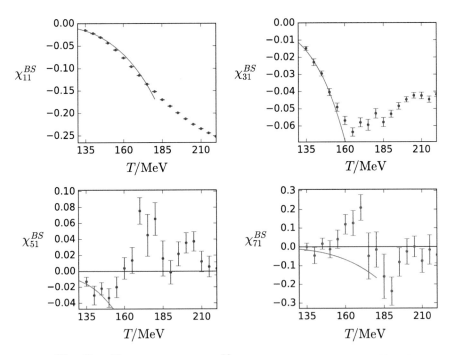

Fig. 3 χ_{11}^{BS}, χ_{31}^{BS}, χ_{51}^{BS} and an estimate for χ_{71}^{BS} as functions of the temperature. The red curves are the HRG model results

5 The Fluctuations from a Simple Model

The high order baryon fluctuations presented in Fig. 2 have a particular pattern, also known from chiral models [35]. While $\chi_2^B(T)$ is a sigmoid function, with an inflection point near $T_c \approx 155$ MeV, $\chi_4^B(T)$ has a pronounced peak. $\chi_6^B(T)$ has one zero crossing, $\chi_8^B(T)$ has two zero crossings. The latter two have vanishing Stefan-Boltzmann limits. This pattern could be easily explained, if higher order fluctuations were T-derivatives of the lower order ones, but it is not the case: the higher order fluctuations are, in fact μ_B/T derivatives.

The structure can be explained, though, if a connection is established between the chemical potential μ_B/T and temperature dependence. The crucial observation here is that some certain quantities show a simple pattern of chemical potential dependence. The most important example is for the χ^B channel. In Fig. 4 we show a particular quantity, the normalized baryon density (χ_1^B/μ_B). The temperature dependence at a finite chemical potential can be reproduced by simply rescaling the temperature axis:

$$\frac{\chi_1^B}{\mu_B} = f\left(T(1 - \kappa\mu_B^2/T^2 + \dots)\right), \tag{9}$$

where $f(T)$ is simply $\chi_2^B(T)$ at zero chemical potential. The $\dots$ mean an approximation, which is the point where our model neglects higher order curvature of the phase diagram.

Equation (9) defines a simple model for all higher order chemical potentials in the transition regime. For this the only input is the curvature of the phase diagram (here we use $\kappa = 0.02$ [36]), and our present result for $\chi_2^B(T)$. We could use this model to reproduce the newly simulated results with imaginary chemical potentials. Instead, we plot the higher order fluctuations in Fig. 5, which contain equivalent information.

In Fig. 5 the agreement is not merely qualitative. At high and low temperatures the small deviation is well understood: at low temperatures the hadron resonance

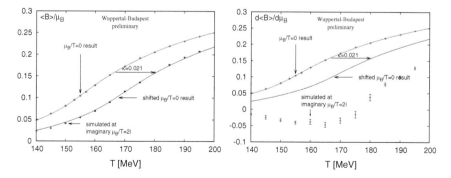

Fig. 4 The temperature dependence of χ_1^B/μ_B is roughly identical for any studied (imaginary) chemical potential, only that the temperature axis is scaled. This behaviour not generic, for example, χ^B (right panel) shows a very different pattern

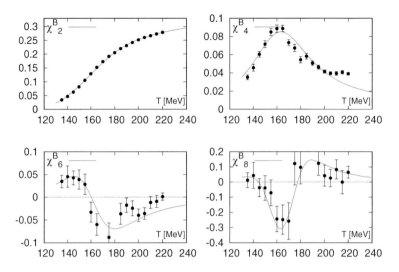

Fig. 5 2nd, 4th, 6th and 8th order baryon fluctuations at zero chemical potentials. These quantities have been proposed to study a nearby critical end point [35]. We find a quantitative agreement with a model which assumes the absence of any critical end-point

gas model governs physics, at high temperature it is described by the (resummed) perturbation theory. In the transition regime, however, we see that the single QCD-predicted sigmoid function and the curvature of the phase diagram alone can describe physics.

6   Phenomenology at Finite Chemical Potential

For a comparison with heavy ion collision experiments, the cumulants of the net-baryon distribution are very useful observables. The first four cumulants are the mean M_B, the variance σ_B^2, the skewness S_B and the kurtosis κ_B. By forming appropriate ratios, we can cancel out explicit volume factors. However, the measured distributions themselves may still depend on the volume, which one should take into account when comparing to experiments.

Heavy ion collisions involving lead or gold atoms at $\mu_B > 0$ correspond to the following situation

$$\langle n_S \rangle = 0 \qquad \langle n_Q \rangle = 0.4 \langle n_B \rangle . \tag{10}$$

For each T and μ_B pair, we have to first calculate μ_Q and μ_S that satisfy this condition. The resulting $\mu_Q(\mu_B)$ and $\mu_S(\mu_B)$ functions, too, can be Taylor expandend [19, 20].

After calculating the Taylor coefficients for $S_B \sigma_B^3 / M_B$ and $\kappa_B \sigma_B^2$, we use these results to extrapolate these quantities to finite chemical potential. They are shown in Fig. 6. In the left panel, $S_B \sigma_B^3 / M_B$ is shown as a function of the chemical potential

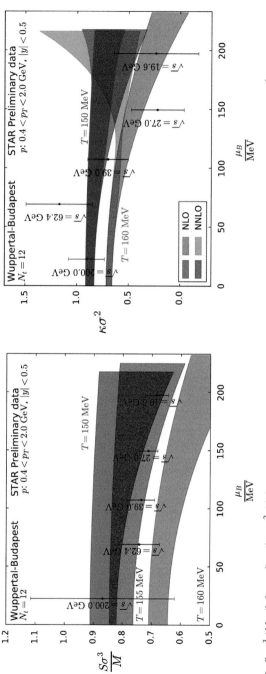

Fig. 6 $S_B \sigma_B^3 / M_B$ (left panel) and $\kappa_B \sigma_B^2$ (right panel) extrapolated to finite chemical potential. The left panel is extrapolated up to $\mathcal{O}(\hat{\mu}_B^2)$. In the right panel, the darker bands correspond to the extrapolation up to $\mathcal{O}(\hat{\mu}_B^2)$, whereas the lighter bands also include the $\mathcal{O}(\hat{\mu}_B^4)$ term

for different temperatures. The Taylor expansion for this quantity is truncated at $\mathcal{O}(\hat{\mu}_B^2)$. The black points in the figure are the experimental results from the STAR collaboration from an analysis of cumulant ratios measured at m id-rapidity, $|y| \le 0.5$, including protons and anti-protons with transverse momenta $0.4\,\mathrm{GeV} \le p_t \le 2.0$ GeV [37, 38]. The beam energies were translated to chemical potentials using the fitted formula of Ref. [39]. Even if we do not quantitatively compare the lattice bands to the measurements t o extract the freeze-out parameters, as experimental higher order fluctuations might be affected by several effects of non-thermal origin and our lattice result s are not continuum extrapolated, we notice that the trend of the data with increasing μ_B can be understood in terms of our Taylor expansion.

In the right panel, we show $\kappa_B \sigma_B^2$ as a function of μ_B/T for different temperatures. The darker bands correspond to the extrapolation up to $\mathcal{O}(\hat{\mu}_B^2)$, whereas the lighter bands also include the $\mathcal{O}(\hat{\mu}_B^4)$ term. Also in this case, the black points are the experimental results from the STAR collaboration with transverse momentum cut $0.4\,\mathrm{GeV} \le p_t \le 2.0\,\mathrm{GeV}$ [37, 38]. Notice that, due to the fact that the $r_{42}^{B,4}$ is positive in the range $160\,\mathrm{MeV} \le T \le 195\,\mathrm{MeV}$, we observe a non-monotonic behavior in $\kappa_B \sigma_B^2$ for $T = 160\,\mathrm{MeV}$ at large chemical potentials. By comparing the two different truncations of the Taylor series we can conclude that, as we increase the temperature, the range of applicability of our Taylor series decreases: while at $T = 150\,\mathrm{MeV}$ the two orders agree in the whole μ_B/T range shown in the figure, at $T = 160\,\mathrm{MeV}$ the central line of the next-to-next-to-leading order bends upwards and is not contained in the next-to-leading order band. To make the NLO prediction precise substantially more computer time would be needed.

7 Conclusions and Outlook

We have presented from high statistics lattice QCD simulations several μ_B-derivatives of observables that account for the confinement-deconfinement transition between the hadron gas and quark gluon plasma phases. These we achieved using imaginary chemical potential runs, which are not hindered by any sign problem. We confronted our results with a simple model assume the absence of any nearby critical end point in the phase diagram. In fact, the simulation results match with the model with no strengthening or weakening of the transition. This makes the experimental finding of a critical end-point below $\mu_B/T < 2$ highly unexpected. The Beam Energy Scan II at RHIC (Brookhaven, USA) starting 2019 will explore the phase diagram above this threshold and produce, among others, the here shown higher order fluctuations.

Our work in Ref. [40] extends these results to various off-diagonal fluctuations and contain more technical details.

In future work we will extend the computations to the equation of state and we will also model the other sectors, studying in detail the role of the strangeness in the QCD transition.

8   Performance on the Hazel Hen System

Much of the presented results have been achieved using the Hazel Hen supercomputer at HLRS, Stuttgart.

Here we used $48^3 \times 12$ lattice. Our lattice code produces results by a factor 1.5 faster than on the Blue Gene Q system, although the scaling for this small lattice is not optimal.

The most challenging numerical effort is to calculate the continuum limit. For that one optimally extends our existing data set with a lattice with a size of $64^3 \times 16$. These lattices have to be generated, such that an ensemble of gauge configurations is collected for each studied temperature and chemical potential.

The performance of the code can be measured by the wall-clock time of one HMC update, and by the conjugate (CG) performance in gigaflop/seconds.

Our generic MPI-based code has been ported to the CRAY system in several steps of optimizations. Because of the multiple NUMA nodes that share a network interface card we use the hybrid approach, having 4 parallel processes per node, each using 6 threads. The communication goes entirely over DMAPP. The scaling of the code is mostly limited by the job placement mechanism of HAZELHEN. To code runs optimally if all nodes share a group, though this happens rarely with a larger job. Nevertheless, higher per-core performance can be achieved on HAZELHEN than on a BG/Q system.

The Cray system's AVX instructions are analogous to those of the A2 core of BG/Q, thus we use the same site fusion technique. The four dimensional lattice is communicated in three dimensions, the fourth dimension is used in the site fusion.

Parallel I/O: we use DMAPP to do a peer-to-peer communication to reorganize data from simulation layout to disk layout in chunks of 64 MB. These chunks are written by several nodes in parallel. Reads work similarly, data files are read in chunks in parallel and distributed by communication.

For our $64^3 \times 16$ lattices the optimal performance is reached with 64 nodes (1536) cores. There we reach 7 Tflops/sec in the multi-precision inverter.

For the full update algorithm we find the following strong scaling behaviour on HAZELHEN:

# cores	192	384	768	1536	3072
one update on a $64^3 \times 16$ lattice [s]	1372	658	413	173	152

The analysis of the generated configurations requires several conjugate gradient inversions. These configurations can be analyzed independently. Due to the large number of the configurations a straightforward parallelization could be used, however, due to cache limitations, the code runs more efficiently in parallel than in serial. The most optimal use of the resources we achieved by selecting a larger partitions with several sub-partitions, each sub-partition (of e.g. 64 nodes) being in charge of a distinct list of configurations.

We noticed, that the performance of the same code on the same data in the same submission depends very deterministically on which sub-partition it was running. If the nodes in one sub-partition was all from one group the performance was 25% higher than when inter-group links were also used. By selecting a larger partition and letting our own code to select the nodes that belong to one sub-partition we achieved better overall performance than by using several small jobs (each with one sub-partition only), because in the latter case the job was almost always fragmented. Our code defines the sub-partitions to minimize fragmentation.

Here we give examples of our timings for the single-source conjugate gradient solver.

256 nodes divided into 4 sub-partitions:

layout spread over # of groups	run time [s]	Gflops/s
1	8.2	3694.2
2	10.3	2952.4
4	10.4	2930.4
4	10.9	2799.3

When all 256 nodes were given in the same group then each sub-partition performed equally well. The best performance we achieved on 64 nodes with a local lattice of $4 \times 4 \times 16 \times 64$, at 5287.9 Gflops/s. The top CG performance we experienced was as follows (note that this performance was significantly fluctuating, probably depending on the traffic coming from nearby applications).

# Hazel Hen cores	Gigaflops/s
3072	6483.8
1536	5287.9
768	1945.0
384	621.8

We note that in this 128 nodes were not in one group, in fact we almost never received enough nodes in the same group such that a scalability to 128 nodes would be practical.

Acknowledgements This project was funded by the DFG grant SFB/TR55. This work was supported by the Hungarian National Research, Development and Innovation Office, NKFIH grants KKP126769 and K113034. An award of computer time was provided by the INCITE program. This research used resources of the Argonne Leadership Computing Facility, which is a DOE Office of Science User Facility supported under Contract DE-AC02-06CH11357. The authors gratefully acknowledge the Gauss Centre for Supercomputing e.V. (www.gauss-centre.eu) for funding this project by providing computing time on the GCS Supercomputer JUQUEEN [41] at Jülich Supercomputing Centre (JSC) as well as on HAZELHEN at HLRS Stuttgart, Germany. This material is based upon work supported by the National Science Foundation under grants no. PHY-1654219 and OAC-1531814 and by the U.S. Department of Energy, Office of Science, Office of Nuclear

Physics, within the framework of the Beam Energy Scan Theory (BEST) Topical Collaboration. C.R. also acknowledges the support from the Center of Advanced Computing and Data Systems at the University of Houston. Financial Support by the German Ministry of Research and Education (Grant 05P18PXFCA) is gratefully acknowledged.

References

1. Y. Aoki, G. Endrodi, Z. Fodor, S. Katz, K. Szabo, Nature **443**, 675 (2006). https://doi.org/10.1038/nature05120
2. Y. Aoki, Z. Fodor, S. Katz, K. Szabo, Phys. Lett. B **643**, 46 (2006). https://doi.org/10.1016/j.physletb.2006.10.021
3. Y. Aoki, S. Borsanyi, S. Durr, Z. Fodor, S.D. Katz et al., JHEP **0906**, 088 (2009). https://doi.org/10.1088/1126-6708/2009/06/088
4. S. Borsanyi et al., JHEP **1009**, 073 (2010). https://doi.org/10.1007/JHEP09(2010)073
5. T. Bhattacharya, M.I. Buchoff, N.H. Christ, H.T. Ding, R. Gupta et al., "QCD Phase Transition with Chiral Quarks and Physical Quark Masses," Phys. Rev. Lett. **113**(8), 082001 (2014). https://doi.org/10.1103/PhysRevLett.113.082001
6. A. Bazavov, T. Bhattacharya, M. Cheng, C. DeTar, H. Ding et al., Phys. Rev. D **85**, 054503 (2012). https://doi.org/10.1103/PhysRevD.85.054503
7. C. Allton, S. Ejiri, S. Hands, O. Kaczmarek, F. Karsch et al., Phys. Rev. D **66**, 074507 (2002). https://doi.org/10.1103/PhysRevD.66.074507
8. C. Allton, M. Doring, S. Ejiri, S. Hands, O. Kaczmarek et al., Phys. Rev. D **71**, 054508 (2005). https://doi.org/10.1103/PhysRevD.71.054508
9. R.V. Gavai, S. Gupta, Phys. Rev. D **78**, 114503 (2008). https://doi.org/10.1103/PhysRevD.78.114503
10. S. Basak et al., PoS **LATTICE2008**, 171 (2008)
11. O. Kaczmarek, F. Karsch, E. Laermann, C. Miao, S. Mukherjee et al., Phys. Rev. D **83**, 014504 (2011). https://doi.org/10.1103/PhysRevD.83.014504
12. Z. Fodor, S. Katz, Phys. Lett. B **534**, 87 (2002). https://doi.org/10.1016/S0370-2693(02)01583-6
13. P. de Forcrand, O. Philipsen, Nucl. Phys. B **642**, 290 (2002). https://doi.org/10.1016/S0550-3213(02)00626-0
14. M. D'Elia, M.P. Lombardo, Phys. Rev. D **67**, 014505 (2003). https://doi.org/10.1103/PhysRevD.67.014505
15. Z. Fodor, S. Katz, JHEP **0203**, 014 (2002). https://doi.org/10.1088/1126-6708/2002/03/014
16. Z. Fodor, S. Katz, JHEP **0404**, 050 (2004). https://doi.org/10.1088/1126-6708/2004/04/050
17. C. Bonati, M. D'Elia, F. Negro, F. Sanfilippo, K. Zambello, Phys. Rev. D **98**(5), 054510 (2018). https://doi.org/10.1103/PhysRevD.98.054510
18. F. Karsch, Central Eur. J. Phys. **10**, 1234 (2012). https://doi.org/10.2478/s11534-012-0074-3
19. A. Bazavov, H. Ding, P. Hegde, O. Kaczmarek, F. Karsch et al., Phys. Rev. Lett. **109**, 192302 (2012). https://doi.org/10.1103/PhysRevLett.109.192302
20. S. Borsanyi, Z. Fodor, S. Katz, S. Krieg, C. Ratti et al., Phys. Rev. Lett. **111**, 062005 (2013). https://doi.org/10.1103/PhysRevLett.111.062005
21. S. Borsanyi, Z. Fodor, S. Katz, S. Krieg, C. Ratti et al., Phys. Rev. Lett. **113**, 052301 (2014). https://doi.org/10.1103/PhysRevLett.113.052301
22. C. Ratti, "Lattice QCD and heavy ion collisions: a review of recent progress," Rept. Prog. Phys. **81**(8), 084301 (2018). https://doi.org/10.1088/1361-6633/aabb97
23. A. Bazavov, et al., "Skewness and kurtosis of net baryon-number distributions at small values of the baryon chemical potential," Phys. Rev. D. **96**(7), 074510 (2017). https://doi.org/10.1103/PhysRevD.96.074510

24. M.A. Stephanov, K. Rajagopal, E.V. Shuryak, Phys. Rev. D **60**, 114028 (1999). https://doi.org/10.1103/PhysRevD.60.114028

25. M. Cheng, N. Christ, S. Datta, J. van der Heide, C. Jung et al., Phys. Rev. D **77**, 014511 (2008). https://doi.org/10.1103/PhysRevD.77.014511

26. J. Gunther, R. Bellwied, S. Borsanyi, Z. Fodor, S.D. Katz, A. Pasztor, C. Ratti, EPJ Web Conf. **137**, 07008 (2017). https://doi.org/10.1051/epjconf/201713707008

27. M. D'Elia, G. Gagliardi, F. Sanfilippo, Phys. Rev. D **95**(9), 094503 (2017). https://doi.org/10.1103/PhysRevD.95.094503

28. R. Bellwied, S. Borsanyi, Z. Fodor, S.D. Katz, A. Pasztor, C. Ratti, K.K. Szabo, Phys. Rev. D **92**(11), 114505 (2015). https://doi.org/10.1103/PhysRevD.92.114505

29. A. Roberge, N. Weiss, Nucl. Phys. B **275**, 734 (1986). https://doi.org/10.1016/0550-3213(86)90582-1

30. V. Vovchenko, A. Pasztor, Z. Fodor, S.D. Katz, H. Stoecker, Phys. Lett. B **775**, 71 (2017). https://doi.org/10.1016/j.physletb.2017.10.042

31. J.I. Kapusta, C. Gale, *Finite-Temperature Field Theory*, 2nd edn. (Cambridge University Press, Cambridge, 2006). https://doi.org/10.1017/CBO9780511535130. Cambridge Books Online

32. A. Vuorinen, Phys. Rev. D **67**, 074032 (2003). https://doi.org/10.1103/PhysRevD.67.074032

33. C. McNeile, C. Davies, E. Follana, K. Hornbostel, G. Lepage, Phys. Rev. D **82**, 034512 (2010). https://doi.org/10.1103/PhysRevD.82.034512

34. H. Akaike, IEEE Trans. Autom. Control **19**, 716 (1974)

35. B. Friman, F. Karsch, K. Redlich, V. Skokov, Eur. Phys. J. **C71**, 1694 (2011). https://doi.org/10.1140/epjc/s10052-011-1694-2

36. P. Cea, L. Cosmai, A. Papa, Phys. Rev. D **93**(1), 014507 (2016). https://doi.org/10.1103/PhysRevD.93.014507

37. X. Luo, PoS **CPOD2014**, 019 (2014)

38. J. Thäder, "Higher Moments of Net-Particle Multiplicity Distributions," Nucl. Phys. A. **956**, 320 (2016). https://doi.org/10.1016/j.nuclphysa.2016.02.047

39. A. Andronic, P. Braun-Munzinger, J. Stachel, Nucl. Phys. A **772**, 167 (2006). https://doi.org/10.1016/j.nuclphysa.2006.03.012

40. S. Borsanyi, Z. Fodor, J.N. Guenther, S.K. Katz, K.K. Szabo, A. Pasztor, I. Portillo, C. Ratti, JHEP **10**, 205 (2018). https://doi.org/10.1007/JHEP10(2018)205

41. JUQUEEN: IBM Blue Gene/Q Supercomputer System at the Jülich Supercomputing Centre. Technical Report 1 A1, Jülich Supercomputing Centre (2015). https://doi.org/10.17815/jlsrf-1-18

Exploring Many-Body Physics with Bose-Einstein Condensates

O. E. Alon, V. S. Bagnato, R. Beinke, S. Basu, L. S. Cederbaum,
B. Chakrabarti, B. Chatterjee, R. Chitra, F. S. Diorico, S. Dutta, L. Exl,
A. Gammal, S. K. Haldar, S. Klaiman, C. Lévêque, R. Lin, N. J. Mauser,
P. Molignini, L. Papariello, R. Roy, K. Sakmann, A. I. Streltsov,
G. D. Telles, M. C. Tsatsos, R. Wu and A. U. J. Lode

Abstract Bose-Einstein condensates (BECs) offer a fruitful, often uncharted ground for exploring physics of many-particle systems. In the present year of the MCTD-HB project at the HLRS, we maintained and extended our investigations of BECs and interacting bosonic systems using the MultiConfigurational Time-Dependent Hartree for Bosons (MCTDHB) method and running the MCTDHB and MCTDH-X software packages on the Cray XC40 system Hazel Hen. The results we disseminate within this report comprise: (i) Entropies and correlations of ultracold bosons in a lattice; (ii) Crystallization of bosons with dipole-dipole interactions and its detection in single-shot images; (iii) Management of correlations in ultracold gases; (iv) Pulverizing a BEC; (v) Dynamical pulsation of ultracold droplet crystals by laser light; (vi) Two-component bosons interacting with photons in a cavity;

O. E. Alon (✉) · S. K. Haldar
Department of Mathematics, University of Haifa, 3498838 Haifa, Israel
e-mail: ofir@research.haifa.ac.il

Haifa Research Center for Theoretical Physics and Astrophysics, University of Haifa,
3498838 Haifa, Israel

V. S. Bagnato · G. D. Telles · M. C. Tsatsos
São Carlos Institute of Physics, University of São Paulo, P.O. Box 369, São Carlos,
São Paulo 13560-970, Brazil

R. Beinke · L. S. Cederbaum · S. Klaiman · A. I. Streltsov
Theoretische Chemie, Physikalisch-Chemisches Institut, Universität Heidelberg,
Im Neuenheimer Feld 229, 69120 Heidelberg, Germany
e-mail: lorenz.cederbaum@pci.uni-heidelberg.de

S. Basu · S. Dutta
Department of Physics, Indian Institute of Technology Guwahati,
Guwahati 781039, Assam, India

B. Chakrabarti · R. Roy
Department of Physics, Presidency University, 86/1 College Street,
Kolkata 700073, India

© Springer Nature Switzerland AG 2019
W. E. Nagel et al. (eds.), *High Performance Computing in Science
and Engineering '18*, https://doi.org/10.1007/978-3-030-13325-2_6

(vii) Quantum dynamics of a bosonic Josephson junction and the impact of the range of the interaction; (viii) Trapped bosons in the infinite-particle limit and their exact many-body wavefunction and properties; (ix) Angular-momentum conservation in a BEC and Gross-Pitaevskii versus many-body dynamics; (x) Variance of an anisotropic BEC; and (xi) excitation spectrum of a weakly-interacting rotating BEC showing enhanced many-body effects. These are all basic and appealing, sometimes unexpected many-body results put forward with the generous allocation of computer time by the HLRS to the MCTDHB project. Finally, expected future developments and research tasks are prescribed, too.

1 Opening Remarks

The multiconfigurational time-dependent Hartree for bosons (MCTDHB) method [1–11] has been developed, implemented, optimized, and benchmarked for more than twelve years now, allowing one to integrate efficiently and reliably the time-independent and time-dependent many-boson Schrödinger equation. Numerous applications were made and a wealth of results on the many-body physics of Bose-Einstein condensates (BECs) reported, see, e.g., [12–18] for recent ones. We have been describing and summarizing annually [19–23] these investigations, many of

B. Chatterjee
Department of Physics, Indian Institute of Technology-Kanpur,
Kanpur 208016, India

R. Chitra · R. Lin · P. Molignini · L. Papariello
Institute for Theoretical Physics, ETH Zürich, 8093 Zürich, Switzerland

F. S. Diorico · C. Lévêque · K. Sakmann · R. Wu · A. U. J. Lode
Vienna Center for Quantum Science and Technology, Atominstitut, TU Wien,
Stadionallee 2, 1020 Vienna, Austria
e-mail: axel.lode@univie.ac.at

A. Gammal
Instituto de Fisica, Universidade de São Paulo, P.O. Box 66318, São Paulo,
CEP 05315-970, Brazil

A. I. Streltsov
Institut für Physik, Universität Kassel, Heinrich-Plett-Str. 40, 34132 Kassel, Germany

L. Exl · C. Lévêque · N. J. Mauser · A. U. J. Lode
Wolfgang Pauli Institute c/o Faculty of Mathematics, University of Vienna,
Oskar-Morgenstern Platz 1, 1090 Vienna, Austria

them were made possible with the generous allocation of high-performance computer time by the HLRS to the MCTDHB project. We aim at extending this practice in the present report. In the sections below, we summarize our theoretical findings given in [24–34], for which the MCTDHB, MCTDHB-LAB, and MCTDH-X software packages [35–37] were used.

2 Entropies and Correlations of Ultracold Bosons in a Lattice

Already a few interacting ultracold atoms in a periodic potential feature the various facets of the collective behaviors that emerge in the case of many particles [38–42]. In periodic potentials, which can be experimentally realized by counterpropagating laser beams, the collective behavior of ultracold bosonic atoms can be categorized into three classes: (1) the state in which all atoms behave alike and simultaneously occupy many wells of the periodic potential, called a *superfluid*. (2) The state where the particles attain individuality by localizing in seperate wells of the lattice.

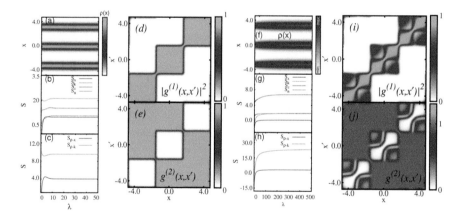

Fig. 1 The collective behavior of ultracold bosons in lattices characterized by correlations and many-body entropies as a function of their repulsion λ. At a filling factor of $\nu = 1$ particles per well, the density **a** does not reveal the transition from superfluid to Mott insulator (SF $\to$ MI); many-body entropies (**b**) and (**c**), and correlation functions (**d**) [$|g^{(1)}(x, x')|^2 = |\frac{\rho^{(1)}(x,x')}{\sqrt{\rho(x)\rho(x')}}|^2$] and (**e**) [$g^{(2)}(x, x') = \frac{\rho^{(2)}(x,x',x,x')}{\rho(x)\rho(x')}$], however, allow to identify the SF $\to$ MI transition. At a filling factor of $\nu = 2$ particles per well, the density (**f**), entropies (**g**), (**h**), and the correlation functions (**i**), (**j**), allow to identify the transition to fermionization at large repulsion λ between the bosons. The formation of intra-well structure as a hallmark of fermionization is manifest in the density [see dashed black line at $\rho(x) = 0.22$ in **f**] and the correlation functions [patterns form for regions where x is in the vicinity of the same minimum of the potential as x' in panels **i** and **j**]. The quantities shown are dimensionless. Figure material (panels) reprinted from [24]

This occurs due to, both, an increased laser intensity and thus a deeper periodic potential and also the increased repulsion between the particles. This state is known as a *Mott insulator*. (3) The state where the repulsion between the particles is so strong that it forces them to localize even within a single minimum of the potential. The formed structure is then referred to as a fermionized state [43].

In our paper [24], we use MCTDH-X [7, 11–14, 37, 44–49] to demonstrate that the collective behaviors, (1)–(3), of bosons in periodic potentials can be unequivocally classified by looking either at their correlation functions or, equivalently, at their many-body entropy measures [50–59] (Fig. 1). For details on how to compute the entropy measures, see Ref. [24].

3 Crystallization of Bosons with Dipole-Dipole Interactions and Its Detection in Experiment

Ultracold bosonic quantum gases are remarkably well under experimental control. In contrast to condensed matter systems, not only the confinement of, but even the interactions between the ultracold atoms can be tuned at will. Further, ultracold atoms have the advantage of direct observation; *the wavefunction of the many-body state* can be photographed in so-called single-shot images [12, 60]. Nowadays ultracold atoms thus serve as almost universal quantum simulators for condensed matter systems [38, 39]. Ultracold quantum gases with dipole-dipole interactions have recently received considerable interest [61–66]. The long-range and anisotropic dipole-dipole interactions entail rich quantum many-body physics, distinct from the effects in ultracold atoms with contact interactions. In our work [25] we use MCTDH-X [7, 11–14, 37, 44–49] to investigate the remarkable crystallization process that takes place when the dipole-dipole interactions are strong enough to dominate the physics of the system [67–73]. In the crystal state, the dipolar atoms localize due to their long-range interactions and form a crystal structure irrespective of the confining potential.

Here, we investigate the ground state of a one-dimensional sample of ultracold bosonic atoms with long-range dipolar interparticle interactions as a function of the interaction strength (Fig. 2a–e). We define an order parameter, $\Delta = \sum_k \left(\frac{\lambda_k}{N}\right)^2$, that is a function of the eigenvalues of the reduced one-body density matrix, λ_k (Fig. 2f–h). We show that this order parameter characterizes all the emergent phases of the system, the superfluid, the Mott insulator, attempted fermionization, and crystallization (Fig. 2h). Importantly, Δ can be inferred from the variance of single-shot images, i.e. by quantifying the fluctuations in the "photos" taken of the wavefunction (Fig. 2i).

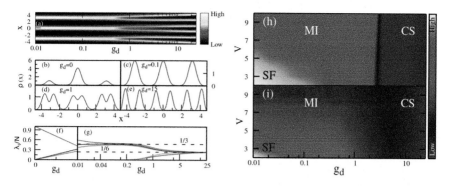

Fig. 2 Tracing the crystallization of bosons with dipole-dipole interactions as a function of the interaction strength. **a** Density as a function of interactions. Densities representative for the emergent phases of dipolar bosons: **b–e** show the superfluid, the Mott insulator, attempted fermionization, and the crystal state, respectively. **f** (**g**) The eigenvalues λ_k of the reduced density matrix on a linear (logarithmic) scale allow to characterize all emergent phases illustrated in **a–e**. **h** Phase diagram obtained using the order parameter $\Delta = \sum_k \left(\frac{\lambda_k}{N}\right)^2$. **i** Variance $\mathcal{V}$ of $N_{\text{shots}} = 10{,}000$ single-shot simulations $\mathcal{B}_j(x)$, for every point in the phase diagram in **h**. The variance is $\mathcal{V} = \int dx \frac{1}{N_{\text{shots}}} \sum_{j=1}^{N_{\text{shots}}} \left[\mathcal{B}_j(x) - \bar{\mathcal{B}}(x)\right]^2$, where $\bar{\mathcal{B}}(x) = \frac{1}{N_{\text{shots}}} \sum_{j=1}^{N_{\text{shots}}} \mathcal{B}_j(x)$. The quantities shown are dimensionless. Figure material (panels) reprinted from [25]

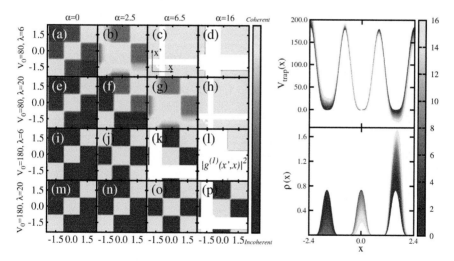

Fig. 3 Right: Potential (top) and density (bottom) of $N = 90$ atoms in a three-well lattice as a function of tilt (see colorbox on the right). Left: The correlations $|g^{(1)}(x', x)|$ between bosons in a three-well potential as a function of tilt α (see column labels), barrier height V_0 and inter-particle repulsion λ (see row labels). The correlations between the particles can be configured almost at will – even long-range correlations are accessible, see patches where $|g^{(1)}| \approx 1$ for x corresponding to the leftmost well and x' to the rightmost well in panels **j** and **o**. The quantities shown are dimensionless. Figure material (panels) reprinted from [26]

4   Managing the Collective Behavior of Ultracold Gases

Matter at the quantum level behaves not totally deterministically but rather proba-bilistically. When many bosons are trapped and cooled down to ultralow tempera-tures (almost absolute zero) they clump up together and condense to a *Bose-Einstein condensate* (BEC) [74]. A BEC is the ultimate quantum state where all external clas-sical "noise" is switched off and the quantum-mechanical "whisper" of nature can be heard. This collective state of many-particles is maximally deterministic, reflecting a maximum of coherence and a minimum of correlations. This means that all particles behave alike; knowing the motion of one translates to knowing the motion of the whole. However, many real-life situation can shatter coherence, either purposefully or not. In our numerical experiment [26] we use MCTDH-X [7, 11–14, 37, 44–49] to simulate a system of 90 bosons where the magnetic or optical fields responsible for trapping the particles have a given desired shape. By controllably varying the shape, position and tilt of the trapping fields, we are able to bring the system from one extreme—being totally coherent—to the other—being totally incoherent (see Fig. 3). This simple, yet effective and unprecedented, method is directly applicable to present-day experiments and allows for a control over the positions and momenta of the particles as well as of the quantum correlations between them.

Studying coherence and the disruption of it is key in understanding the quantum or classical, i.e. the probabilistic or deterministic, behavior of larger systems, which are routinely being produced and engineered nowadays.

5   Pulverizing A Bose-Einstein Condensate

Why is a pile of sand stable? Why does it not fall via its own gravity like ordinary liquids do? Granular matter, like sand, is unique in that it combines properties of all other states of matter: grains can behave as solids, liquids or gases depending on the circumstances, like the externally applied forces. Granulation is the process of transforming a continuous medium to a granular substance of particles, whose sizes can obey specific or random distributions. This happens by triturating, which requires pouring enough energy into the system.

In our paper [27] we use MCTDH-X [7, 11–14, 37, 44–49] and raise the ques-tion of what happens if we pour a sufficient amount of energy in a trapped ultracold gas. Can we also "pulverize" it? We found that, under specific conditions: "Yes we can!" [27]. Indeed, an interesting situation—closely resembling granulation—emerges. When periodically perturbed with high-frequency oscillatory magnetic fields the interaction between the particles modulates periodically. Then the gas appears to break down forming isolated islets of random sizes and positions. Even though the gas initially is a Bose-Einstein condensate and thus not correlated, at the end of the process it becomes highly correlated and fluctuates quantum-mechanically. This means that, the positions and shapes of the islets are highly random and not

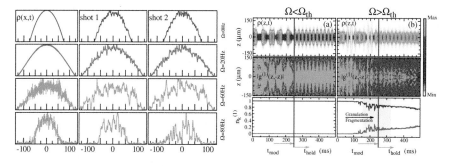

Fig. 4 Left: Granulation of a BEC is manifest in the absorption images (columns labeled "shot 1" and "shot 2") above a critical modulation frequency $\Omega_{th} \approx 40$ Hz, but not seen in the density [column labeled "$\rho(x, t)$"]. Right: Comparison of quasi-adiabatic [column **a**] and granulated [column **b**] time-evolutions of BECs. For the quasi-adiabatic driving, $\Omega < \Omega_{th}$, the density exhibits breathing-like dynamics [**a**, top], the coherence $|g^{(1)}(z, -z)| = |\frac{\rho^{(1)}(z, -z)}{\sqrt{\rho(z)\rho(-z)}}|$ is maintained [$|g^{(1)}(z, -z)| \approx 1$, **a**, middle] and fragmentation is absent [$n_1^{(1)} \approx 1$, **a**, bottom]. For the case of emerging granulation for $\Omega > \Omega_{th}$, the density exhibits a noisy pattern in addition to the breathing dynamics [**b**, top]. The coherence is lost on very short length scales [$|g^{(1)}(z, -z)| \approx 0$, **b**, middle], and fragmentation [$n_1^{(1)} < 1$ and $n_2^{(1)} > 0$] emerges hand-in-hand with granulation [**b**, bottom]. The quantities shown are dimensionless. Figure material (panels) reprinted from [27]

determined—as in ordinary BECs—from the shape of its container (see Fig. 4). Interestingly, we see that the simulated state shows characteristics of various other states too: turbulence [75], localization and correlations appear hand-in-hand [27]. Our theoretical findings are supported by experimental evidence from ongoing studies in the laboratories of Rice, Texas and São Carlos, Brazil.

6 Dynamical Pulsation of Ultracold Droplet Crystals by Laser Light

When certain atoms are cooled to ultracold ($\sim$mK) temperatures, they condense into a new state of matter called a Bose-Einstein condensate (BEC) [76–78]. In this state, a macroscopic portion of the atoms occupy the same lowest-energy state, like droplets of water sliding down to the bottom of a sink to form a unique big drop.

The exotic physics of BECs can be further explored with lasers. When a laser photon collides with the BEC, the atoms can absorb its energy and re-emit it at a later time along an orthogonal direction. If the BEC is placed in between highly reflective mirrors, the re-emitted light can be trapped and amplified in the process. Above a certain threshold of the laser power, the BEC starts to coherently scatter a macroscopic number of photons in the cavity: the system undergoes a phase transition to a so-called superradiant state [79–81]. The scattered photons in turn create an optical lattice—a

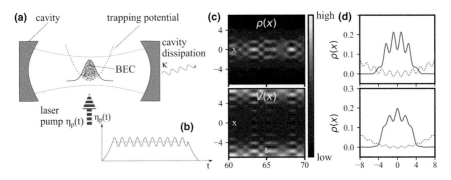

Fig. 5 Many-body parametric resonance of a driven Bose-Einstein condensate in between two mirrors. **a** A one-dimensional BEC in a cavity. **b** Time-dependent pump power to explore resonances in the system as a function of drive amplitude and frequency. **c** Density and potential as a function of time for the many-body parametric resonance. **d** Example densities representative for the dynamical pulsation of the crystal between the different possible lattice configurations. The quantities shown are dimensionless. Figure material (panels) reprinted from [28]

sort of egg-carton-shaped landscape where the BEC droplets arrange themselves to form a crystal structure in two possible configurations.

In our work [28] we use MCTDH-X [7, 11–14, 37, 44–49] to explore a resonance between a BEC and photons by modulating the laser power. The atoms experience a parametric drive by the photons, analogous to what happens for a child on a swing who increases and sustains the motion by stretching and bending the legs at the right frequency [82, 83]. Similarly, in the driven BEC setup, the parametric resonance induces a dynamical pulsation of the droplet crystal between the two configurations (Fig. 5). Heating characteristics confirm that the dynamical phase is a new state of matter: the normal and the superradiant BEC do not heat up when driven, while the dynamical phase does.

7   Two-Component Atoms Dancing with Photons Between Mirrors

Imagine yourself standing between two mirrors facing each other. You get that eerie infinite reflections of yourself. In physics, this is known as an optical cavity. As the mirrors approach perfection, the same particles of light (photons) will bounce back off those mirrors almost indefinitely. If the photon can "talk" to atoms which are also placed inside this optical cavity it will be repeatedly absorbed by the atoms and re-emitted back into the optical cavity. Many of such atoms may form a so-called BEC in the optical cavity—a state of matter where these atoms lose their identity and behave like one giant atom (Fig. 6). The different ways how these atoms talk to the bouncing light in the optical cavity give rise to rich physics. For instance, jacking

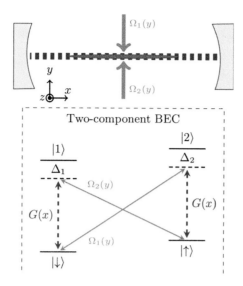

Fig. 6 One dimensional quasi-condensate in a high-finesse optical cavity. The ultracold two-component bosons are pumped by two transverse pump beams that are coupled to the transitions between the atomic states $|\downarrow\rangle$ and $|2\rangle$ and between $|\uparrow\rangle$ and $|1\rangle$ with Rabi frequencies Ω_1 and Ω_2, respectively (see dashed box). The coupling to the cavity is given by $G(x) = g_0\cos(k_c x)$ for both transitions, where g_0 is the single atom coupling strength, see text. In the self-organized phase, the cavity mode is populated (blue dashes) and acts on the atoms as a potential and effective coupling between the $\uparrow$ and $\downarrow$ component. The quantities shown are dimensionless. Figure material (panels) reprinted from [29]

up the amount of light within those mirrors can manipulate the atoms' internal state as if they were *dancing within themselves*, forced by the light.

In our work [29] we extend our numerical package MCTDH-X [7, 11–14, 37, 44–49] to deal with a generalization of the system described in Ref. [84]: we tackle the light-matter interaction within mirror cavities on a many-body level. We show that, when the amount of the scattered light trapped between the mirrors is sufficiently large, the atoms will split into distinct groups because of the interactions between the atoms and between the atoms and the light.

The formation of these groups is called "fragmentation" and the conglomeration due to the light is referred to as "superradiance". We have also shown that increasing the light pumped into the cavity can also lead to the breakdown of the so-called Dicke model [85], a model widely used for describing atoms fixed in space talking to light within a cavity built by mirrors. These are all illustrated in Fig. 7. Correlated atoms coupled to the light in an optical cavity are a hybrid quantum system, because light and matter are interfaced with each other. A detailed understanding of this interface might allow its use as a building block of future quantum computers which need to transfer information from photons to atoms and back.

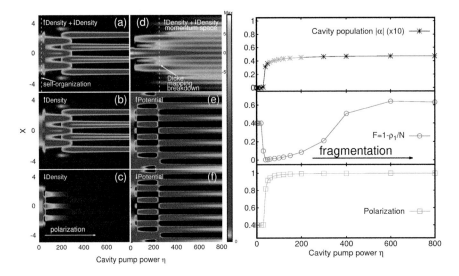

Fig. 7 Left: **a–c** Density of the two-component Bose-Einstein condensate in a cavity as a function of pump power η. The density exhibits self-organization [multiple maxima in **a**] and polarization [density is non-zero in **b**, but zero in **c**], as η becomes larger. The self-organization is triggered by the cavity-mediated one-body potential in **e, f**. The formation of secondary peaks in the momentum distribution, **d**, is a hallmark of the breakdown of the Dicke model. Right: The cavity population (top), fragmentation (middle), and polarization (bottom) are shown as a function of the pump power η and quantify further the findings in **a–f**, see Ref. [29] for further details. The quantities shown are dimensionless. Figure material (panels) reprinted from [29]

8   Quantum Dynamics of a Bosonic Josephson Junction: Impact of the Range of the Interaction

Bosonic Josephson junctions with short-range (contact) interactions have drawn much interest [5, 86–98]. In the context of ultra-cold systems, long-range interactions occur in dipolar [61–66, 99] and Rydberg [100, 101] systems. In this work, we would like to bring the topics of bosonic Josephson junctions and long-range interactions together, and investigate what role the range of the interaction plays in the dynamics of the bosons in the junction.

Having the many-boson wavefunction at hand, we can examine the impact of the range of the interaction on properties of the junction, i.e., the density oscillations and their collapse, self trapping, depletion and fragmentation, as well as the position variance. We find competitive effect between the interaction and the confining trap. The presence of the tail part of the interaction basically enhances the effective repulsion as the range of the interaction is increased starting from a short, finite range. But, as the range becomes comparable with the trap size, the system approaches a

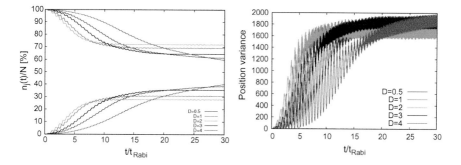

Fig. 8 *Impact of the range of the interaction on a bosonic Josephson junction.* Fragmentation (occupation numbers; left panel) and position variance per particle $\frac{1}{N}\Delta_{\hat{X}}^2$ (right panel) as a function of time for various ranges D. The interaction parameter $\Lambda = 0.1$ and $N = 1000$ bosons. There is an optimal range of the interaction in which properties of the junction are attaining their extreme values. The quantities shown are dimensionless. Figure material (panels) reprinted from [30]

situation where all the atoms feel a constant potential and the impact of the tail on the dynamics diminishes. All in all, there is an optimal range of the interaction in which physical quantities of the junction are attaining their extreme values, see Fig. 8.

9 Trapped Bosons in the Infinite-Particle Limit: Exact Many-Body Wavefunction and Properties

In spite of the fact that the Gross-Pitaevskii equation provides in the infinite-particle limit the exact energy per particle and density per particle [102, 103] (for respective time-dependent results see [104–106]), the overlap of the Gross-Pitaevskii and the exact many-boson wavefunctions is always smaller than 1 and can even vanish [15]. In fact, already the variance of many-particle operators can differ largely when computed at the mean-field and many-body levels in this limit [107].

In [31], we aim at solving the many-body problem of bosons in a trap in the infinite-particle limit. As mentioned above, the mean-field theory has been found to yield the exact density per particle in the infinite-particle limit, implying that the depletion per particle vanishes (i.e., the condensate fraction is 100%). We thus base our theory on the full Hamiltonian expressed in the basis of creation and annihilation operators defined by the mean-field Hamiltonian. We show that the theory formulated in this manner is a priori particle-number conserving.

As an application and test of the theory, we considered the harmonic-interaction model which can be solved analytically [108–111]. It is rather cumbersome to express the exact analytically-known wavefunction of the harmonic-interaction model in the basis of the mean-field orbitals used in our theory. Thus, to have a comparison with the reconstructed wavefunction, we have performed numerical calculations

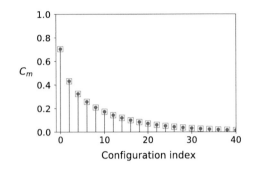

Fig. 9 *Ground-state wavefunction of the harmonic-interaction model.* The reconstructed wavefunc-tion (at the infinite-particle limit) using the phonon Hamiltonian: The coefficients C_m are computed employing an analytic formula (red squares) [31]. Note that the first coefficient C_0 is the overlap of the mean-field and exact wavefunctions. The same coefficients computed numerically for the full many-body Hamiltonian using the MCTDHB method for $N = 10^7$ bosons (blue dots). The values of the coefficients practically (numerically) do not change if N is lowered to $N = 10^6$. The interaction parameter $\Lambda = 100$. The quantities shown are dimensionless. Figure material (panel) reprinted from [31]

(imaginary-time propagation) employing MCTDHB. We remind that MCTDHB has been benchmarked with the harmonic-interaction model [6, 11]. The expansion co-efficients computed numerically for $N = 10^7$ bosons are shown in Fig. 9 for the same coupling strength as used for the reconstructed wavefunction. It is clearly seen that the reconstructed wavefunction excellently reproduces the numerically determined particle-conserving wavefunction.

10   Angular-Momentum Conservation in a BEC: Gross-Pitaevskii Versus Many-Body Dynamics

Conservation of angular momentum is a fundamental symmetry of nature in rotationally-invariant systems. In the field of BECs, the many-body Schrödinger equation is the basic equation that describes the microscopic dynamics. For most BEC experiments the interparticle interaction is a function of the distance between particles only and therefore rotationally invariant. When the external trapping po-tential is rotationally symmetric, the many-body Schrödinger equation conserves angular momentum for any initial state.

In our work [32], we have shown analytically and numerically that the Gross-Pitaevskii mean-field approximation violates the conservation of angular momentum, and have investigated its dependence on parameters. The violation scales proportional to the interaction strength and does not vanish in the infinite-particle particle limit. We have concentrated on the second moment (variance) $\langle \hat{L}_z^2 \rangle(t)$ to illustrate this fact for a system where the exact many-body Schrödinger dynamics conserves all

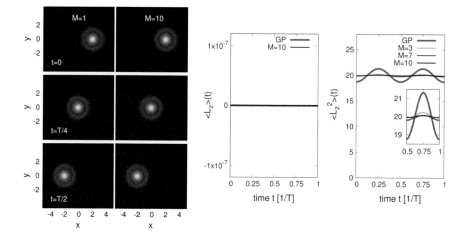

Fig. 10 *Sloshing of a BEC in a two-dimensional isotropic harmonic trap: Angular-momentum conservation and variance.* Left panel: Shown is the single-particle density using $M = 1$ (Gross-Pitaevskii, left column) and $M = 10$ orbitals (MCTDHB, right column). There is no visible difference between the $M = 1$ and $M = 10$ results. Interaction parameter $\Lambda = 0.9$. Number of bosons $N = 10$. Oscillation period $T = 2\pi$. Right panel: Shown are the expectation values $\langle \hat{L}_z \rangle(t)$ (left plot) and $\langle \hat{L}_z^2 \rangle(t)$ (right plot) for the dynamics in the left panel. The Hamiltonian commutes with $\hat{L}_z$ and hence the exact many-body Schrödinger dynamics conserves $\langle \hat{L}_z \rangle(t)$ and $\langle \hat{L}_z^2 \rangle(t)$. Whereas Gross-Pitaevskii mean-field theory conserves $\langle \hat{L}_z \rangle(t)$, it violates the conservation of $\langle \hat{L}_z^2 \rangle(t)$: Here, it oscillates with an amplitude of about 13% of the absolute value instead of being constant. Many-body MCTDHB simulations using $M = 3, 7, 10$ orbitals approach the exact constant result for increasing numbers of orbitals. The oscillations for the $M = 10$ MCTDHB result are only about 0.5% of the absolute value. The quantities shown are dimensionless. Figure material (panels) reprinted from [32]

moments $\langle \hat{L}_z^n \rangle(t)$. We have provided an explanation for this violation based on the time-dependent variational principle [112–114]. The violation is due the fact that the Gross-Pitaevskii ansatz wavefunction is too restrictive and therefore its associated space of allowed variations is not large enough to conserve any moment higher than the first, $\langle \hat{L}_z \rangle(t)$. Using the more general MCTDHB wavefunction, we have shown how going from a mean-field ansatz to a many-body ansatz increases the space of allowed variations. In the example shown in Fig. 10 this leads to a good approximation to the conservation of $\langle \hat{L}_z^2 \rangle(t)$ as required by the the exact dynamics [32].

Summarizing, conservation of angular momentum is a symmetry which generally requires to go beyond the realm of Gross-Pitaevskii mean-field theory. The degree of violation depends on the system's parameters. It will be particularly interesting to investigate the implications of this result for the mechanism of creation of vortices in BECs.

11   Variance of an Anisotropic BEC

In quantum mechanics, the variance of an observable $\hat{a}(x)$ for a particle described by the wavepacket $\phi(x)$ is often used to interpret and quantify the physical behavior of the particle [115]. For instance, a wider wavepacket has a larger position variance in comparison with a narrower wavepacket which has a smaller position variance. Physically, the two situations are associated, respectively, with de-localization and localization of the quantum particle. In different words, we are often used to infer the position and momentum variance from the shape of a particle's density $|\phi(x)|^2$. This intuitive or visual picture may vary in a system made of interacting particles. The purpose of our work [33] is to demonstrate and investigate such many-body effects using a trapped anisotropic BEC as an instrumental example.

Fragmentation of BECs has drawn much attention [1, 41, 116–127]. In particular, bosons with long-range interaction exhibit intriguing fragmentation physics when held in a single-trap potential [48, 128–132]. We consider bosons with long-range

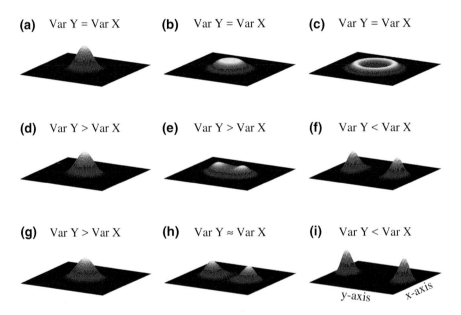

Fig. 11 *Density versus position variance of an anisotropic BEC.* Shown is the density per particle of $N = 10$ bosons for the interaction strengths $\lambda_0 = 0.01, 0.10$, and 0.20 (left to right columns) and trap anisotropies $\beta = 0\%$ (isotropic), 10%, and 20% (upper to lower rows). Increasing the strength of the interaction, the ground state changes its shape and the reduced one-body density matrix fragments. In panels **a–c** the system is isotropic and the position variances in the y and x directions are therefore equal. In panels **d–f** and **g–i** the system is anisotropic and, as the density in the y direction becomes wider than the density in the x direction, the respective variances intriguingly behave in an opposite manner. Namely, the variance in the y direction becomes smaller than that in the x direction. $M = 10$ time-adaptive (self-consistent) orbitals are used. The quantities shown are dimensionless. Figure material (panels) reprinted from [33]

interaction held in a single-well anisotropic potential and show how condensate fragmentation can conspire—together with the trap's anisotropy—to localize the center-of-mass more along the wider y direction than along the narrower x direction, see Fig. 11.

12 Excitation Spectrum of a Weakly-Interacting Rotating BEC: Enhanced Many-Body Effects

Ultracold bosonic gases under rotation have drawn considerable attention [13, 133–148]. Most studies offering analytical and numerical results for the low-energy spectra were carried out by utilizing the Bogoliubov-de Gennes mean-field equation [149, 150] which can be seen as the linear-response theory atop the Gross-Pitaevskii equation, see [151–156]. Examples for a many-body analysis of the low-energy spectra in rotating BECs are generally rare and include the Yrast spectra in a harmonic confinement obtained by exact diagonalization using the lowest Landau level approximation [157–160].

In our work [34], we consider many-body effects in the low-energy excitation spectrum of a weakly-interacting rotating BEC in a regime where the mean-field theory (Bogoliubov-de Gennes equation) is supposed to accurately describe the physics. Going beyond previous works, we study bosons in an anharmonic external confinement [161–163], where one can no longer rely on the validity of the lowest Landau level approximation. It has been recently shown that for a rotating repulsive trapped BEC under certain conditions [164] the excitation energies in the infinite-particle limit converge towards the Bogoliubov-de Gennes spectrum (the same holds for a non-rotating BEC [165]). However, it remains unclear if many-body effects in the excitation spectrum can be observed for mesoscopic and large BECs, typically of the experimentally relevant order of $10^3 - 10^6$ bosons.

As a main result, we show that once the rotation leads to a quantized vortex in the ground state, substantial many-body effects in the low-energy excitation spectrum occur, see Fig. 12. These effects grow with the vorticity of the ground state, which means they can be enhanced by stronger rotation. In addition, we demonstrate for a low-lying excited state which is also accessible within mean-field theory, that these effects clearly persist when the number of bosons is increased up to the experimentally relevant regime, despite the BEC being almost fully condensed [34].

The present work reports on accurate many-body excitation energies of a two-dimensional BEC obtained by LR-MCTDHB [166–168] (the linear-response theory atop MCTDHB), and goes well beyond previous investigations of one-dimensional systems [17, 18], reported upon last year [23].

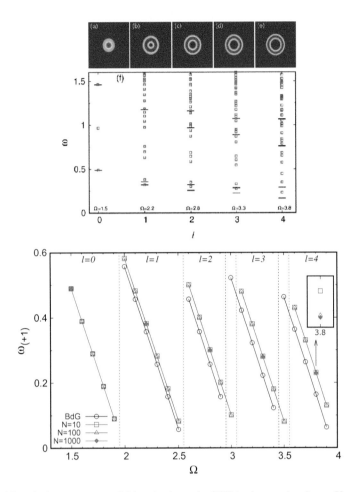

Fig. 12 *Many-body versus mean-field excitations of a BEC in the rotating frame.* Upper panel: **a–e** Many-body ground-state densities of a rotating BEC with $N = 10$ bosons in the anharmonic trap for vorticities $l = 0$, 1, 2, 3, and 4. The interaction parameter $\Lambda = 0.5$. For $l > 0$ the ground state is a single vortex whose size grows with l. The mean-field densities are alike (not shown). **f** Corresponding low-energy excitation spectra for the ground states in the upper raw (**a–e**). Thick lines indicate mean-field results, i.e. with $M = 1$ orbitals (Bogoliubov-de Gennes), and the squares denote the many-body results with $M = 7$ orbitals (LR-MCTDHB). Symbols in blue refer to the (+1) excitation. The many-body spectra show a very rich structure which cannot be accounted for within the mean-field picture. Lower panel: The excitation energies $\omega_{(+1)}$ are shown as a function of the rotation frequency Ω for different number of bosons N. Vertical dotted lines indicate the transition from ground-state vorticity l to $l + 1$ between two adjacent analyzed rotation frequencies (mean-field in black and many-body in red). It is seen that the many-body effects increase with growing vorticity. Even the assignment of the vorticity to the mean-field and many-body ground states does not match as the vorticity grows. The inset shows a magnified view for $\Omega = 3.8$. The quantities shown are dimensionless. Figure material (panels) reprinted from [34]

13 Concluding Remarks and Future Plans

We have summarized in the above sections our recent works on the many-body physics of BECs obtained within the MCTDHB project supported by the HLRS. We have investigated entropies and correlations of ultracold bosons in a lattice; explored crystallization of bosons with dipole-dipole interactions and its detection in single-shot images; studied management of correlations in ultracold gases; discovered the pulverization of a BEC; simulated dynamical pulsation of ultracold droplet crystals by laser light; analyzed two-component bosons interacting with photons in a cavity; researched the impact of the range of the interaction on the quantum dynamics of a bosonic Josephson junction; developed the exact many-body wavefunction and derived properties of trapped bosons in the infinite-particle limit; showed how angular-momentum conservation in a BEC manifests in the Gross-Pitaevskii and many-body dynamics; followed the variance of an anisotropic BEC; and computed enhanced many-body effects in the excitation spectrum of a weakly-interacting rotating BEC.

As future research agenda, we plan to investigate among others: (i) How many-body excitations and position, momentum, and angular-momentum variances depend on the trap's geometry in two dimensions and possibly also in three dimensions; (ii) Out-of-equilibrium many-body dynamics of BECs in two-dimensional cavities; (iii) Efficient new implementation of the coefficient part in MCTDHB equations of motion. Furthermore, keeping track of related developments and applications using restricted-active-space [169, 170] and multi-layer [171–175] theories will be instrumental. All in all, MCTDHB, LR-MCTDHB, and MCTDH-X will continue to explore the increasingly exciting many-body physics of BECs.

Acknowledgements Financial support by the Deutsche Forschungsgemeinschaft (DFG) is gratefully acknowledged. OEA acknowledges funding by the Israel Science Foundation (Grant No. 600/15). RB acknowledges financial support by the IMPRS-QD (International Max Planck Research School for Quantum Dynamics), the Landes-graduiertenförderung Baden-Württemberg, and the Minerva Foundation. We thank Jörg Schmiedmayer for numerous insightful discussions and comments. We acknowledge financial support by the Austrian Science Foundation (FWF) under grant No. F65 (SFB "Complexity in PDEs"), grant No. F40 (SFB "FOQUS"), grant No. F41 (SFB "ViCoM"), grant No. P 31140-N32 ('ROAM'), and the Wiener Wissenschafts- und Technologie-Fonds (WWTF) project No. MA16-066 ("SEQUEX"). We acknowledge financial support from FAPESP, the Department of Science and Technology, Government of India under DST Inspire Faculty fellowship, a FAPESP (grant No. 2016/19622-0), a UFC fellowship, the hospitality of the Wolfgang-Pauli-Institute, and financial support from the Swiss National Science Foundation and Mr. G. Anderheggen.

References

1. A.I. Streltsov, O.E. Alon, L.S. Cederbaum, Phys. Rev. A **73**, 063626 (2006)
2. A.I. Streltsov, O.E. Alon, L.S. Cederbaum, Phys. Rev. Lett. **99**, 030402 (2007)
3. O.E. Alon, A.I. Streltsov, L.S. Cederbaum, J. Chem. Phys. **127**, 154103 (2007)

4. O.E. Alon, A.I. Streltsov, L.S. Cederbaum, Phys. Rev. A **77**, 033613 (2008)
5. K. Sakmann, A.I. Streltsov, O.E. Alon, L.S. Cederbaum, Phys. Rev. Lett. **103**, 220601 (2009)
6. A.U.J. Lode, K. Sakmann, O.E. Alon, L.S. Cederbaum, A.I. Streltsov, Phys. Rev. A **86**, 063606 (2012)
7. A.U.J. Lode, Phys. Rev. A **93**, 063601 (2016)
8. H.-D. Meyer, F. Gatti, G.A. Worth (eds.), *Multidimensional Quantum Dynamics: MCTDH Theory and Applications* (Wiley-VCH, Weinheim, 2009)
9. K. Sakmann, *Many-Body Schrödinger Dynamics of Bose-Einstein Condensates* (Springer Theses. Springer, Heidelberg, 2011)
10. N.P. Proukakis, S.A. Gardiner, M.J. Davis, M.H. Szymanska (eds.), *Quantum Gases: Finite Temperature and Non-equilibrium Dynamics*, Cold Atoms Series, vol. 1 (Imperial College Press, London, 2013)
11. A.U.J. Lode, *Tunneling Dynamics in Open Ultracold Bosonic Systems*, Springer Theses (Springer, Heidelberg, 2015)
12. A.U.J. Lode, C. Bruder, Phys. Rev. Lett. **118**, 013603 (2017)
13. S.E. Weiner, M.C. Tsatsos, L.S. Cederbaum, A.U.J. Lode, Sci. Rep. **7**, 40122 (2017)
14. A.U.J. Lode, C. Bruder, Phys. Rev. A **94**, 013616 (2016)
15. S. Klaiman, L.S. Cederbaum, Phys. Rev. A **94**, 063648 (2016)
16. S. Klaiman, A.I. Streltsov, O.E. Alon, J. Phys.: Conf. Ser. **826**, 012020 (2017)
17. M. Theisen, A.I. Streltsov, Phys. Rev. A **94**, 053622 (2016)
18. R. Beinke, S. Klaiman, L.S. Cederbaum, A.I. Streltsov, O.E. Alon, Phys. Rev. A **95**, 063602 (2017)
19. A.U.J. Lode, K. Sakmann, R.A. Doganov, J. Grond, O.E. Alon, A.I. Streltsov, L.S. Cederbaum, In high performance computing, in *Science and Engineering '13: Transactions of the High Performance Computing Center, Stuttgart (HLRS) 2013*, ed. by W.E. Nagel, D.H. Kröner, M.M. Resch (Springer, Heidelberg, 2013), pp. 81–92
20. S. Klaiman, A.U.J. Lode, K. Sakmann, O.I. Streltsova, O.E. Alon, L.S. Cederbaum, A.I. Streltsov, In high performance computing, in *Science and Engineering '14: Transactions of the High Performance Computing Center, Stuttgart (HLRS) 2014*, ed. by W.E. Nagel, D.H. Kröner, M.M. Resch (Springer, Heidelberg, 2015), pp. 63–86
21. O.E. Alon, V.S. Bagnato, R. Beinke, I. Brouzos, T. Calarco, T. Caneva, L.S. Cederbaum, M.A. Kasevich, S. Klaiman, A.U.J. Lode, S. Montangero, A. Negretti, R.S. Said, K. Sakmann, O.I. Streltsova, M. Theisen, M.C. Tsatsos, S.E. Weiner, T. Wells, A.I. Streltsov, In high performance computing, in *Science and Engineering '15: Transactions of the High Performance Computing Center, Stuttgart (HLRS) 2015*, ed. by W.E. Nagel, D.H. Kröner, M.M. Resch (Springer, Heidelberg, 2016), pp. 23–50
22. O.E. Alon, R. Beinke, L.S. Cederbaum, M.J. Edmonds, E. Fasshauer, M.A. Kasevich, S. Klaiman, A.U.J. Lode, N.G. Parker, K. Sakmann, M.C. Tsatsos, A.I. Streltsov, In high performance computing, in *Science and Engineering '16: Transactions of the High Performance Computing Center, Stuttgart (HLRS) 2016*, ed. by W.E. Nagel, D.H. Kröner, M.M. Resch (Springer, Heidelberg, 2016), pp. 79–96
23. O.E. Alon, R. Beinke, C. Bruder, L.S. Cederbaum, S. Klaiman, A.U.J. Lode, K. Sakmann, M. Theisen, M.C. Tsatsos, S.E. Weiner, A.I. Streltsov, In high performance computing, in *Science and Engineering '17: Transactions of the High Performance Computing Center, Stuttgart (HLRS) 2017*, ed. by W.E. Nagel, D.H. Kröner, M.M. Resch (Springer, Heidelberg, 2018), pp. 93–115
24. R. Roy, A. Gammal, M.C. Tsatsos, B. Chatterjee, B. Chakrabarti, A.U.J. Lode, Phys. Rev. A **97**, 043625 (2018)
25. B. Chatterjee, A.U.J. Lode, Phys. Rev. A **98**, 053624 (2018)
26. S. Dutta, M.C. Tsatsos, S. Basu, A.U.J. Lode, arXiv:1802.02407v1 [cond-mat.quant-gas] to appear in New J. Phys. (2019)
27. J. H. V. Nguyen, M. C. Tsatsos, D. Luo, A. U. J. Lode, G. D. Telles, V. S. Bagnato, R. G. Hulet, Phys. Rev. X **9**, 011052 (2019)
28. P. Molignini, L. Papariello, A. U. J. Lode, R. Chitra, Phys. Rev. A **98**, 053620 (2018)

29. A.U.J. Lode, F.S. Diorico, R. Wu, P. Molignini, L. Papariello, R. Lin, C. Lévêque, L. Exl, M.C. Tsatsos, R. Chitra, N.J. Mauser, New J. Phys. **20**, 055006 (2018).
30. S.K. Haldar, O.E. Alon, Chem. Phys. **509**, 72 (2018).
31. L.S. Cederbaum, Phys. Rev. A **96**, 013615 (2017)
32. K. Sakmann, J. Schmiedmayer, arXiv:1802.03746v1 [cond-mat.quant-gas]
33. S. Klaiman, R. Beinke, L.S. Cederbaum, A.I. Streltsov, O.E. Alon, Chem. Phys. **509**, 45 (2018)
34. R. Beinke, L.S. Cederbaum, O.E. Alon, Phys. Rev. A **98**, 053634 (2018)
35. A.I. Streltsov, L.S. Cederbaum, O.E. Alon, K. Sakmann, A.U.J. Lode, J. Grond, O.I. Streltsova, S. Klaiman, R. Beinke, *The Multiconfigurational Time-Dependent Hartree for Bosons Package*, version 3.x, http://mctdhb.org, Heidelberg/Kassel (2006-Present)
36. A.I. Streltsov, O.I. Streltsova, *The multiconfigurational time-dependent Hartree for bosons laboratory*, version 1.5, http://MCTDHB-lab.com
37. A.U.J. Lode, M.C. Tsatsos, S.E. Weiner, E. Fasshauer, R. Lin, L. Papariello, P. Molignini, C. Lévêque, MCTDH-X: The time-dependent multiconfigurational Hartree for indistinguishable particles software, http://ultracold.org (2018)
38. D. Jaksch, C. Bruder, J.I. Cirac, C.W. Gardiner, P. Zoller, Phys. Rev. Lett. **81**, 3108 (1998)
39. M. Greiner, O. Mandel, T. Esslinger, T.W. Hänsch, I. Bloch, Nature (London) **415**, 39 (2002)
40. E. Haller, R. Hart, M.J. Mark, J.G. Danzl, L. Reichsöllner, M. Gustavsson, M. Dalmonte, G. Pupillo, H.-C. Nägerl, Nature (London) **466**, 597 (2010)
41. O.E. Alon, L.S. Cederbaum, Phys. Rev. Lett. **95**, 140402 (2005)
42. O.E. Alon, A.I. Streltsov, L.S. Cederbaum, Phys. Rev. Lett. **95**, 030405 (2005)
43. M.D. Girardeau, E.M. Wright, J.M. Triscari, Phys. Rev. A **63**, 033601 (2001)
44. E. Fasshauer, A.U.J. Lode, Phys. Rev. A **93**, 033635 (2016)
45. S. Klaiman, A.U.J. Lode, A.I. Streltsov, L.S. Cederbaum, O.E. Alon, Phys. Rev. A **90**, 043620 (2014)
46. A.U.J. Lode, M.C. Tsatsos, J.L. Temp, Phys. **181**, 171 (2015)
47. T. Wells, A.U.J. Lode, V.S. Bagnato, M.C. Tsatsos, J.L. Temp, Phys. **180**, 133 (2015)
48. U.R. Fischer, A.U.J. Lode, B. Chatterjee, Phys. Rev. A **91**, 063621 (2015)
49. A.U.J. Lode, B. Chakrabarti, V.K.B. Kota, Phys. Rev. A **92**, 033622 (2015)
50. D.M. Collins, Z. Naturforsch. **48a**, 68 (1993)
51. P. Ziesche, Int. J. Quantum Chem. **56**, 363 (1995)
52. R.O. Esquivel, A.L. Rodriguez, R.P. Sagar, M. Ho, V.H. Smith, Phys. Rev. A **54**, 259 (1996)
53. P. Gersdorf, W. John, J.P. Perdew, P. Ziesche, Int. J. Quantum Chem. **61**, 935 (1997)
54. P. Ziesche, O. Gunnarsson, W. John, H. Beck, Phys. Rev. B **55**, 10270 (1997)
55. P. Ziesche, V.H. Smith Jr., M. Ho, S. Rudin, P. Gersdorf, M. Taut, J. Chem. Phys. **110**, 6135 (1999)
56. R. Paškauskas, L. You, Phys. Rev. A **64**, 042310 (2001)
57. N.L. Guevara, R.P. Sagar, R.O. Esquivel, J. Chem. Phys. **122**, 084101 (2005)
58. R.P. Sagar, N.L. Guevara, J. Chem. Phys. **123**, 044108 (2005)
59. K.D. Sen, J. Chem. Phys. **123**, 074110 (2005)
60. K. Sakmann, M. Kasevich, Nature Phys. **12**, 451 (2016)
61. M.A. Baranov, Phys. Rep. **464**, 71 (2008)
62. T. Lahaye, C. Menotti, L. Santos, M. Lewenstein, T. Pfau, Rep. Prog. Phys. **72**, 126401 (2009)
63. A. Griesmaier, J. Werner, S. Hensler, J. Stuhler, T. Pfau, Phys. Rev. Lett. **94**, 160401 (2005)
64. Q. Beaufils, R. Chicireanu, T. Zanon, B. Laburthe-Tolra, E. Maréchal, L. Vernac, J.-C. Keller, O. Gorceix, Phys. Rev. A **77**, 061601 (2008)
65. M. Lu, N.Q. Burdick, S.H. Youn, B.L. Lev, Phys. Rev. Lett. **107**, 190401 (2011)
66. K. Aikawa, A. Frisch, M. Mark, S. Baier, A. Rietzler, R. Grimm, F. Ferlaino, Phys. Rev. Lett. **108**, 210401 (2012)
67. S. Zöllner, G.M. Bruun, C.J. Pethick, S.M. Reimann, Phys. Rev. Lett. **107**, 035301 (2011)
68. S. Zöllner, Phys. Rev. A **84**, 063619 (2011)
69. G.E. Astrakharchik, Lozovik, E. Yu, Phys. Rev. A **77**, 013404 (2008)
70. A.S. Arkhipov, G.E. Astrakharchik, A.V. Belikov, Y.E. Lozovik, JETP Lett. **82**, 39 (2005)

71. G.E. Astrakharchik, G.E. Morigi, G. De Chiara, J. Boronat, Phys. Rev. A **78**, 063622 (2008)
72. F. Deuretzbacher, J.C. Cremon, S.M. Reimann, Phys. Rev. A **81**, 063616 (2010)
73. B. Chatterjee, I. Brouzos, L. Cao, P. Schmelcher, J. Phys. B **46**, 085304 (2013)
74. J.R. Anglin, W. Ketterle, Nature (London) **416**, 211 (2002)
75. M.C. Tsatsos, P.E.S. Tavares, A. Cidrim, A.R. Fritsch, M.A. Caracanhas, F.E.A. dos Santos, C.F. Barenghi, V.S. Bagnato, Phys. Rep. **622**, 1 (2016)
76. C.J. Pethick, H. Smith, *Bose-Einstein Condensation in Dilute Gases*, 2nd edn. (Cambridge University Press, Cambridge, 2008)
77. M.H. Anderson, J.R. Ensher, M.R. Matthews, C.E. Wieman, E.A. Cornell, Science **269**, 198 (1995)
78. K.B. Davis, M.-O. Mewes, M.R. Andrews, N.J. van Druten, D.S. Durfee, D.M. Kurn, W. Ketterle, Phys. Rev. Lett. **75**, 3969 (1995)
79. P. Domokos, H. Ritsch, Phys. Rev. Lett. **89**, 253003 (2002)
80. D. Nagy, G. Szirmai, P. Domokos, Eur. Phys. J. D **48**, 127 (2008)
81. K. Baumann, C. Guerlin, F. Brennecke, T. Esslinger, Nature (London) **464**, 1301 (2010)
82. E. Mathieu, J. Phys. Theor. Appl. **2**, 32 (1883)
83. L.D. Landau, E.M. Lifshitz, *Mechanics*, 3rd edn. (Elsevier, Amsterdam, 1982)
84. F. Mivehvar, F. Piazza, H. Ritsch, Phys. Rev. Lett. **119**, 063602 (2017)
85. R.H. Dicke, Phys. Rev. **93**, 99 (1954)
86. G.J. Milburn, J. Corney, E.M. Wright, D.F. Walls, Phys. Rev. A **55**, 4318 (1997)
87. A. Smerzi, S. Fantoni, S. Giovanazzi, S.R. Shenoy, Phys. Rev. Lett. **79**, 4950 (1997)
88. S. Raghavan, A. Smerzi, V.M. Kenkre, Phys. Rev. A **60**, R1787 (1999)
89. C. Orzel, A.K. Tuchman, M.L. Fenselau, M. Yasuda, M.A. Kasevich, Science **291**, 2386 (2001)
90. A. Vardi, J.R. Anglin, Phys. Rev. Lett. **86**, 568 (2001)
91. M. Albiez, R. Gati, J. Fölling, S. Hunsmann, M. Cristiani, M.K. Oberthaler, Phys. Rev. Lett. **95**, 010402 (2005)
92. T. Schumm, S. Hofferberth, L.M. Andersson, S. Wildermuth, S. Groth, I. Bar-Joseph, J. Schmiedmayer, P. Krüger, Nature Phys. **1**, 57 (2005)
93. S. Levy, E. Lahoud, I. Shomroni, J. Steinhauer, Nature (London) **449**, 579 (2007)
94. M. Trujillo-Martinez, A. Posazhennikova, J. Kroha, Phys. Rev. Lett. **103**, 105302 (2009)
95. T. Zibold, E. Nicklas, C. Gross, M.K. Oberthaler, Phys. Rev. Lett. **105**, 204101 (2010)
96. K. Sakmann, A.I. Streltsov, O.E. Alon, L.S. Cederbaum, Phys. Rev. A **89**, 023602 (2014)
97. H. Veksler, S. Fishman, New J. Phys. **17**, 053030 (2015)
98. S. Klaiman, A.I. Streltsov, O.E. Alon, Phys. Rev. A **93**, 023605 (2016)
99. J. Stuhler, A. Griesmaier, T. Koch, M. Fattori, T. Pfau, S. Giovanazzi, P. Pedri, L. Santos, Phys. Rev. Lett. **95**, 150406 (2005)
100. J.E. Johnson, S.L. Rolston, Phys. Rev. A **82**, 033412 (2010)
101. N. Henkel, R. Nath, T. Pohl, Phys. Rev. Lett. **104**, 195302 (2010)
102. E.H. Lieb, R. Seiringer, J. Yngvason, Phys. Rev. A **61**, 043602 (2000)
103. E.H. Lieb, R. Seiringer, Phys. Rev. Lett. **88**, 170409 (2002)
104. L. Erdős, B. Schlein, H.-T. Yau, Phys. Rev. Lett. **98**, 040404 (2007)
105. L. Erdős, B. Schlein, H.-T. Yau, Invent. Math. **167**, 515 (2007)
106. Y. Castin, R. Dum, Phys. Rev. A **57**, 3008 (1998)
107. S. Klaiman, O.E. Alon, Phys. Rev. A **91**, 063613 (2015)
108. L. Cohen, C. Lee, J. Math. Phys. **26**, 3105 (1985)
109. J. Yan, J. Stat. Phys. **113**, 623 (2003)
110. M. Gajda, Phys. Rev. A **73**, 023603 (2006)
111. C. Schilling, R. Schilling, Phys. Rev. **93**, 021601(R) (2016)
112. P.A.M. Dirac, Proc. Cambridge Philos. Soc. **26**, 376 (1930)
113. J. Frenkel, *Wave Mechanics: Advanced General Theory* (Clarendon Press, Oxford, 1934)
114. P. Kramer, M. Saraceno, *Geometry of the time-dependent variational principle* (Springer, Berlin, 1981)
115. C. Cohen-Tannoudji, B. Diu, F. Laloë, *Quantum Mechanics*, vol. 1 (Wiley, New York, 1977)

116. R.W. Spekkens, J.E. Sipe, Phys. Rev. A **59**, 3868 (1999)
117. A.I. Streltsov, L.S. Cederbaum, N. Moiseyev, Phys. Rev. A **70**, 053607 (2004)
118. S. Zöllner, H.-D. Meyer, P. Schmelcher, Phys. Rev. A **74**, 063611 (2006)
119. P. Noziéres and D. Saint James, J. Phys. (Paris) **43**, 1133 (1982)
120. P. Noziéres, in *Bose-Einstein Condensation*, ed. by A. Griffin, D.W. Snoke, S. Stringari (Cambridge University Press, Cambridge, 1996), p. 15
121. E.J. Mueller, T.-L. Ho, M. Ueda, G. Baym, Phys. Rev. A **74**, 033612 (2006)
122. U.R. Fischer, P. Bader, Phys. Rev. A **82**, 013607 (2010)
123. Q. Zhou, X. Cui, Phys. Rev. Lett. **110**, 140407 (2013)
124. Y. Kawaguchi, Phys. Rev. A **89**, 033627 (2014)
125. S.-W. Song, Y.-C. Zhang, H. Zhao, X. Wang, W.-M. Liu, Phys. Rev. A **89**, 063613 (2014)
126. H.H. Jen, S.-K. Yip, Phys. Rev. A **91**, 063603 (2015)
127. A.R. Kolovsky, Phys. Rev. A **95**, 033622 (2017)
128. P. Bader, U.R. Fischer, Phys. Rev. Lett. **103**, 060402 (2009)
129. A.I. Streltsov, Phys. Rev. A **88**, 041602(R) (2013)
130. M.-K. Kang, U.R. Fischer, Phys. Rev. Lett. **113**, 140404 (2014)
131. O.I. Streltsova, O.E. Alon, L.S. Cederbaum, A.I. Streltsov, Phys. Rev. A **89**, 061602(R) (2014)
132. U.R. Fischer, M.-K. Kang, Phys. Rev. Lett. **115**, 260404 (2015)
133. D.A. Butts, D.S. Rokhsar, Nature (London) **397**, 327 (1999)
134. M.R. Matthews, B.P. Anderson, P.C. Haljan, D.S. Hall, C.E. Wieman, E.A. Cornell, Phys. Rev. Lett. **83**, 2498 (1999)
135. K.W. Madison, F. Chevy, W. Wohlleben, J. Dalibard, Phys. Rev. Lett. **84**, 806 (2000)
136. F. Chevy, K.W. Madison, J. Dalibard, Phys. Rev. Lett. **85**, 2223 (2000)
137. K.W. Madison, F. Chevy, V. Bretin, J. Dalibard, Phys. Rev. Lett. **86**, 4443 (2001)
138. I. Coddington, P. Engels, V. Schweikhard, E.A. Cornell, Phys. Rev. Lett. **91**, 100402 (2003)
139. V. Schweikhard, I. Coddington, P. Engels, V.P. Mogendorff, E.A. Cornell, Phys. Rev. Lett. **92**, 040404 (2004)
140. A. Aftalion, X. Blanc, J. Dalibard, Phys. Rev. A **71**, 023611 (2005)
141. N. Regnault, T. Jolicoeur, Phys. Rev. Lett. **91**, 030402 (2003)
142. C.-C. Chang, N. Regnault, T. Jolicoeur, J.K. Jain, Phys. Rev. A **72**, 013611 (2005)
143. S. Stock, B. Battelier, V. Bretin, Z. Hadzibabic, J. Dalibard, Laser Phys. Lett. **2**, 275 (2005)
144. M. Roncaglia, M. Rizzi, J. Dalibard, Sci. Rep. **1**, 43 (2011)
145. I. Bloch, J. Dalibard, W. Zwerger, Rev. Mod. Phys. **80**, 885 (2008)
146. N.R. Cooper, Adv. Phys. **57**, 539 (2008)
147. S. Viefers, J. Phys.: Cond. Matt. **20**, 123202 (2008)
148. A.L. Fetter, Rev. Mod. Phys. **81**, 647 (2009)
149. N.N. Bogoliubov, J. Phys. **11** (USSR), 23 (1947)
150. P.-G. de Gennes, *Superconductivity of Metals and Alloys* (Benjamin, New York, 1966)
151. T. Mizushima, M. Ichioka, K. Machida, Phys. Rev. Lett. **90**, 180401 (2003)
152. T.P. Simula, K. Machida, Phys. Rev. A **82**, 063627 (2010)
153. L.O. Baksmaty, S.J. Woo, S. Choi, N.P. Bigelow, Phys. Rev. Lett. **92**, 160405 (2004)
154. F. Chevy, Phys. Rev. A **73**, 041604 (2006)
155. A. Collin, Phys. Rev. A **73**, 013611 (2006)
156. F. Ancilotto, F. Toigo, Phys. Rev. A **89**, 023617 (2014)
157. S. Viefers, T.H. Hansson, S.M. Reimann, Phys. Rev. A **62**, 053604 (2000)
158. S.M. Reimann, M. Koskinen, Y. Yu, M. Manninen, New J. Phys. **8**, 59 (2006)
159. J.C. Cremon, G.M. Kavoulakis, B.R. Mottelson, S.M. Reimann, Phys. Rev. A **87**, 053615 (2013)
160. J.C. Cremon, A.D. Jackson, E.Ö. Karabulut, G.M. Kavoulakis, B.R. Mottelson, S.M. Reimann, Phys. Rev. A **91**, 033632 (2015)
161. U.R. Fischer, G. Baym, Phys. Rev. Lett. **90**, 140402 (2003)
162. V. Bretin, S. Stock, Y. Seurin, J. Dalibard, Phys. Rev. Lett. **92**, 050403 (2004)
163. M. Correggi, F. Pinsker, N. Rougerie, J. Yngvason, Eur. Phys. J. Spec. Top. **217**, 183 (2013)
164. P.T. Nam, R. Seiringer, Arch. Rational Mech. Anal. **215**, 381 (2015)

165. P. Grech, R. Seiringer, Commun. Math. Phys. **322**, 559 (2013)
166. J. Grond, A.I. Streltsov, A.U.J. Lode, K. Sakmann, L.S. Cederbaum, O.E. Alon, Phys. Rev. A **88**, 023606 (2013)
167. O.E. Alon, A.I. Streltsov, L.S. Cederbaum, J. Chem. Phys. **140**, 034108 (2014)
168. O.E. Alon, J. Phys.: Conf. Ser. **594**, 012039 (2015)
169. C. Lévêque, L.B. Madsen, New J. Phys. **19**, 043007 (2017)
170. C. Lévêque, L.B. Madsen, J. Phys. B **51**, 155302 (2018)
171. H. Wang, M. Thoss, J. Chem. Phys. **131**, 024114 (2009)
172. S. Krönke, L. Cao, O. Vendrell, P. Schmelcher, New J. Phys. **15**, 063018 (2013)
173. L. Cao, S. Krönke, O. Vendrell, P. Schmelcher, J. Chem. Phys. **139**, 134103 (2013)
174. U. Manthe, T. Weike, J. Chem. Phys. **146**, 064117 (2017)
175. L. Cao, V. Bolsinger, S.I. Mistakidis, G.M. Koutentakis, S. Krönke, J.M. Schurer, P. Schmelcher, J. Chem. Phys. **147**, 044106 (2017)

Part II
Molecules, Interfaces, and Solids

During this reporting period the field of molecules, interfaces and solids has profited enormously again from the computational resources provided by the High Performance Computing Center Stuttgart and the Steinbuch Centre for Computing Karlsruhe. For this anthology, we selected one third only of the broadly diversified and greatly successful projects in this area to demonstrate the scientific progress that can be achieved by high performance computing in chemistry, physics, material science and nanotechnology.

One outstanding example of this is the work by K. Schulz and M. Sudmans from the Institute for Applied Materials of the Karlsruhe Institute of Technology who addressed the long-standing, notoriously difficult dislocation-dynamics problem by a original numerical approach, which is based on the dislocation density within a kinematic (continuum) framework and balances the need for a micro-mechanical modeling and the reduction of the numerical effort. The study is promoted by the German Research Foundation (DFG) through the research group FOR 1650 'Dislocation based Plasticity'. Here, the authors focus on materials that are dominated by interfaces such as bi-crystals and metal matrix composite structures and analyze the impact of the dislocation stress interaction for these systems in order to discuss the interplay of physical and numerical accuracy during the the coarse graining at interfaces. Thereby the modeling of the interfaces (being impenetrable to dislocations) is challenging since the dislocations can form pile-ups that induce significant stress gradients. The problem is solved using an MPI parallelization in the finite-element software M++ in combination with an extended discretization in the discontinuous Galerkin scheme for the transport properties. The article gives a detailed discussion of the (plastic) slip profiles and the geometrically necessary dislocations, as well as of the influence of fibers particulary with respect to the stress-strain characteristics. Thereby the size effect of the material is observed to be in good agreement with the discrete dislocation dynamics simulations.

Applying the electron paramagnetic resonance method to semiconductors, the spin-spin zero-field splitting (ZFS) serves as a fingerprint for the identification of high-spin defects in the host material. The Theoretical Material Physics Group at the University Paderborn headed by W.G. Schmidt has inspected the precision and performance of an MPI-based implementation for the DFT-based computation of this important physical quantity on the Cray XC40. In particular, the authors investigated how the accuracy of the calculations depends on the supercell size, imposing periodic boundary conditions. Thereby the algorithms were implemented in the GIPAW and Quantum ESPRESSO software and the negatively charged nitrogen-vacancy defect center NV^- in diamond was chosen as an ideal test system. The results presented impressively demonstrate that any *ab initio* evaluation of the ZFS requires a extreme level of precision. Furthermore, the k-point grid used in Brillouin sampling has a major effect on the efficiency: If it is chosen sufficiently dense the ZFS calculations can be accurate even for small supercells. However, larger unit cells are indispensable in order to reach very high precision of the data, requiring the use of supercomputers. The related scientific investigations of this group were supported by the DFG through the priority program SPP1601, the research unit FOR1700, and the transregio collaborative research center SFB TRR 142.

The contribution of Lesnicki and Sulpizi investigates the dissociation of acids dissolved in water at the liquid/solid interface. Pyruvic acid at the water/silica interface was chosen since pyruvic acid plays an important role in atmospheric chemistry, and the water/silica interface models the situation encounter for a water droplet at a mineral dust particle. The dissociation constant (pK_a value) depends on the free energy change during dissociation and is therefore difficult to calculate. In this contribution, thermodynamic integration along an artificial path was used, a path along which protons are slowly created (in bulk water) and annihilated (near the water/silica interface). In this way, reliable pK_a differences for pyruvic acid in the bulk and near the interface were calculated, and it was found that interfacial acid molecules are more acidic by nearly 2 pK_a units. Since acids tend to become *less* acidic at other interfaces (e.g. water/air), the microscopic origin of the results was analyzed, and it was found that hydrogen bonds to silanol groups stabilize the deprotonated pyruvate anion at the interface. Long simulation runs (25 ps) on large model systems with 1433 atoms were necessary. These molecular dynamics runs on a Born-Oppenheimer surface from density functional theory were conducted with the CP2K/QuickStep software.

The work of Pieck, Pecher, Luy and Tonner describes simulations that give insight into processes that take place when silicon surfaces are functionalized. The experimental method of choice here is metal-organic vapour phase epitaxy (MOVPE), where atoms with organic ligands approach the surface, then only atoms of interest are deposited on the surface. For the adsorption of *tert*-butyl phosphine on silicon, the question is what happens on the surface. It is observed experimentally that these molecules chemisorb in specific patterns. The calculations reported here demonstrate how the first chemisorbed molecule (a silicon-phosphorous bond is formed) directs the adsorption of a second one. Another calculation reported in this contribution targets at the band gap of a GaAs semiconducter where some of the arsenic atoms are replaced by nitrogen. Here it turns out that one has to use large supercells to avoid artificial interactions between nitrogen atoms and their periodic images. It can further be noted that replacing an As atom by a much smaller N one induces large geometric distortions in its vicinity. The calculated decrease of the band gap with increasing nitrogen content nicely fits experimental data and a theoretical continuum model (with empirical parameters from experiment). The calculations reported here have been conducted with the VASP program that scales very well on modern supercomputing hardware such that thousands of CPU core can be used.

Finally the work of Jones reports calculations that uncover ethylene oxidation on silver surfaces. Ethylene oxide is a very important substance produced from ethylene in huge amounts. The terminal oxidant is molecular oxygen and the process takes place at the surface of a silver catalyst. Although this process is known since 80 years, it remained a bafflement until recently. For example, it only works for ethylene and not for higher homologues such as propylene, and the nature of the oxygen species responsible for selective oxidation remained unknown. Selectivity is a key issue here: it is easy to oxidize ethylene but the favoured product is of course carbon dioxide, but we do not want to burn ethylene, we want to make something useful out of it! In this project, it has been demonstrated earlier that sulfate is probably the species responsible for selective oxidation. In the present report, this finding is further elaborated on. First, a complete catalytic cycle including the re-oxidation of the sulfur species is investigated, and second, side reactions leading to formation of acetaldehye are explored. Since acetaldehyde formation is the first step towards complete oxidation (combustion), it is important to know why and how this path is disfavoured (compared to ethylene oxid formation). Modeling chemical reactivity is a difficult task since entiere reaction pathways have to be calculated, here this is done using the nudged elastic band model and variants thereof. In this model, a number of replicas of the system between educts and products are calculated together with an articifial force taking care that the images do not become too different – otherwise they would all fall down either to the product or educt side. By a judicious choice how to distribute the linear algebra workload, the Quantum ESPRESSO program used in this study showed good scaling up to 5000 CPU cores, together with the inherent (and trivial) parallelism of the nudged elastic band model (10-20 replicas), the code could make good use of well beyond 50 thousand CPU cores.

In summary, almost all projects supported during the period under review, have in common, besides a high scientific quality, the strong need for computers with high performance to achieve their results. Therefore, the supercomputing facilities (HLRS and ForHLR) have been essential for their success.

Holger Fehske
Institut für Physik, Lehrstuhl Komplexe Quantensysteme
Ernst-Moritz-Arndt-Universität Greifswald
Felix-Hausdorff-Str. 6, 17489 Greifswald, Germany
e-mail: fehske@physik.uni-greifswald.de

Christoph van Wüllen
Fachbereich Chemie, Technische Universität Kaiserslautern
Erwin-Schrödinger-Str. 52, 67663 Kaiserslautern, Germany
e-mail: vanwullen@chemie.uni-kl.de

Mesoscale Simulation of Dislocation Microstructures at Internal Interfaces

Katrin Schulz and Markus Sudmanns

Abstract The need for predicting the behavior of crystalline materials on small-scales has led to the development of physically based descriptions of the motion of dislocations. Several dislocation-based continuum theories have been introduced, but only recently rigorous techniques have been developed for performing meaningful averages over systems of moving, curved dislocations, yielding evolution equations based on a dislocation density tensor. Those evolution equations provide a physically based framework for describing the motion of curved dislocations in three-dimensional systems. However, a meaningful description of internal interfaces and the complex mechanistic interaction of dislocations and interfaces in a dislocation based continuum model is still an open task. In this paper, we address the conflict between the need for a mechanistic modeling of the involved physical mechanisms and a reasonable reduction of model complexity and numerical effort. We apply a dislocation density based continuum formulation to systems which are strongly affected by internal interfaces, i.e. grain boundaries and interfaces in composite materials, and focus on the physical and numerical realization. Particularly, the interplay between physical and numerical accuracy is pointed out and discussed.

1 Introduction

The modeling and simulation of dislocation motion as the cause of plastic deformation has been of growing interest in recent years. A proper understanding of the physical principles has led to reasonable descriptions of physically motivated small-scale models of plasticity. This has induced renewed efforts to formulate bottom-up formulations of continuum theories of dislocation dynamics. The classical dislocation based continuum theory, which is based on the definition of a second-rank dislocation density tensor, was derived by Kröner [10] and Nye [13] in the 1950s. However, introduced as a measure of the averaged plastic deformation state to link

K. Schulz (✉) · M. Sudmanns
Institute for Applied Materials (IAM-CMS), Karlsruhe Institute of Technology,
Kaiserstr. 12, 76131 Karlsruhe, Germany
e-mail: katrin.schulz@kit.edu

© Springer Nature Switzerland AG 2019
W. E. Nagel et al. (eds.), *High Performance Computing in Science
and Engineering '18*, https://doi.org/10.1007/978-3-030-13325-2_7

the microscopically discontinuous deformation state to a macroscopically continuous deformation state, the classical density tensor just considers geometrically necessary dislocations (GNDs). It does not account for the total dislocation content in an averaging volume and therefore can not properly account for dislocation motion, see [11]. An extended model, which eliminates this shortcoming has been introduced by Hochrainer et al. [9]. They give the mathematical foundations for using the methods of statistical mechanics consistently with three-dimensional systems of curved dislocation lines. Having formulated the kinematic problem, one can combine the model with a constitutive relation for the dislocation velocity, as e.g. proposed by Groma et al. [5] and Hochrainer [8] or discussed in the context of numerical coarse graining by Schmitt et al. [17] and Schulz et al. [18].

The introduction of interfaces, which constrain the dislocation motion or are completely impenetrable for dislocations, requires a precise description of the interface areas including the continuity constraints [14]. The correct representation of the internal stress fields and dislocation microstructure is essential for analyzing structures with internal dislocation density gradients. In the context of a continuum theory, dislocation pile-ups at impenetrable walls have been studied for one dimensional systems e.g. by Friedman and Chrzan [3], Liu et al. [12], Schulz et al. [18] and for single slip systems by Hirschberger et al. [6], Schmitt et al. [17]. Van der Giessen and Needleman [25] analyzed a two phase composite using a discrete dislocation dynamics model in order to understand the complex material behavior of solids with more than one phase. They simplified the system as a 2D plane strain problem considering edge dislocations on parallel slip planes only. Slightly different systems based on discrete dislocation dynamics have been analyzed by Cleveringa et al. [1, 2]. Using a dislocation based continuum approach, a composite material has been investigated by Schulz et al. [20] and Yefimov et al. [28] for a simplified 2D model considering single slip configurations. The deposition of dislocation loops around particles is analyzed by Schwarz et al. [22].

In this paper, we apply a numerical, dislocation density based continuum formulation based on the kinematic formulation of [9] to structures dominated by internal interfaces—a composite material and a bi-crystal structure. Considering the full set of fcc slip systems, a 3D metal matrix composite material according to [4] is investigated. We show that we can reproduce 'discrete dislocation dynamics' (DDD) results for different composite structures including occurring size effects. Additionally, we analyze the impact of dislocation stress interaction for the composite structure as well as the setup of a bi-crystal in order to discuss the interplay of physical and numerical accuracy during numerical coarse graining at interfaces.

2 Numerical Method

2.1 Dislocation Density Based Continuum Model

We consider a two-scale approach for the formulation of the elasto-plastic behavior of metals with face-centered cubic (fcc) structure. The formulation is based on the classical decomposition of the distortion tensor into an elastic and a plastic part

$$\mathbf{Du} = \boldsymbol{\beta}^{\text{pl}} + \boldsymbol{\beta}^{\text{el}}. \tag{1}$$

The plastic distortion $\boldsymbol{\beta}^{\text{pl}}$ is determined as the sum of the plastic slip over all slip systems (defined by the index s) according to the fcc microstructure as

$$\boldsymbol{\beta}^{\text{pl}} = \sum_{s=1}^{N} \gamma_s \mathbf{d}_s \otimes \mathbf{m}_s . \tag{2}$$

Herein, the plastic slip γ_s is assumed to be the result of the dislocation motion on the respective slip system considering the orthonormal basis $\{\mathbf{d}_s, \mathbf{l}_s, \mathbf{m}_s\}$.

The dislocation microstructure is characterized by three dislocation density measures: Total dislocation density ρ_s, the vector of the geometrically necessary dislocations (GND) density $\boldsymbol{\kappa}_s$, and the curvature density q_s. Using theses measures, the dislocation motion is given by the evolution equations according to [9]:

$$
\begin{aligned}
&\partial_t \rho_s = -\nabla \cdot (v_s \boldsymbol{\kappa}_s^{\perp}) + v_s q_s \quad \text{with} \quad \boldsymbol{\kappa}_s^{\perp} = \boldsymbol{\kappa}_s \times \mathbf{m}_s \\
&\partial_t \boldsymbol{\kappa}_s = \nabla \times (\rho_s v_s \mathbf{m}_s) \\
&\partial_t q_s = -\nabla \cdot \left(\frac{q_s}{\rho_s} \boldsymbol{\kappa}_s^{\perp} v_s + \frac{1}{2|\boldsymbol{\kappa}_s|^2} \left((\rho_s + |\boldsymbol{\kappa}_s|) \boldsymbol{\kappa}_s \otimes \boldsymbol{\kappa}_s - (\rho_s - |\boldsymbol{\kappa}_s|) \boldsymbol{\kappa}_s^{\perp} \otimes \boldsymbol{\kappa}_s^{\perp} \right) \nabla v_s \right),
\end{aligned}
\tag{3}
$$

where v_s denotes the velocity on the individual slip system. Thus, the evolution of the plastic slip can be written by the Orowan equation and the Burger's vector $\mathbf{b}_s = b_s \mathbf{d}_s$ as

$$\partial_t \gamma_s = v_s b_s \rho_s. \tag{4}$$

The evolution equations have to be closed by the formulation of the velocity law, which characterize the dislocation flux on the slip systems, respectively. We assume a linear dependency on the resolved shear stress on each slip system τ_s, thus the velocity can be written as $v_s = (b_s/B)\tau_s$ with B denoting a friction stress of 5×10^{-5} Pa s. Herein, the resolved shear stress represents the superposition of shear stresses due to external loading and the internal stress fields due to the dislocation microstructure given as $\tau_s = \tau_{\text{ext},s} + \tau_{\text{int},s}$. Particularly for the consideration of internal interfaces, the internal stresses $\tau_{\text{int},s}$ induced by the interaction of the dislocation stress fields play a significant role since the representation of dislocation pile-ups mainly affects the material behavior.

While the stress field of a single dislocation is known, e.g. according to [7], the challenge of the continuous modeling by dislocation density is the derivation of a formulation which is independent on the choice of the averaging volumes, which is assumed to be equivalent to the element size of a numerical discretization. In order to optimize between numerical resolution and accuracy, we distinguish between long- and short-range stresses in the formulation. The 'long-range stress fields' are considered to resolve the mean dislocation stress field according to the numerical resolution of the system. Using a mean field approach as given in [17], this introduces a mesh dependent stress component. On the one hand, this component looses accuracy during coarse graining of the considered discretization and disappears at all in configurations consisting solely of statistically stored dislocations. On the other hand, it can be shown, that the formulation yields a direct transition to small scale approaches as e.g. DDD models for a discretization yielding the resolution of the single dislocations in the structure [18, 19].

Since the goal of an efficient numerical consideration has to be to choose a discretization as coarse as possible, the loss of accuracy in the dislocation stress representation has to be balanced in order to yield a formulation which is independent on the discretization. This is derived by adding an interaction term accounting for 'short-range stresses' as introduced in [18]. As shown in [17], this term is able to balance the loss of accuracy in the dislocation stress homogenization for varying averaging volumes and particularly in systems with high GND gradients as it is expected in interface dominated structures.

In addition, a 'friction stress' which reduces the dislocation density motion, e.g. due to forest dislocations on other slip systems, is considered, which can be interpreted as a local yield stress. Based on [24], this yield stress τ_y is incorporated into the velocity law as

$$
v_s = \begin{cases} \frac{b_s}{B}(|\tau_s| - \tau_y)\mathrm{sign}\tau_s & \text{if } |\tau_s| > \tau_y \\ 0 & \text{if } |\tau_s| \leq \tau_y \end{cases} \tag{5}
$$

The dislocation density based microstructural evolution is incorporated into the numerical elasto-plastic framework described in [21] using the finite element code M++, which is based on a parallel multigrid method [26, 27]. We consider the fully three dimensional setup of the evolution of the dislocation microstructure coupled with the elastic problem as a two-scale approach. For the numerical solution of the evolution equations (Eq. 3), we combine the evolution equations of the dislocation densities to solve them by the system of equations

$$
\partial_t \begin{pmatrix} \rho_s \\ \kappa_s \end{pmatrix} + \nabla \cdot \boldsymbol{F}_s \begin{pmatrix} \rho_s \\ \kappa_s \end{pmatrix} = \boldsymbol{G}_s \tag{6}
$$

with the flux $\boldsymbol{F}$ and the source $\boldsymbol{G}_s$ given as

$$F_s \begin{pmatrix} \rho_s \\ \kappa_s \end{pmatrix} = v_s \begin{pmatrix} \mathbf{l}_s \cdot \kappa_s \mathbf{d}_s - \mathbf{d}_s \cdot \kappa_s \mathbf{l}_s \\ (\mathbf{d}_s \otimes \mathbf{l}_s - \mathbf{l}_s \otimes \mathbf{d}_s)\rho_s \end{pmatrix} \text{ and } G_s = \begin{pmatrix} v_s q_s \\ 0 \end{pmatrix}. \tag{7}$$

Thereby, the evolution of the dislocation densities and the curvature density q_s are treated as separate transport problems. The right hand side in Eq. (6) represents a production term depending on the curvature density q_s, which is considered a conserved quantity for the solution of the dislocation density evolution equations. Thus, the curvature density q_s couples both parts of the transport problem.

Using a discontinuous Galerkin scheme, the equations are solved separately for each slip system. For the integration of both parts in the discontinuous Galerkin scheme, we apply an upwind flux formulation using an implicit midpoint rule. The stresses are computed by the finite element method. Hereby, the the same mesh is applied to the computation of the stress as well as the internal variables of the microstructure. However we use linear shape functions for the finite element method and quadratic shape functions for the microstructural evolution equations.

2.2 Grain Boundaries and Internal Interfaces

The representation of internal interfaces plays a key role in polycrystalline as well as heterogeneous materials. The modeling of interfaces which are impenetrable to dislocations are a challenge since dislocation can form pile-ups that induce strong stress concentrations with significant stress gradients, cp. [23]. Two kinds of internal interfaces are considered in this investigation: Grain boundaries as interface between two crystallographic grains with tilted crystal orientation and internal interfaces separating two phases of a composite material, for which only one phase shows a plastic activity. The latter may induce further numerical challenges if the stiffness of the composed materials significantly varies. Therefore, we limit the present analyses to small strain configurations.

Both kinds of interfaces are modeled by a zero flux condition, thus no dislocation density transport is allowed across the boundaries. With the interface unit normal vector $\mathbf{n}$, it holds

$$\mathbf{v}_s = \frac{v_s}{\rho_s} \kappa_s^\perp \qquad \mathbf{n} \cdot \mathbf{v}_s = 0. \tag{8}$$

The stress fields of the crystallographic lattice distortion due to the dislocation pile-ups at the interfaces are covered by the mean field stress and solved incorporated into the elastic problem. However, the accumulation of dislocation density at internal interfaces does not imply that the dislocation flux is reduced in either case. For a composite material, which poses no interconnected inclusions, the dislocation transport can continue aligned to the interface with ongoing production of dislocation line length. Thus, a moving dislocation line, which encounters an inclusion in a composite will continue its motion by surrounding the inclusion.

2.3 System Setup

For the consideration of internal interfaces, we model two different system setups:
A bi-crystal and a composite structure.

The bi-crystal consists of two neighboring grains separated by an elastic tilt grain
boundary which is modeled impenetrable to dislocations, see Fig. 1a. The crystal
orientation of the different grains can be chosen independently. However, analyzed
as a benchmark system of a symmetric tilt boundary, here only one representative
adjacent slip system per grain is considered, respectively. The cubic system is defined
by an edge length of 4 μm and subjected to a constant external stress of 15 MPa
on each slip system. An initial dislocation microstructure is given by a uniform
distribution of dislocation loops with a constant radius yielding an initial dislocation
density of $\rho = 7 \times 10^{12}$ 1/m^2. The elastic material constants are given by the Young's
modulus $E = 70$ GPa and the Poisson's ratio $\nu = 0.3$.

The considered composite structure is composed of SiC-fibers oriented in z-
direction and homogeneously distributed in the xy plane incorporated into an alu-
minum matrix, see Fig. 1b. For the matrix material, the full fcc set of 12 slip
systems is considered and initially populated with a dislocation density of $\rho_0 =
1.75 \times 10^{13}$ 1/m^2. The system is subjected to a tensile loading in y-direction transver-
sal to the fiber orientation. The material constants are given as $E_{Al} = 71.3$ GPa and
$E_{SiC} = 373$ GPa as well as the Poisson's ratio $\nu_{Al} = 0.375$ and $\nu_{SiC} = 0.265$.

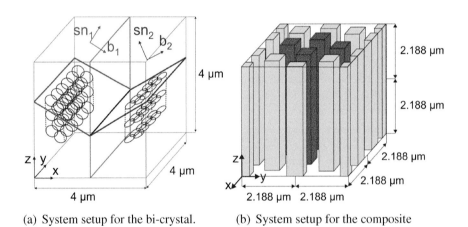

(a) System setup for the bi-crystal.   (b) System setup for the composite

Fig. 1 Setup of considered interfacial systems—bi-crystal (**a**) and composite with metal matrix (**b**)

2.4 Performance and Scalability

The two scale approach of the elastic boundary value problem coupled with the transport system of dislocation density flow is solved using MPI parallelization in the finite element software M++. Regarding the transport problem, an extended discretization scheme according to [16] is used in the discontinuous Galerkin scheme. For the time integration, we apply a Runge-Kutta scheme with an implicit midpoint rule which conserves quadratic energy functionals, so that no Courant-Friedrichs-Lewy (CFL) condition is required. The linear system is solved by the GMRES method according to [15].

A comparison of the computation time of the first two time steps of a benchmark simulation is performed to show the scalablilty, see Fig. 2. Since the dislocation microstructure is determined by the evolution equations of the internal variables (Eqs. 3 and 4) and the boundary value problem is solved by a standard FE solver in each time step, it is sufficient to evaluate the computational effort by analyzing single time steps independent of the present dislocation configuration and the specific microstructural problem considered. The refinement of the spatial numerical discretization is achieved by subdividing each volume element in each direction. This leads to a total numbers of elements, which differs by a factor of 8 between each refinement step. It can be seen in Fig. 2 that the computation time scales very well with the degree of parallelization given by the number of CPU cores, if a sufficiently high refinement is used. For high core numbers, a significant further reduction in computation time is only observed for systems with sufficiently high element numbers i.e. numbers of degrees of freedom. For less refinement, the scaling becomes less efficient and the reduction of computation time seems to converge to stagnation for increasing numbers of used cores (and can even increase again as observes for the

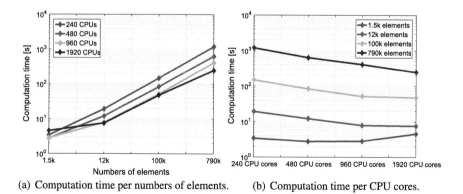

(a) Computation time per numbers of elements.   (b) Computation time per CPU cores.

Fig. 2 Comparison of computation times for different refinements of the system and different degrees of parallelization plotted over the number of elements in the system (**a**) and the number of CPU cores (**b**)

lowest refinement in Fig. 2b). In contrast, simulations with high numbers of elements need a distinctive higher computational time if used with very low CPU numbers.

Considering the setups introduced in Sect. 2.3, the following system data have been measured: The bi-crystal structure has been discretized by about 790 k quadratic tetragonal elements, which results in about 31 Mio. degrees of freedom in total. The simulations have been conducted on 260 cores for 48 h each, resulting in about 20 GB of data for each simulation. For the composite system, a discretization by about 100 k elements has been applied, which corresponds to about 3.8 Mio. degrees of freedom. Using 500 cores for 12 h each, simulation output data of about 20 GB for each simulation have been produced. Additionally, simulations on a higher refinement were carried out for convergence and accuracy studies, using 790 k elements on 1960 cores for 34 h, resulting in about 40 GB of simulation data.

Issues of mesh refinement in the context of numerical and physical accuracy have been studied as shown for two exemplary setups in Fig. 3. Here, the plastic slip for one representative slip system affecting the metallic matrix material is shown for two different refinements of the system. The results of the internal variables are achieved by averaging over all node values in the respective element. Therefore only the average values of the respective elements are shown. It can be seen in Fig. 3, that the dislocation free inclusions appear larger in the higher refinement, than in the lower. This is an artifact of setting arbitrary internal boundaries into a regular grid, which creates elements that are cut by the internal boundaries. Thus, the averaging procedure of the internal variables produces a non-zero average value, although at least one node value is zero. However, the evaluation of the internal variables is conducted on each element node and is therefore not affected by the averaging.

(a) Simulation mesh with 100k elements.   (b) Simulation mesh with 790k elements.

Fig. 3 Comparison of the plastic slip in the composite system on one slip system with a moderate resolution using 100 k elements (**a**) and a high resolution using 790 k elements (**b**). Only the lower half of the system is shown for visualization purposes

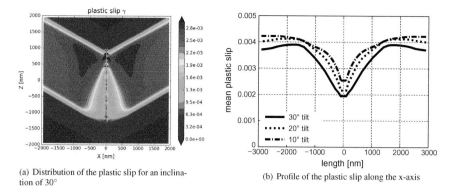

(a) Distribution of the plastic slip for an inclination of 30°

(b) Profile of the plastic slip along the x-axis

Fig. 4 Distribution of the plastic slip on a plane normal to the y-axis for 30° (**a**) and plastic slip profile in x-direction averaged over the y- and z-direction for three different inclinations of the slip planes (**b**)

3 Results

3.1 Bicrystal

As a first example, the bi-crystal structure presented in Sect. 2.3 is analyzed. Figure 4a shows a contourplot of the plastic slip on both sides of the grain boundary plotted on a plane normal to the y-axis for an inclination of 30° with respect to the grain boundary. It can be seen, that the slip profile forms a characteristic "V-shape" with a higher slip in the upper and a lower slip in the lower part of the bi-crystal. A strong gradient is formed near the grain boundary in the lower part of the bi-crystal. Figure 4b shows the plastic slip averaged over thin slices in the y-z-plane and plotted along the x-axis for different misorientations. It can be seen, that for a lower misorientation, the average plastic slip is higher in the vicinity of the grain boundary.

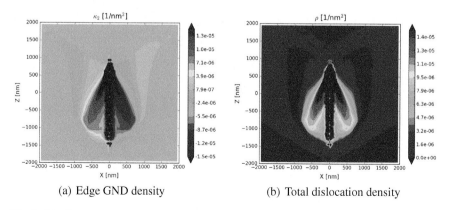

(a) Edge GND density

(b) Total dislocation density

Fig. 5 Distribution of the edge GND density (**a**) and the total dislocation density (**b**)

The distribution of the edge GND-density is shown in Fig. 5a. Comparing the GND-density distribution with the plastic slip in Fig. 4a, it can be clearly seen, that the characteristic "V-shape" is observed in both distributions, in which the shape of the GND-density correlates with the shape of the plastic slip. The distribution of the total dislocation density shown in Fig. 4b corresponds to the absolute value of the GND density. It therefore can be concluded, that the system at the converged state is only affected by GND density.

3.2 Composite

As a second example, we analyze the deformation behavior of a composite system as described in Sect. 2.3. Here, we increase the complexity by using a full fcc configuration of 12 slip systems. Figure 6 show the distribution of the plastic slip and GND density on one exemplary slip system visualized by 3D iso-surfaces. It can be seen, that the plastic slip concentrates on the areas between the fibers and shows a tendency towards a certain direction (Fig. 6a). The plastic slip is lower near the internal interfaces, which is due to the accumulation of GND density around the fibers (Fig. 6b). In both, the plastic slip and the GND density, no significant gradient is observed in the direction longitudinal to the fiber axis.

Therefore, we use contour plots taken from the x-y-plane in the middle of the system in the following to analyze the behavior of the internal variables. Figure 7 shows contourplots of the plastic slip for two different slip systems for a fiber volume fraction of 20%. Figure 7b, which is the same configuration as in Fig. 6a, clearly shows lower areas of plastic slip, where the distance between fibers is the smallest. However, the distribution of plastic slip can be significantly different depending on the slip system considered, as shown in Fig. 7a.

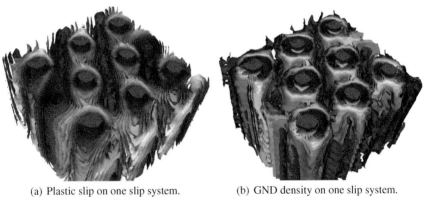

(a) Plastic slip on one slip system.   (b) GND density on one slip system.

Fig. 6 Plastic slip and GND density on one slip system in the composite material visualized by 3D iso-surfaces

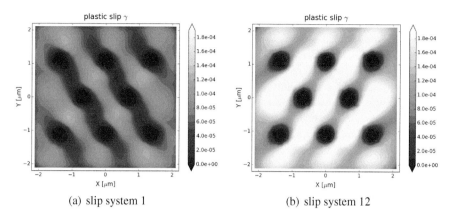

(a) slip system 1 (b) slip system 12

Fig. 7 Contourplot of the plastic slip on two different slip systems taken from a x-y-plane in the middle of the system for the configuration with a higher distance between fibers. The fiber volume fraction is 20%

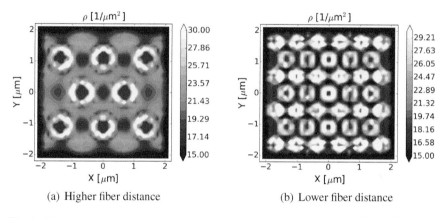

(a) Higher fiber distance (b) Lower fiber distance

Fig. 8 Total dislocation density summarized over all slip systems comparing higher (**a**) and an lower distance between fibers (**b**). In both systems, the fiber volume fraction is 20%, resulting in different sizes of the fibers

To further analyze the influence of the fibers, we now compare the total dislocation density and overall plastic slip in configurations with different fiber distances, but the same fiber volume fraction. This is achieved by an increase of the total number of fibers in the system with the same volume fraction of 20%, which leads to a decrease in the fiber cross section and distance. Figure 8 shows contourplots of the total density summarized over all slip systems for both fiber configurations. In both cases, the total density accumulates around the fibers similar to the GND density (cf. Fig. 6b). However, for the smaller fiber distance, the dislocation density between the fibers outside of the pile-ups is significantly lower compared to the wider fiber distance (cf. Fig. 8a, b). Thus, for the smaller fiber distance almost the complete dislocation density is accumulated around the fibers.

The consequence of this behavior can be seen in the contourplots of the overall plastic slip summarized over all slip systems (Fig. 9). Here, the plastic slip is largely heterogeneous for the wider fiber distance with higher magnitudes in certain areas between fibers. In contrast, reducing the distance between fibers yields a plastic slip with a significantly lower magnitude overall in the system, as shown in Fig. 9b.

Finally, Fig. 10a shows the stress-strain curves for different fiber volume fractions using transversal tensile loading compared to DDD-results [4]. In this loading orientation, the deformation behavior of the system is dominated by the plastic behavior of the matrix material, instead of the elastic properties of the fibers. In both, the continuum approach and DDD, a distinctive size effect with stronger hardening is observed for a higher volume fraction. However, the continuum formulation has difficulties to reproduce the same yield points as observed in DDD. This comes in due to different initial configurations with a given initial discrete structure that can

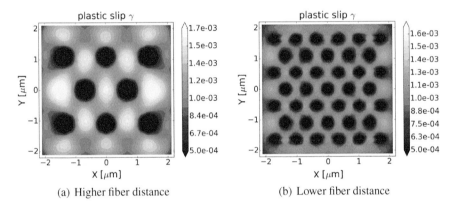

(a) Higher fiber distance   (b) Lower fiber distance

Fig. 9 Plastic slip summarized over all slip systems, corresponding to Fig. 8

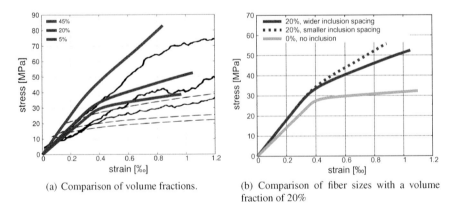

(a) Comparison of volume fractions.   (b) Comparison of fiber sizes with a volume fraction of 20%

Fig. 10 Stress-strain curve for tensile loading transversal to the fibers in y-direction for different fiber volume fractions compared to DDD-results [4] (**a**) and comparison of fiber sizes for a constant volume fraction of 20%

not be modeled by the continuum approach. The size effect observed in the continuum model as depicted in Fig. 10a can be explained by the fact that with a higher fiber volume fraction, less matrix material—and thus less line length overall—can produce plasticity. This ambiguous behavior can be overcome by varying the fiber distance with a constant volume fraction, as shown before in Figs. 8b and 9b. The corresponding stress-strain curves, shown in Fig. 10b, show a higher hardening for a smaller distance between fibers, which is in accordance with the distribution of the overall plastic slip, cf. Fig. 9.

4 Discussion

A dislocation density based continuum formulation has been applied to a bi-crystal and a composite structure in order to analyze the ability of the formulation for the representation of internal interfaces in a physically and numerically reasonable manner. The investigation of the bi-crystal showed that the stress fields originating from dislocation pile-ups close to the grain boundary can be well reproduced. As theoretically expected, the dislocation density topology in the surrounding of the boundary yields the characteristic shape of a small angle grain boundary due to the stress interaction inside the individual grains as well as the interaction between the two grains. Despite the fact that the grain boundary is impenetrable to dislocations, the grain boundary is transparent to dislocation stress fields. Therefore, the observation, that the extent of the pile-up—and thus the plastic slip at the grain boundary—is pronounced differently depending on the misorientation of the aligning grains (as shown in Fig. 4b) is physically reasonable and well resolved by the numerical formulation.

The considered composite structure has been studied in comparison with DDD simulations from the literature and shows qualitatively good accordance to the reference DDD solution given in [4]. It has been shown, that the internal boundaries lead to a very heterogeneous evolution of the plastic slip within as well as between the slip systems (Figs. 6 and 7).

Is has been found, that the cross section as well as the mutual distance of the fibers have a significant effect on the microstructural behavior (see Figs. 8 and 9). It is observed, that the accumulation of dislocation density around the fibers leads to a closing of the free channels between fibers due to the formation of a back-stress on the remaining dislocations in the channel when the fibers are thinner, but closer to each other (see Fig. 8). This imposes a geometrical constraint with ongoing dislocation accumulation around the fibers. Furthermore, it is shown in Fig. 10b, that this yields a significantly different hardening behavior.

Looking at the results of performance and scalability, it can be concluded that the numerical formulation scales well in the considered parameter space. However, an increase of the considered system sizes is a central objective in order to reach critical volumes to extract fundamental physical behavior from the simulations. Therefore, further analyses on performance acceleration as well as a smart algorithmic framework for the abstraction of physical mechanisms and structure formation has to be envisaged in further investigations.

5 Conclusion

A dislocation density based continuum formulation has been applied to a bi-crystal as well as a metal matrix composite structure. Both systems are highly dependent on internal interfaces. Although the grain boundary in the bi-crystal is impenetrable to dislocations, it is transparent to the dislocation stress fields. This yields differently pronounced dislocation pile-ups close to the boundary dependent on the misorientation of the neighboring grains. For the metal matrix composite structure, the dislocation pile-ups at the internal interfaces directly affects the dislocation flux through the structure, since the channel width is reduced with ongoing dislocation motion. This yields a pronounced size effect of the material which is observed in good accordance to discrete dislocation dynamics simulations.

Acknowledgements The Financial support for the research group FOR1650 *Dislocation based Plasticity* funded by the German Research Foundation (DFG) under the contract number GU367/36-2 as well as the support by the European Social Fund and the state of Baden-Württemberg is gratefully acknowledged. This work was performed on the computational resource ForHLR II funded by the Ministry of Science, Research and the Arts Baden-Württemberg and DFG ("Deutsche Forschungsgemeinschaft").

References

1. H. Cleveringa, E. Van der Giessen, A. Needleman, A discrete dislocation analysis of bending. Int. J. Plast. **15**, 837–868 (1999)
2. H.H.M. Cleveringa, E. VanderGiessen, A. Needleman, Comparison of discrete dislocation and continuum plasticity predictions for a composite material. Acta Mater. **45**(8), 3163–3179 (1997)
3. L. Friedman, D. Chrzan, Continuum analysis of dislocation pile-ups: influence of sources. Phil. Mag. A **77**(5), 1185–1204 (1998)
4. S. Groh, B. Devincre, L. Kubin, A. Roos, F. Feyel, J.L. Chaboche, Size effects in metal matrix composites. Mater. Sci. Eng. A **400**, 279–282 (2005)
5. I. Groma, F. Csikor, M. Zaiser, Spatial correlations and higher-order gradient terms in a continuum description of dislocation dynamics. Acta Mater. **51**, 1271–1281 (2003)
6. C. Hirschberger, R. Peerlings, W. Brekelmans, M. Geers, On the role of dislocation conservation in single-slip crystal plasticity. Model. Simul. Mater. Sci. Eng. **19**(085002) (2011)
7. J. Hirth, J. Lothe, *Theory of Dislocations* (Wiley, New York, 1982)
8. T. Hochrainer, Thermodynamically consistent continuum dislocation dynamics. J. Mech. Phys. Solids **88**, 12–22 (2016)
9. T. Hochrainer, S. Sandfeld, M. Zaiser, P. Gumbsch, Continuum dislocation dynamics: towards a physical theory of crystal plasticity. J. Mech. Phys. Solids **63**, 167–178 (2014)
10. E. Kröner, *Kontinuumstheorie der Versetzungen und Eigenspannungen* (Springer, 1958)
11. E. Kröner, Benefits and shortcomings of the continuous theory of dislocations. Int. J. Solids Struct. **38**, 1115–1134 (2001). https://doi.org/10.1016/S0020-7683(00)00077-9

12. D. Liu, Y. He, B. Zhang, Towards a further understanding of dislocation pileups in the presence of stress gradients. Phil. Mag. 1–23 (2013). https://doi.org/10.1080/14786435.2013.774096
13. J. Nye, Some geometrical relations in dislocated crystals. Acta Metall. **1**, 153–162 (1953)
14. T. Richeton, G. Wang, C. Fressengeas, Continuity constraints at interfaces and their consequences on the work hardening of metal-matrix composites. J. Mech. Phys. Solids **59**(10), 2023–2043 (2011)
15. Y. Saad, M.H. Schultz, Gmres: a generalized minimal residual algorithm for solving nonsymmetric linear systems. SIAM J. Sci. Stat. Comput. **7**(3), 856–869 (1986)
16. S. Sandfeld, E. Thawinan, C. Wieners, A link between microstructure evolution and macroscopic response in elasto-plasticity: formulation and numerical approximation of the higher-dimensional continuum dislocation dynamics theory. Int. J. Plast. **72**, 1–20 (2015)
17. S. Schmitt, P. Gumbsch, K. Schulz, Internal stresses in a homogenized representation of dislocation microstructures. J. Mech. Phys. Solids **84**, 528–544 (2015)
18. K. Schulz, D. Dickel, S. Schmitt, S. Sandfeld, D. Weygand, P. Gumbsch, Analysis of dislocation pile-ups using a dislocation-based continuum theory. Model. Simul. Mater. Sci. Eng. **22**(2), 025,008 (2014)
19. K. Schulz, S. Schmitt, Discrete-continuum transition: a discussion of the continuum limit. Tech. Mech. **38**(1), 126–134 (2018)
20. K. Schulz, M. Sudmanns, P. Gumbsch, Dislocation-density based description of the deformation of a composite material. Model. Simul. Mater. Sci. Eng. **25**(6), 064,003 (2017)
21. K. Schulz, L. Wagner, C. Wieners, A mesoscale approach for dislocation density motion using a runge-kutta discontinuous Galerkin method. PAMM **16**(1), 403–404 (2016)
22. C. Schwarz, R. Sedláček, E. Werner, Plastic deformation of a composite and the source-shortening effect simulated by the continuum dislocation-based model. Model. Simul. Mater. Sci. Eng. **15**, S37–S49 (2007)
23. M. Stricker, J. Gagel, S. Schmitt, K. Schulz, D. Weygand, P. Gumbsch, On slip transmission and grain boundary yielding. Meccanica **51**(2), 271–278 (2016)
24. G.I. Taylor, The mechanism of plastic deformation of crystals. Part I. Theoretical. Proc. R. Soc. Lond. Ser. A, Containing Papers of a Mathematical and Physical Character **145**(855), 362–387 (1934)
25. E. Van der Giessen, A. Needleman, Discrete dislocation plasticity: a simple planar model. Model. Simul. Mater. Sci. Eng. **3**, 689–735 (1995). https://doi.org/10.1088/0965-0393/3/5/008
26. C. Wieners, A geometric data structure for parallel finite elements and the application to multigrid methods with block smoothing. Comput. Vis. Sci. **13**(4), 161–175
27. C. Wieners, Distributed point objects. a new concept for parallel finite elements, in *Domain Decomposition Methods in Science and Engineering* (Springer, 2005), pp. 175–182
28. S. Yefimov, I. Groma, E. van der Giessen, A comparison of a statistical-mechanics based plasticity model with discrete dislocation plasticity calculations. J. Mech. and Phys. Solids **52**(2), 279–300 (2004). https://doi.org/10.1016/S0022-5096(03)00094-2, http://www.sciencedirect.com/science/article/B6TXB-49JPKTK-1/2/5df57c08baa877d5ebfd601f33e50933

Accurate and Efficient Spin-Spin Zero-Field Splitting Calculations for Extended Periodic Systems

T. Biktagirov, C. Braun, S. Neufeld, U. Gerstmann and W. G. Schmidt

Abstract Spin-spin zero-field splitting (ZFS) is a sensitive spectroscopic signature accessible by electron paramagnetic resonance. It is the key fingerprint used for the identification of high-spin defect centers in semiconducting host materials and for the characterization of their electronic structure. In recent years, much progress has been made in developing an efficient first-principles methodology for ZFS calculations that help in the interpretation of experimental data. Here we address the negatively charged nitrogen-vacancy center (NV^-) in diamond. It is used as a test system to explore the accuracy and efficiency of spin-spin zero-field splitting calculations on massively parallel computing systems.

1 Introduction

High-spin ($S \geq 1$) defect centers in semiconducting host materials are promising candidates for quantum information processing and sensing applications [1–3]. They rely on efficient detection and manipulation of their spin states. The first system for which such manipulations were realized is the negatively charged nitrogen-vacancy center (NV^-) in diamond. This defect exhibits a large ground-state transverse spin-relaxation time and optically induced spin polarization in a broad range of wavelengths at ambient conditions. Nowadays, there are other spin centers under investigation in diamond and silicon carbide (SiC) that possess similar spin properties, and the search for further structures continues.

The zero-field splitting (ZFS) accessible by electron-paramagnetic resonance (EPR) spectroscopy [4–6] is one of the key fingerprints used in the search for high-spin centers. It describes the interactions between the unpaired electrons in the absence of external magnetic fields and is extremely sensitive to the electronic structure and microscopic configuration of the defects. Within second order perturbation theory, the ZFS can be parametrised via the phenomenological spin-Hamiltonian

T. Biktagirov · C. Braun · S. Neufeld · U. Gerstmann · W. G. Schmidt (✉)
Lehrstuhl Für Theoretische Materialphysik Universität Paderborn,
33095 Paderborn, Germany
e-mail: W.G.Schmidt@upb.de

© Springer Nature Switzerland AG 2019
W. E. Nagel et al. (eds.), *High Performance Computing in Science and Engineering '18*, https://doi.org/10.1007/978-3-030-13325-2_8

and split into spin-spin and spin-orbit contributions [7]. For a number of important systems, such as spin centers in diamond and SiC, the spin-spin part is believed to be dominant [8, 9].

There exists a well-established framework for spin-spin ZFS calculation within all-electron density functional theory (DFT) for finite size systems [10–12]. It is frequently used for molecular systems. This methodology is, however, not directly applicable to extended periodic systems, due to unavoidable finite-size effects. This is in particular the case for defects with strongly delocalized spin density, typical for spin centers in semicoductors. While it is in principle possible to approximate the semiconductor host material by finite clusters [13, 14], the numerical convergence with respect to the cluster size is typically slow and requires extensive computational resources.

Therefore, the more suitable way is to adopt the supercell approach with explicitly imposed periodic boundary conditions. It relies on Bloch's theorem and involves Fourier transformation of the basic quantities into the reciprocal space. This strategy is routinely employed in computational materials science for the calculations of various properties, often in combination with pseudopotentials. The supercell approach can be effectively complemented with Blöchl's projector augmented wave (PAW) formalism [16] for EPR parameter calculations. In this method, all-electron wave functions are substituted by smooth pseudo-wave functions expanded in plane waves, while the true shape of the wave functions within the atomic core region is preserved. This allows for addressing all-electron properties with the efficiency of pseudopotential calculations.

The gauge-including extension of the projector augmented wave formalism (GIPAW) [17], which ensures the translational invariance of the wave-functions in external magnetic fields, has already proven highly successful for the ab initio calculation of EPR parameters. Recently, we presented an implementation of the spin-spin contribution to ZFS within the PAW formalism [18] that allows for their efficient calculation using periodic boundary conditions. The comparison with all-electron calculations, shown in Fig. 1 for a series of high-spin defects in hydrogen-terminated

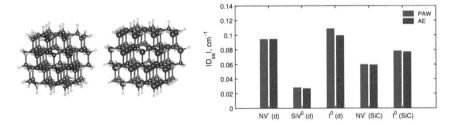

Fig. 1 Comparison of the spin-spin ZFS calculated with the present GIPAW implementation using the Quantum ESPRESSO code (*blue*) with results obtained using the all-electron (AE) approach implemented in the ORCA package [15] (*red*) for the same hydrogen-terminated clusters cut from optimized supercells (*left*). Shown are data for the nitrogen-vacancy center (NV$^-$) in diamond and SiC, the neutral silicon-vacancy (SiV^0) center in diamond, and the neutral carbon self-interstitial (I^0) in diamond and SiC

clusters of diamond and cubic SiC, demonstrates that this implementation reaches chemical accuracy of the calculated spin-spin ZFS.

The successful application of the implemented algorithms requires the control over all technical factors that affect the precision of the calculated ZFS. The spurious dipole-dipole interaction between the periodic replicas and the strain-induced redistribution of the spin density within the supercell are among the main sources of possible errors in this context. In principle, both factors are related to the size of the supercell used for calculations. This relation is obvious for dipole-dipole interaction, since it explicitly depends on spatial separation between the periodic copies of the defect. Furthermore, the calculated ZFS of the defect can sense local strain fields produced by the replicas, which should vanish upon increasing the supercell size. In either case, the results for larger supercells are expected to be more reliable. The computational costs, however, scale unfavorably with the size of the supercell.

Therefore, in this report we investigate how the accuracy of spin-spin ZFS calculations depends on the supercell size. The negatively charged nitrogen-vacancy center (NV$^-$) in diamond is chosen as an ideal candidate for the numerical tests. Particular emphasis is placed (i) on the possibility to obtain sufficiently accurate results with smaller cells upon assuring the convergence with respect to the Brillouin zone sampling in the reciprocal space (i.e. the number of used k-points) and (ii) on the performance of our implementation on the HLRS computing resources.

We mention that a similar study for spin centers in SiC was recently reported in Ref. [19]. There other essential spectroscopic fingerprints, i.e., the zero-phonon line and the hyperfine coupling were explored. We believe that the present results are a valuable contribution to the understanding and optimization of computational schemes for the comprehensive ab initio characterization of spin centers.

2 Methodology

The algorithms implemented in the GIPAW module of the Quantum ESPRESSO software [20, 21] are based on the the theoretical framework thoroughly described in Ref. [18]. Here, we only provide a brief outline of the methods with an extra focus put on technical details of their implementation.

Conventionally, the spin-spin ZFS is defined in terms of a phenomenological spin Hamiltonian [22] as a symmetric 3×3 tensor $\mathbf{D}$:

$$\hat{H}_{SS} = \frac{\alpha^2}{2} \sum_{m,n} \left[\frac{\hat{\mathbf{s}}_m \cdot \hat{\mathbf{s}}_n}{r_{mn}^3} - 3 \frac{(\hat{\mathbf{s}}_m \cdot \mathbf{r}_{mn})(\hat{\mathbf{s}}_n \cdot \mathbf{r}_{mn})}{r_{mn}^5} \right] = \hat{\mathbf{S}} \cdot \mathbf{D} \cdot \hat{\mathbf{S}} . \tag{1}$$

In Eq. (1), α is the fine-structure constant, $\mathbf{r}_{mn}$ is the vector between electrons m and n, and $\hat{\mathbf{S}}$ is the total effective spin of the system. To further address the $\mathbf{D}$ tensor from first-principles calculations, it is convenient to reformulate it by separating the spin and spatial degrees of freedom ($a, b = x, y, z$) [23]:

$$D_{ab} = \sum_{m,n} f_m f_n \sigma_{mn} D_{ab,mn} \, , \qquad (2)$$

where f_m, f_n are the occupations of the orbitals m and n, and σ_{mn} originates from the matrix elements of spin-operators: $\sigma_{mn} = 1$ for parallel spins and $\sigma_{mn} = -1$ in case of antiparallel spins. In Eq. (2) the space-dependent part of the spin-spin ZFS contribution from two orbitals, m and n, is given by McWeeny and Mizuno [23]

$$D_{ab,mn} = \frac{\alpha^2}{4S(2S-1)} \int d^3\mathbf{r} d^3\mathbf{r}' \frac{|\mathbf{r} - \mathbf{r}'|^2 \delta_{ab} - 3(\mathbf{r} - \mathbf{r}')_a (\mathbf{r} - \mathbf{r}')_b}{|\mathbf{r} - \mathbf{r}'|^5}$$
$$\times \left[n_{mm}(\mathbf{r}) n_{nn}^*(\mathbf{r}') - n_{mn}(\mathbf{r}) n_{mn}^*(\mathbf{r}') \right] , \qquad (3)$$

which consists of Coulomb-like and exchange-like contributions. In the above expression we define the charge density, $n_{mn}(\mathbf{r}) = \psi_m^*(\mathbf{r}) \psi_n(\mathbf{r})$, from the space distribution of the spin orbitals.

In case of all-electron calculations, the densities $n_{mn}(\mathbf{r})$ are readily available from the self-consistent field (SCF) calculations. However, when the pseudopotential approach is utilized, the resulting charge densities are devoid of important details about the true shape of the spin orbitals in the core region. As already mentioned in the introduction, this problem can be circumvented within the PAW framework. Here, the two-electron density is presented by a superposition

$$n_{mn}(\mathbf{r}) = \tilde{n}_{mn}(\mathbf{r}) + \sum_R (n_{mn}^R(\mathbf{r}) - \tilde{n}_{mn}^R(\mathbf{r})) \qquad (4)$$

of the smooth pseudo-density $\tilde{n}_{mn}(\mathbf{r}) = \langle \tilde{\Psi}_m | \mathbf{r} \rangle \langle \mathbf{r} | \tilde{\Psi}_n \rangle$ treated on a plane-wave grid, and the difference between on-site expansions of the all-electron density n^R and the pseudo-density $\tilde{n}^R$ defined on a radial grid within the augmentation sphere around an atomic site R.

As justified in Ref. [18] for norm-conserving pseudopotentials, the spin-spin ZFS also can be constructed in a form of three non-interacting contributions,

$$D_{ab} = \tilde{D}_{ab} + \sum_R \left[D_{ab}^R - \tilde{D}_{ab}^R \right] = \tilde{D}_{ab} + \Delta D_{ab} \, , \qquad (5)$$

where the sum ΔD_{ab} of the on-site terms constitutes the essence of the PAW reconstruction.

Thereby the treatment of ΔD_{ab} is based on evaluation of the following four-partial-wave two-electron integral [18],

$$W_{ab,ijkl}^R = \int_{\Omega_R} d^3\mathbf{r} d^3\mathbf{r}' \langle \phi_i | \mathbf{r} \rangle \langle \mathbf{r} | \phi_j \rangle \langle \phi_l | \mathbf{r}' \rangle \langle \mathbf{r}' | \phi_k \rangle$$
$$\times \left\{ (\frac{1}{3} \delta_{ab} \nabla^2 - \partial_a \partial_b)_{r'} \frac{1}{|\mathbf{r} - \mathbf{r}'|} \right\} , \qquad (6)$$

where the index $i \in R$ refers to angular momentum quantum numbers l_i and m_i for the atom R. The all-electron partial wave ϕ_i is a solution of the radial Schrödinger (or Dirac) equation for the reference atom. The pseudo partial wave $\tilde{\phi}_i$ is equivalent to ϕ_i outside the augmentation sphere. In $W^R_{ab,ijkl}$, angular integration can be further performed analytically making the evaluation of ΔD_{ab} computationally very efficient.

The implementation of the plane-wave pseudo-density contribution, $\tilde{D}_{ab}$, is based on the work by Rayson and Briddon [24, 25], who formulate it in reciprocal space as

$$\tilde{D}_{ab,mn} = \frac{\alpha^2 \Omega \pi}{S(2S-1)} \sum_{\mathbf{G}} \left(\frac{G_a G_b}{G^2} - \frac{\delta_{ab}}{3} \right) \left[\tilde{n}_{mm}(\mathbf{G}) \tilde{n}_{nn}(-\mathbf{G}) - |\tilde{n}_{mn}(\mathbf{G})|^2 \right]. \quad (7)$$

In the above expression, $\tilde{n}_{mm}(\mathbf{G})$ and $\tilde{n}_{nn}(-\mathbf{G})$ are the Fourier coefficients of the two-electron densities at the points $\mathbf{G}$ and $-\mathbf{G}$, and Ω is the volume of the unit cell. Note that Eq. (7) corresponds to the situation where the Brillouin zone is sampled with only one single k point. It can be generalized to a denser sampling by substituting $\mathbf{G}$ by $\mathbf{G} + \mathbf{k}$, where the summation is to be made over $\mathbf{G}$ as well as $\mathbf{k}$.

In practice, numerical evaluation of $\tilde{D}_{ab,mn}$ involves a two-step procedure. First, the pseudo-wave functions are Fourier transformed to the plane-wave grid in real space, where the two-electron densities are subsequently constructed. Then the densities are transferred again to the reciprocal space, and the summation in Eq. (7) is computed. The procedure has to be carried out for each orbital pair m and n. This makes $\tilde{D}_{ab,mn}$ the computationally most demanding term in Eq. (5). To assure reasonable efficiency of the implementation, our code was thus designed to support parallelization with respect to the plane-wave coefficients. This is achieved by redistributing the coefficients across the processors in such a way that for each processor both $\tilde{n}_{mm}(\mathbf{G})$ and the corresponding $\tilde{n}_{nn}(-\mathbf{G})$ are addressable [(cf. Eq. (7)].

The overall procedure of computing D_{ab} is thus as follows. At the first step, we determine the electronic ground state in the PWSCF module of the Quantum ESPRESSO software. We restrict ourselves to norm-conserving pseudopotentials (generated with the Troullier-Martins approach [26]) and the PBE exchange-correlation functional [27] in the numerical tests presented below. The convergence criterion for the geometry optimization was set to 10^{-4} Ry/Bohr. At the next step, the numerically converged ground-state orbitals are taken over by the GIPAW module, where the $\tilde{D}_{ab}$ and ΔD_{ab} contributions are evaluated separately.

3 Results

For the numerical tests, we have chosen the NV$^-$ defect in diamond (see Fig. 2). In this so-called color center, the nitrogen impurity occupies a substitutional site neighboring the carbon vacancy. The nitrogen-vacancy pair is formed in such a way

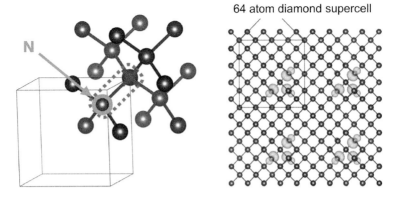

Fig. 2 Schematics of the microscopic model of the NV$^-$ center in diamond (*left*) and an illustration of the 64-atom diamond supercell containing the center (*right*). Yellow isosurfaces represent the spin-density distribution of the defect over the three carbon dangling bonds surrounding the vacancy

that the symmetry axis of the complex coincides with the [111] axis of the crystal. At the ground state ($^3A_2$, $S = 1$) the center manifests C_{3v} point group symmetry with the spin density mostly localized on the three carbon dangling bonds surrounding the vacancy. Its ZFS (of about 2.88 GHz [28]) is supposed to be almost entirely caused by the spin-spin contribution [8], which makes it a perfect choice for the current study.

We begin by incorporating NV$^-$ into a series of cubic diamond supercells with increasing size. It is expected that the small supercells are most affected by the spurious influence from the periodic replicas of the defect. The ideal case scenario would imply isolating a defect center by using as large supercells as possible within the available computational resources. However, also in the experiment the color center is never entirely isolated, because of unavoidable presence of different kinds of other defects and impurities in the host material [29]. Therefore, the essential question is, whether it is possible to achieve *sufficient* accuracy of the calculated spin-spin ZFS without pushing the size of supercell beyond the limits of a reasonable computational cost.

Thus, for each of the considered supercells we had to assure tight convergence of the results. In particular, we study the convergence behaviour of the ZFS with respect to the level of Brillouin zone sampling (i.e. the number of k points) used in the SCF cycle. For another convergence parameter, the plane-wave cutoff, the same (and sufficiently large [18]) value of 50 Ry was used for all of the supercell sizes. Therefore, its influence is not further discussed.

The convergence of D_{SS} with respect to the dimensions of a Monkhorst-Pack [30] k-point grid is illustrated in Fig. 3 for different supercell sizes. The first conclusion is that a single k point for Brillouin zone sampling is in general not sufficient. Aside from that, the smaller supercells demonstrate slower convergence, and require denser k-point grids. This becomes particularly obvious in the extreme case of a 8-atom supercell. Here, the calculated D_{SS} value is converged within a 10×10^{-4} cm^{-1}

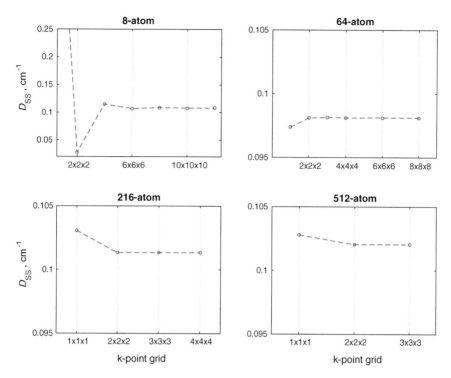

Fig. 3 Spin-spin ZFS (D_{SS}) calculated for the NV$^-$ center in diamond supercells of varying size: Convergence behavior with respect to the Brillouin zone sampling, i.e. the number of k-points generated according to the Monkhorst-Pack scheme [30]

window only for k-point grids as large as at least $6 \times 6 \times 6$. Of course, such a small supercell is considered here only for the purpose of numerical exercise and should not be used in practice, when agreement with experimental data is desired.

With this being said, we can then carry out a comparative analysis of the converged results obtained with different supercell sizes. The major conclusion which can be derived from the data presented in Fig. 4 is that even comparably small supercells can provide results within a reasonable level of error, provided k-point convergence is ensured. Surprisingly, even the D_{SS} value calculated for the 8-atom supercell is not too far off. However, we cannot exclude that this is due to fortuitous error cancellation. The 216-atom supercell provides a much more reasonable choice, and the corresponding D_{SS} deviates by only 2% from the value obtained for the largest system considered here.

These results provide an obvious guideline for practical calculation of spin-spin ZFS: One strategy is to use a reasonably small supercell with sufficiently dense k-point mesh. Alternatively, the supercell size might be increased and the Brillouin zone sampling restricted to a single k-point. The latter strategy provides more accurate D values and is mandatory for spin centers that show very small ZFS, which is easily

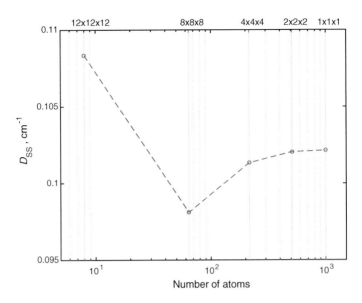

Fig. 4 Dependence of the spin-spin ZFS (D_{SS}) calculated for the NV$^-$ center in diamond on the supercell size. For each supercell, the presented value was previously converged with respect to the number of k-points (the required dimensions of the k-point grid are shown on the topmost horizontal axis.)

masked by numerical noise. However, the usage of large unit cells is computationally expensive.

Therefore, an important question is how well do the underlying algorithms perform on massively parallel systems such as the Cray XC40 (Hazel Hen) for a relatively large supercells. As mentioned above in the Methodology section, our implementation supports parallelization over plane-wave expansion coefficients (whose number increases upon increasing of the supercell size). This leads to efficient speed up of the calculations and is expected to provide efficient scaling. Figure 5 illustrates the performance of the Cray XC40 for spin-spin ZFS calculation for NV$^-$ in a 216-atom supercell (with the plane-wave cutoff increased up to 100 Ry). It can be seen that the parallelization which distributes the plane-wave expansion coefficients indeed leads to an appreciable saving of computation time with increasing number of cores. However, the speed up saturates already for a few hundred cores. Therefore it is essential to improve the scalability further. This is presently done by exploring the parallelization over **k** points as well as over electronic states. However, as in case of exact exchange functionals [31], the "sum-over-pairs of states" nature of the ZFS hinders their straightforward implementation.

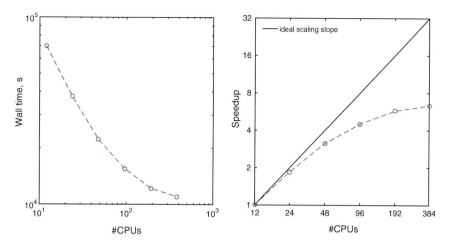

Fig. 5 Scaling of spin-spin ZFS calculation for the NV$^-$ center in 216-atom diamond supercell with high plane-wave cutoff of 100 Ry on the HLRS Cray XC40. MPI parallelization is restricted to the plane waves

4 Conclusions

To summarize, in the present report we addressed the accuracy and performance of our recently developed MPI-based implementation for DFT based calculations of spin-spin ZFS with periodic boundary condition. The zero-field splitting is a sensitive spectroscopic signature, and its ab initio evaluation requires a high level of precision. The number of k points used for the Brillouin zone sampling is found to have a crucial influence on the calculated D values. Provided the k-point grid is chosen sufficiently dense, the ZFS calculations may be accurate even for relatively small supercells. In case of the NV$^-$ center in diamond considered here, cubic 216 and 512 atom supercells appear to be appropriate for routine calculations. On the other hand, when very high precision is required (e.g. to identify defects with extremely small D values) larger unit cells may be necessary. This requires the usage of massively-parallel computing systems, such as the Cray XC40. However, the scalability of our code still needs to be improved.

Acknowledgements The Deutsche Forschungsgemeinschaft is acknowledged for financial support (SPP1601, FOR1700 and SFB TRR 142). We thank the Höchstleistungs-Rechenzentrum Stuttgart (HLRS) and the Paderborn Center for Parallel Computing (PC^2) for grants of high-performance computer time. Images of molecular structures were prepared using VESTA [32].

References

1. D. DiVincenzo, Nat. Mater. **9**, 468–469 (2010)
2. D.D. Awschalom, L.C. Bassett, A.S. Dzurak, E.L. Hu, J.R. Petta, Science **339**, 1174–1179 (2013)
3. H.J. Mamin, M. Kim, M.H. Sherwood, C.T. Rettner, K. Ohno, D.D. Awschalom, D. Rugar, Science **339**, 557–560 (2013)
4. H.J. Von Bardeleben, J.L. Cantin, E. Rauls, U. Gerstmann, Phys. Rev. B **92**, 064104 (2015)
5. V.A. Soltamov, B.V. Yavkin, D.O. Tolmachev, R.A. Babunts, A.G. Badalyan, V.Yu. Davydov, E.N. Mokhov, I.I. Proskuryakov, S.B. Orlinskii, P.G. Baranov, Phys. Rev. Lett. **115**, 247602 (2015)
6. A. Csóré, H.J. von Bardeleben, J.L. Cantin, A. Gali, Phys. Rev. B **96**, 085204 (2017)
7. J.E. Harriman, *Theoretical Foundations of Electron Spin Resonance* (Academic press, New York, 1978)
8. V. Ivády, T. Simon, J.R. Maze, I.A. Abrikosov, A. Gali, Phys. Rev. B **90**, 235205 (2014)
9. H.J. Von Bardeleben, J.L. Cantin, A. Csóré, A. Gali, E. Rauls, U. Gerstmann, Phys. Rev. B **94**, 121202 (2016)
10. F. Neese, J. Am. Chem. Soc. **128**, 10213–10222 (2006)
11. S. Sinnecker, F. Neese, J. Phys. Chem. A **110**, 12267–12275 (2006)
12. F. Neese, J. Chem. Phys. **127**, 164112 (2007)
13. T.T. Petrenko, T.L. Petrenko, V. Ya Bratus, J. Phys. Condens. Matter **14**, 12433 (2002)
14. A. Komarovskikh, A. Dmitriev, V. Nadolinny, Y. Palyanov, Diam. Relat. Mater. **76**, 86–89 (2017)
15. F. Neese, Wiley Interdiscip. Rev. Comput. Mol. Sci. **2**, 73–78 (2012)
16. P.E. Blöchl, Phys. Rev. B **50**, 17953 (1994)
17. C.J. Pickard, F. Mauri, Phys. Rev. B **63**, 245101 (2001)
18. T. Biktagirov, W.G. Schmidt, U. Gerstmann Phys. Rev. B **97**, 115135 (2018)
19. J. Davidsson, V. Ivdy, R. Armiento, N.T. Son, A. Gali, I.A. Abrikosov, New J. Phys. **20**, 023035 (2018)
20. P. Giannozzi, O. Andreussi, T. Brumme et al., J. Phys. Condens. Matter **29**, 465901 (2017)
21. P. Giannozzi, S. Baroni, N. Bonini, M. Calandra et al., J. Phys. Condens. Matter **21**, 395502 (2009)
22. R. McWeeny, *Methods of Molecular Quantum Mechanics* (Academic press, 1992)
23. R. McWeeny, Y. Mizuno, in *Proceedings of the Royal Society of London A: Mathematical, Physical and Engineering Sciences*, vol. 259 (The Royal Society, 1961), pp. 554–577
24. M.J. Rayson, P.R. Briddon, Phys. Rev. B **77**, 035119 (2008)
25. Z. Bodrog, A. Gali, J. Phys. Condens. Matter **26**, 015305 (2013)
26. N. Troullier, J.L. Martins, Phys. Rev. B **43**, 1993 (1991)
27. J.P. Perdew, K. Burke, M. Ernzerhof, Phys. Rev. Lett. **77**, 3865 (1996)
28. J. Loubser, J. van Wyk, Rep. Prog. Phys. **41**, 1201 (1978)
29. T. Biktagirov, A.N. Smirnov, V.Yu. Davydov, M.W. Doherty, A. Alkauskas, B.C. Gibson, V.A. Soltamov, Phys. Rev. B **96**, 075205 (2017)
30. H.J. Monkhorst, J.D. Pack, Phys. Rev. B **13**, 5188 (1976)
31. T.A. Barnes, T. Kurth, P. Carrier, N. Wichmann, D. Prendergast, P.R. Kent, J. Deslippe, Comput. Phys. Commun. **214**, 52 (2017)
32. K. Momma, F. Izumi, J. Appl. Crystallogr. **44**, 1272 (2011)

Enhanced Acid Dissociation at the Solid/Liquid Interface

Dominika Lesnicki and Marialore Sulpizi

Abstract In this chapter we review some recent results from *first principles* molecular dynamics simulations which show how molecular properties, such as proton dissociation, can be influenced upon adsorption at a solid/liquid interface. In particular, we discuss in details the increased acidity of pyruvic acid at the quartz /water interface, which is of relevance for the chemistry of the atmosphere. Our simulations unveil the special role of the microsolvation at interface, as well as the role of the silanols in stabliziing the deprotonated form of the acid. The enhanced acidity at the hydrophilic quartz/water interface is at odd with what typically found at the water/air interface where acidity is normally reduced, and the associated form of the acidity stabilized.

1 Introduction

In this contribution we review some recent work from our group on the calculation of acidity constants for adsorbed molecules at interfaces [1]. In particular we will discuss in details the case of dissociation of pyruvic acid, the simplest α-keto acid, at silica/water interface. This choice is motivated by the relevance of such system in the chemistry of the atmosphere. As recently shown in the work of Vaida and coworkers [2], pyruvic acid plays a key role in the intricate chemistry of the atmosphere as intermediate in the oxidative channels of isopropene and of secondary organic aerosols. On the other hand, silica is the main component of mineral dust particles in the atmosphere, and reactions on its surface, in presence of variable levels of humidity, have a direct impact, e.g. on the radical concentration and the ability of dust particles to serve as cloud condensation nuclei [3]. An up to date review can be found in the recent review from the group of Grassian [3].

D. Lesnicki · M. Sulpizi (✉)

Johannes Gutenberg University Mainz, Staudinger Weg 7, 55099 Mainz, Germany
e-mail: sulpizi@uni-mainz.de

© Springer Nature Switzerland AG 2019

W. E. Nagel et al. (eds.), *High Performance Computing in Science and Engineering '18*, https://doi.org/10.1007/978-3-030-13325-2_9

Acid dissociation constants, namely pK_a, for molecules in bulk solution can be routinely determined using standard techniques like potentiometric titration, voltammetry, and electrophoresis [4]. However, in the case of molecules localized at an interfaces, as e.g. a solid/liquid interface, the pK_a measurements become more complicate, since the signal originating from the molecules in the bulk typically overwhelms the signal coming from the interfacial molecules. Progresses in selectively probing interfacial properties have been recently made thanks to the advent of interface selective vibrational spectroscopy techniques, capable to probe the molecular structure of the nanometric layer of a liquid interface. This has recently permitted also to selectively probe the degree of proton dissociation at an interface. In the case of the water/air interface spectroscopic studies of acid/base pairs have shown that the surface favors the neutral form of the acid/base pair in comparison to the bulk [5]. For instance, using second harmonic generation (SHG) and vibrational Sum Frequency Generation (VSFG), it has been shown for phenol and carboxylic acids that the neutral acid species is favored over the anionic conjugate base [6–8]. A similar result was also found for molecules containing an acid ammonium group: also in this case the neutral amine base species is favored with respect to the cationic quaternary ammonium acid [9, 10]. In the case of a strong acid, such as nitric acid, the acid dissociation does not occurs at the water surface, but it is more likely to happens in the first and second surface layers below the surface [11–14]. Recently, a VSFG study has shown that some amino acids, such as L-alanine and L-proline present a reduced degree of acid dissociation at the water surface. Namely, higher pH values are required to induce the dissociation of carboxylic acid groups of molecule at surface [15]. Information on molecules adsorbed at buried interfaces, such as the silica/water interface are more difficult to obtain, thefore also information on how acidity constants are affected at an interface are not easily available. The group, e.g. of Grassian has been extensively studying adsorption and dissociation of strong and weak acids on silica particle [16, 17], however it is challenging for such experiments to quantify a change in the acidity upon adsorption at the solid/liquid interface.

At the solid/liquid interface the situation may be different which respect to that at the water/vapor interface, since the solid surface may contain hydrophilic groups (silanols in the case of quartz and silica), as well as ions, which can influence the dissociation constant. Some recent experimental and computational work has shown that ions at interfaces can shift the acidity constants of a sizeable amount [18].

In this report we will discuss in details the case of pyruvic acid at the quartz/water interface. The dissociation constant of pyruvic acid in bulk solution is 2.5. Such value is close to the point of zero charge of α-quartz surface [19, 20], therefore pyruvic acid and silanols may compete for protons at the quartz surface. In some works from few years ago, we have shown that on the quartz (0001) surface two species of silanols exist, namely the out-of-plane silanols, presenting a pK_a of 5.6 [20], and the in-plane silanols with an higher pK_a. Similarly, at the amorphous silica/water interface, we also found that some convex geminals groups, as well as some type of vicinals can be very acidic with a pK_a of 2.9 and 2.1, respectively [21]. These pK_a values are very close to the pK_a of bulk pyruvic acid of 2.5.

In our recent work [1] we have shown that, in the case of pyruvic acid, acidity is enhanced at the quartz/water interface, of almost two units. This result goes in the opposite direction of what normally found for acids at the water/vapor interface, since in this case the interface favors the deprotonation. We will review here, in this report, how such increased acidity is the results of the specific micro-solvation around the acid/base groups and, in particular, of the stabilization of the conjugated base by the silanols groups on the quartz surface. The report is structured as follows: in the methods section we present the proton insertion/deletion method to calculate acidity constants and the details on the computational setup. In the results we discuss how the method of insertion and removal of proton can give an estimate of relative pK_a values from the vertical energy gaps. A detail analysis of the structure, through radial distribution functions around the protonated and deprotonated pyruvic acid permits to understand the enhanced acidity at the surface.

2 Methods

In order to obtain a detailed microscopic understanding of the interface pK_a variations, molecular modeling at the atomistic level is required. Using *ab initio* molecular dynamics simulations, the acidity constants can be computed using the reversible proton insertion/deletion method [22, 23] that we already applied to a wide range of chemical groups, also at solid/liquid interface [18, 20, 21, 24]. The advantage of our approach is that it permits to explicitly take into account the solvent and its specific interaction with the investigated molecule as well as with the surface. This permits us to pin down the origin of the pK_a changes in terms of molecular orientation and local hydrogen bond strength. Moreover, the inclusion of the full electronic structure permits us to include anharmonicity and polarization effects, which cannot be easily incorporated in a simpler, empirical force field approach.

2.1 The Proton Insertion/Deletion Method

The scheme to compute the difference in pK_a's between the aqueous value for the pyruvic acid and that at quartz/water interface is presented in Fig. 1. Our approach is based on the vertical energy gap method [22, 23]: the proton of the acid at interface marked $(AH)_I$ is gradually transformed into a dummy $(Ad^-)_I$ (d is a particle with no charge) while the dummy atom in bulk water marked $(Ad^-)_B$ is gradually transformed into a proton $(AH)_B$. The free energy change of this process is calculated using the thermodynamic integration technique:

$$\Delta A = \int_0^1 d\eta \langle \Delta E \rangle_\eta \qquad (1)$$

where $\Delta E = E_1 - E_0$ is the vertical energy gap, defined as the potential energy difference between product P: $\left((\text{Ad}^-)_\text{I}, (\text{AH})_\text{B}\right)$ and reactant R: $\left((\text{AH})_\text{I}, (\text{Ad}^-)_\text{B}\right)$ configurations extracted from the molecular dynamics trajectory. The subscript η indicates that the averages are evaluated over the restrained mapping Hamiltonian

$$\mathcal{H}_\eta = (1 - \eta)\mathcal{H}_R + \eta\mathcal{H}_P + V_r \tag{2}$$

where η is a coupling parameter which gradually increased from 0 $\left((\text{AH})_\text{I}, (\text{Ad}^-)_\text{B}\right)$ to 1 $\left((\text{Ad}^-)_\text{I}, (\text{AH})_\text{B}\right)$ and $\mathcal{H}_R$ and $\mathcal{H}_P$ stand for the reactant and product states, respectively. The restrained harmonic potential V_r, which is used to keep the dummy position close to where it was in the protonated state,

$$V_r = \sum_{bonds} \frac{k_r}{2}(r - r_{eq})^2 + \sum_{angles} \frac{k_\theta}{2}(\theta - \theta_{eq})^2$$
$$+ \sum_{dihedrals} \frac{k_\phi}{2}(\phi - \phi_{eq})^2 \tag{3}$$

consists of the bonding, angle bending and dihedral torsion terms whose equilibrium values are r_{eq}, θ_{eq}, ϕ_{eq} respectively. We assume that we can treat separately the deprotonation of the acid in the bulk (labeled B) and at the interface (labeled I), respectively, considering along the trajectory that the two acids do not interact with each other. Thus the model presented on the left side of Fig. 1 can then be transformed into the one shown on the right side Fig. 1. One can then write the following equations for the free energy deprotonation of the pyruvic acid at the interface and in the bulk, respectively,

$$(\Delta_{dp} A_{AH})_I = \int_0^1 d\eta \left\langle (\Delta_{dp} E_{AH})_I \right\rangle_\eta \tag{4}$$

and

$$(\Delta_{dp} A_{AH})_B = \int_0^1 d\eta \left\langle (\Delta_{dp} E_{AH})_B \right\rangle_\eta \tag{5}$$

with $(\Delta_{dp} E_{AH})_I = E_{1I} - E_{0I}$ and $(\Delta_{dp} E_{AH})_B = E_{1B} - E_{0B}$.

The difference in pK$_a$'s is then given by the following equation:

$$\Delta p K_a = p K_{a_I} - p K_{a_B} = \frac{(\Delta_{dp} A_{AH})_I - (\Delta_{dp} A_{AH})_B}{\ln 10 k_B T} \tag{6}$$

where we have assumed that the de-solvation free energy of the aqueous proton, the translational entropy generated from the acid dissociation and the free energy for the quantum correction for the nuclear motion are similar in the bulk and at the interface. It can be written, using the thermodynamic integral Eq. 1 and the assumption $E_1 - E_0 = (E_{1I} + E_{0B}) - (E_{0I} + E_{1B})$, as

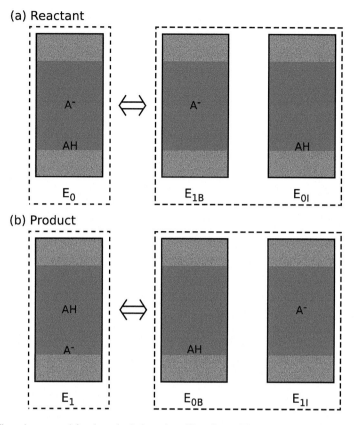

(a) Reactant

(b) Product

Fig. 1 The scheme used for the calculation of ΔpK_a value. **a** The reactants, or initial state. The pyruvic acid at the interface is in the protonated form, while that in the bulk is in the deprotonated one. **b** The products, or final state: pyruvic acid at the in the interface is deprotonated, while pyruvic acid in the bulk is protonated

$$\Delta pK_a = \frac{1}{ln\,10 k_B T} \int_0^1 d\eta \langle \Delta E \rangle_\eta \qquad (7)$$

The final ΔpK_a reads as

$$\Delta pK_a = \frac{1}{6 ln\,10 k_B T} (\langle \Delta E \rangle_0 + \langle \Delta E \rangle_1)$$
$$+ \frac{2}{3 ln\,10 k_B T} \langle \Delta E \rangle_{0.5} \qquad (8)$$

using the 3-point Simpson rule for the integral in Eq. 7.

2.2   Simulation Details

In the following we present the details on the system setup. A water box of 19.64 × 17.008 × 30 Å was created with 348 molecules using packmol. The box was equilibrated for 3.87 ps in a canonical NVT ensemble with a timestep of 0.5 fs using the CSVR thermostat and a target temperature of 298 K. This equilibrated water box was placed on the fully hydroxylated (0001) α-quartz interface along the z direction. The pyruvic acid with a configuration *Tce* was placed on top of a geminal group, parallel to the quartz surface. Water molecules near the acid were moved away to avoid any overlapping. The so obtained quartz/pyruvic acid/water system contained 1438 atoms with the box dimensions of 19.64 × 17.008 × 43.5 Å. We performed a NVT simulation of 10 ps in the canonical NVT ensemble with a timestep of 0.5 fs using the CSVR thermostat and a target temperature of 330 K. The equilibrium values used for the two pyruvic acids in Eq. 3, r_{eq} the distance of the O–H bond, θ_{eq} the angle C–O–H and ϕ_{eq} the dihedral torsion angle of C–C–O–H, were extracted from this simulation and are summarized in Table 1.

To the previous system, a second *Tce* pyruvic acid was added in the bulk water, which required 5 water molecules to be removed. The resulting box (see on the left side of Fig. 2) contained 1433 atoms with the same dimensions (19.64 × 17.008 × 43.5 Å). The charge is compensated by a uniform neutralizing background. This system was used to calculate the pK_a difference between the pyruvic acid at the interface and in the bulk.

DFT based Born-Oppenheimer molecular dynamics (DFTMD) simulations were performed using Becke exchange [25] and Lee et al. [26] correlation functionals. All calculations have been carried out with the freely available DFT package CP2K/Quickstep which is based on the hybrid Gaussian and plane wave method [27]. Analytic Goedecker-Tecter-Hutter pseudopotentials [28, 29], a TZV2P level basis set for the orbitals and a density cutoff of 280 Ry were used. Dynamics were conducted in the canonical NVT ensemble with a timestep of 0.5 fs using the CSVR thermostat with the target temperature of 330 K to avoid the glassy behavior of BLYP water and a time constant of 1 ps. A periodic orthorhombic box of dimensions 19.64 × 17.008 × 43.5 Å, containing 1438 atmos is considered. For each simulation, an equilibrium trajectory of 5 ps and an additional production run of 20 ps was used for the ΔpK_a calculation. Typical run used 64 nodes of Hazelhen (Cary XC40) with a production time 58 s per molecular dynamics step.

Table 1 Force constants and structural parameters for the restraining potential Eq. 3. k_r, k_θ, k_ϕ are the force constants in kcal/mol for bonds (Å), angles (°) and torsion angles (°) respectively

k_r	r_{eq}	k_θ	θ_{eq}	k_ϕ	ϕ_{eq}
40	1.01956	10	113.543	10	−5.71599

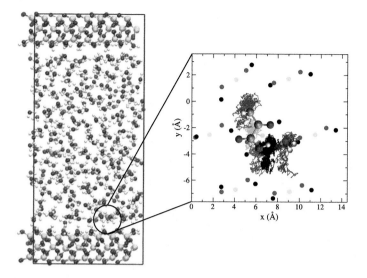

Fig. 2 On the left side a side view (xz side) of the (0001) α-quartz interface is reported. The unit cell in the x and z directions is marked with a blue line. The three dimensional periodic simulation cell contains 6 O–Si–O layers, with two fully hydroxylated surfaces (in total 96 Si, 224 oxygens and 64 hydrogen atoms) and 343 water molecules. On the right: zoom on the top view of the (0001) α-quartz interface with adsorbed pyruvic acid. The coloured dots represent the average position of Si (yellow), O (red) and H (black) atoms from the quartz interface. The lines represent instead the time traces (position as function of time) of pyruvic acid center of mass (cyan), the methyl group (magenta), the oxygen (red) and hydrogen (black) atoms from the pyruvic acid and the oxygen atom (blue) from the water bridge

3 Results

For both bulk and interface the most stable *Tce* isomer was considered. When at the interface, pyruvic acid is stabilized in a configuration with the heavy atoms plane parallel to the quartz surface. During the simulation time (few tenths of ps) the acid remains localized within the first adsorbed layer of water, establishing hydrogen bonds with water molecules in such first layer. In the protonated form the binding between pyruvic acid and the surface silanols is not direct, but mediated by a water bridge (see for example a snapshot from the simulation in Fig. 2). Additional stabilization is provided by the methyl group of pyruvic acid, which accommodates on the hydrophobic patch on the quartz surface (see Fig. 2 for a top view perspective). The described interfacial configuration is maintained along the simulation trajectory as shown by the time traces reported in Fig. 2. Essentially, what we observe is that pyruvic acid cannot freely diffuse at the interface, but remains anchored to the surface.

The pK_a difference between the acid at the interface and the acid in the bulk is calculated using the thermodynamics integration method described in Sect. 2, transforming the acid into its conjugated base.

From our calculation we find that the interfacial pyruvic acid is more acidic than the one in the bulk by 1.84 pK_a units [1].

The thermodynamic integral in Eq. 7 was computed by determining the average energy gap using 3 values of the coupling parameter η: 0, 0.5 and 1. Each value of η shows a different mixing state and requires a separate DFTMD simulation. The thermal average of the vertical energy gaps $\langle \Delta_{dp} E \rangle_\eta$ and the corresponding standard deviations obtained from DFTMD simulations are summarized in Table 2.

Figure 3 shows the vertical energy gap for different values of η as a function of time. Using Eq. 8 it turns out that the interfacial pyruvic acid is more acidic than the one in the bulk by pK_a units of 1.84.

The increase tendency to deprotonate at the quartz/water interface can be understood through an analysis of the solvation structure around both protonated and deprotonated species [1]. In particular, in the following, we will discuss the solvation structure in terms of radial distribution functions and coordination numbers, specifically defined for the solid/liquid interface as described in our recent work [1].

When analyzing the protonated form, we can observe that the solvation structure around the OH group is similar when comparing bulk and interface. In both cases the OH group donates a relatively strong hydrogen bond. At the interface such hydrogen bond is slightly shorter (red line in Fig. 4, position of the first peak at 1.61 Å) with respect to the corresponding one in the bulk (black line in Fig. 4, position of the first

Table 2 Energy gap $\langle \Delta_{dp} E \rangle_\eta$ (in eV) and standard deviation σ_η for different values of η. $\langle \Delta_{dp} E \rangle_{\eta 1}$ and $\langle \Delta_{dp} E \rangle_{\eta 2}$ denotes the averaged energy gap calculated over first and second half of the production trajectory

η	$\langle \Delta E \rangle_{\eta 1}$	$\langle \Delta E \rangle_{\eta 2}$	$\langle \Delta E \rangle_\eta$	σ_η
0	4.73	4.82	4.78	0.69
0.5	−0.09	−0.35	−0.22	0.63
1	−4.47	−4.64	−4.55	0.80

Fig. 3 Accumulative averages of vertical energy gaps with $\eta = 0, 0.5$ and 1 as a function of time (plain). The dashed lines represent the accumulative value for the second half of production trajectory

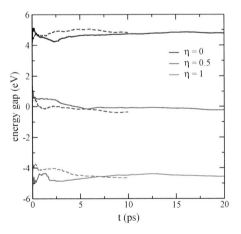

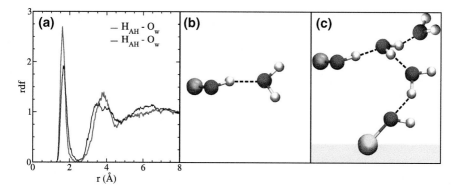

Fig. 4 a Radial distribution functions between the hydrogen atom of the protonated pyruvic acid (H_{AH}) and the oxygen atoms of the water solvent (O_w) (acid in bulk in black, acid at the interface in red), the H-bonds formed between the protonated pyruvic acid and the surrounding atoms: **b** protonated acid in the bulk water, **c** protonated acid at the interface

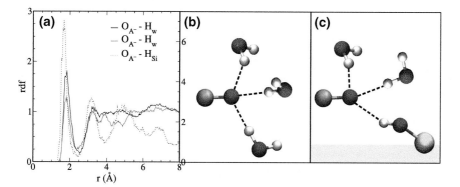

Fig. 5 a Radial distribution functions between the oxygen atom of the deprotonated pyruvic acid (O_{A-}) and the hydrogen atoms of the water solvent (H_w) (conjugated base in bulk in black, at the interface in red. The rdf between O_{A-} and the silanol proton (H_{Si}) is also reported (red dashed line, y axis values readable on the right). The H-bonds formed between the deprotonated pyruvic acid and the surrounding atoms: **b** deprotonated acid in the bulk water, **c** deprotonated acid at the interface

peak at 1.71 Å). This small difference could suggest that at the interface the proton is more keen to leave its oxygen. The observed difference in the local solvation structure is the result of the different water behavior in the bulk and at the interface. The first adsorbed water layer shows slower orientational and diffusion dynamics when compared to the bulk. As a result the hydrogen bond established between the interfacial water molecule and the OH group of pyruvic acid has a life time, which is larger than 20 ps, actually exceeding the simulation length.

It is also interesting to compare the behavior of the conjugated base at the interface and in the bulk as they also present some interesting differences. In the bulk the conjugated base is stabilized accepting hydrogen bonds from water (with an average number of Hbonds in the first shell of 2.64). At the interface, instead, it is stabilized by hydrogen Hbonds from water (on average 1.87), but additionally also by accepting hydrogen bonds from a surface silanol (on average 0.58). The hydrogen bond established with the silanol (dashed red line in Fig. 5) is shorter (1.69 Å) than that accepted by water (continuous red line in Fig. 5b, position of the first peak at 1.85 Å). Such a shorter hydrogen bond is due to the specific properties of the OH group in the silanols. As we have already shown in our previous work [20, 21], silanols may have acidity constants as low as 5, substantially lower than water, in turn establishing stronger hydrogen bonds than water. The silanols play therefore a special role in stabilizing the conjugated base and therefore in favoring the deprotonated form of pyruvic acid at the interface.

4 Conclusions

In this report we have reviewed some recent work addressing molecular acidity at the quartz/water interface. We have shown that in the case of pyruvic acid the molecular acidity at the interface is enhanced as a result of the peculiar microsolvation structure which is established at the interface. The behavior observed for pyruvic acid at the quartz/water interface is possibly not an isolated case, and indeed a similar mechanism for the stabilization of the conjugated base could be in place also for other acids at hydroxilated interfaces. Both the propensity to localize at the interface, as well as the structure of the local solvation will strongly depend on the degree of hydrophilicity and hydrophobiticity of the acid as well as that of the solid surface.

Acknowledgements This work was supported by the Deutsche Forschungsgemeinschaft (DFG) TRR146, project A4. All the calculation were performed on the supercomputer of the High Performance Computing Center (HLRS) of Stuttgart (grant 2DSFG).

References

1. S. Parashar, D. Lesnicki, M. Sulpizi, Increased acid dissociation at the quartz/water interface. J. Phys. Chem. Lett. **9**, 2186, Apr 2018
2. A.E. Reed Harris, A. Pajunoja, M. Cazaunau, A. Gratien, E. Pangui, A. Monod, E.C. Griffith, A. Virtanen, J.-F. Doussin, V. Vaida. Multiphase photochemistry of pyruvic acid under atmospheric conditions. J. Phys. Chem. A **121**(18), 3327–3339 (2017). PMID: 28388049
3. D.M. Cwiertny, M.A. Young, V.H. Grassian, Chemistry and photochemistry of mineral dust aerosol. Ann. Rev. Phys. Chem. **59**(1), 27–51 (2008). PMID: 18393675
4. J. Reijenga, A. van Hoof, A. van Loon, Development of methods for the determination of pka values. Anal. Chem. Insights **8**, 53 (2013)
5. K. Eisenthal, Liquid interfaces probed by second-harmonic and sum-frequency spectroscopy. Chem. Rev. **96**, 1343–1360 (1996)
6. P. Miranda, Q. Du, Y. Shen, Interaction of water with a fatty acid langmuir film. Chem. Phys. Lett. **286**, 1–8 (1998)
7. Y. Rao, M. Subir, E.A. McArthur, N.J. Turro, K.B. Eisenthal, Organic ions at the air/water interface. Chem. Phys. Lett. **477**, 241–244 (2009)

8. C.Y. Tang, Z. Huang, H.C. Allen, Binding of mg2+ and ca2+ to palmitic acid and deprotonation of the cooh headgroup studied by vibrational sum frequency generation spectroscopy. J. Phys. Chem. B **114**, 17068–17076 (2010)

9. H. Wang, X. Zhao, K.B. Eisenthal, Effects of monolayer density and bulk ionic strength on acid-base equilibria at the air/water interface. J. Phys. Chem. B **104**, 8855–8861 (2000)

10. X. Zhao, S. Ong, H. Wang, K.B. Eisenthal, New method for determination of surface pka using second harmonic generation. Chem. Phys. Lett. **214**, 203–207 (1993)

11. M.C. Kido Soule, P.G. Blower, G.L. Richmond, Nonlinear vibrational spectroscopic studies of the adsorption and speciation of nitric acid at the vapor/acid solution interface. J. Phys. Chem. A **111**(17), 3349–3357 (2007). PMID: 17419597

12. E.S. Shamay, V. Buch, M. Parrinello, G.L. Richmond, At the water's edge: Nitric acid as a weak acid. J. Am. Chem. Soc. **129**(43), 12910–12911 (2007). PMID: 17915872

13. S. Wang, R. Bianco, J.T. Hynes, Depth-dependent dissociation of nitric acid at an aqueous surface: car-parrinello molecular dynamics. J. Phys. Chem. A **113**(7), 1295–1307 (2009). PMID: 19173580

14. S. Wang, R. Bianco, J.T. Hynes, Dissociation of nitric acid at an aqueous surface: large amplitude motions in the contact ion pair to solvent-separated ion pair conversion. Phys. Chem. Chem. Phys. **12**, 8241–8249 (2010)

15. S. Strazdaite, K. Meister, H.J. Bakker, Reduced acid dissociation of amino-acids at the surface of water. J. Am. Chem. Soc. **139**(10), 3716–3720 (2017). PMID: 28177623

16. Y. Fang, M. Tang, V.H. Grassian, Competition between displacement and dissociation of a strong acid com pared to a weak acid adsorbed on silica particle surfaces: the role of adsorbed water. J. Phys. Chem. A **120**(23), 4016–4024 (2016). PMID: 27220375

17. M. Tang, W.A. Larish, Y. Fang, A.N. Gankanda, V.H. Grassian, Heterogeneous reactions of acetic acid with oxide surfaces: effects of mineralogy and relative humidity. J. Phys. Chem. A **120**(28), 5609–5616 (2016). PMID: 27322707

18. M. Pfeiffer-Laplaud, M.-P. Gaigeot, M. Sulpizi, pka at quartz/electrolyte interfaces. J. Phys. Chem. Lett. **7**(16), 3229–3234 (2016). PMID: 27483195

19. K. Leung, I.M.B. Nielsen, L.J. Criscenti, Elucidating the bimodal acid-base behavior of the water-silica interface from first principles. J. Am. Chem. Soc. **131**(51), 18358–18365 (2009). PMID: 19947602

20. M. Sulpizi, M.-P. Gaigeot, M. Sprik, The silica-water interface: how the silanols determine the surface acidity and modulate the water properties. J. Chem. Theory Comp. **8**(3), 1037–1047 (2012)

21. M. Pfeiffer-Laplaud, D. Costa, F. Tielens, M.-P. Gaigeot, M. Sulpizi, Bimodal acidity at the amorphous silica/water interface. J. Phys. Chem. C **119**(49), 27354–27362 (2015)

22. F. Costanzo, M. Sulpizi, R.G.D. Valle, M. Sprik. The oxidation of tyrosine and tryptophan studied by a molecular dynamics normal hydrogen electrode. J. Chem. Phys. **134**, 244508 (2011)

23. M. Sulpizi, M. Sprik, Acidity constants from vertical energy gaps: density functional theory based molecular dynamics implementation. Phys. Chem. Chem. Phys. **10**, 5238–5249 (2008)

24. S. Tazi, B. Rotenberg, M. Salanne, M. Sprik, M. Sulpizi, Absolute acidity of clay edge sites from ab-initio simulations. Geochim. Cosmochim. Acta **94**, 1–11 (2012)

25. A.D. Becke, Density-functional exchange-energy approximation with correct asymptotic behavior. Phys. Rev. A **38**, 3098–3100 (1988)

26. C. Lee, W. Yang, R.G. Parr, Development of the colle-salvetti correlation-energy formula into a functional of the electron density. Phys. Rev. B **37**, 785–789 (1988)

27. J. VandeVondele, M. Krack, F. Mohamed, M. Parrinello, T. Chassaing, J. Hutter, Quickstep: fast and accurate density functional calculations using a mixed gaussian and plane waves approach. Comput. Phys. Commun. **167**(2), 103–128 (2005)

28. S. Goedecker, M. Teter, J. Hutter, Separable dual-space gaussian pseudopotentials. Phys. Rev. B **54**, 1703–1710 (1996)

29. C. Hartwigsen, S. Goedecker, J. Hutter, Relativistic separable dual-space gaussian pseudopotentials from h to rn. Phys. Rev. B **58**, 3641–3662 (1998)

Inorganic and Organic Functionalisation of Silicon Studied by Density Functional Theory

Fabian Pieck, Lisa Pecher, Jan-Niclas Luy and Ralf Tonner

Abstract The present report deals with different approaches to functionalize the silicon surface studied by density functional theory. With the examples of the adsorption of *tert*-butylphosphine (TBP) on a silicon surface and the prediction of the band gap for the diluted nitride system Ga(N, As), elemental questions in the growth of inorganic semiconductors via the chemical vapour deposition (CVD) technique are addressed. The organic functionalisation is presented on the basis of the monofunctional cyclooctyne and its bifunctional derivate. Whereby, the adsorption dynamics and the lifetime of intermediate structures are studied with the aid of ab initio molecular dynamic simulations. The scaling of large simulations on the resources of the High-Performance Computing Center Stuttgart is addressed in a final section.

1 Introduction

Up to now silicon is one of the most fundamental materials for our everyday technology, as it is the most important material for the microelectronics industry [1]. The idea behind the functionalisation of silicon is to start with a well established material and to add new functionalities by combining a silicon substrate with another material. For example, to overcome the shortcoming of an indirect band gap in silicon, III/V semiconductors, i.e. materials containing elements from the groups 13 and 15, are deposited on a silicon substrate, so that the composite material is suitable to be used in lasers or solar cells [2–6]. The following sections focus on two different material classes, namely inorganic semiconductors (Sects. 2 and 3) and organic molecules (Sects. 4 and 5), which are combined with silicon substrates. Key aspects in the synthesis of the composite materials are studied by density functional theory (DFT), which has been proven to be a valuable theoretical method for these kinds of systems [7, 8].

A frequently used experimental method to grow III/V semiconductors on silicon is metalorganic vapour phase epitaxy (MOVPE) [9]. In this process, volatile precursors

F. Pieck · L. Pecher · J.-N. Luy · R. Tonner (✉)
Philipps-Universität Marburg, Marburg, Germany
e-mail: tonner@chemie.uni-marburg.de

© Springer Nature Switzerland AG 2019
W. E. Nagel et al. (eds.), *High Performance Computing in Science and Engineering '18*, https://doi.org/10.1007/978-3-030-13325-2_10

are used, which can partially decompose in the gas phase. The residual molecules and fragments adsorb on the substrate surface and deposit the desired elements.

In previous studies, the gas phase decomposition of precursor molecules for the deposition of B, Ga, N, As and P atoms was investigated [10–14]. In Sect. 2 another elemental step is addressed with the example of the adsorption of *tert*-butylphosphine (TBP) on the Si(001) surface. TBP is used as a P precursor in the growth of GaP [10] and Ga(N, As, P) [11, 12] and is known to reach the surface intact [15–17]. Therefore, no further gas phase decomposition products are considered. One should mention that the Si(001) surface does not show the common tilted dimers [18–20] under MOVPE conditions but is rather saturated with H atoms (H/Si(001)) [21–23], albeit some H vacancies exist [23]. Experimental studies concerning the adsorption of TBP on Si(001) are also available [24–27].

Beside the reactivity of the precursor molecules, the properties of the grown material are of interest, as they motivate any future applications. Ga(N, As), GaAs with a small amount of N, exhibits an unexpected band structure [28–30]. A reliable ab initio description of that band structure (Sect. 3) is therefore highly desirable, since it would allow an accurate prediction of other complex systems. For an accurate description of the Ga(N, As) band gap large and thus computationally demanding supercells are needed [31–34].

While Sects. 2 and 3 focused on the reactivity of the precursor and the material properties, further studies are presented that deliver information about the lifetime of structures or the magnitude of reaction rate constants, since these dynamical aspects are quite important. For the organic functionalisation of semiconductors, the adsorption dynamics of precursors controls the formation of well-ordered structures on the surface [35, 36]. Cyclooctyne, which is known to form dense and well-ordered structures on the Si(001) surface [37, 38], is an exemplary system for dynamical studies (Sect. 4). An advantage of cyclooctyne is, that further functional groups can be added [39–42] leading to a bifunctional organic building block, which is needed for the layer-by-layer synthesis [38, 43]. An example for such a bifunctional molecule is 5-ethoxymethyl-5-methylcyclooctyne (EMC) (Sect. 5).

2 Adsorption of *tert*-Butylphophine on Si(001)

The reactivity of precursors on the surface consists in a first approximation of several elemental steps like adsorption, diffusion and decomposition of a single precursor and reactions between several precursors, which are commonly studied separately. For the adsorption of TBP on an ideal H/Si(001) surface only physisorbed structures, structures bonded mainly by dispersion interactions, with an bonding energy up to -34 kJ mol^{-1} are observed. If H vacancies are available, TBP can form a chemisorbed structure where a $H_2(g)$ molecule is released. This exothermic reaction ($\Delta H \geq -78$ kJ mol^{-1}) leads to three different conformers shown in Fig. 1. The bonding energy of the chemisorbed structures is with up to -281 kJ mol^{-1} clearly lower than that of the physisorbed structures and the dispersion energy plays a minor role.

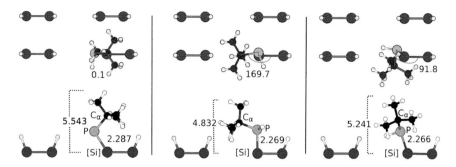

Fig. 1 Adsorption modes found for PH(t–C_4H_9) on a vacancy site of H/Si(001) ($\Theta = 0.125$ ML) in top and side view. Conformation of the *tert*-butyl group parallel (left), anti parallel (middle), or perpendicular (right) to the Si dimer at the surface. Structural parameters $< (C_\alpha - P - [Si] - [Si])$ (top-view), d(Si–P) and adsorbate height relative to the surface-plane (side view) are given in $^\circ$ and Å, respectively. Reprint with permission from Ref. [44]

The conformers differ in the dihedral angle between the organic ligand and the surface dimer. The most stable structure Fig. 1 (middle) is observed for an angle of 169.7° with the smallest distance between the topmost hydrogen and the surface dimer plane of 4.832 Å and with 2.269 Å one of the smallest P–C bond length. The study of several TBP precursors showed, that a second precursor preferably adsorbs on a diagonally shifted dimer atom in comparison to the first precursor, because of an attractive interaction between the precursors leading to an bonding energy for the second precursor of up to -311 kJ mol^{-1}. The adsorption on the same dimer is hindered, because of the steric repulsion between the organic ligands. These results alone were not able to explain the experimental findings of a non-statistical adsorption across the dimer rows [44]. To explain the experimental observations a second approach was considered, in which the ordered adsorption is explained by the stabilization of the H vacancies at a certain position. Figure 2 shows the relative energies for H vacancies at different Si atoms.

For the most stable adsorbate structure the vacancy position across the dimer row is favored by 7.5 kJ mol^{-1} in comparison to all other positions. The defect is stabilized at this position because of charge donation from the C–H groups. The following reaction mechanism is therefore predicted (Scheme 1).

The first precursor adsorbs randomly on a H vacany site and stabilizes the created defect. A stabilized defect leads to a higher detention time of the vacancy across the dimer row and a higher probability for another precursor to adsorb on that position. A small diffusion of the vacancy, which is equivalent to a diffusion of the H atom, is possible [45], but does not change the favorable adsorption parallel to the dimer rows. In the end, this mechanism is suitable to explain the experimental findings of a non-statistical adsorption across and a statistical adsorption along the dimer rows [44].

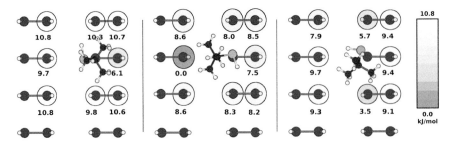

Fig. 2 Vacancy formation energies around an adsorbate TBP on H/Si(001) in top view. Relative energies are given and visualized by a color code (bar to the right) w.r.t the most stable site in kJ mol^{-1}. Reprint with permission from Ref. [44]

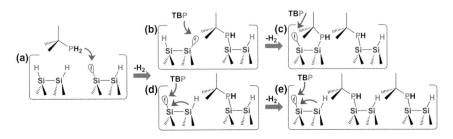

Scheme 1 Summary of the proposed adsorption mechanism, leading to a preferred agglomeration of TBP molecules in chains perpendicular to the dimer rows of the H/Si(001) surface. **a** The first TBP adsorbs on a hydrogen vacancy site. The next TBP molecule can either adsorb on the **b** stabilized vacancy site and **c** stabilize a further vacancy on the same dimer or **d** TBP adsorbs after vacancy migration on one dimer (red arrow) leading to chain growth **e** as well. Reprint with permission from Ref. [44]

3 Accurate ab initio Calculation of the Ga(N, As) Band Gap

Besides the understanding of the growth of III/V materials the prediction of their properties is highly desirable. Ga(N, As), a diluted nitride material, presents a challenge for the commonly used density functionals in their predictive power for band gaps as discussed in the following. For band gap calculations the choice of an accurate functional is crucial as shown in Table 1. The functionals differ considerably in the obtained band gap for the binary compounds GaAs, GaN and GaP. PBE shows the well-known underestimation of band gaps. LDA-1/2 and the computationally more demanding hybrid functionals (PBE0 + HSE06) perform better than the GGA functional. The meta-GGA TB09 outperforms all the other functionals and is even close to the much more demanding GW method. The usage of TB09 comes only with a minor increase in cost in comparison to a GGA functional and is overall perfectly suited for the calculation of band gaps for III/V semiconductors.

Table 1 Direct band gap (eV) of III/V semiconductors in zinc blende structure with various methods. G_0W_0 is based on PBE and without spin-orbit coupling. Lattice constants (GaAs: 5.689 Å, GaN: 4.580 Å, GaP: 5.477 Å) were theoretically optimized (PBE-D3). The root mean square deviation (RMSD) w.r.t. experimental reference values is given for each method. Data from Ref. [46]

	PBE	LDA-1/2	PBE0	HSE06	TB09	G_0W_0	Ref. [47]
GaAs	0.32	1.17	1.71	1.11	1.44	1.41	1.52
GaN	1.58	3.16	3.48	2.78	3.03	3.09	3.28
GaP	1.74	2.58	3.31	2.66	2.95	2.86	2.86
RMSD	1.36	0.27	0.30	0.39	0.16	0.13	

Table 2 Displacement relative to ideal lattice site of selected atoms in $Ga_{125}(As_{124}N_1)$ compared to $Ga_{125}(As_{124}Bi_1)$ in Å. Labeling of position relative to atom E (E = N, Bi) and according to Fig. 3. Data from Ref. [46]

	$Ga_{125}(As_{124}N_1)$	$Ga_{125}(As_{124}Bi_1)$
Ga_{NN}	−0.379	0.190
As_{NNN}	−0.110	0.053
Ga_{3N}(on)	−0.067	0.036
Ga_{3N}(off)	−0.017	0.008
As_{4N}(on)	−0.032	0.019
Ga_{5N}(on)	−0.005	0.003

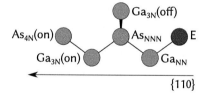
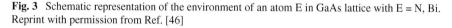

Fig. 3 Schematic representation of the environment of an atom E in GaAs lattice with E = N, Bi. Reprint with permission from Ref. [46]

Before the band gap of Ga(N, As) is discussed, we want to address the problems one is facing when studying this system. One major issue is that N is significantly smaller than As, which leads to a huge distortion in the lattice. The distortion of one N atom in a 5 × 5 × 5 GaAs supercell is shown in Table 2 in comparison to Bi, the other extreme of group V.

The nomenclature follows the connectivity to atom E (E = N, Bi) along the chemical bonds. While the displacement for the nearest neighbors (NN) and the next-nearest neighbors (NNN) is isotropic, a distinction is observed for the third-nearest neighbors ($3N$), which sit along the {110} chains [Ga_{3N}(on)] and off these chains [Ga_{3N}(off)]. Overall the distortion for N is around four times larger than for Bi and the distortion along the {110} chains is twice as large as the distortion off these chains even when the atom is further away from atom E. The distortion is halved in every step along the {110} direction leading to a distortion smaller than 1 pm at the fifth atom.

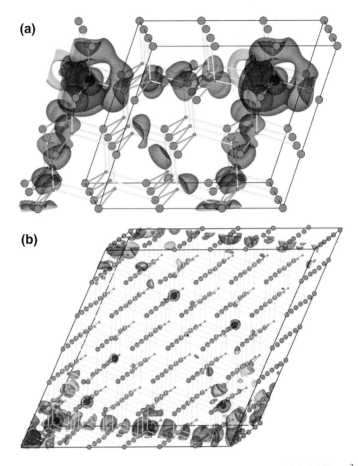

Fig. 4 Partial charge density of the first conduction band of Ga(N, As) with 3.7% N. **a** 3^3 supercell and **b** 6^3 supercell. Isosurface at 10% of the respective maximal value. Ga: light blue, As: orange, N: blue; Ga and As atoms are shown in smaller size. Adapted reprint with permission from Ref. [46]

The second issue corresponds to the electronic interaction of N atoms with their periodic images. Figure 4 shows the charge density of the first conduction band for a 3^3 and a 6^3 supercell both with 3.7% nitrogen content. In Fig. 4a it is clearly visible that a too small cell leads to interactions between N and its images. Similar to the distortion, the charge density is localized along the {110} chains. In the larger 6^3 unit cell some charge density is still observed along the {110} chains, but the largest amount is localized at the N atoms. For an accurate description of the structural distortion and the electronic states a computationally demanding 6^3 unit cell containing 432 atoms is needed. Smaller supercells lead to an artificial interaction of N atoms with their periodic images and by this to an incorrect description of the band gaps.

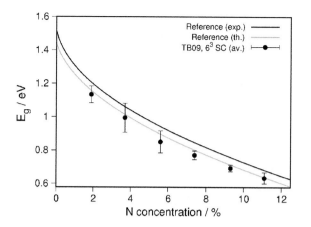

Fig. 5 Band gaps of Ga(N, As) for a 6^3 supercell averaged over four arrangements each for the three concentrations up to 5.6% N, three arrangements each for 7.4 and 9.3% N and two for 11.1%. The references are based on the CBAC model parameters as given by Ref. [48] with the experimentally derived band gap (exp.) and with the TB09 band gap (th.) of GaAs (see Table 1). Reprint with permission from Ref. [46]

Finally, the band gap of Ga(N, As) for N concentration of up to 11.1% is calculated with the 6^3 unit cell and the TB09 functional (Fig. 5). For every N concentration the band gap was averaged over different N arrangements. This is necessary because different arrangements can lead to a spread in band gap over up to ±100 meV (3.4% N content). The calculated band gaps are compared to the conduction band anticrossing (CBAC) model [48–50], which is used with the experimental (exp.) GaAs band gap of 1.52 eV or the theoretical (th.) band gap of 1.44 eV. While the calculated band gaps perfectly match with CBAC (th.) they are always slighty smaller than CBAC (exp.). The trend of smaller band gaps with higher N contents is preserved. It can be concluded that a 6^3 unit cell in combination with the TB09 functional is well suited to predict band gaps of diluted nitrides.

4 Adsorption Dynamics of Cyclooctyne on Si(001)

The calculation of static systems and properties as shown in the previous sections is a valuable approach to understand and to predict experiments. Nevertheless, for a thorough understanding of the chemical reactions an important component is missing, namely the dynamical aspect. So far the lifetime of a structure, which correlates with the probability to experimentally observe that structure, is unknown. These information can be addressed in a statistical approach, for example by the quasi-quantum harmonic transition state theory (qq-hTST) [51, 52], or in an explicit approach as in ab initio molecular dynamics (AIMD) [53]. From a computational point of view these dynamical aspect is even for small systems (~100 atoms) very demanding. The difficulty is to ensure a reliable time scale (at least some ps), while the fastest movements in the system, which are commonly the vibrations containing the dis-

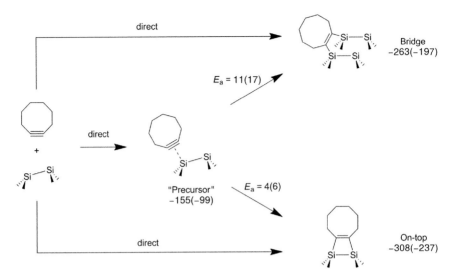

Scheme 2 Adsorption and surface reactivity map of cyclooctyne on Si(001). Bonding energies E^{bond} and activation energies E_a calculated at PBE-D3/PAW, given in kJ mol^{-1}. Gibbs energies (in parentheses) calculated at $T = 300\,K$, $p = 1$ bar. Reprint with permission from Ref. [54]

placement of hydrogen atoms, have to be accurately described leading to time steps of ~1 fs. As a consequence every calculation has to perform several thousand steps.

Scheme 2 shows adsorption structures for cyclooctyne. In the gas phase, cyclooctyne prefers to orient close to the Si_{down} atom, which is attributed to the attractive interaction of the electron rich triple bond and the electrophile Si_{down} atom. The adsorption to the On-top and Bridge structure can proceed via an intermediate state or a direct path. The observation of an intermediate state is unexpected because experiments predicted a direct adsorption, only [38]. In concordance with the experiment the On-top structure is thermodynamically favored by $-45\,kJ\,mol^{-1}(-40\,kJ\,mol^{-1})$ over the Bridge structure [37, 55].

In order to predict the lifetime of the precursor state 20 AIMD calculations were performed at 300 K. Figure 6 shows two exemplary trajectories. An intermediate state is assumed if several criteria are fulfilled:

- An energy plateau around $-150\,kJ\,mol^{-1}$
- An asymmetric C–Si distance [d(C–Si)], with a d(C–Si) value close to 2 Å
- A slight elongation of the C–C triple bond [d(C^1–C^2)] up to a bond length of 1.3 Å

Dotted lines indicate the largest change in the C^1–C^2 bond length and are used to estimate the lifetime of the intermediate state. The trajectory in Fig. 6a fulfills all the mentioned criteria and is assigned to show an adsorption via the intermediate state, while the trajectory in Fig. 6b shows a direct adsorption. The 15 trajectories showing an intermediate state result in an averaged lifetime for that state of 1.0 ps. With the help of the lifetime and the reaction barrier the pre-exponential factor of

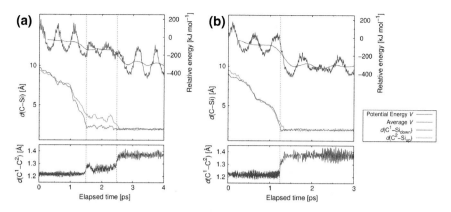

Fig. 6 Potential energy V and selected interatomic distances indicative for the progress of the adsorption, taken from two exemplary AIMD trajectories: **a** adsorption via intermediate; **b** direct adsorption. Averaging of V done over 600 fs, the approximate oscillation periodicity of the Nosé thermostat. The zero point of the potential energy was set to the sum of averaged (over 6 ps) potential energies of separated molecule and slab simulations in equilibrium. Reprint with permission from Ref. [54]

Table 3 Estimation of the lifetime τ of the intermediate at different temperatures T using the Gibbs activation energy G_a and the Arrhenius pre-exponential factor A_0 (in THz) taken from the AIMD calculations or, alternatively, qq-hTST. Data from Ref. [54]

		PBE-D3		HSE06-D3	
T (K)	A_0 (THz)	G_a (kJ mol^{-1})	τ (ps)	G_a (kJ mol^{-1})	τ (ps)
AIMD					
50	12 ± 8	3.8	720 ± 500	2.7	60 ± 40
150	12 ± 8	4.5	3.0 ± 2.0	3.4	1.3 ± 0.9
300	12 ± 8	6.2	1.0 ± 0.7	5.1	0.7 ± 0.5
qq-hTST					
50	1.8	3.8	4700	2.7	400
150	2.2	4.5	17	3.4	7.3
300	2.6	6.2	4.5	5.1	3.0

an Arrhenius equation is estimated. This allows to calculate the lifetime for different temperatures as shown in Table 3. Beside the values extracted from the AIMD calculations, Table 3 contains lifetimes calculated with the qq-hTST method. Albeit the absolute values change for the different methods and functionals, the overall statement of a lifetime smaller than some nanoseconds is the same. In conclusion the adsorption of cyclooctyne is shown to proceed via pseudodirect adsorption, i.e. adsorption via an ultra-short-lived intermediate state, which is in good agreement with the experiments.

5 Competitive Bonding of a Bifunctional Cyclooctyne on Si(001)

The description of static and dynamical properties becomes even more demanding if a second functional group, an ether group in EMC, is added to the cyclooctyne. Beside the understanding of the adsorption via one functional group the key question dealing with the competitive adsorption arises.

Scheme 3 shows the singly bonded states of the EMC molecule. Three different cycloadducts are obtained if the molecule bonds via the triple bond to the Si(001) surface. The On-Top (OT) position is favored over the Bridge (B) position by at least $44 \, kJ \, mol^{-1}$. For the adsorption via the ether group two structures showing a dative bond (DB) from the oxygen to the Si_{down} atom are obtained, from which the adsorbate can further react via bond cleavage (C). The DB1 structure has with $-132 \, kJ \, mol^{-1}$ a smaller absolute bonding energy than the OT structure with $-352 \, kJ \, mol^{-1}$, but the products C1 and C2 are with -301 to $-304 \, kJ \, mol^{-1}$ of similar magnitude. Doubly bonded states, which are either a combination of OT and DB2 (OT + DB) or OT and C2 (OT + C), with an bonding energy of -370 to $-474 \, kJ \, mol^{-1}$ are observed, too. The interaction between the two functional groups can be studied further if the reactivity and the chemical bonds with the Si surface are compared to the isolated functional groups in cyclooctyne and diethylether. It becomes evident that the reactivity of the triple bond in EMC is unaffected, while the ether group is hindered by the large ring, but the bond strength of the dative O–Si bond is at the same time increased by around 12% as shown by periodic energy decomposition analysis (pEDA) [56].

Since the bonding energies alone allow no reasoning concerning the competitive adsorption, 10 AIMD simulations at 300 K were performed. An exemplary trajectory is shown in Fig. 7. The trajectories were evaluated according to the rules of Sect. 4. Out of the 10 simulations, 4 runs ended in the OT, 3 in the OT + DB and 1 in the B1, B2 and in a physisorbed structure respectively. The dative bond via the ether group was not observed. Similar to cyclooctyne an ultra-short-lived stationary state with a lifetime of 1.38 ps for the OT adsorption was observed. The similar magnitude of the lifetime in comparison to the adsorption of cyclooctyne confirms that the triple bond is unaffected by the second functional group. The fact that no adsorption via a dative O–Si bond was observed indicates that the adsorption via the triple bond is favored. Additionally, it can be concluded that OT is the most probable position. The observation of OT + DB adsorption structures does not contradict with that conclusion, because it is reasonable that both structures constantly interconvert at 300 K as supported by the negligible difference in the Gibbs energy of $2 \, kJ \, mol^{-1}$ (or see Fig. 7). In conclusion, the EMC molecule is suitable for the functionalisation of silicon, because a controlled adsorption via the triple bond is possible, leading to the OT structure, which is in agreement with the experiment [38]. Based on these results the focus is now shifted to the reactions of further organic building blocks with the remaining ether group and how that influences the properties of the novel materials.

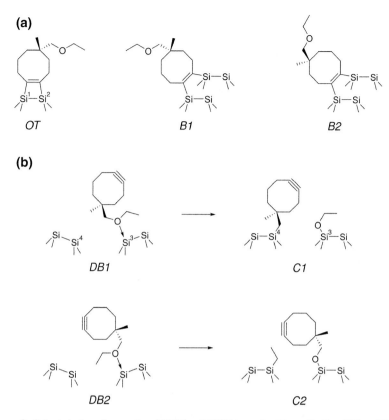

Scheme 3 Selected adsorption modes of EMC on Si(001): **a** cycloadducts On-Top (OT) and Bridge (B). Dative bond (DB) formation in two possibel orientations and the corresponding C–O cleavage (C) products. Reprint with permission from Ref. [56]

6 Scaling and Resources

Sections 2–5 introduced the computationally demanding problems of a large system or a long trajectory, which need to be handled in an efficient way. The scaling for the main code VASP [57–60] was previously shown to be excellent up to several thousand cores [61, 62]. Futhermore, the calculations can be sped up by carefully considering the parallelisation options of VASP. Most critical is the parallelisation over K-points and bands controlled by the keywords KPAR and NCORE, respectively. With NCORE equal to the number of cores per node and KPAR equal to the number of K-points of the system, 5000 steps (5 ps simulation time) within 24 h on 240 cores are achievable in a AIMD simulation describing a system with up to 144 atoms. To reach a simulation time of up to 20 ps or more, the number of cores has to be multiplied at least with the same factor, which leads to a requirement of 1000 to 1500 cores. If the system becomes larger, $\sim 3\times$ more atoms in Ga(N, As), the

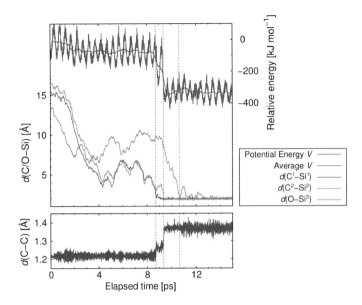

Fig. 7 Evolution of the potential energy V and selected interatomic distances over time in simulation 6. Vertical lines denote a change in the adsorption mode: Physisorbed (until 8.71 ps), intermediate state (8.71–9.36 ps), OT (9.36–10.65 ps), OT + DB (from 10.65 ps). Reprint with permission from Ref. [56]

number of performed steps reduces significantly. In the worst case, the code scales with N^4, whereby N is the number of basis functions or roughly the number of atoms. In order to treat larger system with multiple adsorbates or to reach experimentally meaningful timescales, further performance improvements are inevitable. To do so, we plan to either reduce the accuracy of the calculations by using for example the density functional tight binding approach [63], or use more sophisticated sampling approaches like metadynamic calculations [64].

References

1. *International Technology Roadmap for Semiconductors* (2015), http://www.itrs2.net. Accessed 2 May 2018
2. K. Volz, A. Beyer, W. Witte, J. Ohlmann, I. Németh, J. Cryst. Growth **315**, 37 (2011)
3. B. Kunert, K. Volz, I. Németh, W. Stolz, J. Lumin. **121**, 361 (2006)
4. B. Kunert, K. Volz, W. Stolz, Phys. Status Solidi B **244**, 2730 (2007)
5. D. Thomson, A. Zilkie, J. E. Bowers, T. Komljenovic, G. T. Reed, L. Vivien, D. Marris-Morini, E. Cassan, L. Virot, J.-M. Fédéli, J.-M. Hartmann, J. H. Schmid, D.-X. Xu, F. Boeuf, P. O' Brien, G. Z. Mashanovich, M. Nedeljkovic, J. Opt. **18**, 073003 (2016)
6. B. Kunert, S. Reinhard, J. Koch, M. Lampalzer, K. Volz, W. Stolz, Phys. Status Solidi C **3**, 614 (2006)
7. J. Neugebauer, T. Hickel, Wiley Interdiscip. Rev.-Comput. Mol. Sci. **3**, 438 (2013)

8. H. Pedersen, S.D. Elliott, Theor. Chem. Acc. **133**, 1476 (2014)
9. G.B. Stringfellow, Mater. Sci. Eng. B **87**, 97 (2001)
10. K. Volz, A. Beyer, W. Witte, J. Ohlmann, I. Németh, B. Kunert, W. Stolz, J. Cryst. Growth **315**, 37 (2011)
11. N. Hossain, S.R. Jin, S. Liebich, M. Zimprich, K. Volz, B. Kunert, W. Stolz, S.J. Sweeney, Appl. Phys. Lett. **101**, 011107 (2012)
12. S. Liebich, M. Zimprich, A. Beyer, C. Lange, D.J. Franzbach, S. Chatterjee, N. Hossain, S.J. Sweeney, K. Volz, B. Kunert, W. Stolz, Appl. Phys. Lett. **99**, 071109 (2011)
13. M. Imam, L. Souqui, J. Herritsch, A. Stegmüller, C. Höglund, S. Schmidt, R. Hall-Wilton, H. Högberg, J. Birch, R. Tonner, H. Pedersen, J. Phys. Chem. C **121**, 26465 (2017)
14. M. Imam, K. Gaul, A. Stegmüller, C. Höglund, J. Jensen, L. Hultman, J. Birch, R. Tonner, H. Pedersen, J. Mater. Chem. C **3**, 10898 (2015)
15. A. Stegmüller, R. Tonner, Chem. Vap. Deposition **21**, 161 (2015)
16. A. Stegmüller, P. Rosenow, R. Tonner, Phys. Chem. Chem. Phys. **16**, 17018 (2014)
17. A. Stegmüller, R. Tonner, Inorg. Chem. **54**, 6363 (2015)
18. R.E. Schlier, H.E. Farnsworth, J. Chem. Phys. **30**, 917 (1959)
19. R.M. Tromp, R.J. Hamers, J.E. Demuth, Phys. Rev. Lett. **55**, 1303 (1985)
20. J. Dąbrowski, M. Scheffler, Appl. Surf. Sci. **56–58**, 15 (1992)
21. H. Döscher, P. Kleinschmidt, T. Hannappel, Appl. Surf. Sci. **257**, 574 (2010)
22. S. Brückner, H. Döscher, P. Kleinschmidt, T. Hannappel, Appl. Phys. Lett. **98**, 211909 (2011)
23. P. Rosenow, R. Tonner, J. Chem. Phys. **144**, 204706 (2016)
24. Y. Fukuda, T. Kobayashi, S. Mochizuki, Appl. Surf. Sci. **212–213**, 329 (2003)
25. R.L. Woo, U. Das, S.F. Cheng, G. Chen, K. Raghavachari, R.F. Hicks, Surf. Sci. **600**, 4888 (2006)
26. Y. Fukuda, T. Kobayashi, S. Mochizuki, Appl. Surf. Sci. **175–176**, 218 (2001)
27. G. Kaneda, N. Sanada, Y. Fukuda, Appl. Surf. Sci. **142**, 1 (1999)
28. M. Weyers, M. Sato, H. Ando, Jpn. J. Appl. Phys. **31**, L853 (1992)
29. M. Kondow, K. Uomi, K. Hosomi, T. Mozume, Jpn. J. Appl. Phys. **33**, L1056 (1994)
30. K. Uesugi, N. Morooka, I. Suemune, Appl. Phys. Lett. **74**, 1254 (1999)
31. L. Bellaiche, S.-H. Wei, A. Zunger, Phys. Rev. B **54**, 17568 (1996)
32. L. Bellaiche, S.-H. Wei, A. Zunger, Appl. Phys. Lett. **70**, 3558 (1997)
33. P.R.C. Kent, A. Zunger, Phys. Rev. B **64**, 115208 (2001)
34. A. Zunger, Phys. Status Solidi B **216**, 117 (1999)
35. C.T. Rettner, D.J. Auerbach, J.C. Tully, A.W. Kleyn, J. Phys. Chem. **100**, 13021 (1996)
36. R. Díez Muiño, H.F. Busnengo (eds.) *Dynamics of Gas-Surface Interactions* (Springer, Berlin, Heidelberg, 2013)
37. G. Mette, M. Dürr, R. Bartholomäus, U. Koert, U. Höfer, Chem. Phys. Lett. **556**, 70 (2013)
38. M. Reutzel, N. Münster, M.A. Lipponer, C. Länger, U. Höfer, U. Koert, M. Dürr, J. Phys. Chem. C **120**, 26284 (2016)
39. N.J. Agard, J.M. Baskin, J.A. Prescher, A. Lo, C.R. Bertozzi, ACS Chem. Biol. **1**, 644 (2006)
40. J.A. Codelli, J.M. Baskin, N.J. Agard, C.R. Bertozzi, J. Am. Chem. Soc. **130**, 11486 (2008)
41. Y. Sun, X. Ma, K. Cheng, B. Wu, J. Duan, H. Chen, L. Bu, R. Zhang, X. Hu, Z. Deng, Angew. Chem., Int. Ed. **54**, 5981 (2015)
42. N. Münster, P. Nikodemiak, U. Koert, Org. Lett. **18**, 4296 (2016)
43. M.Z. Hossain, Y. Yamashita, K. Mukai, J. Yoshinobu, Chem. Phys. Lett. **388**, 27 (2004)
44. A. Stegmüller, K. Werner, M. Reutzel, A. Beyer, P. Rosenow, U. Höfer, W. Stolz, K. Volz, M. Dürr, R. Tonner, Chem. - Eur. J. **22**, 14920 (2016)
45. M. Dürr, U. Höfer, Prog. Surf. Sci. **88**, 61 (2013)
46. P. Rosenow, L.C. Bannow, E.W. Fischer, W. Stolz, K. Volz, S.W. Koch, R. Tonner, Phys. Rev. B. **97**, 075201 (2018)
47. I. Vurgaftman, J.R. Meyer, L.R. Ram-Mohan, J. Appl. Phys. **89**, 5815 (2001)
48. I. Vurgaftman, J.R. Meyer, J. Appl. Phys. **94**, 3675 (2003)
49. W. Shan, W. Walukiewicz, J. Ager, E. Haller, J. Geisz, D. Friedman, J. Olson, S. Kurtz, Phys. Rev. Lett. **82**, 1221 (1999)

50. L.C. Bannow, O. Rubel, S.C. Badescu, P. Rosenow, J. Hader, J.V. Moloney, R. Tonner, S.W. Koch, Phys. Rev. B **93**, 205202 (2016)
51. G. Henkelman, A. Arnaldsson, H. Jónsson, J. Chem. Phys. **124**, 044706 (2006)
52. L. Árnadóttir, E.M. Stuve, H. Jónsson, Surf. Sci. **606**, 233 (2012)
53. M.E. Tuckerman, J. Phys.: Condens. Matter **14**, R1297 (2002)
54. L. Pecher, S. Schmidt, R. Tonner, J. Phys. Chem. C **121**, 26840 (2017)
55. J. Pecher, C. Schober, R. Tonner, Chem. - Eur. J. **23**, 5459 (2017)
56. L. Pecher, R. Tonner, Theor. Chem. Ac. **137**, 48 (2018)
57. G. Kresse, J. Hafner, Phys. Rev. B **47**, 558 (1993)
58. G. Kresse, J. Hafner, Phys. Rev. B **49**, 14251 (1994)
59. G. Kresse, J. Furthmüller, Phys. Rev. B **54**, 11169 (1996)
60. G. Kresse, J. Furthmüller, Comput. Mater. Sci. **6**, 15 (1996)
61. A. Stegmüller, R. Tonner, in *High Performance Computing in Science and Engineering' 13* (Springer International Publishing, 2013), pp. 185–199
62. P. Rosenow, A. Stegmüller, J. Pecher, R. Tonner, in *High Performance Computing in Science and Engineering' 15* (Springer International Publishing, 2016), pp. 199-213
63. G. Seifert, J.O. Joswig, Wiley Interdiscip. Rev. Comput. Mol. Sci. **2**, 456 (2012)
64. A. Barducci, M. Bonomi, M. Parrinello, Wiley Interdiscip. Rev.-Comput. Mol. Sci. **1**, 826 (2011)

Progress Report on: Sulfur in Ethylene Epoxidation on Silver (SEES2)

Travis Jones

Abstract The primary goal of the "Sulfur in ethylene epoxidation on silver" (SEES2) project is to elucidate the mechanism(s) by which catalytic ethylene epoxidation occurs over silver surfaces. There is a particular focus on the role of sulfur. The program involves using density functional theory to predict stable surface phases under various conditions by way of ab initio atomistic thermodynamics. Once identified, the spectroscopic properties of candidate phases are computed to enable experimental verification. Minimum energy paths associated with the (re)formation and reaction of the identified phases are then computed to determine their possible roles in ethylene epoxidation. Through this approach we identified a novel $Ag(SO_4)$ phase and showed it selectively transfers oxygen to ethylene to form the epoxide during temperature programed reaction. In the last year we have shifted the focus to the behavior of surface species under catalytic conditions, focusing on the 0 K minimum energy paths of the surface reaction network before moving on to finite temperature effects. In this effort we have identified additional phases and studied the competition between them. Of the studied species the novel SO_4 phase, where sulfur is present as $S(V+)$, is the only silver one capable of selectively reacting with ethylene to form the epoxide. We further found the SO_4 species is rapidly regenerated through reaction with oxygen, which suppresses the coverage of an adsorbed SO_3 that appears to be selective in total oxidation. The presence of SO_x species is also found to reduce the EO:AcH branching ratio associated with the reaction of ethylene with atomic oxygen on the unreconstructed $Ag(111)$. Thus, it appears under conditions that are not artificially clean EO is produced in large part by oxygen transfer from the novel SO_4 phase. These new insights are only possible due to the use of various levels of parallelization to extend the scaling of our code on the Cray XC40 system Hazel Hen, which has allowed us to compute the minimum energy paths of a complex network of surface reactions.

T. Jones (✉)
Department of Inorganic Chemistry, Fritz-Haber-Institut der Max-Planck-Gesellschaft,
Faradayweg 4-6, 14195 Berlin, Germany
e-mail: trjones@fhi-berlin.mpg.de

© Springer Nature Switzerland AG 2019
W. E. Nagel et al. (eds.), *High Performance Computing in Science
and Engineering '18*, https://doi.org/10.1007/978-3-030-13325-2_11

1 Introduction

Ethylene oxide (EO) is the simplest cyclic ether. It is formed by adding a single oxygen atom across the C–C double bond of ethylene, see schematic in Fig. 1. Despite its simplicity the epoxide is one of the most important industrial chemicals with a worldwide production in excess of 15×10^6 t per annum [1]. Today virtually all of the EO produced each year is made by the catalytic partial oxidation of ethylene over silver under an oxygen atmosphere [1–3]. Although epoxidation may appear formally trivial, see Fig. 1, the mechanism under catalytic conditions is unclear [1, 4–7]. The difficulty lies in the competition between partial and total oxidation in the reaction triangle shown in Fig. 1, that is selectivity to the epoxide.

Thermodynamic considerations of Fig. 1 show, with an enthalpy of reaction an order of magnitude higher than EO formation [1], CO_2 production is favored under typical epoxidation conditions. Thus, it is only kinetics that drive EO formation. Silver is unique in that under a mixture of ethylene and oxygen at atmospheric pressure these kinetics already lead to a $\approx$50% selectivity to the epoxide at $\approx$500 K; with the judicious use of promoters the selectivity can reach above 80% [1, 7–9]. Understanding why and how EO can be produced selectively over silver is a significant challenge in heterogeneous catalysis. Scientifically it is an interesting question as to how both EO and CO_2 are produced over silver as it is one of the simplest and most successful examples of selectivity in heterogeneously catalyzed reactions. If the understanding of such studies can be translated to an industrial setting the results may be profound; a 1% increase in selectivity would result in tens of millions of dollars in annual savings for EO producers [2]. Beyond these aspects a deep understanding of the direct gas-phase epoxidation of ethylene may open the door to the rational design of new catalysts for reactions that are currently out of reach. By way of example, unlike the tremendous success of the direct partial oxidation of ethylene under oxygen, the equally useful epoxide of propylene cannot be produced economically by this route and is instead manufactured through one of several less appealing processes [10]. To move towards these goals, however, we must first understand the catalytic production of EO over (unpromoted) silver.

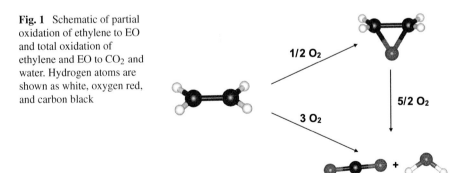

Fig. 1 Schematic of partial oxidation of ethylene to EO and total oxidation of ethylene and EO to CO_2 and water. Hydrogen atoms are shown as white, oxygen red, and carbon black

1/2 O$_2$

5/2 O$_2$

3 O$_2$

Plausible models to explain ethylene epoxidation over silver emerged already in the 1940s [9]. These early models posited a special form of adsorbed oxygen reacts with ethylene to form EO while a second type of oxygen reacts to form CO_2 [9]. Temperature programmed reaction (TPR) was eventually used to show two types of oxygen could be prepared under vacuum and only one is involved in EO production [7, 11]. This selective species became known as *electrophilic oxygen*, while the oxygen involved only in combustion is often referred to as *nucleophilic oxygen* [7, 11], in reference to an earlier proposal that the EO-selective species has a lower formal negative charge than the CO_2-selective species [12]. Such arguments found support from X-ray photoelectron spectroscopy (XPS), with nucleophilic oxygen appears at 528.4 eV in the O 1s spectrum and electrophilic oxygen appears at 530.5 eV [7, 11].

The O 1s binding energy of nucleophilic oxygen is consistent with that of oxygen in oxygen induced surface reconstructions known to form on Ag(111) [13–15] and Ag(110) [16, 17] surfaces. It was thus assumed the oxygen in these reconstructions can only produce CO_2 upon reaction with ethylene, an assumption verified as part of the SEES2 project by way of density functional theory (DFT) calculations combined with TPR [4]. Electrophilic oxygen, however, resisted isolation under ultra-high vacuum (UHV) and was long assumed to be atomic oxygen adsorbed on unreconstructed silver surfaces [18]. This assumption led to a number of theoretical studies aimed at uncovering and understanding the reaction mechanism of epoxidation by focusing on the possible reactivity of atomic oxygen on the unreconstructed surface [19–24]. While these DFT studies predicted such an oxygen species could participate in non-selective epoxidation, they failed to find a highly-selective species. Moreover, the picture that atomic oxygen adsorbed on the unreconstructed surface is the electrophilic oxygen species was cast into doubt when an oxygen species consistent with electrophilic oxygen was prepared on a Ag(111) single crystal surface under UHV and found to give rise to a $(7 \times \sqrt{3})$rect low energy diffraction (LEED) pattern [25].

As part of the SEES2 project we solved the structure of the $(7 \times \sqrt{3})$rect reconstruction attributed to electrophilic oxygen and found it to be a parent phase of the oxygen species active in EO production during TPR, for full details see [5]. Briefly, by first noting the spectroscopic properties of electrophilic oxygen require the presence of covalently bound oxygen [26] we were able to determine the $(7 \times \sqrt{3})$rect reconstruction contains sulfur, as verified by XPS [5]. An extensive structural search based on DFT at the Perdew, Burke, and Ernzerhof [27] level was then carried out on Hazel Hen. This search identified a low-energy $Ag_9(SO_4)_3$ overlayer which a LEED-*I(V)* analysis confirmed is the structure of the $(7 \times \sqrt{3})$rect reconstruction previously attributed to electrophilic oxygen. However, minimum energy path (MEP) calculations performed on Hazel Hen suggested the oxygen in the $(7 \times \sqrt{3})$rect reconstruction is inactive in TPR. Further analysis showed that the $(7 \times \sqrt{3})$rect reconstruction is lifted when a surface partially covered in it is oxidized. This transformation leads to the appearance of SO_4 moieties adsorbed on the unreconstructed (111) surface, $SO_{4,ads}$. A wide range of additional MEP calculations suggested this $SO_{4,ads}$ can selectively transfer an oxygen atom to ethylene to produce EO and adsorbed SO_3. As shown in Fig. 2, EO can be produced from adsorbed ethylene in a process

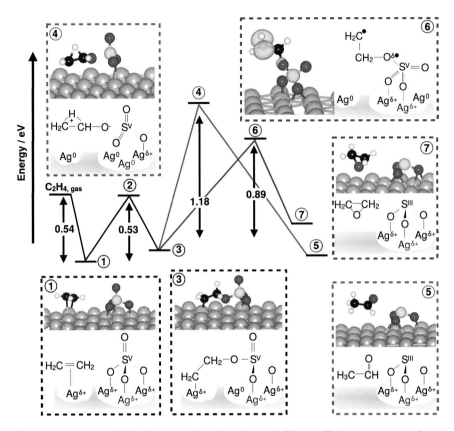

Fig. 2 Lowest energy MEPs for the reaction of ethylene with $SO_{4,ads}$. Hydrogen atoms are shown as white spheres, oxygen red, carbon black, sulfur yellow, and silver grey. The black line shows the formation of an adsorbed C_2H_4–O–SO_3 intermediate. This intermediate decomposes into either AcH (red path) or EO (blue path)

activated by 0.96 eV, in close agreement with the apparent activation energy for EO desorption during TPR [28], whereas acetaldehyde (AcH) formation, a CO_2 precursor, is activated by 1.5 eV relative to adsorbed ethylene. TPR experiments confirm the computed behavior [5].

With the oxygen phase active in TPR fully characterized we have now turned to its behavior under reaction conditions. To do so we identified other phases that may be present and studied their competition with $SO_{4,ads}$. To begin we have focused on the fate of the SO_3 produced in the production of EO as shown in Fig. 2. As shown below, our MEP calculations suggest reoxidation is fast with respect to further reduction when the surface coverage of adsorbed ethylene is no more than 3–4 orders of magnitude higher than that of adsorbed molecular oxygen. In cases where SO_3 reduction occurs we find it is selective towards AcH, and hence total combustion. The competing reactions of atomic oxygen on the unreconstructed surface were also found to favor AcH production when dispersion corrections are accounted for, especially in the presence of adsorbed SO_x. To test the generality of the dispersion interactions we also explored the MEPs for the reaction of ethylene with oxygen on the unreconstructed Ag(110), which support the conclusion that atomic oxygen on the unreconstructed surfaces produces AcH, in many cases selectively. Of the studied species only SO_4 has MEPs consistent with selective EO formation.

2 Methods

DFT calculations were performed with the Quantum ESPRESSO package [29]. All calculations employed the Perdew, Burke, and Ernzerhof [27] exchange and correlation potential, and, unless otherwise noted the exchange-hole dipole moment (XDM) dispersion corrections [30, 31]. Through extensive testing of the computational setup [4, 5] we found the calculated energies were converged when using ultrasoft pseudopotentials with non-linear core corrections with a kinetic energy (charge density) cutoff of 30 (300) Ry.

MEPs on the Ag(111) were computed with the climbing image nudged elastic band (NEB) method where the MEPs were discretized into 8–20 replicas with a single climbing image along each path. All calculations employed four-layer (4 × 4) silver slabs where the bottom two layers were held fixed at the computed lattice constant (4.15 Å) during ionic relaxation. Slabs were separated by ≈15 Å of vacuum. The Brillouin zone was sampled with a Monkhorst-Pack **k**-point mesh equivalent to (12 × 12 × 1) for the surface unit (111) cell. Marzari-Vanderbilt cold smearing was used with a smearing parameter of 0.02 Ry [32]. A similar computational setup was employed for the Ag(110) surface, see [6] for full details.

In this portion of the study the MEP calculations were the most computationally demanding, each of which required the accurate calculation of forces on the atoms in the 8–20 images used to discretize the reaction path. These computed forces tangent to the MEP are used along with a (variable) fictitious spring force parallel to the path to converge the initially guessed images to the true MEP, where the fictitious forces prevents the images from collapsing into a local minimum [33]. As the images converge to the true MEP the full force is employed on the highest energy replica with the component normal to the MEP inverted to drive the geometry to the transition state. As these calculations can take several hundred force evaluations per image we worked to identify strategies for efficient parallelization.

First we tested the serial efficiency of the code using a one-atom fcc cell of silver with 182 **k**-points, 10 Kohn-Sham states (N_b), and a kinetic energy (charge density) cutoff of 30 (300) Ry. Here we found good serial performance with 2.3 instructions/cycle using 9 GB/s of the total 160 GB/s L2 bandwidth.

The systems of interest contain approximately 75 atoms, 450 Kohn-Sham states, 30 **k**-points, and 8–20 replicas. For scalability testing we employed one replica, a 4-layer (4 × 4) slab with a single SO_4 and C_2H_4 adsorbed on the surface. The 30 **k**-points were distributed over 30 pools of CPUs as the communication requirement between pools is low. Wave-function coefficients were then distributed over the n_{PW}

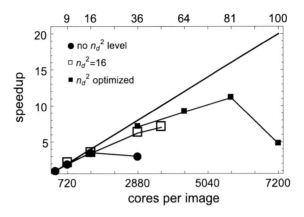

Fig. 3 Testing parallelization levels for a system composed of 4-layer (4 × 4) silver slab with adsorbed SO_4 and C_2H_4. The filled circles show results without linear algebra parallelization. The empty squares show results with $n_d^2 = 16$. The filled squares show results with the optimized values of n_d^2: 36, 64, 81, and 100 for 2880, 4320, 5760, and 7200 cores per image respectively

CPUs working on each **k**-point. As shown in Fig. 3 this parallelization is somewhat disappointing, saturating before 2880 cores per replica. To further extend the scaling we distributed the linear algebra operations required for subspace diagonalization on $\approx N_b \times N_b$ square matrices on a square grid of n_d^2 cores, where $n_d^2 \leq n_{PW}$.

Using this linear algebra parallelization level we are able to extend the scaling to 2880 cores per replica using $n_d^2 = 16$, see Fig. 3. A further increase of n_d^2 to 36 in resulted in a minor decrease in the time to solution, while further increases in n_d^2 with 2880 cores led to decreased performance. With $n_d^2 = 16$, however, scaling does not extend significantly past 2880 cores. To maintain good scalability to a larger number of cores n_d^2 must be careful chosen. Doing so we find we can employ up to 5760 cores per image with $n_d^2 = 81$. As the typical simulation was performed with at least 8 replicas, with parallelization over replicas trivial, scaling to over 40k cores was possible.

3 Results and Discussion

3.1 Reoxidation of $SO_{3,ads}$

Figure 2 shows EO can be produced selectively through the reaction of ethylene with $SO_{4,ads}$. For the process to be catalytic the resultant $SO_{3,ads}$ must be reoxidized to $SO_{4,ads}$. Two general mechanism can be envisaged for this process on the unreconstructed surface, reaction of $SO_{3,ads}$ with adsorbed atomic oxygen (O_{ads}) or molecular oxygen ($O_{2,ads}$). (Note that the notation O_{ads} is used to differentiate atomic oxygen adsorbed on an unreconstructed surface from atomic oxygen in the well known surface reconstructions [13–17].)

To begin, consider the reaction with atomic oxygen. While O_{ads} is not observed under UHV at elevated temperatures—low-energy electron microscopy established an upper coverage of 0.03 monolayer (ML) at ≈ 500 K on Ag(111) [34]—it can be found at low coverage under an oxygen atmosphere. In situ XPS results show oxygen is present on the unreconstructed Ag(110) up to a coverage of ≈ 0.02 ML at 423 K [6] and up to ≈ 0.06 ML on the Ag(111) at 423 K [35]. Thus, O_{ads} is a relevant species to consider at low coverages. To capture such low coverages and avoid spurious interactions between periodic images we employed a (4×4) Ag(111) surface when studying reoxidation of $SO_{3,ads}$ by O_{ads}.

The first step investigated in the reoxidation of $SO_{3,ads}$ by atomic oxygen was the formation of O_{ads} through the dissociation of $O_{2,ads}$. As seen in state 1 of Fig. 4, dissociation begins with molecular oxygen adsorbing on a bridge site where each oxygen atom is bound to one surface silver atom. Adsorption increases the gas-phase bond length from 1.23 to 1.29 Å and reduces the magnetic moment from its gas phase value (2.0 μ_B) to 1.5 μ_B. These changes are indicative of slight charge transfer to the $O_{2,ads}$, showing it adsorbs as a superoxo species (O_2^-) as opposed to the peroxo (O_2^{2-}) adsorption seen on the Ag(110) [6]. Adsorption as a superoxo species is further supported by the low (0.12 eV) adsorption energy. From this adsorbed state the O–O bond can be broken to form 2 O_{ads} when $O_{2,ads}$ moves such that both oxygen atoms are located at Ag–Ag bridge sites, see state 2 of Fig. 4. The activation energy for dissociation is 1.21 eV and the heat of reaction is 0.06 eV, where the positive

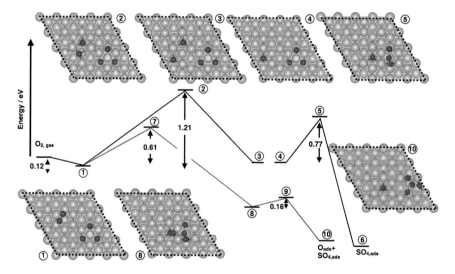

Fig. 4 MEPs for $SO_{4,ads}$ regeneration from $SO_{3,ads}$ and gas phase O_2. The black MEPs show the pathway for $SO_{4,ads}$ formation with atomic oxygen. States 1–3 show O_2 dissociation in the presence of $SO_{3,ads}$. States 4–6 show the subsequent reaction of O_{ads} with $SO_{3,ads}$. The red MEPs show the competing pathway to $SO_{4,ads}$ by the reaction of $SO_{3,ads}$ and $O_{2,ads}$. This MEP passes through the $SO_{5,ads}$ intermediate in state 8. Coloring of atomic spheres matches Fig. 2

sign indicates $O_{2,ads}$ dissociation is weakly endothermic in the presence of 1/16 ML $SO_{3,ads}$.

After this slow $O_{2,ads}$ dissociation step the formed O_{ads} can react with the adsorbed SO_3 to reform $SO_{4,ads}$; note competing reactions are discussed below. This reaction is 1.30 eV exothermic and has a barrier of 0.77 eV, see states 4–6 in Fig. 4. As seen in the figure, the $SO_{3,ads}$ is only coordinated to the surface through two Ag–O bonds at the transition state, which causes the sulfur center to develop radical character. However, because the adsorbed oxygen atom remains formally O(II-) along the MEP, spin-polarization has a negligible effect on the energy of the transition state; ignoring spin-polarization increases the barrier to 0.79 eV.

Conversely, we can consider the direct reaction of $O_{2,ads}$ with $SO_{3,ads}$. While an accurate estimate of the coverage of $O_{2,ads}$ is difficult to obtain from in situ XPS, the low adsorption energy suggests the coverage of $O_{2,ads}$ will also be low. As, such we again employed a (4×4) Ag(111) surface for studying reoxidation of $SO_{3,ads}$ directly by $O_{2,ads}$.

States 1–8 in Fig. 4 show reoxidation of $SO_{3,ads}$ by $O_{2,ads}$ proceeds by first forming an adsorbed SO_5 species from the same initial state as $O_{2,ads}$ dissociation. Now, however, because both reactants have significant radical character at the transition state, spin-polarization is essential to recover the correct barrier, which we find it to be 0.61 eV, significantly less than the 1.21 eV found for the competing $O_{2,ads}$ dissociation. Formation of $SO_{5,ads}$ is 0.66 eV exothermic, and once formed, it will quickly decompose into $SO_{4,ads}$ and O_{ads} by crossing a barrier of 0.17 eV in a step that is 0.59 eV exothermic, see states 8–10 in Fig. 4. The remaining O_{ads} formed in this process can then participate in either reoxidation of $SO_{3,ads}$, formation of an oxygen induced surface reconstruction, or (partial) oxidation of ethylene.

Assuming the prefactors associated with the elementary steps in Fig. 4 are the same, a simple Arrhenius equation can be used to express the ratio of the frequency of O_2 dissociation through the two routes. Doing so while ignoring coverage one finds that at 500 K the rate of direct $O_{2,ads}$ dissociation is $\approx 10^6$ times slower than SO_5 formation on this surface. Furthermore, under the same assumptions the ratio of $SO_{3,ads}$ reoxidation (through state 7 in Fig. 4) to EO and $SO_{3,ads}$ formation from $C_2H_{4,ads}$ and $SO_{4,ads}$ (through state 6 in Fig. 2) is 7×10^3 at 500 K. Thus we expect reoxidation of $SO_{3,ads}$ will be fast relative to its formation. Our experimental results support this conclusion, with a majority of the SO_x on the surface during epoxidation at 503 K present as $SO_{4,ads}$ [5].

3.2 $SO_{3,ads}$ Reduction

In addition to the previously discussed $SO_{4,ads}$ regeneration step, during catalytic epoxidation the $SO_{3,ads}$ formed through oxygen transfer to ethylene, Fig. 2, may also be reduced to $SO_{2,ads}$ by further reaction with ethylene. Such a reaction is expected to lower the SO_x coverage, as $SO_{2,ads}$ rapidly desorbs at elevated temperatures [36].

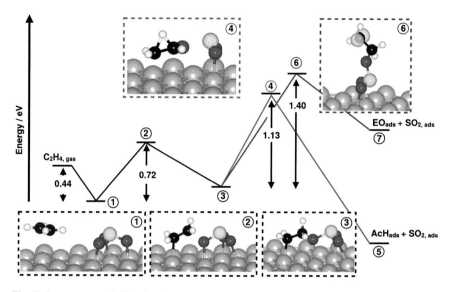

Fig. 5 Lowest energy MEPs for the reaction of ethylene with $SO_{3,ads}$. The black line shows the formation of an adsorbed C_2H_4–O–SO_2 intermediate. This intermediate decomposes into either AcH (red path) or EO (blue path). The coloring of the atomic spheres is identical to Fig. 2

Like the MEPs for the reaction of $SO_{4,ads}$ with ethylene, we find that $SO_{3,ads}$ can react with $C_2H_{4,ads}$ to form a C_2H_4–$OSO_{2,ads}$ intermediate, see Fig. 5. The first step in the process is the adsorption of ethylene, which is 0.44 eV exothermic in the presence of 1/16 ML $SO_{3,ads}$. This values is slightly (0.1 eV) smaller than that seen for ethylene adsorption in the presence of 1/16 ML $SO_{4,ads}$, Fig. 2. The activation energy associated with the formation of C_2H_4–$OSO_{2,ads}$, 0.72 eV, can also be seen to be $\approx$0.2 eV higher than that associated with the formation of the C_2H_4–$OSO_{3,ads}$ intermediate in the case of $SO_{4,ads}$. Consistent with the higher barrier, formation of C_2H_4–$OSO_{2,ads}$ is 0.18 eV endothermic, as opposed to 0.08 eV for formation of C_2H_4–$OSO_{3,ads}$. While these difference in activation energies for the elementary steps associated with C_2H_4–$OSO_{x,ads}$ formation are non-negligible, a more striking difference between the reactivity of $SO_{3,ads}$ and $SO_{4,ads}$ comes from the decomposition of the C_2H_4–$OSO_{x,ads}$ intermediates.

Analogous to C_2H_4–$OSO_{3,ads}$, once the C_2H_4–$OSO_{2,ads}$ intermediate is formed it can decompose into either AcH and $SO_{2,ads}$ or EO and $SO_{2,ads}$. Figure 5 shows the barrier to AcH is 1.13 eV in a step that is 0.70 eV exothermic. The activation energy for this elementary step is nearly identical to the 1.18 eV found for the decomposition of C_2H_4–$OSO_{3,ads}$ shown in Fig. 2. The MEP to EO, however, is activated by 1.40 eV in the case of C_2H_4–$OSO_{2,ads}$, as opposed to 0.89 eV for the $SO_{4,ads}$ analogue. Part of the reason for the discrepancy between the two reactions is the $SO_{3,ads}$ moieties inability to effectively stabilize radical character on the oxygen atom involved in ring closure. In the case of C_2H_4–$OSO_{3,ads}$ spin-polarization reduces the activation energy to EO by $\approx$0.5 eV, whereas for C_2H_4–$OSO_{2,ads}$ the reduction is only $\approx$0.3 eV.

Thus, this difference accounts for 40% of the roughly 0.5 eV difference in activation energy for EO formation through the $SO_{4,ads}$ and $SO_{3,ads}$ moieties.

Overall the results then predict $SO_{3,ads}$ will selectively produce AcH, and hence CO_2. To understand the possible importance of this reaction we can contrast it with the competing $SO_{3,ads}$ reoxidation. Comparing Figs. 4 and 5 reveals $SO_{3,ads}$ reoxidation is associated with lower barriers than $SO_{3,ads}$ reduction. The ratio of a simple Arrhenius equation with identical prefactors for the slowest elementary step for $SO_{3,ads}$ reoxidation through reaction with $O_{2,ads}$ and that for AcH formation is $\approx 10^5$. Thus, when the coverage of both adsorbed ethylene and molecular oxygen are within 3–4 orders of magnitude we expect $SO_{3,ads}$ reduction to be slow relative to its reoxidation. This finding is also supported by our experimental results, see [5]. Here we found the $SO_{4,ads}$ coverage increases on a silver powder catalyst as the system moves towards steady-state in a 1 : 1 or 2 : 1 mixture of $O_{2,gas}$ to $C_2H_{4,gas}$ by way of in situ XPS. $SO_{3,ads}$, however, could only be transiently accumulated by titrating the active surface with ethylene.

3.3 Competing Reactions

With the MEPs for the consumption and reoxidation of $SO_{3,ads}$ now identified we can turn to competing reactions to complete the picture of the MEPs associated with reactions on a surface partially covered in adsorbed SO_x under catalytic conditions. Perhaps the most prominent competing reaction under low-pressure conditions is that of the oxygen induced surface reconstructions. These species are active only in combustion, where, for example, the $p(4 \times 4)$-O reconstruction of the Ag(111) produces AcH through an oxametallacycle (OMC) mechanism with a barrier of only 0.28 eV, making it active at room temperature [4]. As mentioned earlier, under epoxidation conditions the oxygen in this $p(4 \times 4)$-O reconstruction may come either from direct O_2 dissociation or $SO_{5,ads}$ decomposition, see Fig. 4. In both cases O_{ads} is formed initially, and this O_{ads} may also react with ethylene before reconstructing the silver surface.

The reaction of O_{ads} with ethylene is one of the most popularized mechanisms proposed for EO production. Numerous authors have shown using PBE DFT without dispersion corrections that, if present, O_{ads} can react with adsorbed ethylene to form an OMC intermediate where this C_2H_4O intermediate is bound to the surface through the oxygen and terminal carbon [19, 20, 22, 23, 37], see state 3 in Fig. 6. These studies find the OMC intermediate can decompose into either AcH or EO. The activation energy to EO (AcH) is found by various authors to be, for example, 0.73 eV (0.75 eV) [37] and 0.79 eV (0.84 eV) on the Ag(111) [23]. Hence the EO:AcH branching ratio, which represents the upper limit to EO selectivity through the OMC mechanism, is thought to be near 1:1 for this type of reaction on the Ag(111). On other surfaces the branching ratio is argued to generally increase towards EO [19, 20, 22, 23, 37], where these slight variations in activation energy as a function of surface termination can be well described as a competition between the Ag–C and Ag–O bond energies

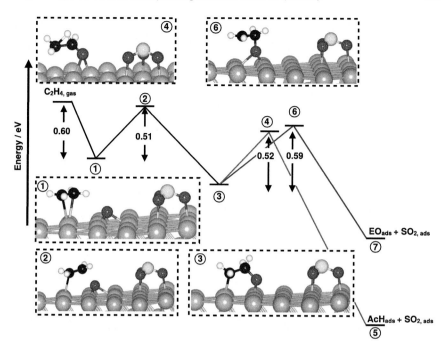

Fig. 6 MEPs for the reaction of ethylene with O_{ads} in the presence of $SO_{3,ads}$. Coloring matches Fig. 2. An OMC intermediate is formed in step 3 and decomposes into AcH via step 4 or EO via step 6

of the OMC; a stronger Ag–O bond relative to the Ag–C bond favors EO production relative to AcH production and the converse [22].

The OMC mechanism in the presence of 1/16 ML $SO_{3,ads}$ shows some subtle differences compared to the reaction on the clean surface. The first is that the adsorption energy of ethylene (0.60 eV) on the surface with 1/16 ML $SO_{3,ads}$ is higher than adsorption on a clean surface (0.32 eV) or a surface with adsorbed oxygen (0.51 eV). This difference can be attributed to the $Ag^{\delta+}$ formed in response to the $SO_{3,ads}$, which tends to bind the π system of ethylene more strongly [38, 39]. Once adsorbed, ethylene can react with O_{ads} on the surface with 1/16 ML $SO_{3,ads}$ to form an OMC by crossing a barrier of 0.51 eV. Once formed this OMC can decompose into either EO or AcH. However, in the presence of 1/16 ML $SO_{3,ads}$ AcH production is predicted to be favored by almost 0.1 eV, with a 0.59 eV (0.52 eV) activation energy to EO (AcH). While this value is within the typical PBE error [40], it is instructive to consider its difference with previously reported results and study the trends in activation energies.

There are two competing factors when comparing to previous results, the inclusion of dispersion corrections and the presence of 1/16 ML $SO_{3,ads}$. We find in the absence of dispersion interactions the activation energy for the OMC to decompose into EO (AcH) is 0.72 eV (0.74 eV) on the clean surface, in agreement with previous results. Inclusion of dispersion interactions reduces both barriers to 0.64 eV, showing van der

Waals corrections change the overall barrier but only subtly increase the branching ratio towards AcH in this case. On a surface with 1/16 ML $SO_{3,ads}$, ignoring dispersion interactions leads to an activation energy to EO (AcH) of 0.65 eV (0.61 eV) compared to the 0.59 eV (0.52 eV). Again, dispersion corrections reduce the magnitude of the barriers and slightly increase the branching ratio towards AcH. Thus, it appears both dispersion interactions and the 1/16 ML $SO_{3,ads}$ drive the branching ratio towards AcH. The underlying reason for this behavior can be understood when considering the aforementioned competition between the Ag–C and Ag–O bond energies of the OMC [22]. The $SO_{3,ads}$ produces $Ag^{\delta+}$ sites which bind carbon more strongly than the clean surface and the van der Waals C_6 coefficients are typically larger for carbon than for oxygen [41]. The increase in Ag–C bond energy relative to Ag–O then increases the branching ratio to AcH.

This issue is exacerbated on the Ag(110) surface, as the surface silver atoms are more under coordinated than on the Ag(111) surface. The Ag(110) is also of particular interest as we were able to show experimentally that O_{ads} can be isolated on it at low-temperature, where computed XPS spectra and phase diagrams inspired the pursuit and eventual identification of the O_{ads} species on this surface [6]. Reaction of O_{ads} with ethylene through an OMC mechanism is, however, predicted to favor AcH formation. Figure 7 shows the MEPs for this reaction. OMC formation is weakly

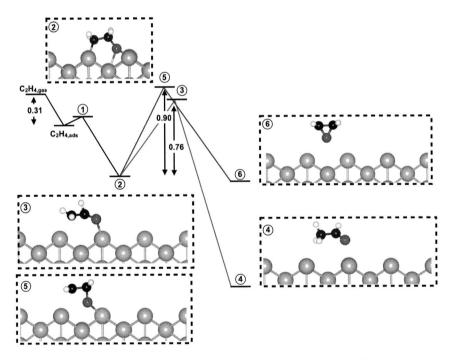

Fig. 7 MEPs for the reaction of ethylene with O_{ads} on the Ag(110) surface. Coloring matches Fig. 2. An OMC intermediate is formed in step 2 and decomposes into AcH via step 3 or EO via step 5

activated (0.09 eV), and once formed, the OMC can decompose into AcH (EO) by crossing a barrier of 0.76 eV (0.90 eV).

Similar behavior is also seen when considering the reaction of O_{ads} with ethylene in the presence of 1/16 ML $SO_{4,ads}$. When considering the earlier discussion that a fraction of the $SO_{4,ads}$ is likely reformed through an $SO_{5,ads}$ intermediate it is clear such a situation may be relevant during catalytic epoxidation. Though not shown, the MEPs for OMC mechanism in the presence of 1/16 ML $SO_{4,ads}$ qualitatively match those shown in Fig. 6. In this case, however, the activation energy to form AcH and EO from the OMC intermediate are somewhat lower, 0.44 eV (0.53 eV without dispersion corrections) and 0.54 eV (0.58 eV without dispersion corrections), respectively. Despite these differences the EO:AcH branching ratio is similar to the case of 1/16 ML $SO_{3,ads}$ shown in Fig. 6.

It is worth noting that in all the aforementioned cases the reaction of O_{ads} with ethylene tends to have lower associated barriers than its reaction with $SO_{3,ads}$. Thus, it appears likely the reaction between ethylene and atomic oxygen may still limit the EO selectivity as the reaction through the OMC is predicted to be non-selective to EO, that is, with a EO:AcH branching ratio less than 1. And while this does not imply the famous 6/7 upper limit to selectivity [1], it does suggest oxygen management strategies remain an essential aspect of achieving high selectivity.

4 Summary

In summary, by leveraging the excellent scalability of Quantum ESPRESSO on Hazel Hen we have been able to develop a full picture of the 0 K MEPs for the catalytic epoxidation of ethylene with surface oxygen and SO_x species on the unpromoted Ag(111). From these results we have found $SO_{4,ads}$ selectively transfers oxygen to ethylene through an elementary step with an activation energy lower than that associated with direct O_2 dissociation. The $SO_{3,ads}$ formed in this process can be reduced by ethylene in a slow reaction leading to AcH, and eventually CO_2, production. The competing reoxidation of $SO_{3,ads}$, however, is predicted to be faster than either SO_3 formation or reduction when the coverage of adsorbed ethylene is no more than 4 order of magnitude higher than the coverage or adsorbed O_2, both of which will be low owing to their low adsorption energies. We also found that competing reactions of adsorbed atomic oxygen with ethylene on the (un)reconstructed Ag(111) and Ag(110) have low-barriers relative to the SO_x routes. However, by including both dispersion corrections and/or a small coverage of $SO_{x,ads}$ we find these competing reactions are non-selective to EO, and in some cases selective to AcH, and hence CO_2. Of the studied species only electrophilic oxygen, $SO_{4,ads}$, is predicted to selectively epoxidize ethylene.

Acknowledgements Dr. M. Greiner and Dr. E. Carbonio from the FHI offered valuable discussions. T. E. Jones acknowledges the Alexander-von-Humboldt foundation for financial support of the work.

References

1. S. Rebsdat, D. Mayer, *Ethylene Oxide* (Wiley-VCH Verlag GmbH & Co. KGaA, Weinheim, 2012)
2. C. Stegelmann, N. Schiødt, C. Campbell, P. Stoltze, J. Catal. **221**, 630 (2004)
3. J. Serafin, A. Liu, S. Seyedmonir, J. Mol. Catal. A Chem. **131**, 157 (1998)
4. T.E. Jones, R. Wyrwich, S. Böcklein, T.C.R. Rocha, E.A. Carbonio, A. Knop-Gericke, R. Schlögl, S. Günther, J. Wintterlin, S. Piccinin, J. Phys. Chem. C **120**, 28630 (2016)
5. T.E. Jones, R. Wyrwich, S. Böcklein, E.A. Carbonio, M.T. Greiner, A.Y. Klyushin, W. Moritz, A. Locatelli, T.O. Mentes, M.A. Nino et al., ACS Catal. **8**, 3844 (2018)
6. E.A. Carbonio, T.C.R. Rocha, A.Y. Klyushin, I. Pis, E. Magnano, S. Nappini, S. Piccinin, A. Knop-Gericke, R. Schlogl, T.E. Jones, Chem. Sci. **9**, 990 (2018)
7. V.I. Bukhtiyarov, A. Knop-Gericke, *Nanostructured Catalysts: Selective Oxidations* (The Royal Society of Chemistry, 2011), pp. 214–247
8. R.V. Santen, H. Kuipers, Adv. Catal. **35**, 265–321 (1987). Academic Press
9. W.M.H. Sachtler, C. Backx, R.A. Van Santen, Catal. Rev. **23**, 127 (1981)
10. H. Baer, M. Bergamo, A. Forlin, L.H. Pottenger, J. Lindner, *Propylene Oxide* (Wiley-VCH Verlag GmbH & Co. KGaA, Weinheim, 2012)
11. V. Bukhtiyarov, I. Prosvirin, R. Kvon, Surf. Sci. **320**, L47 (1994)
12. R.B. Grant, R.M. Lambert, J. Catal. **92**, 364 (1985)
13. R. Reichelt, S. Günther, J. Wintterlin, W. Moritz, L. Aballe, T.O. Mentes, J. Chem. Phys. **127** (2007)
14. M. Schmid, A. Reicho, A. Stierle, I. Costina, J. Klikovits, P. Kostelnik, O. Dubay, G. Kresse, J. Gustafson, E. Lundgren et al., Phys. Rev. Lett. **96**, 146102 (2006)
15. J. Schnadt, A. Michaelides, J. Knudsen, R.T. Vang, K. Reuter, E. Lægsgaard, M. Scheffler, F. Besenbacher, Phys. Rev. Lett. **96**, 146101 (2006)
16. S.R. Bare, K. Griffiths, W. Lennard, H. Tang, Surf. Sci. **342**, 185 (1995)
17. M. Pascal, C. Lamont, P. Baumgärtel, R. Terborg, J. Hoeft, O. Schaff, M. Polcik, A. Bradshaw, R. Toomes, D. Woodruff, Surf. Sci. **464**, 83 (2000)
18. V.I. Bukhtiyarov, M. Hävecker, V.V. Kaichev, A. Knop-Gericke, R.W. Mayer, R. Schlögl, Phys. Rev. B **67**, 235422 (2003)
19. S. Linic, M.A. Barteau, J. Am. Chem. Soc. **124**, 310 (2002)
20. S. Linic, M.A. Barteau, J. Am. Chem. Soc. **125**, 4034 (2003)
21. J. Greeley, M. Mavrikakis, J. Phys. Chem. C **111**, 7992 (2007)
22. A. Kokalj, P. Gava, S. de Gironcoli, S. Baroni, J. Catal. **254**, 304 (2008)
23. M. Ozbek, I. Onal, R. Santen, Top. Catal. **55**, 710 (2012)
24. M.L. Bocquet, D. Loffreda, J. Am. Chem. Soc. **127**, 17207 (2005)
25. S. Böcklein, S. Günther, J. Wintterlin, Angew. Chem. Int. Ed. **52**, 5518 (2013)
26. T.E. Jones, T.C.R. Rocha, A. Knop-Gericke, C. Stampfl, R. Schlögl, S. Piccinin, ACS Catal. **5**, 5846 (2015)
27. J.P. Perdew, K. Burke, M. Ernzerhof, Phys. Rev. Lett. **77**, 3865 (1996)
28. R.B. Grant, R.M. Lambert, J. Chem. Soc. Chem. Commun. **355**, 662 (1983)
29. P. Giannozzi, S. Baroni, N. Bonini, M. Calandra, R. Car, C. Cavazzoni, D. Ceresoli, G.L. Chiarotti, M. Cococcioni, I. Dabo et al., J. Phys. Condens. Matter **21**, 395502 (2009). http://www.quantum-espresso.org
30. E.R. Johnson, A.D. Becke, J. Chem. Phys. **128**, 124105 (2008)
31. A.O. de-la Roza, E.R. Johnson, J. Chem. Phys. **136**, 174109 (2012)
32. N. Marzari, D. Vanderbilt, A. De Vita, M.C. Payne, Phys. Rev. Lett. **82**, 3296 (1999)
33. G. Henkelman, B.P. Uberuaga, H. Jónsson, J. Chem. Phys. **113**, 9901 (2000)
34. S. Günther, T.O. Mentes, M.A. Niño, A. Locatelli, W.J. Böcklein, Nat. Commun. **5**, 3853 (2014)
35. T.E. Jones, T.C.R. Rocha, A. Knop-Gericke, C. Stampfl, R. Schlogl, S. Piccinin, Phys. Chem. Chem. Phys. **17**, 9288 (2015)

36. H.-W. Wassmuth, J. Ahner, M. Höfer, H. Stolz, Prog. Surf. Sci. **42**, 257 (1993)
37. P. Christopher, S. Linic, J. Am. Chem. Soc. **130**, 11264 (2008)
38. A. Kokalj, A.D. Corso, S. de Gironcoli, S. Baroni, Surf. Sci. **532–535**, 191 (2003)
39. C. Backx, C.D. Groot, P. Biloen, W. Sachtler, Surf. Sci. **128**, 81 (1983)
40. K. Duanmu, D.G. Truhlar, J. Chem. Theory Comp. **13**, 835 (2017)
41. S. Grimme, J. Comp. Chem. **27**, 1787 (2006)

Part III
Reactive Flows

For the section "Reactive Flows" we have a two contributions:

"Towards clean propulsion with synthetic fuels" by

Mathis Bode, Marco Davidovic, Heinz Pitsch

and

"Improved vectorization for efficient chemistry computations in OpenFoam for large scale combustion simulations" by

Thorsten Zirwes, Feichi Zhang, Jordan A. Denev, Peter Habisreuther, Henning Bockhorn, Dimosthenis Trimis.

In the first contribution the authors consider clean propulsion with synthetic fuels. Still today most of the propulsions are based on liquid fuels and unhealthy emissions. It was found that synthetic fuels have the potential to burn soot free and therefore can reduce the pollution. The used mathematical model is based on the compressible Navier-Stokes equations with additional tabulated chemistry models. The authors consider up to 822 species and 4597 reactions.

The numerical method is based on the in-house code CIAO. They use WENO schemes, local mesh refinement, load balancing, the volume of fluid and level set methods for trekking interfaces. The Lagrangian particle methods are used for studying sprays.

Good scaling is obtained up to 12288 cores. Comparison of the numerical results with physical experiments are very convincing. This approach is unique in the sense that it does not require any tuning, resolves the small skales during primary break up and they can consider detailed mechanisms with hundreds of species.

In the second contribution the authors develop a software which automatically generates a C++ code for computing the reaction rates. The generated code reduces the CPU time for serial computing chemical reaction rates by up to 50% and for the total simulation by up to 25%. Compared to the performance of the OpenFOAM implementation of the reaction rate computation, the new code is faster by a factor of 10. The numerical results of the new code are the same as before, since the same equations are solved more efficiently. The new code has been applied to a large scale numerical simulation of a turbulent flame on a grid with 150 million cells and 14400 processors. The results are in good agreement with measurements. This project was supported by the Helmholtz Association and the DFG.

Dietmar H. Kröner

Institut für Angewandte Mathematik, Albert-Ludwig-Universität Freiburg
Hermann-Herder-Straße 10, 79104 Freiburg im Breisgau, Germany
e-mail: dietmar.kroener@mathematik.uni-freiburg.de

Towards Clean Propulsion with Synthetic Fuels: Computational Aspects and Analysis

Mathis Bode, Marco Davidovic and Heinz Pitsch

Abstract In order to support sustainable powertrain concepts, synthetic fuels show significant potential to be a promising solution for future mobility. It was found that CO_2 emissions during the combustion process of synthetic fuels can be reduced compared to conventional fuels and that sustainable fuel production pathways exists. Furthermore, it is possible to burn some synthetic fuels soot-free, which indirectly also eliminates the well-known soot-NO_x tradeoff. However, in order to use the full potential of the new fuels, optimization of currently used injection systems needs to be performed. This is still challenging since fundamental properties are not known and pollutant formation is a multi-physics, multi-scale process. Therefore, the high-fidelity simulation framework CIAO is improved and optimized for predictive simulations of multiphase, reactive injections in complex geometries. Due to the large separation of scales, these simulations are only possible with current supercomputers. This work discusses the computational performance of the high-fidelity simulations especially focusing on vectorization, scaling, and input/output (I/O) on Hazel Hen (Cray XC40) supercomputer at the High Performance Computing Center Stuttgart (HLRS). Moreover, the impact of different internal nozzle flow initial conditions is shown, the effect of different chemical mechanisms studied, and the predictability of soot emissions investigated. The Spray A case defined by the Engine Combustion Network (ECN) is used as the target case due to the availability of experimental data for this injector.

Electronic supplementary material The online version of this chapter (https://doi.org/10.1007/978-3-030-13325-2_12) contains supplementary material, which is available to authorized users.

M. Bode (✉) · M. Davidovic · H. Pitsch
Institute for Combustion Technology, RWTH Aachen University,
Templergraben 64, 52056 Aachen, Germany
e-mail: m.bode@itv.rwth-aachen.de

1   Introduction

Propulsion based on liquid fuels provides a large fraction of today's transportation energy. Since alternative technologies, such as electric mobility, are not able to substitute propulsion technology based on liquid fuel within the next decades, processes of conventional propulsion such as injection of a liquid fuel into a gaseous environment and energy conversion in a turbulent environment need to be improved in order to deal with stricter emission regulations and the finiteness of fossil fuels. Additionally, the usage of alternative fuels can potentially overcome classical limitations of conventional energy conversion machines. For example, it was found that synthetic fuels, such as oxymethylenethers, have the potential to burn soot-free and thus eliminate the well-known soot-NO_x tradeoff of classical diesel engines [23]. Also, CO_2 emissions are potentially lower than in gasoline engines and an efficient production from renewable resources seems possible leading to a closed CO_2 loop. However, the optimization of these processes is very challenging and details are still not completely understood due to their complexity and the difficulty of performing experiments characterizing the flow features inside the nozzle and the atomization and chemistry processes outside the nozzle.

The performance of a particular injection system/fuel combination depends on a cascade of physical processes, originating from the nozzle internal flow driven by the injector geometry and operation conditions, potential cavitation, turbulence, the mixing of the coherent liquid stream with the gaseous ambient environment during the atomization process outside of the injector orifice, and the chemical properties of the fuel [20]. It is well known that the primary breakup, the first phase of the atomization process, is especially critical for the overall engine performance in terms of emission and pollutant formation [26], and the transfer occurring between liquid and gas has to be governed by an interface topology [17]. In this regard, the objective of an injector design could be to maximize the interface surface density in order to ensure homogeneous mixing, small droplets, fast evaporation, and emission-free ignition. However, how design parameters influence surface density and ultimately droplet size distribution is not clear, and simple predictive models do not exist. Therefore, the Spray A case defined by the Engine Combustion Network (ECN) [1] is used as the target case in this work for gaining more insight into the underlying physical mechanisms and the development of predictive models. Since the Spray A case uses a single hole injector, simulations of the internal nozzle flow are computationally cheaper and issues such as asymmetric mass flow rates are avoided compared to multi-hole injectors as e. g. used for the Spray B case. Furthermore, the broad range of available experimental data for the Spray A case allows to validate the simulations and isolate single aspects in the atomization/ignition process. The conditions of the Spray A case are summarized in Table 1.

Beside the numerical problems of such simulations descending from the discontinuities at the interface, complex boundary conditions, and ill-conditioned systems of equations, the large range of relevant scales in an injection system is a major issue preventing the realization of simulations governing both nozzle internal flow

Table 1 Details of the Spray A case [3, 19, 25]

Injector configuration	Single hole
Outlet diameter	0.09 mm
Nozzle length	1.03 mm
Velocity coefficient	0.96
Fuel	n-dodecane
Fuel temperature at nozzle	363 K
Fuel injection pressure	150 MPa
Fuel density	703.82 Kgm^{-3}
Fuel dynamic viscosity	6.09×10^{-4} kg/ms
Ambient gas temperature	900 K
Ambient gas pressure	6 Mpa
Ambient gas dynamic viscosity	3.81×10^{-5} kg/ms
Ambient gas velocity	Near-quiescent
Surface tension	0.019 kgs^{-2}

and atomization/ignition process. Coupled simulations can reduce the computational cost by simultaneously performing simulations with different resolutions and considered physics. As successfully demonstrated by the authors [8, 10], the usage of a large-eddy simulation (LES) for the nozzle internal flow, a direct numerical simulation (DNS) for the primary breakup, and a Lagrangian particle based LES (LP-LES) describing the far-field spray evolution is a promising solution for the coupling. It is also employed for this work and the resulting simulation setup is sketched in Fig. 1. An overview of the the simulation features and the resulting scales for the Spray A case are given in Table 2. Due to the much higher spatial resolution required for the nozzle internal flow LES and the primary breakup DNS, the corresponding simula-

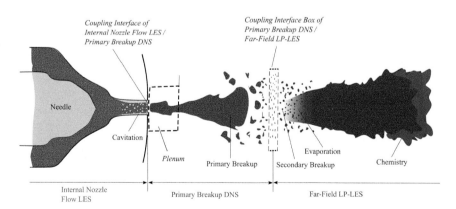

Fig. 1 Overview of the simulation setup

Table 2 Overview of simulation features and resulting scales for Spray A with Δx as highest grid resolution and Δt as resulting simulation time step size

Nozzle internal flow LES	Primary breakup DNS	Far-field LP-LES
Compressible	Incompressible	Compressible
	Interface captured	LP droplets
3 phases (liquid/vapor/gas)	2 phases (liquid/gas)	3 phases (liquid/vapor/gas)
$\Delta x = 5\,\mu\text{m}$	$\Delta x = 5\,\mu\text{m}$	$\Delta x = 5\,\mu\text{m}$
$\Delta t = 10\,\text{ns}$	$\Delta t = 1\,\text{ns}$	$\Delta t = 500\,\text{ns}$

tion cost are higher than that for the far-field LP-LES. Therefore, the nozzle internal flow LES and primary breakup DNS are run simultaneously by executing both solvers at the same time and accessing data at the coupling interface in the memory in order to significantly save computing time as described by Bode et al. [5]. In contrast, the LP-LES is run separately and coupled via data which were written by the primary breakup DNS to the file system. This does not only allow to run the LP-LES with fewer cores and therefore enough data per core to scale well but also enables cheap parameter variations with respect to the used chemistry model.

Another challenge while computing injection systems is the high number of species involved in the chemical processes during ignition and energy conversion. A direct consideration of all of these species in spray simulations is not possible as it would end up in more than trillions of degrees of freedom, which would exceed the limit of current supercomputers. The use of optimized, reduced chemical mechanisms, which significantly lowers the considered number of species, is one common way to overcome this issue but has a significant effect on the accuracy of the results as will be shown. Instead, the multiple representative interactive flamelet (MRIF) model, in which the chemistry is conditioned on the mixture fraction field and solved interactively to the flow, is mainly used in this work. Since the chemical reactions only need to be solved over multiple one-dimensional (1-D) grids and chemical time scales do not need to be resolved in the flow computation, the computational cost can be drastically reduced such that detailed chemical mechanisms can be applied. This is evaluated by considering two reduced mechanisms denoted as RM I/II and one detailed mechanism denoted as DM II in this work.

The described coupling and chemistry approaches are used to compute a consistent and predictive data set for the Spray A case, which includes shot-to-shot nozzle internal flow variations, the consideration of details chemical mechanisms, and soot. It will be shared with the ECN before the next ECN workshop and thus serve as basis for many analyses in the ECN for better understanding injection systems. Furthermore, the developed methods are general and predictive enough to apply them to arbitrary injector/fuel combinations and suggest alternatives for clean propulsion. This paper focuses on the computational aspects of the simulations, such as input/output (I/O) and vectorization, the effect of shot-to-shot nozzle internal flow variations, the impact of chemical mechanisms, and the predictability of soot.

2 Numerical Methods and Models

All simulations in this work were performed using the code framework CIAO, which has been developed during the last several years. CIAO solves the Navier-Stokes equations in compressible or low Mach formulation along with multi-physics effects. It is a structured, arbitrary order, finite difference code [15], which enables the coupling of multiple domains and their simultaneous computation. The compressible solver uses advanced equation of states [11] or tabulated fluid properties, a transport equation for internal or total energy, and a low-storage five-stage, explicit Runge-Kutta method for time integration [18, 28]. The momentum equations are discretized using central differences resulting in low numerical dissipation. All scalars are discretized by WENO schemes ensuring bounded solutions. Spatial staggering of flow variables is used in order to increase the accuracy of the numerical stencil. For LES, a dynamic Smagorinsky model [16] is used as a subfilter stress model employing weighted averaging backward in time along Lagrangian trajectories of fluid particles. A reduced Cartesian communicator (RCC) has been developed for reducing load imbalances of structured code in complex geometries and mesh blanking can be used for moving geometries. An overset mesh method is available for local mesh refinement and more flexible load balancing. Both, interface capturing/tracking methods based on volume of fluid (VOF) and level set [8] and Lagrangian particle (LP) methods can be used for studying sprays. In case of LP simulation, the Kelvin-Helmholtz/Rayleigh-Taylor model [24] can be applied for secondary breakup and evaporation can be modeled according to Miller et al. [21]. The source terms in the gas phase resulting from evaporating liquid droplets are distributed using Gaussian distribution kernels [4] reducing mesh dependence and increasing numerical stability. The MRIF model conditioning the chemistry on the mixture fraction field is employed for computing chemistry in sprays. Thus, time-histories of the scalar dissipation rate, which is used to model turbulence-chemistry interactions, and the species composition is captured more precisely compared to tabulated chemistry models. Additionally, all chemical time scales are resolved within the flamelet domain and hence, no special treatment of gas phase pollutants like CO or NO_x is required as it is required for steady flamelet models. Furthermore, the hybrid method of moments (HMOM) model [22] has been coupled to the MRIF model in order to compute the resulting soot formation. Recently, several studies of sprays were published that used the CIAO simulation framework [7–9, 11, 14].

3 Computational Methods

CIAO uses MPI for communication between cores and nodes and has been successfully run on SuperMUC (LRZ), JUQUEEN (JSC), and Mare Nostrum III (Barcelona) among others. As it was shown to scale on the full JUQUEEN (458,752 cores), it is member of the High-Q club [2, 6].

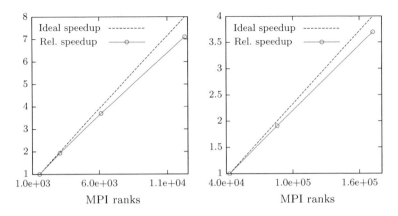

Fig. 2 Scaling of the coupled LES/DNS setup on Hazel Hen. Left: Domain with 900 millions cells and scaling up to 12,288 cores. Right: Domain size of 2560^3 cells and scaling up to full machine

The left plot in Fig. 2 shows a scaling plot for the coupled nozzle internal flow LES/primary breakup DNS with 900 million cells on Hazel Hen. Good scaling can be observed until 12,288 cores. Additionally, the right plot in Fig. 2 shows that the code is also able to efficiently use the full Hazel Hen machine as long as the problem size is large enough.

Besides scaling, node performance is also important for efficiently using super-computers. Due to the limited accessibility of hardware performance counters on Haswell-CPUs (such as Hazel Hen), the node performance of CIAO was analyzed using the Skylake-CPUs of Cray's SWAN cluster. Since both clusters use the same Cray environment, conclusions are also possible for CIAO's performance on Hazel Hen. The analysis was done using Cray's Perftools-lite and focused on peak performance, memory transfer rate, and cache and data-translation lookaside buffer (DTLB) utilization. In order to access particular hardware performance counters, the environment variable 'PAT_RT_PERFCTR' needs to be set to particular values and exported. An overview of used values for this study is given in Table 3. The information is written by the executable at the end of execution and can be viewed with the terminal command 'pat_report <ap2-files>'.

The results are summarized in Table 4. The achieved peak performance is 0.8 GFlop/s per core, which corresponds to 2% peak performance of the system.

Table 3 Overview of used 'PAT_RT_PERFCTR' values

Overall performance (e. g. peak performance)	export PAT_RT_PERFCTR=1
Cache information	export PAT_RT_PERFCTR=2
DTLB information	export PAT_RT_PERFCTR=dtlb_l3_misses
Memory performance	export PAT_RT_PERFCTR=mem_bw

Table 4 Overview of CIAO performance on SWAN using 448 MPI ranks and 512^3 cells (Cray's Perftools-lite notation used)

HW FP Ops / User time	0.804 G/s
Computational intensity	0.32 ops/cycle, 0.67 ops/ref
MFLOPS (aggregate)	360,809.94 M/s
Memory traffic GBytes	2.440 G/s
TLB utilization	4,667.55 refs/miss, 9.12 avg uses
D1 cache hit, miss ratios	97.7% hits
D1 cache utilization (misses)	43.70 refs/miss, 5.46 avg hits
D2 cache hit, miss ratio	95.2% hits
D1+D2 cache hit, miss ratio	99.9% hits
D1+D2 cache utilization	917.18 refs/miss, 114.65 avg hits
D2 to D1 bandwidth	0.719 GiB/s
L3 cache hit, miss ratio	7.3% hits
DTLB miss overhead cycles	7.3% cycles
L3 miss overhead cycles	9.1% cycles

The memory transfer rate is 2.44 GB/s, which is 1.9% of the memory bandwidth. As can be seen, one reason for this is the bad utilization of DTLB and L3 cache. While D1 and D2 caches have a combined hit/miss ratio of 99.9%, the L3 cache has a hit/miss ratio of only 7.3%, which, together with suboptimal DTLB usage, account for 16.4% overhead cycles.

Due to the findings with respect to cache utilization and peak performance, the following subsections focus on corresponding improvements mainly achieved on the "Optimization of Scaling and Node-level Performance on Hazel Hen (Cray XC40)" workshop. Furthermore, details on I/O issues and improvements are given.

3.1 Vectorization

One way to improve the peak performance is to better utilize the vector units of the cores. Perftools showed that a large amount of time is spent in the routines calculating the WENO5 coefficients (39.1%) and evaluating the WENO5 residuals (15.2%). Analyzing these routines using the detailed optimization reports (enabled with compilation flag '-qopt-report=5'), it was found that the compiler does not vectorize these routines because it assumes ANTI and FLOW dependencies between pointers. Since CIAO's multi-domain feature only works with pointers but not allocatables, changing the pointers to allocatables was not possible. Thus, OpenMP instructions ('!$OMP SIMD') were added as exemplarily shown in the code in Listing 1. Additionally, OpenMP was enabled by compiling the code with '-openmp'. In

the code example, the compiler assumed ANTI and FLOW dependencies between
sc_grad_p/sc_grad_c/sc_grad_m and weno5_xp/weno5_xm.

```
   do  k = sd%kmin_ -st1,sd%kmax_+st2
       do  j=sd%jmin_ -st1,sd%jmax_+st2
!$OMP SIMD
          do  i=sd%imin_ -st1,sd%imax_+st2
             !  Direction  x  -  U>0
             a0  =  1.0_WP/(epsilon+sc_grad_p(i-2,j,k))
             a1  =  6.0_WP/(epsilon+sc_grad_c(i-1,j,k))
             a2  =  3.0_WP/(epsilon+sc_grad_m(i,j,k))
             do  s  =  -3,1
                weno5_xp(s,i,j,k)  =  (a0*S0_xp_1d(s,i)  &
                +a1*S1_xp_1d(s,i)  &
                +a2*S2_xp_1d(s,i))/(a0+a1+a2)
             end  do

             !  Direction  x  -  U<0
             a0  =  1.0_WP/(epsilon+sc_grad_m(i+1,j,k))
             a1  =  6.0_WP/(epsilon+sc_grad_c(i,j,k))
             a2  =  3.0_WP/(epsilon+sc_grad_p(i-1,j,k))
             do  s  =  -2,2
                weno5_xm(s,i,j,k)  =  (a0*S0_xm_1d(s,i)  &
                +a1*S1_xm_1d(s,i)  &
                +a2*S2_xm_1d(s,i))/(a0+a1+a2)
             end  do
          end  do
       end  do
   end  do
```

Listing 12.1 Code example of a loop for computing WENO5 coefficients in one physical direction

Using the OpenMP instructions, CIAO was totally 19.2% faster for the coupled
LES/DNS case due to a distinct reduction of cycles. Interestingly, the DTLB miss
overhead cycles remained 7.3% cycles as the misses reduce and the number of cycles
also reduces, but the L3 miss overhead cycles increased to 11.1% cycles because the
increase in L3 cache misses outpace the reduction in overall cycles. The D1 cache
hit/miss ratios and the D2 cache hit/miss ratios slightly reduced to 96.1% and 93.9%,
respectively. The D1+D2 cache hit, miss ratio minimally reduced to 99.8%. Thus,
even though the additional time due to L3 cache misses and bad DTLB utilization
increases to 18.4%, the code runs 19.2% faster.

3.2   LP-LES Scaling

The LP-LES involves three computational domains for gaseous flow, liquid droplets,
and chemistry, which are advanced sequentially in time. In the flow solver, the
3-D numerical grid is split into Cartesian partitions and communication for gradient
evaluation is performed between neighbors using the MPI Cartesian communicator.
The liquid droplets are modeled as Lagrangian point particles, which are solved along

their trajectory and continuously interact with the gaseous phase. A droplet is solved by the MPI process that also owns the flow partition on which the droplet is physically located since local information is required for evaluating the source terms. Hence, communication is only needed if a particle moves across a flow domain partition border. However, the computational load is not equally distributed, if the droplets are not homogeneously distributed in physical space, which is typically the case during injection processes.

The chemistry is represented by multiple interactive flamelets i.e. the chemical reactions of individual gaseous species are solved over coordinates, which represent mixing of fuel and oxidizer at different fuel residence times. The flamelets only share information with the flow solver but not with each other and, hence, can be solved independently and simultaneously. The 1-D flamelet grid is decomposed to multiple MPI processes and neighbor-to-neighbor communication is applied in order to compute derivatives for evaluating diffusive transport.

Figure 3 illustrates the domain decomposition for a typical production job. The gaseous flow domain and liquid droplets are shown in physical space in the top, while the first and nth flamelet grids are exemplarily shown in the bottom. The colors indicate the share of individual MPI ranks.

In typical production runs, the number of MPI ranks is much larger than the grid size of a single flamelet. Hence, MPI sub-communicators are created for each flamelet such that the total number of ranks is equally distributed and all flamelets are solved simultaneously. For updating the flow solver domain, the flamelet solution of major species, e. g. chemistry products and pollutants, are broadcast via the global communicator. If the number of MPI processes do not exceed the amount of flamelet grid points, all processes are involved in computing flow and chemistry, while some processes have additional computational load due to solving the Lagrangian particles.

In order to eliminate uncertainties arising from the chemical mechanism reduction process, a detailed chemistry model has been targeted for the final simulations. Consequently, the number of reacting species increases by an order of magnitude. While LP-LES setups in CIAO have been investigated regarding computational performance before, it is the first time that such a detailed chemical mechanism is applied.

The timings and scaling behavior of CIAO and its most expensive modules are shown in Fig. 4 for the revised LP-LES setup. Overall, very good scaling is observed up to 150 nodes. However, it is interesting to note that the three different solvers show entirely different behavior. The scalar module, which is a part of the gaseous flow solver, scales well up to 50 nodes but shows only minor performance improvements beyond this point, since the cost of 3-D communication steadily increases with smaller partitions. The spray module including the Lagrangian particle solver also shows reasonable scaling up to 50 nodes. Beyond 50 nodes, however, no speed up can be observed. In fact, in the 75 and 150 node runs, the computational performance of this module decreases, which is caused by increasing load imbalance. The combustion model scales very well and shows the most significant speed up from 75 to 100 nodes, where it even exceeds ideal scaling. While the scaling of the spray and scalar modules meet the expectations from previous measurements, the super-ideal

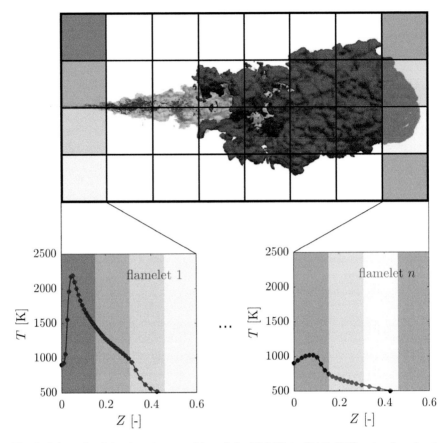

Fig. 3 Schematic of the data decomposition of the LP-LES in CIAO. Different colors denote individual MPI ranks

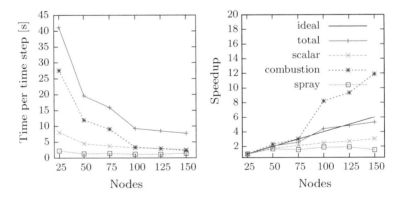

Fig. 4 Scaling of the LP-LES setup on Hazel Hen. Left: Time per time step. Right: Relative speed up

Table 5 Scaling and cache utilization of the chemistry solver on Hazel Hen

MPI ranks	Wall time (s)	Rel. speedup	L3 hit/miss (%)	D2 hit/miss (%)	D1 hit/miss (%)
24	6144	1.0	64.4	46.8	84.9
48	2098	2.9	67.6	47.0	88.2
72	1991	3.1	72.5	47.4	97.4
96	361	17.0	81.2	47.6	90.2

scaling of the combustion module was observed for the first time and thus will be analyzed in more detail.

In order to investigate the computational performance of the chemistry solver, a standalone test case computing an entire ignition event has been set up. The initial conditions are the same as in the production job and the number of grid points has been set to 96. Input variables that would be supplied by the flow solver in a production job have been held constant. The measured timings and relevant performance counters of the standalone chemistry test case are shown in Table 5. All cache levels, except D1 from 72 MPI ranks to 96 MPI ranks, show more effective cache utilization at higher number of MPI ranks, which is caused by the reduced data size that needs to be handled by each rank. Again, overall super-ideal scaling can be observed. However, from 48 MPI ranks to 72 MPI ranks, only minor performance benefits can be seen due to severe load imbalance, since the number of grid points cannot be equally distributed to all processes in the 72 MPI ranks case.

To summarize, the code shows overall good scaling up to 100 nodes for the targeted LP-LES case. While the flow and chemistry solver show typical scaling behavior, the chemistry solver shows super-ideal scaling due to more effective cache utilization at high numbers of MPI ranks. Thus, if the total number of flamelet grid points exceed the number of available processes, it might be more efficient to compute the flamelets partly sequentially instead of complete simultaneously in the LP-LES.

3.3 I/O

I/O can consume a significant amount of the computing time especially for unsteady simulations if multiple data files over time are necessary for the analysis. For I/O optimization, MPI-IO environmental variables were set in order to monitor the reading and writing. An overview of the used flags is given in Table 6.

Two different I/O strategies are implemented in CIAO and have been successfully used on many supercomputers. First, CIAO can use pure MPI-IO for I/O, where all cores write their data in parallel to the file system using 'MPI_FILE_WRITE_ALL'. With pure MPI-IO, writing rates of only 1–2 GB/s were observed even for optimal stripe sizes and the optimal number of object storage targets (OSTs) over which to

Table 6 Overview of used MPICH_MPIIO variables

export MPICH_MPIIO_AGGREGATOR_PLACEMENT_DISPLAY=1
export MPICH_MPIIO_HINTS_DISPLAY=1
export MPICH_MPIIO_TIMERS=1
export MPICH_MPIIO_STATS=1

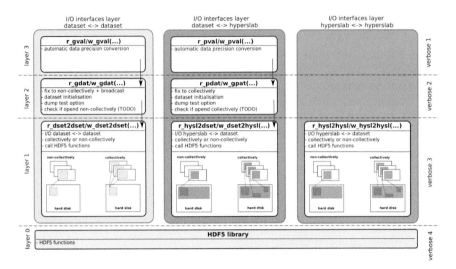

Fig. 5 Overview of parIO

stripe a file. Second, CIAO can use the external library parIO for writing HDF5 data files, which uses the HDF5 library internally. The wrapper library parIO was developed at the Institute for Combustion Technology to keep the I/O for the application simple and to achieve the best read/write performance on different supercomputers at the same time. It currently provides the ability to archive 1-D, 2-D, 3-D, and 4-D arrays of different types distributed over any number of MPI threads and an overview is sketched in Fig. 5. It has shown good read/write performances on various supercomputers and is available from the authors on request. Since I/O via parIO is faster on many supercomputers than pure MPI-IO, it was the focus during the workshop at HLRS. First runs also showed a writing rate of 1–2 GB/s for I/O via parIO. However, doing some small changes to the internal data alignment resulting in larger data chunks, using a striping size of 4 MB, and using as many OSTs as nodes, a performance of 8–12 GB/s was finally achieved on the ws8 file system of Hazel Hen for writing a 4-D array with 20×1024^3 cells via parIO. This number is compared with other writing rates for similar test cases on other clusters in Table 7. As can be seen, the writing rate on Hazel Hen is worst and might require even more tuning in the future. As the I/O of CIAO performs better on other clusters for similar cases, tuning on the hardware and the general file system settings might be desirable. However, due to different numbers of available OSTs, different sizes of the OSTs, different

Table 7 Write rates for a 4-D array with 20×1024^3 cells on various clusters with parIO

	SuperMUC (LRZ)	JUQUEEN (JSC)	CLAIX (ITC Aachen)	Hazel Hen/ws8 (HLRS)
File system	gpfs	gpfs	lustre	lustre
Write rate (GB/s)	45–65	20–30	10–15	8–12

sizes of single HDDs, and different file systems, it cannot be expected that equally good I/O can be achieved on all shown supercomputers in Table 7.

Besides the improved writing rates, two additional findings were made, which might be also of importance for other codes. First, it was observed that HDF5 encounters problems with pre-filling the allocated arrays with zeros with increasing number of cores. Whereas pre-filling with zeros generated an overhead in the order of microseconds for less than 50,000 MPI ranks, it increased for cases using more than 100,000 MPI ranks to the order of 100–1000 s per array. Here, the given numbers of used cores are only guidelines and vary between clusters. This problem can be avoided by using the internal HDF5 variable 'H5D_FILL_TIME_NEVER_F'. More precisely, pre-filling can be avoided by using

'call H5Pset_fill_time_f (pario_noncoll_dcreate_id, H5D_FILL_TIME_NEVER_F, ierr)'.

Second, an issue with respect to file closing was observed on Hazel Hen. While file closing generates almost no overhead on the other used clusters, it added an overhead in the order of 100 s to the write process on Hazel Hen (which was not included in the measurements for Table 7). This issue might be related to problems with 'MPI_File_SET_SIZE' on lustre as pointed out in a recent white paper of the HDF5 developers [12]. However, as written in the paper, a solution does not exist yet and collaboration between the cluster operator and the HDF5 group would be preferable.

4 Results and Discussion

For this work, the nozzle internal flow, primary breakup, and LP-LES were computed for three consecutive injections for the Spray A target case but with partly shorter injection duration in order to save computing cost. Furthermore, the nozzle internal flow results of the first injection process were used as boundary conditions to perform LP-LESs with three different chemical mechanisms. Details about the computational setups and cost are presented in Table 8.

Figure 6[1] shows the nozzle internal flow result including needle opening and closing as well as the result of one corresponding LP-LES. The different processes

---

[1]Videos may only be displayed correctly with Adobe Acrobat Reader in the electronic version of this work. All other readers are kindly referred to the given YouTube links.

Table 8 Performed simulations on Hazel Hen. LP-LES numbers correspond to a single injection realization

	Nozzle LES/primary breakup DNS	LP-LES (RM I/II)	LP-LES (DM II)
Number of grid cells (1e6)	2×900	80	80
Number of LP		400,000	400,000
Memory demand (GB)	7962	1422	1424
Number of nodes	256	60	60
Simulation time step size (ns)	1.6	25.0	25.0
Total simulation cost (core-h)	91,24,522[a]	196,236	223,182

[a]only partially run on Hazel Hen

Fig. 6 Visualization of coupled reactive spray simulation. Left (https://youtu.be/chC9ol8U8Rs): Velocity magnitude resulting from the internal nozzle flow LES (video speed increased). Right (https://youtu.be/2XJ4daFLq68): Liquid droplets (dark blue), stoichiometric mixture fraction (light blue), and iso-temperature-surface 1100 K obtained from a reactive LP-LES. Results of primary breakup DNS are not shown

such as evaporation and ignition can be clearly identified and the results yield good agreement with experimental data in terms of penetration, droplet size distribution, and mixing field. These validation results are not further discussed here for brevity. Instead, the effect of shot-to-shot variations, different chemical mechanisms, and the predictability of soot are presented in the next subsections.

4.1 Nozzle Initial Conditions

The framework and awarded computational time allow to perform multiple injections without resetting the conditions inside the nozzle. This enables the analysis of the effect of shot-to-shot variations in simulations for the first time and can be used to enforce more accurate initial conditions in the nozzle in the future even if only one shot is simulated. One important question is how much fuel vapor and oxidizer are present in the nozzle before start of injection. In state-of-the-art methods, these gases are usually neglected although experiments have shown that gases are formed and sucked into the nozzle during needle closing. This is confirmed by the results shown in Fig. 7. It can be seen that after needle closing, both oxidizer and vapor are present in the nozzle, which affect the next injection. Even though only three shots were computed, deviations of more than 10% in a least square sense were found for the spray characteristics such as interface formation in the early injection phase. This is significant and means that the error due to initial conditions compared to measurements is higher than usually assumed and cannot be neglected.

4.2 Chemistry Modeling

Liquid fuels consist of long-chain hydrocarbon molecules, which react with the oxidizer in extremely complex reaction networks towards the combustion products. During the energy conversion in a compression-ignition (CI) engine, the fuel first undergoes a complex ignition process that involves low-temperature and high-temperature chemistry with a broad range of time scales in a highly turbulent environment at varying pressure. The ignition timing alone drastically affects the entire chemistry history and thus pollutant formation. After ignition, a turbulent non-premixed flame establishes at a certain distance from the injector, the so-called flame lift-off length

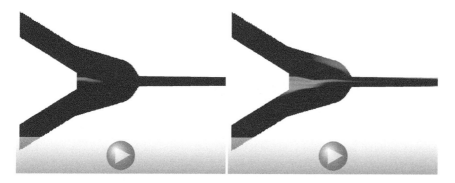

Fig. 7 Simulation results of internal nozzle flow LES. Left (https://youtu.be/-kim2ifpPRM): Vapor mass fraction. Right (https://youtu.be/d8XM5MNoBY0): Oxidizer mass fraction

Table 9 Summary of chemical mechanisms

	Species	Reactions	NO$_x$	PAH
RM I	54	269	No	No
RM II	71	225	Yes	No
DM II	822	4597	Yes	Yes

(FLOL). The pollutants and soot precursors form on longer time scales and are strongly affected by mixing, pressure, and local temperature. In order to capture all effects, the chemistry model requires extremely high accuracy in a broad operating range. Even though very accurate and physically motivated chemistry models exist, their application to full system simulations typically fails due to computational cost. In order to perform those simulations, detailed mechanisms are reduced in size keeping only species and reactions that are assumed to be important for the targeted operating range. In a first study, two reduced mechanisms designed and optimized for Spray A conditions were applied to the Spray A target case in order to quantify the effect of the chemical mechanisms.

The first mechanism by Yao [29], denoted as RM I, is a reduced mechanism that has been extensively used by many researchers within the ECN. However, it does not include NO$_x$ and PAH chemistry, which is required for predicting pollutant formation. The second mechanism, RM II, is a reduced version of the mechanism DM II, which is based on rate rules theory [13]. The mechanism size and pollutants chemistry is given in Table 9.

Figure 8 shows the comparison of the simulation results with experimental Schlieren data [27]. The Schlieren technique visualizes gradients of the refractive index, which correlates with density, in a line-of-sight view. In order to get a comparative quantity from the LES results, the absolute values of the density gradients have been integrated in line-of-sight direction. The color scale has been chosen arbitrarily such that acoustic waves are not visible. It is clearly visible in the experimental results that the Schlieren signal partly softens during the ignition process, e. g. at 347 μs almost the entire spray head vanishes. This phenomenon is caused by heat release arsing from low-temperature chemistry, which locally increases the temperature and thus decreases the density. When the jet density approaches the density of the surrounding gas, the gradients decrease and the Schlieren signal softens.

Overall, the jet structure of both simulations look similar to the experimental data. It is clearly visible that the Schlieren signal level deviates strongly between the two mechanisms. RM II shows stronger signal softening compared to RM I, which seems to be in better agreement with the experimental observation. At the same time, the Schlieren signal softening occurs to early with RM II.

In Fig. 9, the solution of the first flamelet is shown at different time values during the ignition phase in order to gain a more quantitative comparison of the results of the different mechanisms. In the top row, the temperature T is depicted over the mixture fraction coordinate Z. It can be seen that the results already differ significantly at the

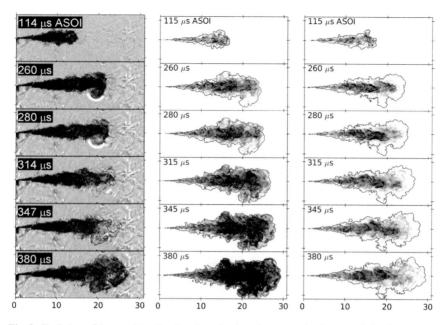

Fig. 8 Evalution of the considered reduced mechanisms by comparing integrated absolute density gradients in line of sight direction with experimental data. Left: Experimental Schlieren visualization [27]. Center: RM I. Right: RM II

early stages (190 μs). At 490 μs, both flamelets are ignited and show similar values at stochiometric mixtures. RM I shows a slightly broader high temperature range, which can be explained by its earlier ignition timing. The CO mass fraction Y_{CO} is shown in the second row. It can be seen that RM I has higher CO compared to RM II before ignition, which also explains the differences in temperature. In the third row, the formaldehyde (CH_2O) mass fraction Y_{CH_2O}, which can be seen as a tracer for low temperature chemistry, is shown. Clearly, both mechanisms differ by an order of magnitude prior to ignition. The last row shows the scalar dissipation rate χ, which is a measure for local mixing.

It can be concluded that even though the ignition delay times of both reduced mechanisms yield reasonable agreement with the experimental values, the ignition process itself shows significant differences in terms of temperature distribution and species composition. However, since no quantitative species or temperature measurements are available, it is unclear, which mechanism performs better. The detailed mechanism DM II, which has been developed based on rate rules and also includes NO_x as well as PAH chemistry for pollutant formation, was therefore applied in a more detailed simulation in order to understand the differences in the results of the simulations using the reduced, optimized mechanisms and achieve more trustful results.

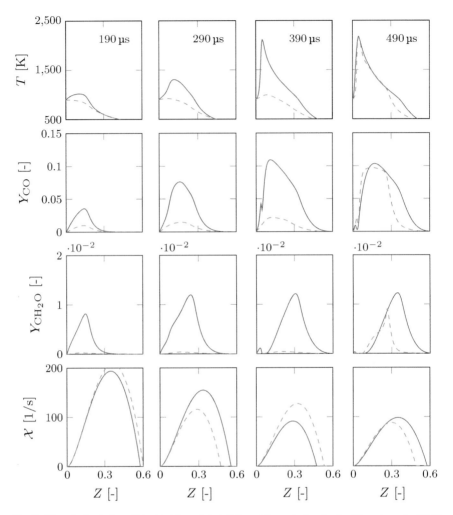

Fig. 9 Chemical solution of the first flamelet at different times during the ignition phase for RM I (solid red) and RM II (dashed green)

4.3 Emission Predictions

The detailed chemical model DM II has been applied to the Spray A case in order to achieve results which are more accurate even for predicting chemical species concentrations. The LES predicts the ignition delay, which has been defined as the time, when the OH mass fraction reaches 2% of its maximum value for the first time, with 390 μs quite accurately compared to the 410 μs, which has been observed in the experiments.

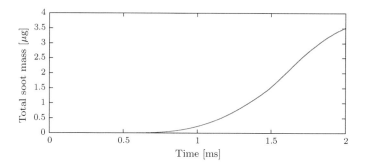

Fig. 10 Snapshot from the LP-LES at end of injection (EOI). Top: Liquid droplets (blue points), stoichiometric mixture (blue line), and soot volume fraction (colorscale) on a plane through the spray axis. Bottom: Liquid droplets (blue points), experimental flame lift-off length (averaged in time), and temperature iso-surface of 1800 K (red)

Figure 10 shows the instantaneous flow field at the end of injection. In the upper picture, the liquid droplets, an iso-line of stoichiometric mixture, and the soot volume fraction are shown. It can be seen that soot is formed in rich regions, which are inside of the mixture fraction iso-line. The highest soot volume fraction is observed in the spray head, where the residence times are the longest. In the lower picture of Fig. 10, a 3-D temperature iso-surface of the value 1800 K is shown that corresponds to the occurrence of high temperature chemistry. Additionally, the experimentally determined time-averaged FLOL is illustrated by the black arrow. The FLOL is the distance from the injector orifice to the location, where a stable flame establishes after ignition. It can be seen that the temperature iso-contour matches well with the experimentally observed value.

The temporal evolution of the total soot mass is given in Fig. 11. The onset timing, i. e. the timing where soot noticeably starts to form, agrees fairly well with the

Fig. 11 Total soot mass evolution

experimental value of 730 μs and the total soot mass is in the same order of magnitude as in the experiment (underestimated by a factor of 5), which is, considering that no tuning was performed for this result, a good agreement compared to other simulations available in literature.

In a next step, this detailed simulation will be used to identify important chemistry pathways which are not accurately considered in the reduced, optimized mechanisms in order to improve these.

5 Outlook

In the future, the model framework will be also applied to a multi-hole injector (Spray B), which involves more features of typical high-pressure injection systems, like cavitation. The effect of cavitation and hydraulic flip inside the injector on the pollutant formation will be numerically investigated. Therefore, cavitation and hydraulic flip models have been already improved. Figure 12 shows the comparison of simulation results with X-ray measurements [8] for a 3-hole injector with a shorter nozzle length, which was used as benchmark for the performance of the cavitation and hydraulic flip model. In the left image, where existing cavitation and hydraulic flip models were used, the penetrating gas is clearly on the opposite site compared to the experimental findings. Thus, the models were improved with respect to their boundary treatment and the consideration of 3-phase thermodynamic in order to achieve better results. As can be seen in the middle figure, the modified methods were able to predict the penetrating gas on the correct side and, thus, the analysis of multi-hole injectors is possible in the future.

Even though, only constant volume spray chambers have been computed in this study up to now, the developed methods are directly applicable to full in-cylinder flows due to detailed chemistry and the variable pressure combustion model. First

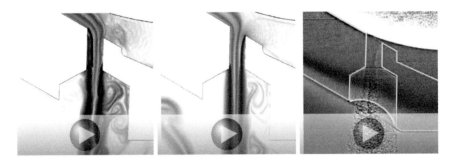

Fig. 12 Results of a shorter injector. Left (https://youtu.be/KHDVURSKY3M): Density gradients of a simulation using existing cavitation/hydraulic flip models. Center (https://youtu.be/MQKjlsv57ow): Density gradients of a simulation using improved cavitation/hydraulic flip models. Right (https://youtu.be/wkivdPoBn3c): X-ray measurements [8]

Fig. 13 Simulation of a single cylinder research engine operated with di-n-buthylether (bio fuel). Red color indicates CO formation. (https://youtu.be/UZJmhHks6Eg)

simulations of a single cylinder engine with multi-hole injector and bio fuels have been performed for a research engine that is located at the Institute for Combustion Engines (VKA) at RWTH Aachen University contributing to the Cluster of Excellence "Tailor-Made Fuels from Biomass" (TMFB) and are shown in Fig. 13. They are used to estimate the performance of newly developed bio or synthetic fuels. This is essential for developing clean propulsion since new fuels might be promising from a chemical point of view but not suitable for practical usage due to their spray characteristics. Thus, the simulations help identifying the best fuel candidates, can be used to adopt and optimize the injection process for promising fuels and thus contribute to clean propulsion based on synthetic fuels.

6 Conclusions

The aim of this work is to develop predictive and affordable methods towards clean propulsion with synthetic fuels. For that, the Spray A target case has been addressed as a first step and as reference for further studies with multi-hole injectors and new fuels. The presented approach is unique in the sense that it does not require any tuning, resolves the small scales during primary breakup, and can consider detailed mechanisms with hundreds of species. As briefly shown, the framework is also flexible enough to focus on more complex spray situations featuring moving pistons and synthetic fuels. Computational aspects have been discussed and improvements with respect to I/O, cache utilization, and vectorization were shown.

Several publications in peer-reviewed journals based on these data are in preparation focusing on different physical aspects:

- Impact of nozzle internal flow on shot-to-shot variations
- Impact of nozzle internal flow thermodynamics on time dependent cone angles
- Analysis of chemistry pathways during ignition and soot formation
- Generating physically more accurate reduced, optimized mechanisms.

Moreover, a joint spray analysis within the ECN is currently prepared. Due to the high quality of the computed data in this work, these data are used as basis of the analysis as well as boundary condition for further simulations and therefore also shared with other scientists in the ECN. Results of the joint investigations will be presented on the next ECN workshop in September 2018.

Acknowledgements The authors gratefully acknowledge support by Stefan Andersson (Cray), Björn Dick (HLRS, University of Stuttgart), Philipp Offenhäuser (HLRS, University of Stuttgart), Andreas Ruopp (HLRS, University of Stuttgart), and Jens Henrik Göbbert (JSC, FZ Jülich). Additionally, funding by the Cluster of Excellence "Tailor-made Fuels from Biomass" and computing time on the national supercomputer Cray XC40 at the HLRS under the grant number GCS-MRES are acknowledged. Data and support provided by Honda R&D and Argonne National Laboratory (Advanced Photon Source) are also kindly acknowledged.

References

1. Engine combustion network (2018). https://ecn.sandia.gov
2. High-Q Club (2018). http://www.fz-juelich.de/ias/jsc/EN/Expertise/High-Q-Club/_node.html
3. National institute of standards and technology (2018). https://www.nist.gov
4. S.V. Apte, K. Mahesh, T. Lundgren, Accounting for finite-size effects in simulations of disperse particle-laden flows. Int. J. Multiph. Flow **34**(3), 260–271 (2008)
5. M. Bode, M. Davidovic, H. Pitsch, Multi-scale coupling for predictive injector simulations, in *High-Performance Scientific Computing*, ed. by E. Di Napoli, M.A. Hermanns, H. Iliev, A. Lintermann, A. Peyser (Springer, 2017), pp. 96–108
6. M. Bode, A. Deshmukh, J.H. Göbbert, H. Pitsch, Ciao: multiphysics, multiscale Navier-Stokes solver for turbulent reacting flows in complex geometries, in *JUQUEEN Extreme Scaling Workshop 2016*, ed. by D. Brömmel, W. Frings, B.J.N. Wylie (Springer, 2016), pp. 15–24
7. M. Bode, F. Diewald, D.O. Broll, J.F. Heyse, V. Le Chenadec, H. Pitsch, Influence of the injector geometry on primary breakup in diesel injector systems. SAE Technical Paper 2014-01-1427 (2014)
8. M. Bode, T. Falkenstein, M. Davidovic, H. Pitsch, H. Taniguchi, K. Murayama, T. Arima, S. Moon, J. Wang, A. Arioka, Effects of cavitation and hydraulic flip in 3-hole GDI injectors. SAE Int. J. Fuels Lubr. **10**(2), 380–393 (2017)
9. M. Bode, T. Falkenstein, V. Le Chenadec, S. Kang, H. Pitsch, T. Arima, H. Taniguchi, A new Euler/Lagrange approach for multiphase simulations of a multi-hole GDI injector. SAE Technical Paper 2015-01-0949 (2015)
10. M. Bode, T. Falkenstein, H. Pitsch, T. Kimijima, H. Taniguchi, T. Arima, Numerical study of the impact of cavitation on the spray processes during gasoline direct injection, in *13th Triennial International Conference on Liquid Atomization and Spray Systems* (Tainan, Taiwan, 2015)
11. M. Bode, S. Satcunanathan, K. Maeda, T. Colonius, H. Pitsch, An equation of state tabulation approach for injectors with non-condensable gases: development and analysis, in *10th International Cavitation Symposium* (Baltimore, USA, 2018)
12. S. Breitenfeld, J. Mainzer, R. Warren, File open, close, and flush performance issues in hdf5. White Paper (2018)

13. L. Cai, H. Pitsch, S.Y. Mohamed, V. Raman, J. Bugler, H. Curran, S. Mani Sarathy, Optimized reaction mechanism rate rules for ignition of normal alkanes. Combust. Flame **173**, 468–482 (2016)
14. M. Davidovic, T. Falkenstein, M. Bode, L. Cai, S. Kang, J. Hinrichs, H. Pitsch, LES of n-dodecane spray combustion using a multiple representative interactive flamelets model. Oil Gas Sci. Technol. Rev. IFP Energ. Nouv. **72**(29) (2017)
15. O. Desjardins, G. Blanquart, G. Balarac, H. Pitsch, High order conservative finite difference scheme for variable density low mach number turbulent flows. J. Comput. Phys. **227**(15), 7125–7159 (2008)
16. M. Germano, U. Piomelli, P. Moin, W.H. Cabot, A dynamic subgrid-scale eddy viscosity model. Phys. Fluids A Fluid Dyn. **3**, 1760–1765 (1991)
17. M. Gorokhovski, M. Herrmann, Modeling primary atomization. Annu. Rev. Fluid Mech. **40**(1), 343–366 (2008)
18. F. Hu, M.Y. Hussaini, J.L. Manthey, Low-dissipation and low-dispersion Runge-Kutta schemes for computational acoustics. J. Comput. Phys. **124**(1), 177–191 (1996)
19. A.L. Kastengren, F.Z. Tilocco, C.F. Powell, J. Manin, L.M. Pickett, R. Payri, T. Bazyn, Engine combustion network (ECN): measurements of nozzle geometry and hydraulic behavior. At. Sprays **22**(12), 1011–1052 (2012)
20. P. Marmottant, E. Villermaux, On spray formation. J. Fluid Mech. **498**, 73–111 (2004)
21. R.S. Miller, K. Harstad, J. Bellan, Evaluation of equilibrium and non-equilibrium evaporation models for many-droplet gas-liquid flow simulations. Int. J. Multiph. Flow **24**(6), 1025–1055 (1998)
22. M.E. Mueller, G. Blanquart, H. Pitsch, Hybrid method of moments for modeling soot formation and growth. Combust. Flame **156**(6), 1143–1155 (2009)
23. A. Omari, B. Heuser, S. Pischinger, Potential of oxymethylenether-diesel blends for ultra-low emission engines. Fuel **209**, 232–237 (2017)
24. M.A. Patterson, R.D. Reitz, Modeling the effects of fuel spray characteristics on diesel engine combustion and emissions. SAE Technical Paper 980131 (1998)
25. L.M. Pickett, C.L. Genzale, G. Bruneaux, L.M. Malbec, L. Hermant, C. Christiansen, J. Schramm, Comparison of diesel spray combustion in different high-temperature, high-pressure facilities. SAE Int. J. Engines **3**(2), 156–181 (2010)
26. J. Shinjo, A. Umemura, Detailed simulation of primary atomization mechanisms in Diesel jet sprays (isolated identification of liquid jet tip effects). Proc. Combust. Inst. **33**(2), 2089–2097 (2011)
27. S.A. Skeen, J. Manin, L.M. Pickett, Simultaneous formaldehyde plif and high-speed schlieren imaging for ignition visualization in high-pressure spray flames. Proc. Combust. Inst. **35**(3), 3167–3174 (2015)
28. D. Stanescu, W.G. Habashi, 2N-storage low dissipation and dispersion Runge-Kutta schemes for computational acoustics. J. Comput. Phys. **143**(2), 674–681 (1998)
29. T. Yao, Y. Pei, B.J. Zhong, S. Som, T. Lu, K. Hong Luo, A compact skeletal mechanism for n-dodecane with optimized semi-global low-temperature chemistry for diesel engine simulations. Fuel **191**, 339–349 (2017)

Improved Vectorization for Efficient Chemistry Computations in OpenFOAM for Large Scale Combustion Simulations

Thorsten Zirwes, Feichi Zhang, Jordan A. Denev, Peter Habisreuther, Henning Bockhorn and Dimosthenis Trimis

Abstract The computation of chemical reaction rates is commonly the performance bottleneck in CFD simulations of turbulent combustion with detailed chemistry. Therefore, an optimization method is used where C++ source code is automatically generated for arbitrary reaction mechanisms. The generated code is highly optimized for the chosen mechanism and contains all routines for computing chemical reaction rates. In this work, the serial performance of the automatically generated source code, which in an earlier work only used ISO C++, is further improved by utilizing two compiler extensions: `restrict` and `__builtin_assume_aligned`. Introducing these two extensions to the generated code reduces the time for computing chemical reaction rates by up to 50% for the investigated reaction mechanism and total simulation time by up to 25%. Compared to OpenFOAM's standard chemistry implementation, the new code is faster by a factor of 10. This work discusses the effect of the two compiler extensions on performance by looking at two specific kernel functions from the automatically generated code and the effect on the assembly generated by the gcc and Intel compilers. The newly optimized code is used to evaluate the performance gain in a large scale parallel case, which simulates an experimentally investigated turbulent flame of laboratory scale on 14,400 CPU cores on the Hazel Hen cluster at HLRS. In the simulation, no combustion models are used and the flame is resolved down to the smallest length scales. With this approach, comparison of measured data with the simulation shows very good agreement. Using the optimized code including compiler extensions, total simulation time decreases by 20% compared to the same code without compiler extensions. A comprehensive database from the simulation results has been assembled and will consist of 10 TB of 3D and 2D transient field variables.

T. Zirwes (✉) · J. A. Denev
Steinbuch Centre for Computing, Karlsruhe Institute of Technology,
Hermann-von-Helmholtz-Platz 1, Karlsruhe, Germany
e-mail: thorsten.zirwes@kit.edu

F. Zhang · P. Habisreuther · H. Bockhorn · D. Trimis
Engler-Bunte-Institute, Karlsruhe Institute of Technology,
Engler-Bunte-Ring 7, Karlsruhe, GermanyF. Zhang
e-mail: feichi.zhang@kit.edu

© Springer Nature Switzerland AG 2019
W. E. Nagel et al. (eds.), *High Performance Computing in Science and Engineering '18*, https://doi.org/10.1007/978-3-030-13325-2_13

1 Introduction

In this work, OpenFOAM [1] is used together with the thermo-physics library
Cantera [2] to simulate turbulent combustion. Using detailed chemical reaction
mechanisms is computationally very expensive, so a good code performance is
mandatory. Because of this, a highly efficient solver for direct numerical simulation (DNS) of turbulent combustion has been developed which includes an optimization method for speeding up the computation of chemical reaction rates [3, 4]. The
method is based on automated code generation and is described in Sect. 3. In this
follow-on work, further performance improvements of the serial performance are
achieved by extending the automatically generated code with additional information
about data alignment and pointer aliasing through compiler extensions. This yields
more efficient code and enables auto-vectorization of loops that would otherwise not
be vectorized by the compiler.

The new code is used to perform a highly-resolved large scale simulation of a
turbulent flame in order to evaluate the performance of the new code. The simulated
burner has been developed to study the effect of inhomogeneous mixing conditions
of fuel and oxidizer on the flame [5]. Such conditions are widely applied in practical
combustion devices, but are difficult to model due to the existence of both premixed
and non-premixed combustion modes. The results from the detailed simulation are
used to create a comprehensive database of the burner. The database can be used
to gain an in-depth understanding of the underlying mechanism of flame-regime
transition and consists of about 10 TB of data including a large number of temporally-
spatially resolved quantities.

Section 2 gives a short overview of previous performance optimizations. Since
parallel performance has been presented in previous works, this work focuses on
optimizing serial performance by including two compiler extensions (`restrict`
and `__builtin_assume_aligned`) in the code base. The benefits in terms of
auto-vectorization from the extensions are discussed. Two kernel functions from
the actual code base are used as an example to illustrate the effect of the compiler
extensions on the machine code that is generated by the gcc and Intel compilers
on the ForHLR II cluster at KIT and the Hazel Hen cluster at HLRS [6]. Section 3
describes the numerical setup of the large scale simulation and presents the results
as well as validation with experimental data.

2 Serial Code Optimization

2.1 General Optimization Concept from Previous Works

In order to achieve accurate results for the flame-turbulence interaction and concentrations of intermediate species, detailed reaction mechanisms have to be employed.
But the computation of chemical reaction rates from these detailed mechanisms is

computationally expensive and is often the performance bottleneck. In order to set up a new combustion simulation in OpenFOAM, the user has to provide the information about which chemical species should be considered and which chemical reactions can occur. These information are generally provided in a reaction mechanism as a text file, which contains the parameters to describe the properties of the chemical species and reactions. This text file serves as input for the CFD solver, which computes the chemical reaction rates based on the information in the reaction mechanism text file during the simulation.

In a previous work [3], an optimization technique was introduced which is based on automatic code generation and is also used in other fields of research [7, 8]. Before a new simulation is set up, the user can use a self-developed converter tool, which takes a specific reaction mechanism file as input and converts it into C++ source code. The GRI 3.0 [9] reaction mechanism is used as an example for the following discussion. The mechanism contains 53 chemical species and 325 chemical reactions. The user executes the converter tool with the GRI 3.0 mechanism in xml format as input. The converter tool generates a new directory called optimizedChemistry_gri30 which contains the source code for the chemical reaction rates. When the user compiles the automatically generated code, a new shared library is created which can directly be used in the simulation due to OpenFOAM's runtime selection mechanism. This method has the advantage that the chemical reaction rates are not computed from general routines, but from routines that are optimized for one specific reaction mechanism. All parameters from the reaction mechanism, like the number of chemical species, are compile-time constants, allowing for better optimizations by the compiler. Additionally, the code is generated in a way that makes it easier for the compiler to generate auto-vectorized loops by changing the order of species and reactions from the reaction mechanism and grouping them by their respective types.

2.2 Compiler Extensions for Improved Vectorization

In this work, two additions to the automatically generated chemistry code are discussed. Up until now, the automatically generated code contained only ISO C++ code. In this work, compiler extensions are used to give the compiler more information about data alignment and pointer aliasing which results in more efficient code. Parallel performance of the code has been shown to scale linearly up to 10,000 CPU cores in a previous work [3]. Therefore, the optimization focus in this work lies on serial performance. This section discusses the benefits of using two compiler extensions: providing alignment information via __builtin_assume_aligned and the restrict keyword. The same information can be given to the compiler also through OpenMP pragmas, but in this work the discussion is limited to compiler extensions. Two functions from the automatically generated chemistry code for the GRI 3.0 mechanism, which previously did not use any compiler extensions, are used to illustrate the effects of using the two compiler extensions.

First, the use of the `restrict` keyword is discussed with the help of the auto-matically generated routine that computes the Gibbs energy G_k° of the species for the GRI 3.0 mechanism, which is needed to compute the equilibrium constants of each chemical reaction (Sect. 2.2.1). In C/C++, different pointers that are passed as func-tion arguments are allowed to point to the same memory location. This is in contrast to programming languages like FORTRAN, where different function arguments are implicitly not allowed to point to the same location in memory. The `restrict` keyword can be used to annotate a pointer and to tell the compiler that the memory a pointer points to is only accessed by this pointer. The `restrict` keyword is part of the ISO C standard since C11 but has not been included in the recent C++ standards. Therefore, compilers offer the same functionally as compiler extension. The exact implementation differs between compilers. In the automatically generated chemistry code, the C preprocessor is used to find out which compiler is used and to define the `restrict` keyword accordingly:

```
#define RP __restrict__
```

In this example, it is assumed for the sake of brevity that the compiler is either gcc or the Intel compiler icpc. In the real code, more compilers are considered. A new keyword RP is defined, which is set to __restrict__. Alternatively, the key-word `restrict` can be used for the icpc if the code is compiled with the flag `-restrict`.

2.2.1 Computation of Thermo-chemical Properties

In order to compute the reaction rates of the reverse reactions, the equilibrium con-stant K_r has to be computed. This requires the computation of the Gibbs energy G°:

$$\frac{G_k^\circ}{\mathscr{R}T} = \frac{\tilde{h}_k}{\mathscr{R}T} - \frac{\tilde{s}_k}{\mathscr{R}} \tag{1}$$

$$\frac{\tilde{h}_k}{\mathscr{R}T} = a_{k,0} + \frac{a_{k,1}}{2}T + \frac{a_{k,2}}{3}T^2 + \frac{a_{k,3}}{4}T^3 + \frac{a_{k,4}}{5}T^4 + \frac{a_{k,5}}{T} \tag{2}$$

$$\frac{\tilde{s}_k}{\mathscr{R}} = a_{k,0} \ln T + a_{k,1}T + \frac{a_{k,2}}{2}T^2 + \frac{a_{k,3}}{3}T^3 + \frac{a_{k,4}}{4}T^4 + a_{k,6} \tag{3}$$

$$\frac{\tilde{c}_{p,k}}{\mathscr{R}} = a_{k,0} + a_{k,1}T + a_{k,2}T^2 + a_{k,3}T^3 + a_{k,4}T^4 \tag{4}$$

The index k denotes the kth species, T is the temperature, $\mathscr{R}$ the universal gas constant, $\tilde{h}_k$ and $\tilde{s}_k$ the molar enthalpy and entropy of the species and $\tilde{c}_{p,k}$ is the molar heat capacity of the species. The $a_{k,0}$ to $a_{k,6}$ are coefficients for the NASA polynomials [10] above.

Figure 1 shows the automatically generated source code for the GRI 3.0 mecha-nism, which computes the species Gibbs enthalpies and the heat capacities from the equations above. The code in Fig. 1 starts with the definition of the function, where

```
1  void compute_thermo_gri30(double T,   double* RP G_RT,
2       double* RP cp_R, double* RP a0, double* RP a1,
3       double* RP a2,   double* RP a3, double* RP a4,
4       double* RP a5,   double* RP a6) noexcept
5  {
6  /*
7  G_RT = (double*)(__builtin_assume_aligned(G_RT, 64));
8  cp_R = (double*)(__builtin_assume_aligned(cp_R, 64));
9  a0 =  (double*)(__builtin_assume_aligned(a0,  64));
10 a1 =  (double*)(__builtin_assume_aligned(a1,  64));
11 a2 =  (double*)(__builtin_assume_aligned(a2,  64));
12 a3 =  (double*)(__builtin_assume_aligned(a3,  64));
13 a4 =  (double*)(__builtin_assume_aligned(a4,  64));
14 a5 =  (double*)(__builtin_assume_aligned(a5,  64));
15 a6 =  (double*)(__builtin_assume_aligned(a6,  64));
16 */
17
18 double T1 = T,  T2 = T*T;
19 double T3 = T1*T2, T4 = T2*T2;
20 double invT = 1./T;
21 double logT = std::log(T);
22
23 for (unsigned int i=0; i<50; ++i)
24 {
25     double ct0 = a0[i];    // a0
26     double ct1 = a1[i]*T;  // a1 * T
27     double ct2 = a2[i]*T2; // a2 * T^2
28     double ct3 = a3[i]*T3; // a3 * T^3
29     double ct4 = a4[i]*T4; // a4 * T^4
30
31     cp_R[i] = ct0 + ct1 + ct2 + ct3 + ct4;
32
33     double h_RT = ct0 + 0.5*ct1 + (1.0/3.0)*ct2 + 0.25*ct3
34                   + 0.2*ct4+ a5[i]*invT;
35     double s_R = ct0*logT + ct1 + 0.5*ct2 + (1.0/3.0)*ct3
36                  + 0.25*ct4 + a6[i];
37     G_RT[i] = h_RT - s_R;
38 }
39 }
```

Fig. 1 Automatically generated code for computing the Gibbs energy and heat capacity of each species from the GRI 3.0 mechanism as shown in Eqs. (1)–(4)

all pointers are annotated with RP. The loop starting in line 23 iterates over the first 50 species of the reaction mechanism. Because this is part of the automatically generated code, the number of species appears as a compile-time constant in the loop, making it easier for the compiler to make decisions about vectorization and loop unrolling. Using RP for the function arguments enables gcc and icpc to generate a vectorized version of the loop. The generated assembly is shown in Fig. 2. The first 48 elements are computed in a vectorized loop, which can be seen from the instructions ending

	gcc 7.2: -Ofast -march=native	icpc 18.0: -fast -restrict

```
 1  .L2:                                             ..B1.2:
 2  vmulpd       ymm3, ymm12, [r15+rax]            movzx        eax,   dl
 3  vmovapd      ymm4, [rcx+rax]                   add          dl,    4
 4  vmulpd       ymm6, ymm11, [r8+rax]             vmovupd      ymm2,  [r13+rax*8]
 5  vmulpd       ymm5, ymm15, [r14+rax]            vmovupd      ymm3,  [r8+rax*8]
 6  vmulpd       ymm2, ymm14, [r13+0+rax]          vmulpd       ymm15, ymm10, [rdi+rax*8]
 7  vaddpd       ymm7, ymm3,  ymm6                 vmulpd       ymm14, ymm12, [rbx+rax*8]
 8  vaddpd       ymm1, ymm2,  ymm5                 vmulpd       ymm1,  ymm13, [r12+rax*8]
 9  vaddpd       ymm1, ymm1,  ymm7                 vmulpd       ymm4,  ymm11, ymm3
10  vmovapd      ymm7, [r12+rax]                   vfmadd213pd  ymm3,  ymm11, ymm2
11  vfmsub213pd  ymm7, ymm13, [rbx+rax]            vfmadd231pd  ymm3,  ymm12, [rbx+rax*8]
12  vaddpd       ymm1, ymm1,  ymm4                 vfmadd231pd  ymm3,  ymm13, [r12+rax*8]
13  vmovapd      [rsi+rax], ymm1                   vaddpd       ymm3,  ymm15, ymm3
14  vmulpd       ymm1, ymm3,  ymm10                vmovupd      [r14+rax*8], ymm3
15  vfmadd231pd  ymm1, ymm6,  ymm9                 vmovupd      ymm3,  [rcx+rax*8]
16  vfmadd231pd  ymm6, ymm0,  ymm4                 vfmadd231pd  ymm3,  ymm8,  ymm14
17  vaddpd       ymm1, ymm7,  ymm1                 vfnmsub231pd ymm3,  ymm0,  ymm2
18  vmulpd       ymm7, ymm2,  .LC3[rip]            vfnmadd231pd ymm2,  ymm6,  ymm15
19  vfnmadd132pd ymm2, ymm4,  ymm8                 vmulpd       ymm15, ymm4,  ymm5
20  vfmadd231pd  ymm7, ymm5,  ymm8                 vfmadd213pd  ymm14, ymm7,  ymm15
21  vmulpd       ymm5, ymm5,  ymm10                vmulpd       ymm15, ymm4,  ymm6
22  vaddpd       ymm1, ymm1,  ymm7                 vfnmsub213pd ymm4,  ymm7,  ymm1
23  vfmadd132pd  ymm3, ymm5,  ymm9                 vfmadd231pd  ymm15, ymm8,  ymm1
24  vsubpd       ymm1, ymm1,  ymm6                 vaddpd       ymm1,  ymm2,  ymm4
25  vsubpd       ymm1, ymm1,  ymm3                 vaddpd       ymm14, ymm15, ymm14
26  vaddpd       ymm1, ymm1,  ymm2                 vaddpd       ymm2,  ymm2,  ymm3
27  vmovapd      [rdi+rax], ymm1                   vaddpd       ymm4,  ymm2,  ymm14
28  add          rax,  32                          vfmadd231pd  ymm4,  ymm9,  [rsi+rax*8]
29  cmp          rax,  384                         vmovupd      [r15+rax*8], ymm4
30  jne          .L2                               cmp          dl,    48
31  noop                                           jb           ..B1.2
```

Fig. 2 Assembler output for the loop over the first 48 elements in Fig. 1 using the `restrict` keyword and alignment information compiled on Hazel Hen

in `pd`, which stands for "packed double". In this case it means the instructions act on four double precision variables at a time. Also, the loop counter in the gcc assembly is increased by 32 bytes in line 28, which corresponds to four double values, and by four elements in line 3 of the icpc output, so that every iteration of the loop acts on four elements simultaneously.

The compiler explorer by Godbolt [11] offers a convenient way to check the assembly output of code depending on the compiler, compiler version and optimization flags. An interactive example of the discussed case can be found in footnote[1] where the output of gcc and icpc are compared. If RP is omitted, then the loop will not be vectorized because the compiler does not know that all pointers point to different memory locations and carry no dependencies across iterations (see footnote[2]).

The second compiler extension that can lead to more efficient code is providing information about data alignment. If the comment around lines 7–15 of Fig. 1 is removed, the compiler is given the information, that each pointer points to a double element, which starts on an address in memory that is divisible by 64. In the case of the investigated loop in Fig. 2, four double values are accessed together as a single

---

[1]https://godbolt.org/g/sUKL2j: vectorized loop due to use of `restrict`.

[2]https://godbolt.org/g/CWobvD: not vectorized without `restrict`.

```
1  void compute_Kc_gri30(double* k_rev, const double* G,
2  double T) noexcept
3  {
4      // G = G/RT
5      double logP_RT = std::log(1.0132e+05/(8.3144621e+03*T));
6      k_rev[0]   = std::exp(-2.0   * G[2] + G[3]  + logP_RT);
7      k_rev[1]   = std::exp(-G[1]  - G[2] + G[4]  + logP_RT);
8      // skip ...
9      k_rev[307] = std::exp( G[3]  - G[6] - G[46] + G[47]);
10     k_rev[308] = std::exp(-G[12] + 2.0 * G[25] - G[46]);
11 }
```

Fig. 3 Automatically generated code from the previous work [3] without compiler extensions for computing the inverse equilibrium constant for all reversible reactions (depicted are 4 out of the 309 reactions) for the GRI 3.0 mechanism

vector unit. Compilers can produce the most efficient code if the data is aligned by the size of the vector unit, which means that the first element starts on an address in memory which is at least divisible by the size of the vector unit. Removing the comments in Fig. 1 leads to more efficient code by gcc, while the icpc output stays nearly the same (see footnote[3]). In case of gcc, using the alignment information reduces the assembly code size by a factor of three, because the compiler does not need to generate checks to test the data alignment at runtime. The effect can be seen for example in line 3 of the gcc output in Fig. 2, where the instruction vmovapd is used to load aligned data. The icpc always generates unaligned loads (vmovupd) and is therefore less effected by additional alignment information. This method requires that the data has been allocated with the correct alignment. Otherwise the CPU will throw an exception if unaligned data is accessed by aligned loads.

2.2.2 Computation of Equilibrium Constants

Another example from the automatically generated code is given by the routine that computes the equilibrium constant K_r. It is computed from

$$K_r = \exp\left(-\frac{\sum_k (v''_{k,r} - v'_{k,r})G^\circ_k}{\mathscr{R}T}\right)\left(\frac{p^\circ}{\mathscr{R}T}\right)^{\sum_k (v''_{k,r} - v'_{k,r})} \tag{5}$$

where G°_k is the Gibbs energy, p° a reference pressure and $v'_{k,r}$ and $v''_{k,r}$ the stoichiometric coefficients of the forward and reverse reactions r. Because typically only two to three species take part in a single reaction, most of the stoichiometric coefficients $v_{k,r}$ are zero. In order to utilize this sparsity, Eq. (5) is generated explicitly for each individual reaction of the reaction mechanism. Figure 3 shows the automatically generated source code of the routine computing the chemical equilibrium constants

---

[3]https://godbolt.org/g/95smzt: smaller code size for gcc when alignment info is given.

```
 1  void compute_Kc_gri30(double* RP k_rev, const double* RP G,
 2                        double T) noexcept
 3  {
 4      k_rev = (double*)__builtin_assume_aligned(k_rev, 64);
 5      G =     (const double*)__builtin_assume_aligned(G,  64);
 6      // G = G/RT
 7      double logP_RT = std::log(1.0132e+05/(8.3144621e+03*T));
 8
 9      k_rev[0]    = -2.0    * G[2] + G[3]   + logP_RT;
10      k_rev[1]    = -G[1]   - G[2] + G[4]   + logP_RT;
11      // skip ...
12      k_rev[307] =  G[3]    - G[6] - G[46] + G[47];
13      k_rev[308] = -G[12]  + 2.0   * G[25] - G[46];
14
15      for (unsigned int i = 0; i!=309; ++i)
16          k_rev[i] = std::exp(k_rev[i]);
17  }
```

Fig. 4 Automatically generated code from this work for the inverse equilibrium constants for all reversible reactions (depicted are 4 out of the 309 reactions) for the GRI 3.0 mechanism

from the previous work [3]. Line 6 of the code corresponds to the equilibrium constant of the first reaction appearing in the GRI 3.0 mechanism $2O \rightleftharpoons O2$:

$$\frac{1}{K_0} = \exp\left(\frac{-2G_2^\circ + G_3^\circ}{\mathscr{R}T}\right)\left(\frac{p^\circ}{\mathscr{R}T}\right)^1 \tag{6}$$

In this work, the routine from Fig. 3 is optimized as shown in Fig. 4. Alignment information is added for the pointers k_rev and G in line 4 and 5, and the restrict keyword is used for the function arguments. Additionally, the computation of the exponential function is done at the end of the routine at line 15. Thereby, the function is split into two parts: the code from line 9 to line 13 cannot be vectorized while the loop at the end can be vectorized. The resulting machine code for the last loop is shown in Fig. 5. Both Intel and gcc generate almost the same code. The loop has been vectorized, as highlighted by the calls to the exponential function from the short vector math library shown in red. The first 308 reactions are computed in the vectorized loop and the 309th element is computed separately. gcc computes the last element with a serial version of exp while icpc uses the vectorized version again. Also, icpc uses unaligned moves while gcc uses the aligned versions.

2.3   Serial Performance Measurements

In this section, two performance measurements are presented: First, the performance of only a single routine from Fig. 4, which computes the equilibrium constants. And

gcc 7.2: -Ofast -march=native

icpc 18.0: -fast -restrict

```
1  .L2:
2      vmovapd ymm0, YMMWORD PTR [r12]
3      add     r12, 32
4      call    _ZGVdN4v___exp_finite
5      vmovapd YMMWORD PTR [r12-32], ymm0
6      cmp     r13, r12
7      jne     .L2
8  vmovsd  xmm0, QWORD PTR [rbx+2464]
9  vzeroupper
10 call    __exp_finite
11 vmovsd  QWORD PTR [rbx+2464], xmm0
```

```
..B1.2:
    mov     r12d, esi
    vmovupd ymm0, YMMWORD PTR [rdi+r12*8]
    call    __svml_exp4_19
    add     esi, 4
    vmovupd YMMWORD PTR [rdi+r12*8], ymm0
    cmp     esi, 308
    jb      ..B1.2
vbroadcastsd ymm0, QWORD PTR [2464+rdi]
call __svml_exp4_19
vmovsd QWORD PTR [2464+rdi], xmm0
```

Fig. 5 Assembler output of gcc and icpc for the code from Fig. 4 compiled on Hazel Hen

second, a simulation with OpenFOAM which uses all of the automatically generated code for computing the reaction rates, including the two functions shown above. All measurements have been performed on a single core on the Hazel Hen and ForHLR II cluster with gcc 7.3 (ForHLR II), gcc 7.2 (Hazel Hen) and icpc 18.0 using the flags -Ofast -march=native for gcc and -fast -restrict for icpc. The target architectures are Intel Xeon E5-2680 v3 on Hazel Hen and Intel Xeon E5-2660 v3 on ForHLR II. The following codes are compared:

- "gcc" and "icpc": The automatically generated code from the previous work [3] without compiler extensions
- "gcc ext" and "icpc ext": The exact same code as "gcc" and "icpc", with the only difference that the compiler extensions restrict and alignment information are used and the code from Fig. 4 instead of Fig. 3 with the vectorizable loop at the end is used.

The performance measurements have been performed with the perf tool from the Linux performance tools. The following measurements have been done:

- CPU time
- number of instructions for fixed problem size
- instructions per cycle
- GFLOPs measured from $4 \cdot r04C6$/time [12].

Figure 6 shows the measurements for only the function that computes the equilibrium constants for the GRI 3.0 mechanism with the code from this work (Fig. 4) and the previous work (Fig. 3). The function is called 10 million times with randomized input. In Fig. 6a, the total computing time is compared between the different compiler versions and clusters. For the code that uses only ISO C++ without compiler extensions ("gcc" and "icpc"), measurements on Hazel Hen and ForHLR II show similar results, with the Intel compiler creating code that is about twice as fast. Also, code on Hazel Hen runs about 10% faster compared with ForHLR II (red vs blue bars). By using the new code from this work ("gcc ext" and "intel ext") including restrict and alignment information, the code from gcc gets a 2.5 speedup on Hazel Hen (red bars "gcc ext" vs. "gcc") while the code by icpc gets about 40% faster ("icpc ext" vs. "icpc" in Fig. 6a). Therefore, using restrict and alignment information also helps to

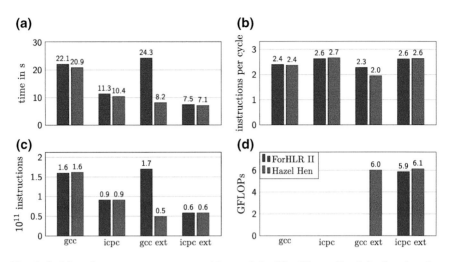

Fig. 6 Serial performance measurements with `perf` for 10 million calls of the function that computes the Gibbs energy for the chemical equilibrium constants from this work (Fig. 4) and the previous work (Fig. 3)

close the performance gap between gcc and icpc, but the Intel compiler still performs about 10% faster than gcc. The only outlier in the CPU times is the code generated by gcc on ForHLR II. The reason is, that ForHLR II uses an old version of the `glibc` library, which does not provide the library `libmvec` containing vectorized versions of transcendental functions like `exp`. Therefore, no loops containing these functions can be vectorized by gcc on ForHLR II.

Figure 6b shows the number of instructions per cycle, which lies between 2.0 and 2.7. The number of instruction required to compute 10 million function calls in Fig. 6c shows that the additional alignment information can decrease the number of alignment checks generated by the compilers, especially gcc, as previously discussed, and more vectorized code also leads to fewer instructions. In Fig. 6d, the performance in terms of GFLOPs is measured, which reaches about 6, compared to a theoretical maximum of 40 GFLOPs. The GFLOPs are only shown for the cases which could be properly vectorized.

The second performance measurement in Fig. 7 has been conducted for a full OpenFOAM simulation on one core in order to provide more realistic performance measurements than the first measurement, which only considered an isolated function from the chemistry code. The test case is a mesh consisting of 1 million cells and the time it takes to compute the chemical reaction rates for all cells is measured. OpenFOAM uses the generated code from this work for the chemistry computations, which includes not only the routines from Figs. 1 and 4, but other functions, too. The results for the CPU times in Fig. 7 show similar trends as the previous one: Code on Hazel Hen performs slightly faster than on ForHLR II and the Intel compiler produces code that is faster than the code by gcc. Using the compiler extensions in the automatically generated code leads to performance gains of about 40–50%.

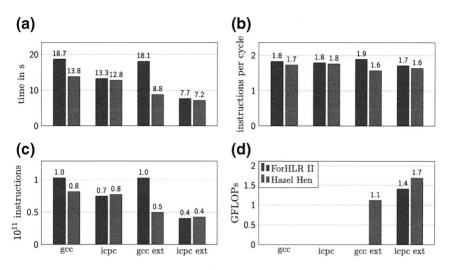

Fig. 7 Serial performance measurements with `perf` for computing the chemical reaction rates with OpenFOAM on a grid with 1 million cells from the GRI 3.0 mechanism

The instructions per cycle are between 1.6 and 1.9 as the code becomes increasingly more memory-bound and peak performance reaches between 1 and 1.7 GFLOPs. In summary, both measurements have shown that the computing time can be decreased by up to 50% by just adding the `restrict` keyword and alignment information to the code. Since the chemistry part typically takes about 50% of the simulation time [4], 25% reduction of total simulation time can be expected with the new code. These savings increase when larger reaction mechanisms are employed [3]. If the same case is run with OpenFOAM's standard solver `reactingFoam` using Open-FOAM's general chemistry implementation yields simulation times that are 10 times slower compared to the new code from this work.

3 Simulation of an Inhomogeneous Turbulent Burner

3.1 Methodology and Numerical Setup

The last performance evaluation for the newly optimized code is done by running a large scale parallel combustion simulation on Hazel Hen. The setup of the simulation is given by a turbulent mixed-mode combustion burner from literature (FJ200-5GP-Lr75-57 [5]) which has been experimentally investigated. The computational domain consists of 150 million cells. Figure 8 shows a two-dimensional cut through the flame from the simulation. The cold fuel-oxidizer mixture enters the domain on the left. A hot pilot flow surrounding the inner fuel-oxidizer stream also enters the domain

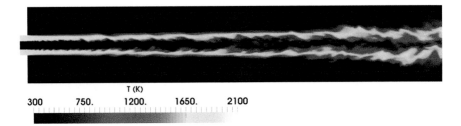

Fig. 8 Two-dimensional cut through the simulated flame showing the temperature distribution

on the left and ignites the unburnt methane-air mixture. Due to the high velocities of the fuel-oxidizer mixture of about 80 m/s, turbulent flow structures emerge which lead to a wrinkling of the flame. The case has been run on 14,400 CPU cores and required about 15 million core hours on the Hazel Hen cluster at the High Performance Computing Center in Stuttgart (HLRS). Post-processing and visualization of the case has been done on the ForHLR II cluster at KIT.

The GRI 3.0 mechanism is used for computing the chemical reaction rates in the performance assessment. The simulation is performed with the code from the previous work without compiler extensions and the new code including the compiler extensions discussed above. The widely used CFD tool OpenFOAM 5.x [1] is used to solve the compressible Navier-Stokes equations [4]. The open-source chemical kinetics library Cantera [2] is used for computing detailed transport coefficients for the diffusion of heat and mass based on the Curtiss-Hirschfelder approximation [13]. The interested reader is referred to [14, 15] for more information about the solver. A second-order implicit time discretization and fourth-order interpolation methods for the discretization of spatial derivatives are used for the numerical solution of the conservation equations for mass, energy, species masses and momentum.

3.2 Simulation Results

Due to the highly resolved nature of the simulation setup and the use of detailed chemical reaction mechanisms as well as detailed molecular diffusion coefficients, comparison with experimental measurements shows good agreement. For example, Fig. 9 presents a comparison between the simulation and the experiment [5] of the time averaged temperature T and Bilger mixture fraction Z values as well as their rms values at a fixed axial position in the flame.

The simulation results allow a detailed view into the underlying processes that govern the turbulent flame. For example, Fig. 10 shows two quantities which are not directly accessible in experiments: The instantaneous heat release rate is depicted on a two-dimensional cutting plane in black-blue-white, which identifies regions of high chemical reactions rates. Additionally, an iso-surface of the vorticity magnitude

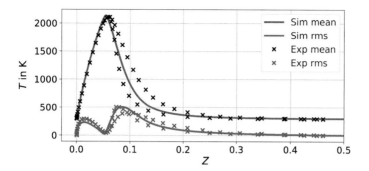

Fig. 9 Mean and rms values of temperature T and mixture fraction Z compared between simulation and experiment [5]

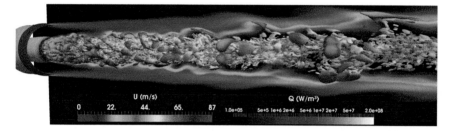

Fig. 10 Heat release rate on a 2D cutting plane and iso-surface of vorticity magnitude colored by fluid velocity

is shown, which visualizes turbulent flow structures in the shear layer between the unburnt fuel-air mixture and the hot pilot flow. The iso-surface is colored by the fluid velocity, which reaches up to 80 m/s in the core region and about 20 m/s in the outer region of the pilot flow. Both quantities—heat release rate and the turbulent flow field—as well as their mutual interaction constitute an important factor for efficient and stable combustion, but can only be measured with limited resolution.

Figure 11 shows an iso-surface of the vorticity magnitude, and therefore depicts the structure of turbulent vortices in the shear layer between the fuel-air flow and the pilot flow. On the left, the fast unburnt fuel-air mixture can be seen leaving the nozzle and forming characteristic vortex tubes in the shear layer. The iso-surface is colored by the flame index $FI \equiv (\nabla Y_{CH_4} \cdot \nabla Y_{O_2}) / |\nabla Y_{CH_4} \cdot \nabla Y_{O_2}|$. Here, Y is the mass fraction of a chemical species. The flame index allows to identify different regimes of combustion in the flame which play an important role in the chosen burner setup [5]. Because the flame index is computed from local gradients of species concentrations, it is not accessible in the experiments. The simulation results have been compiled into a comprehensive data base. Full 3D data sets of turbulent flow variables and chemical scalars for 10 time steps are included, as well as 2D datasets corresponding to experimentally investigated locations in the flame at high time frequencies. In total, 10 TB of data will be provided in the database, which will be made publicly available.

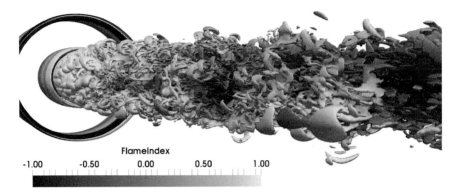

Fig. 11 Iso-surface of vorticity magnitude in the shear layer between unburnt fuel-air and pilot flow colored by the flame index

3.3 Performance Evaluation of the Parallel Case

The parallel scaling of the code has been shown in a previous work [3] to be linear up to about 10,000 MPI processes. Therefore, optimization of serial performance is a reasonable approach to further improve performance. With the given setup of the burner flame, the computation of a single simulation time step in parallel takes about 5.0 s on average with the code without compiler extensions. Of the 5.0 s for the whole time step, 3.5 s are spent on computing the chemical reaction rates. With the new code, which includes the discussed compiler extensions, the time for computing the time step reduces to 3.8 s on average with the computation of chemical reaction rates requiring 2.2 s. Therefore, the performance gains in the serial test cases are also present in the parallel one: the time for chemistry computations is reduced by about 40% and total simulation time is reduced by about 20%.

4 Summary

The performance bottleneck of the combustion simulations using detailed reaction mechanisms is often the computation of chemical reaction rates. In this work, the serial performance of the code for computing chemical reaction rates has been further optimized by utilizing compiler extensions to give the compiler more information about pointer aliasing and data alignment. Adding these compiler extensions to the code base requires little changes and does not affect the numerical results but drastically reduces the size of the machine code, enables additional auto-vectorization and reduces the computation times for chemical reaction rates by up to 50% and total simulation times by up to 25%, which will even increase with increasing complexity of the reaction mechanism. The introduced optimizations also help to close the gap between gcc and icpc performance results. Compared to OpenFOAM's standard implementation of the reaction rate computations, the newly optimized chemistry code performed ten times faster without affecting the numerical results because the same equations are solved but in a more efficient way.

The new code has also been applied to a highly resolved large scale numerical simulation of a turbulent flame on a grid with 150 million cells and 14,400 MPI processes. Applying the new code containing the compiler extensions leads to a reduction in total simulation time of 20%. The simulation is validated with measured data and the results are compiled into a database containing 10 TB of 3D and 2D data.

Acknowledgements This work was supported by the Helmholtz Association of German Research Centres (HGF) through the Research Unit EMR. This work was performed on the national supercomputer Cray XC40 Hazel Hen at the High Performance Computing Center Stuttgart (HLRS) and on the computational resource ForHLR II with the acronym Cnoise funded by the Ministry of Science, Research and the Arts Baden-Württemberg and DFG ("Deutsche Forschungsgemeinschaft").

References

1. OpenCFD, *OpenFOAM: The Open Source CFD Toolbox. User Guide Version 1.4, OpenCFD Limited* (Reading UK, April 2007)
2. D. Goodwin, H. Moffat, R. Speth, Cantera: a object-oriented software toolkit for chemical kinetics, thermodynamics, and transport processes, version 2.3.0b (2017). Software available at www.cantera.org
3. T. Zirwes, F. Zhang, J. Denev, P. Habisreuther, H. Bockhorn, Automated code generation for maximizing performance of detailed chemistry calculations in OpenFOAM, in *High Performance Computing in Science and Engineering '17*, ed. by W. Nagel, D. Kröner, M. Resch (Springer, 2017), pp. 189–204
4. F. Zhang, H. Bonart, T. Zirwes, P. Habisreuther, H. Bockhorn, N. Zarzalis, Direct numerical simulation of chemically reacting flows with the public domain code OpenFOAM, in *High Performance Computing in Science and Engineering '14*, ed. by W. Nagel, D. Kröner, M. Resch (Springer, 2015), pp. 221–236
5. R. Barlow, S. Meares, G. Magnotti, H. Cutcher, A. Masri, Local extinction and near-field structure in piloted turbulent CH4/air jet flames with inhomogeneous inlets. Combust. Flame **162**(10), 3516–3540 (2015)
6. High Performance Computing Center Stuttgart (2018). www.hlrs.de/systems/cray-xc40-hazel-hen
7. V. Damian, A. Sandu, M. Damian, F. Potra, R. Carmichael, The kinetic preprocessor kpp—a software environment for solving chemical kinetics. Comput. Chem. Eng. **26**, 1567–1579 (2002)
8. K. Niemeyer, N. Curtis, pyJac v1.0.4 (2017). https://github.com/slackha/pyJac
9. G. Smith, D. Golden, M. Frenklach, N. Moriarty, B. Eiteneer, M. Goldenberg et al., Gri 3.0 reaction mechanism
10. B. McBride, S. Gordon, M. Reno, *Coefficients for Calculating Thermodynamic and Transport Properties of Individual Species* (1993). National Aeronautics and Space Administration, NASA Technical Memorandum 4513
11. M. Godbolt, Compiler explorer (2018). https://godbolt.org/
12. U. Paul, B. Dick, M. Kellner, Towards efficient codes for sustainable hpc, in *20th Results and Review Workshop*, Stuttgart (2017)

13. T. Poinsot, D. Veynante, *Theoretical and Numerical Combustion* (R.T. Edwards, 2001)
14. F. Zhang, T. Zirwes, P. Habisreuther, H. Bockhorn, Effect of unsteady stretching on the flame local dynamics. Combust. Flame **175**, 170–179 (2017)
15. T. Zirwes, F. Zhang, T. Häber, D. Roth, H. Bockhorn, Direct numerical simulation of ignition by hot moving particles," in *26th International Colloquium on the Dynamics of Explosions and Reactive Systems* (2017)

Part IV
Computational Fluid Dynamics

The following chapter presents a selection of research projects conducted in the field of Computational Fluid Dynamics (CFD) during the reporting period, both on the CRAY XC40 Hazel Hen of the High Performance Computing Center Stuttgart (HLRS) and the ForHLR II of the Steinbuch Centre for Computing (SCC) in Karlsruhe. As in the years before, CFD had the strongest request for supercomputing resources amongst all disciplines.

This year, 38 annual reports were submitted and underwent a peer review process. Many important results and new scientific findings have been achieved and notable progress has been made in the field of CFD. Research focuses were set on the simulation of fluid dynamic phenomena being detectable only through an extremely fine temporal and spatial discretization, often in combination with high-order methods, as well as on the adaptation of the employed numerical codes to the specific hardware architectures making the codes run more efficiently on the respective supercomputers.

Due to limited space, only nine contributions were selected for publication in this book, which means that quite a number of highly qualified reports could, unfortunately, not be considered. Challenging fluid dynamic problems had been looked into, and the results for most complex flow physical problems have only been achievable by the use of High Performance Computing. The spectrum of projects presented in the following covers fundamental research as well as application-oriented problems of industrial relevance. Examples for the latter are liquid jet breakups, helicopter aeromechanics, flows in water turbines, and safety issues in nuclear reactors, just to name a few. Different numerical methods, such as Finite Difference, Finite Volume, Smoothed Particle Hydrodynamics, or Discontinuous Galerkin methods, were employed using in-house or commercial codes.

In the first four contributions, fluid dynamic problems have been addressed by Direct Numerical Simulation (DNS), i.e. no turbulence modelling has been applied. *Wenzel, Peter, Selent, Weinschenk, Rist, and Kloker* of the transition and turbulence working group at the Institute of Aerodynamics and Gas Dynamics, University of Stuttgart, present DNS of compressible subsonic turbulent boundary layers with adverse pressure gradients. The boundary layer profiles are self-similar in their streamwise evolution, which represents the most general canonical form of the adverse pressure gradient case. Only few results are available in literature for incompressible cases, but no results at all can be found for the present problem.

Additionally, first results are shown for turbulent superstructures in the unsteady flow field, which can be denoted to be the most dominant global events in wall-bounded turbulent flows. Significant advancements in the parallel efficiency of the code were achieved by replacing the former OpenMP decomposition in spanwise direction by an MPI decomposition.

A detailed DNS of a primary liquid jet breakup including evaporation was performed by *Reutzsch, Ertl, Baggio, Seck, and Weigand* from the Institute of Aerospace Thermodynamics of the University of Stuttgart. Aiming at enhancing the understanding of the process of evaporation during breakup, which occurs during fuel injection in combustion engines and gas turbines, comparison between an evaporating and a non-evaporating water jet into atmospheric air was made. They use the Volume of Fluid (VoF) method in combination with Piecewise Linear Interface Calculation (PLIC) to reconstruct sharp interfaces. For simulating the evaporation, the energy equation is solved to obtain the phase change at the interface, and a second volume fraction for the vapour phase has been introduced. An enhancement of the computational performance was achieved by replacing large pointer arrays in the code by allocatable arrays.

Stein and Sesterhenn from the Institute of Fluid Dynamics and Technical Acoustics of the TU Berlin performed, for the first time, a three-dimensional DNS of a turbulent flow over a cavity, which acts as a Helmholtz resonator. They used a newly developed multi-block in-house code to predict the non-linear interaction of turbulence and acoustics and compared their results with experiments. Avoiding costly experimental parameter studies in an aeroacoustic wind tunnel, the derived numerical model is aimed to simplifying the design process of acoustic cavity resonators in turbulent flow. The report also includes a detailed description of weak and strong scaling tests for different problem sizes performed on the employed supercomputer Hazel Hen, including a discussion of the information content and value created by such tests.

In their project involving several institutes of the University of *Stuttgart, Föll, Pandey, Chu, Munz, Laurien, and Weigand* have performed DNS of supercritical channel flow using a Discontinuous Galerkin Spectral Element Method (DG-SEM). Supercritical fluids, like the considered carbon dioxide, exhibit an attenuation of turbulence under certain conditions resulting in a deteriorated heat transfer. As this peculiarity is difficult to predict with conventional turbulence models, DNS has to be used. While formerly Finite Volume methods had been applied, results of a DG simulation are shown in the present paper, with the higher order of accuracy leading to an improved fidelity. One focus of the work was to demonstrate the applicability of the DG code for complex compressible flow. An excellent strong scaling was achieved up to 12.000 cores for a problem with 200 Mio. degrees of freedom.

The above mentioned DG-SEM code has been developed by the working group of Munz at the Institute of Aerodynamics and Gas Dynamics in Stuttgart. It has systematically been designed for scale resolving simulations in an HPC context. The next contribution stems from this working group and deals with recent developments of their high-order code for compressible multi-scale flows. *Beck, Bolemann, Flad, Krais, Zeifang, and Munz* report on two extensions, namely the Chimera grid technique for easy handling of curved boundaries and moving grids, and a machine learning approach to provide advanced subgrid closures for Large Eddy Simulations. The latter is very novel and a very promising approach. Various artificial neural network architectures have been investigated and trained with different strategies. First results of the simulation of a challenging high Reynolds number supersonic dual nozzle flow including shock waves are shown.

In the next article, *Chaussonnet Dauch, Braun, Keller, Kaden, Schwitzke, Jacobs, Koch, and Bauer* from the Institute of Thermal Turbomachinery at the Karlsruhe Institute of Technology present predictions of primary breakups of liquids in air-assisted atomization systems using an in-house numerical code based on the Smoothed Particle Hydrodynamics method (SPH). The suitability of this method for breakup processes is successfully demonstrated for two application cases. The first one is the two-phase flow in fuel spray nozzles for aero-engine combustors, and the second one is the atomization of a viscous slurry used in gasifiers for biofuel production. The computations were performed on the ForHLR II tier-2 supercomputer at the SCC in Karlsruhe using up to 2000 cores in parallel.

For a long time, the Helicopter and Aeroacoustics Group at the Institute of Aerodynamics and Gas Dynamics (IAG) at the University of Stuttgart has performed high-fidelity numerical simulations on complete rotorcraft configurations on HLRS' supercomputers using the flow solver FLOWer. This code was originally developed by DLR and has successively been improved for high-performance rotorcraft applications over the previous years at the IAG. The recent efforts of the group are presented by *Frey, Herb, Letzgus, Weihing, Keßler, and Krämer* and concentrate on further enhancements of the code as well as on coupled simulations for Airbus' compound helicopter RACER. The first part covers the implementation and testing of the γ-Re_θ laminar-turbulent transition model and its comparison to experiments. The simulation of the RACER configuration includes application of a 5th order WENO scheme, adaptive mesh refinement, and deflectable flaps on the box-wing and the H-tail surfaces. Flow interaction phenomena between main rotor, propellers and fuselage have been investigated for hover and cruise conditions.

The work of *Mansour, Kaltenbach, and Laurien* from the Institute of Nuclear Technology and Energy Systems at the University of Stuttgart is dedicated to enhancing the safety in nuclear pressurized water reactors. They have studied the applicability of two-phase CFD to simulate the injection of cold water spray into a nuclear containment, which had been filled by hot steam due to a leak in the reactor's primary circuit. For that purpose, models with heat and mass transfer for spray cooling and aerosol particle washout were implemented into the commercial code ANSYS CFX v16.1. In their article, they present the physical mechanisms of spray cooling and particle washout in the frame of a generic incident scenario within a simplified reactor building. Furthermore, they elaborate intensively on the parallel performance for different grid sizes and topologies in order to define guidelines for the optimal use of computational resources.

The last contribution in this chapter comes from *Wack and Riedelbauch* of the Institute of Fluid Mechanics and Hydraulic Machinery, University of Stuttgart. They have investigated cavitation in a water turbine. Firstly, they present results for a

NACA009 hydrofoil test case with a cavitating tip leakage vortex. The aim of this study was to answer the fundamental question whether a homogeneous or a non-homogeneous model should be used in a two-phase modelling approach. It turned out that the latter did not increase the accuracy of the results. In the second part, two-phase simulation were performed for a Francis turbine at a stable full load operating point. For the chosen pressure level, a stable cavitating vortex rope forms in the draft tube cone. While the effect of the cavitation constants can be neglected, the grid resolution, the geometry of the runner nut and the enabling of the curvature correction in the SST turbulence model have a decisive impact on the simulation accuracy. The ANSYS CFX v18.1 code was used for these simulations.

The presented selection of projects reflects the continuous progress in supercomputer simulation in the field of CFD. It illustrates that High Performance Computing is crucial in order to be able to tackle problem classes, which seemed unreachable only a few years ago. Furthermore, it impressively demonstrates the power of modern computer simulation methods to achieve more reliable and more efficient predictions in industrial design processes and a better understanding of complex flow physical phenomena. However, it is obvious that the engineers and natural scientists, in order to fully exploit the potential offered by modern supercomputers, must not focus solely on their physical problem. Moreover, they have to think about appropriate numerical algorithms as well as to get their teeth into the optimization of their codes with respect to the specific hardware architecture they are going to use. Usually, this exceeds the capabilities and skills of one individual and, thus, different experts in the respective fields should work together. The very fruitful collaboration of researchers with experts from the HLRS and the SCC in many of the projects shows how this can work. Gratitude is owed in this respect to the staff of both computing centres for their individual and tailored support. Without their efforts, it would not have been possible and will not be possible in the future to sustain the high scientific quality we see in the projects.

Ewald Krämer
Institute of Aerodynamics and Gas Dynamics, University of Stuttgart
Pfaffenwaldring 21, 70550 Stuttgart, Germany
e-mail: kraemer@iag.uni-stuttgart.de

DNS of Compressible Turbulent Boundary Layers with Adverse Pressure Gradients

Christoph Wenzel, Johannes M. F. Peter, Björn Selent, Matthias B. Weinschenk, Ulrich Rist and Markus J. Kloker

Abstract First direct-numerical-simulation results of compressible subsonic adverse-pressure-gradient turbulent boundary-layers are presented with self-similar boundary-layer profiles in their streamwise evolution, which represents the most general canonical form of the adverse pressure gradient case. Only few results are available in literature for this problem even in the incompressible regime, since the achievement of such canonical flows requires very costly iterative procedures in order to find the correct pressure distribution which has to be prescribed at the top of the simulation domain. Additionally, preliminary result are presented for the detection of turbulent superstructures in the unsteady flow field, which can be denoted to be the most dominant global events in wall-bounded turbulent flows. All results have been calculated with a new version of our numerical in-house code *NS3D*, which now also allows MPI decomposition in the spanwise direction of the simulation domain. A scaling study shows a maximum increase in the efficiency of up to 300% compared to the previous code version where only OpenMP decomposition has been available for the spanwise decomposition.

C. Wenzel (✉) · J. M. F. Peter · B. Selent · M. B. Weinschenk · U. Rist · M. J. Kloker
Institut für Aero- und Gasdynamik, Universität Stuttgart, Pfaffenwaldring 21,
70550 Stuttgart, Germany
e-mail: wenzel@iag.uni-stuttgart.de

J. M. F. Peter
e-mail: johannes.peter@iag.uni-stuttgart.de

B. Selent
e-mail: selent@iag.uni-stuttgart.de

M. B. Weinschenk
e-mail: weinschenk@iag.uni-stuttgart.de

U. Rist
e-mail: rist@iag.uni-stuttgart.de

M. J. Kloker
e-mail: kloker@iag.uni-stuttgart.de

© Springer Nature Switzerland AG 2019
W. E. Nagel et al. (eds.), *High Performance Computing in Science and Engineering '18*, https://doi.org/10.1007/978-3-030-13325-2_14

229

1 Introduction

For many years now the canonical case of the flat-plate zero-pressure-gradient turbu-lent boundary layer (ZPG TBL) forms an essential foundation in turbulence research. Whereas fundamental understanding of the physical behavior of wall-bounded turbu-lent flows was gained already for more than hundred years by experimental measure-ments and analytical investigations, a second and more closer view into turbulence was provided by the very first Direct Numerical Simulation (DNS) of the ZPG TBL case by Spalart in 1988 [22]. Even though the computational power has grown enor-mously in the last thirty years, DNS of TBLs are still strongly restricted by the computational resources and many investigations are therefore often limited to the incompressible zero-pressure-gradient case even today.

For realistic applications, however, where turbulent flows are almost always influ-enced by pressure gradients caused by the presence of curved surfaces, the incom-pressible ZPG TBL case only represents a very small portion in the parametric space and its expansion on studies of canonical TBLs influenced by pressure gradients are of utmost importance [12]. In contrast to the ZPG TBL case, however, pressure-gradient influenced TBLs are very difficult to study systematically since the pressure distributions are arbitrarily changing in the streamwise direction (and also spanwise of course) and thus making the history of every different TBL fairly unique. The dynamical properties of the TBL like its flow structure or its turbulent statistics for instance are strongly influenced by this history even in the ZPG TBL case [21]—often referred to as history effects—and thus strongly depend upon the specific distribution of the streamwise pressure gradient.

A special subgroup in the variety of pressure gradient distributions is represented by so-called self-similar PG TBL, meaning that all terms, or the most important ones, of the flow-describing equations of motion have the same proportionality in their streamwise evolution [3, 5, 19, 25] which minimises the impact of history effects. The idea of self-similar TBLs has been developed by Mellor and Gibson [19] by the determination of the incompressible non-dimensional pressure gradient, often referred to as Clauser parameter: $\beta = \delta_1^* \left(dp_e^*/dx \right) / \tau_w^*$, which must be constant and thus independent from the streamwise position, where δ_1^* is the displacement thickness, τ_w^* the mean shear stress at the wall and dp_e^*/dx the far-field pressure gradient. According to the sign of the the far-field pressure gradient dp_e^*/dx, β classifies the various types of TBL into the ZPGTBL case with $\beta = 0$, favourable pressure gradient cases (FPG) with $\beta < 0$, APG cases (A-adverse) with $\beta > 0$ and the APG case at the separation point with $\beta \to \infty$.

Since the efficiency of many technical applications relies on the assumption that the turbulent flow remains attached over the whole wing of an airplane for instance—which may not be the case if the APG is too strong—self-similar APG TBLs are arguably the most relevant PG cases to study in fundamental research [12]. Concern-ing the self-similar APG TBL, many aspects of scaling, structure and stability remain unclear even in the incompressible case. Thus, if compressibility is additionally taken into account, the problem becomes even more demanding and a lot of fundamental investigations are highly desirable since no compressible DNS results of APG TBLs are known to the authors up to now.

In the present work, preliminary DNS results of compressible APG TBLs are presented for a nearly incompressible inlet Mach number of $M_\infty = 0.5$ for three Clauser parameters $\beta = 0.25, 0.6$ and 1.0. With the main goal of understanding the unsteady flow-field of wall-bounded turbulent flows, an additional section is focused on the detection of turbulent superstructures, which represent major events in the unsteady turbulent flow-field. Additional information about the numerical code as well as its performance are given at the end.

2 Numerical Setup

Numerical method All computations are performed with a revised version of the compressible high-order in-house DNS code $NS3D$ (see Chap. 5 for further information), which has been extensively validated for a broad variety of applications. Fundamentals of the code are described in Babucke [1], Linn and Kloker [16, 17] and Keller and Kloker [9, 10]. It solves the three-dimensional, unsteady, compressible Navier-Stokes equations together with the continuity and energy equation in conservative formulation. Assuming a velocity vector $\mathbf{u} = [u, v, w]^T$ in streamwise, wall-normal and spanwise directions x, y and z, respectively, the dimensionless solution vector is $\mathbf{Q} = [\rho, \rho u, \rho v, \rho w, E]^T$, where ρ and E are the density and the total energy. Velocities and length scales are normalized by the streamwise reference velocity U_∞^* and the reference length L^*, respectively. Thermodynamical quantities are normalised with the reference temperature T_∞^* and the reference density ρ_∞^*. Pressure p is non-dimensionalised by $\rho_\infty^* U_\infty^{*2}$. The specific heats c_p and c_v as well as the Prandtl number Pr are assumed to be constant. Temperature dependence of the viscosity is modeled by Sutherland's law. The equations are solved on a block-structured Cartesian grid spanning a rectangular integration domain, see Fig. 1. All three spatial directions are discretized using 6th-order subdomain-compact finite differences [8] for the present simulations. The classical fourth-order Runge-Kutta scheme is used for time integration.

Numerical setup The main region of the computational box has a dimension of $340\,\delta_{99,0}^* \times 34\,\delta_{99,0}^* \times 8\pi\,\delta_{99,0}^*$ in streamwise, wall-normal and spanwise direction, respectively, where $\delta_{99,0}^*$ is the boundary-layer thickness at the inlet of the simulation domain. The use of an identical setup for all different pressure-gradient cases should allow a maximum of comparability. The numerical grid has been designed to provide a comparable grid resolution as given in [27], where also detailed information about the grid validation can be found. The grid resolution is based on the most restricting $\beta = 0.25$ case with $2500 \times 235 \times 512$ grid points for the main region (red framed in Fig. 1) and $2700 \times 270 \times 512$ grid points for the overall domain resulting in 373,248,000 total grid points. Calculated in viscous wall units, this gives a grid spacing of $\Delta x^+ = 20$, $\Delta y_1^+ = 0.8$ and $\Delta z^+ = 6.9$ at $Re_\theta = 670$ for $M_\infty = 0.5$ and of $\Delta x^+ = 19.5$, $\Delta y_1^+ = 0.75$ and $\Delta z^+ = 6.4$ at the inlet of the

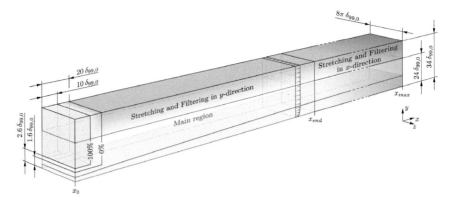

Fig. 1 Simulation domain for present DNS. Yellow colored region represents sponge zone, blue colored regions represent grid-stretched and filtered regions. The red bordered zone represents the main region of the simulation

Table 1 Thermodynamic properties

M_∞	–	0.5
T_∞^*	K	288.15
Pr	–	0.71
ρ_∞^*	kg/m³	1.225
R^*	J/kg	287
κ	–	1.4

domain in the three dimensions. The averaging time after the initial transient is about $\Delta t^+ = \Delta t^* u_\tau^{*2}/v_w^* \approx 7000$ viscous time units for the three cases. The basic thermodynamic flow parameters are shown in Table 1 and equal for all cases.

In analogy to simulations available in literature, the main region (red framed in Fig. 1) of the simulation domain is designed with a height of at least three boundary-layer thicknesses $\delta_{99,end}^*$ and width of about $\pi\,\delta_{99,end}^*$ in spanwise direction, measured at the end of the main region x_{end}^* for the $\beta = 1.0$ case with the thickest boundary layer. The flow field is periodic in the spanwise direction.

At the solid wall, the flow field is treated as fully adiabatic with $(dT/dy)_w = 0$ which suppresses any heat exchange between wall and fluid, whereas the pressure at the wall is calculated by $(dp/dy)_w = 0$ from the interior field. For the velocity components, a no-slip boundary condition is applied.

At the inlet a digital filtering synthetic eddy method (SEM) is used to generate an unsteady, pseudo-turbulent inflow condition [11, 13, 24]. The required distributions for the mean flow field and for the Reynolds fluctuations are used from a small auxiliary simulation using a recycling/rescaling method at $M_\infty = 0.5$ [18, 23, 27, 28] for all cases. Even if the SEM boundary condition provides a pseudo-physical turbulent flow field already at the inlet of the domain, the flow needs about $10\delta_{99,0}$ in the streamwise direction to satisfy equilibrium turbulent flow statistics. In order to prevent the farfield flow from being distorted by this transition process, a sponge region [14] is applied in the inlet region of the simulation domain (yellow colored in Fig. 1), which damps down the flow to an unperturbed free-stream baseflow.

At the top of the domain ρ, T and p are kept constant. The velocity components u and w are specified by $d/dy = 0$, whereas the wall-normal velocity component v is computed from the continuity equation under the assumption of $d\rho/dy = 0$ such that $dv/dy = -1/\rho \left(d \left(\rho u \right) / dx + d \left(\rho w \right) / dz \right)$.

At the outflow, the time derivative is extrapolated with $\partial \mathbf{Q}/\partial t|_N = \partial \mathbf{Q}/\partial t|_{N-1}$.

In both wall-normal and streamwise direction, the numerical grid is stretched and the solution filtered [4, 26] in order to avoid reflections from the boundaries.

Modeling of the pressure distribution prescribed at the free-stream boundary condition According to Mellor and Gibson [19], self-similar APG TBL can be determined by the non-dimensional pressure gradient, $\beta = \delta_1^* \left(dp_e^*/dx \right) / \tau_w^*$ which must be independent from the streamwise position and therefore constant in the nearly incompressible case. Following the idea of Townsend [25], this only can be fulfilled if the streamwise velocity at the edge of the boundary layer follows the power-law relation $u_{APG}(x) = u_\infty(x_0)\left(x - x_{origin}\right)^m$ where m is the pressure gradient strength exponent and x_{origin} the virtual origin of the boundary layer. It should be mentioned however that it is not an easy task to define the pressure distribution at the top of the simulation domain yielding to a requested β-value, since both the direct correlation of m and β as well as the exact value of x_0 are unknown a priori. Additionally, the streamwise velocity u is varying through the farfield in the wall-normal direction caused by non-constant values for $\partial v/\partial x$ in order to fulfill the condition of vanishing vorticity $\omega_z = \partial v/\partial x - \partial u/\partial y \approx 0$ in the farfield. Thus the exact velocity distribution which has to be prescribed initially at the top of the domain has to be defined iteratively. The resulting parameters for the present simulations are summarized in Table 2 for the three cases.

Based on the precalculated velocity distribution $u_{APG}(x)$, the dimensional distributions of the temperature and pressure calculated from the one-dimensional isentropic gas equations

$$T_{APG}^*(x) = T_t^* - \frac{\frac{\kappa-1}{2} u_{APG}^*(x)^2}{\kappa R^*}, \quad p_{APG}^*(x) = \frac{p_t^*}{\left(1 + \frac{\kappa-1}{2} M_{APG}(x)^2\right)^{\frac{\kappa}{\kappa-1}}}, \quad (1)$$

Table 2 Initial values for the three different APG cases at $\beta = 0.25$, $\beta = 0.6$ and $\beta = 1.0$. Given parameters are the starting point of the APG region in boundary-layer thicknesses at the inlet of the domain $\delta_{99,0}^*$, the virtual origin of the velocity distribution x_0 and the pressure gradient strength exponent m

	$\beta = 0.25$	$\beta = 0.6$	$\beta = 1.0$
Start APG zone in $\delta_{99,0}^*$	30	30	30
x_0 in $\delta_{99,0}^*$	−1.2	−1.0	−1.0
m	−0.07	−0.15	−0.2

with the Mach number $M_{APG}^2 = u_{APG}^2 / (\kappa R^* T_{APG}^*)$, the total temperature T_t^* and total pressure p_t^* as well as the isentropic exponent κ calculated at the inlet of the simulation domain. The distribution of the far-field density is calculated from the equation of state $\rho_{APG}^* = p_{APG}^* / (R^* T_{APG}^*)$. The velocity as well as the forced pressure distributions at the top of the domain are given in Fig. 3a, b respectively.

3   Results

This section is mainly focused on the streamwise evolution of basic flow parameters which are related to the self-similarity of the resulting flow fields. A deeper look into the physics of turbulence will be the result of ongoing investigations.

In order to give the reader an impression of the resulting flow field, iso-surfaces of the λ_2-criterion are depicted in Fig. 2 for the $\beta = 0.6$ case at one snapshot in time for the main region of the simulation domain. Color indicates the height of the rapidly thickening boundary layer, which increases from $\delta_{99,0}^*$ at the inlet up to eleven $\delta_{99,0}^*$ at the end of the main region.

Next to the velocity and pressure distributions $u_{APG}(x)$ and $p_{APG}(x)$, the resulting development of the boundary-layer thickness $\delta_{99}^*(x)/\delta_{99,0}^*$ as well as the Clauser-parameter distributions β are given in Fig. 3c, d respectively. Related to the self-similarity of the APG cases, which is the utmost prerequisite for future studies, almost perfectly constant Clauser parameters can be found for the $\beta = 0.25$ case for $x^*/\delta_{99,0}^* \gtrsim 100$, for the $\beta = 0.6$ case for $x^*/\delta_{99,0}^* \gtrsim 130$ and for the $\beta = 1.0$ case for $x^*/\delta_{99,0}^* \gtrsim 150$. However, as mentioned in the introduction, the constant values of the Clauser parameter do not inevitably guarantee perfect self-similarity of the local flow field in the streamwise direction, since the flow fields are still influenced by the

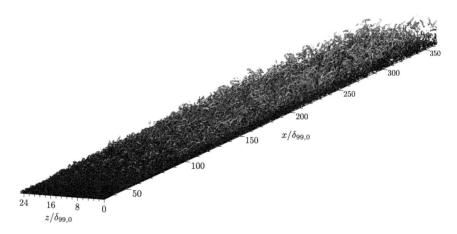

Fig. 2 Iso-surfaces of λ_2-criterion for the $\beta = 0.6$ case. Color indicates height of the boundary layer, which approximately thickens up eleven times in the long domain, see also Fig. 3c

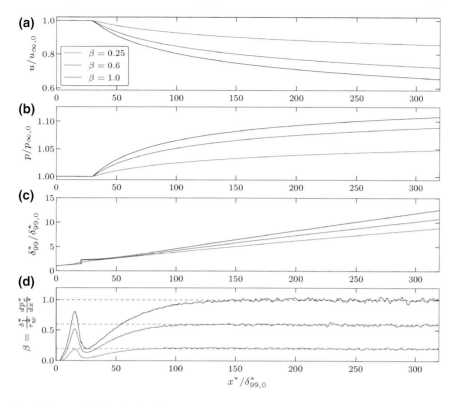

Fig. 3 Streamwise distribution of **a** farfield velocity $u/u\infty$, **b** farfield pressure p/p_∞, **c** boundary-layer thickness ratio $\delta_{99}^*(x)/\delta_{99,0}^*$ and **d** Clauser parameter β for three different cases at $\beta = 0.25$, 0.6 and 1.0

history occurring in the run-up distance where β is not yet constant. Nevertheless, investigations of local velocity profiles have shown that regions of significant length can be found in the long domain where true self-similarity can be achieved.

In comparison to the ZPG TBL case (not shown here), these velocity profiles are much more influenced by "high Reynolds-number effects" yielding to a rapid separation of smaller and larger turbulent length-scales. Whereas the smaller scales are mainly located at the wall, the larger ones are located somewhat further outside in the boundary layer. This effect is mainly caused by the rapid increase of the boundary layer thickness δ_{99}^*, which is approximately twice the one of the ZPG case at the end of the simulation domain for the $\beta = 1.0$ case and almost one and a half of the $\beta = 0.25$ case, as shown in Fig. 3c. The step in the distribution of the boundary-layer thickness located at about $x^*/\delta_{99,0}^* \approx 20$ is caused by the implementation of the inlet sponge (see Fig. 1) and the sensitivity of the weakly defined boundary-layer thickness $\delta_{99,0}^*$ to such small mean-flow distortions itself. Nevertheless, since these steps are located far away from the usable region of the simulations, they do not affect the quality of the ongoing investigations.

4 Postprocessing Tools—Fast and Adaptive Bideminsional Empirical Mode Decomposition

Next, we apply the Fast and Adaptive Bidimensional Empirical Mode Decomposition algorithm (FABEMD) to separate turbulent scales. The FABEMD is a modification of the Bidimensional Empirical Mode Decomposition (BEMD) [2]. The FABEMD–algorithm uses 2D input data and decomposes the data into its characteristic bidimensional intrinsic mode functions (BIMFs) and a residue by using an order-statistics filter, which uses different window sizes depending on the distance between maxima and minima of the data (note that the BIMFs are locally orthogonal among each other). In order to obtain the BIMFs, FABEMD applies a filtering process which separates and arranges the different BIMFs depending on the scales/fluctuations. In the case of a turbulent flow the first BIMF contains the smallest scales of the flow and the final BIMF contains the largest scales whereas the residue contains the mean flow.

In order to identify 'Turbulent Superstructures' in our case we apply the FABEMD–algorithm to instantaneous snapshots in a flat-plate boundary layer in planes approximately parallel to the wall at constant relative height with respect to the local boundary-layer thickness, for instance at 70%. The domain is periodic in spanwise direction and finite in streamwise direction. Therefore, the order-statistics filter window size in spanwise direction remains constant whereas in streamwise direction the filter window size becomes smaller near the boundaries. The filtering process produces N BIMFs and a residue. Now a separating process is applied where all BIMFs up to N are added, this represents the small scales, whereas the residue represents the large scales. This procedure is repeated for several planes at constant relative boundary-layer thickness, and the newly generated planes are layered on top of each other and iso-surfaces are built. Figure 4 represents isosurfaces of the streamwise velocity component u of the original instantaneous flow field without any filtering for comparison with the flow field after applying the FABEMD–algorithm (large scales are depicted, separated at the 4th BIMF). It can be seen that FABEMD removes the small fluctuations from the flow field. What remains are large coherent structures as represented in Fig. 4b. These coherent structures consist of large positive (blue) and negative (yellow) instantaneous streaks. The negative streaks can be called 'Turbulent Superstructures'. They are generated by the main–shear region of the turbulent boundary layer at the wall and are lifted up to the edge of the boundary layer in their spatial evolution. The positive ones are characterized by more laminar behaviour since they mainly consist of laminar fluid which is engulfed from the laminar far-field.

In order to quantify the different BIMFs a Fourier-Transform (FT) in streamwise direction is applied to the original data, the small scales and the large scales. The FT is performed for each plane relative to the boundary-layer thickness and each separate spanwise z-position. Subsequently the mean value over the spanwise direction of the absolute value of $|u'_\alpha|$ is computed. In Fig. 5 the mean value over z of $|u'_\alpha|$ at the edge of the boundary layer is depicted versus the streamwise wave number α at different separation points ($N = 2, 4, 6$). Whereas the original data is represented by the solid black line, the small scales are represented by the solid lines and the large

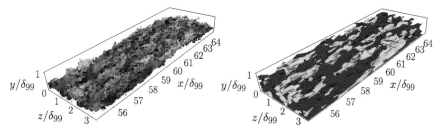

| (a) Original instantaneous flow field. | (b) Large scales separated at the 4th BIMF. |

Fig. 4 Comparison of original instantaneous flow field with the large scales (separated at the 4th BIMF) of the streamwise velocity component u

Fig. 5 Fourier-Transform of original instantaneous data and the large scales (separated at different BIMFs) located at the edge of the boundary layer

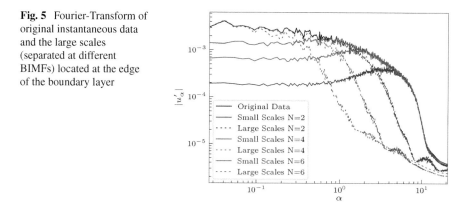

scales are represented by the dashed lines. It can be seen that the small scales contain the fluctuations at high α-values (small wavelengths) whereas the fluctuations at low α-values are filtered out. In contrast the large scales contain fluctuations at lower wave numbers (large wavelengths) and the fluctuations at high α are filtered out. The summation of small and large scales yields again to the original data. The 4th BIMF is chosen to be the separating point since it separates the scales in the middle of the spectrum as can be seen in Fig. 5. Scale-separation at $N = 2$ means that large scales contain too much small scales, whereas scale-separation at $N = 6$ means that small scales contain too much large scales. As a result, FABEMD appears well suited to separate scales in a data/flow field but the separation point is empirical and depends on the data/flow field.

5 Numerical Code

The numerical simulations described in the previous sections were all performed using a newly developed program which is heavily based on the formulations and algorithms used by the well established existing code *NS3D* [1]. Nonetheless it contains quite a number of new and improved features which account for both the evolution in HPC resources as well as changes with respect to scientific and numerical demands.

Implementation The revised *NS3D* is a tool to perform direct numerical simulations. According to the previous version, either an explicit 8th-order alternating [6] or a 6th-order subdomain-compact finite difference scheme [8] can be used to discretize the spatial derivatives whereas a 4th-order explicit Runge-Kutta time integration scheme is utilized to obtain the necessary high-fidelity approximation of the governing equations. The solver allows for a very efficient parallelization because any computations of physical quantities can be performed locally and only some limited ghost-point information has to be shared. Originally the former *NS3D* was designed and optimized for a vector computing architecture. Therefore domain decomposition, i.e. distributed memory parallelization, was only possible in x and y coordinate directions whereas in z direction shared-memory parallelization was used. To overcome the limitations of this approach and to account for the development towards massively parallel computer clusters the new *NS3D* features fully three-dimensional domain decomposition. An example layout is shown in Fig. 6. The computational domain is decomposed in blocks by splitting it in any direction. These blocks communicate along their internal boundaries via MPI routines [20] to exchange neighbouring data. Additionally the spanwise direction is parallelized within each block by means of openMP directives. The new *NS3D's* memory management is fully dynamic and therefore allows for varying block sizes which again is an improvement over its former version as the gain in domain-decomposition flexibility outperforms the penalty of the dynamic memory management. The communication's overhead of the larger explicit stencil and additional decomposition is more than compensated for by the fact that no more linear systems are solved and the communication is hidden behind computational routines. This is especially true for communication in z-direction

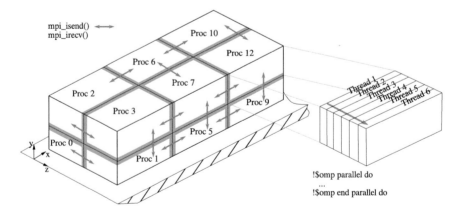

Fig. 6 Domain decomposition in x, y, z-direction and parallelization features in the new *NS3D* program

which commences before any derivatives or fluxes are computed. The new *NS3D* has three-dimensional transformations implemented and therefore allows for the use of arbitrary curvilinear meshes. Grid transformation and the weaker damping of the explicit finite differences makes filtering necessary at times. This now leads to a slight breakage of the all explicit approach because filtering for high fidelity simulations is most effectively done by applying compact filter schemes [15].

Performance The deliberate use of its well established ancestor and the implementation of the enhancements described above make the new *NS3D* highly efficient and very well behaved on the HLRS cluster. To investigate its parallel performance in comparison to the former version a scaling study is performed. For the former version, domain decomposition using MPI is only possible in the *x*- and *y*-direction where explicit 8th-order or sub-domain compact 6th-order finite differences are used for the spatial discretization. The Fourier-spectral ansatz in the spanwise direction *z* is parallelized using a shared-memory OpenMP approach, limiting the number of CPU cores used for the spanwise discretization to the 24 available cores per node (then with one MPI process per node) for the current Cray XC40 at HLRS. The missing domain decomposition in this direction limits the number of usable nodes and has been the main reason for advancing *NS3D*.

The test case used for the study consists of a laminar boundary-layer flow over a flat plate. The flow is two-dimensional but a fully three-dimensional calculation is performed to check the code's performance numbers. The domain both for the former and new *NS3D* is discretized using $6912 \times 600 \times 256$ grid points in the *x*-, *y*- and *z*-direction, respectively, leading to a total amount of 1.06 billion grid points. Whereas all three dimensions with the new *NS3D* and the *x*- and *y*-directions with the former *NS3D* are discretized using the sub-domain compact finite differences, the *z*-direction with the former *NS3D* is discretized using the Fourier-spectral ansatz. The domain is split into 384, 768 or 1536 total sub-domains in the *x*-*y*-plane. The *z*-direction parallelization with the old *NS3D* code is done using a pure OpenMP approach and with the new *NS3D* code a mixed MPI-OpenMP-approach is used: the number of OpenMP threads is hereby capped to never exceed 4, meaning that 4 cores for the spanwise direction are split into one MPI process with 4 OpenMP threads, whereas 6 cores are split into 2 MPI processes with 3 OpenMP threads each for instance. This yields a total number of cores used from $384 \times 1 = 384$ up to $1536 \times 24 = 36864$ cores for both code versions.[1] In order to allow an honest comparison between the former and new version of the code, also a pure OpenMP *z*-parallelization test is performed using 1 up to 24 cores. Each employed node is always fully used. Hence for 4 OpenMP threads 6 MPI processes are placed on one node, for example. If the product of OpenMP threads and MPI processes is smaller than 24, then MPI with the *x*-*y*-domains are used to fill up the nodes.

Figure 7 shows the speed-up and efficiency as function of the cores used for the spanwise decomposition. Note that the base-speed for both code versions is almost

---

[1]Note that the number of cores used for the *z*-parallelization is limited to 24 cores only for the former *NS3D* version due to the pure OpenMP approach. The revised version does allow higher scaling which was not investigated here.

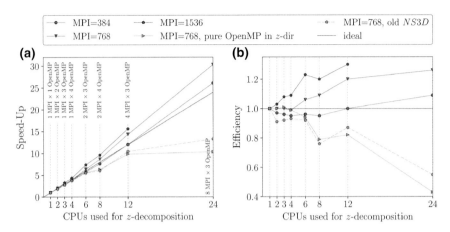

Fig. 7 a Speed-up and **b** efficiency for a variation of both MPI and OpenMP threads used for the decomposition of the spanwise z-direction (1, 2, 3, 4, 6, 8, 12 and 24). The decomposition of the domain in the x-y-plane varies between 384, 768 and 1536 MPI threads. Note that the speedup by the various MPIs is not explicitly shown in the plot

identical and the scaling by increasing the number of subdomains in the x-y-plane shows linear behavior. Therefore, the speed-up and the efficiency directly relate to absolute performance. If the results for the pure OpenMP parallelization in spanwise direction are discussed first (cyan lines for the old and new version of the code), both code versions show comparable trends and almost perfect scaling can be achieved for up to 6 OpenMP threads, staying well above 90% efficiency. For the case with 8 OpenMP threads the efficiency drops below 80%. This is due to the ccNUMA architecture of the Cray XC40 cluster, where one node consists of two sockets, each housing a 12-core CPU with a memory controller and "socket-local" memory. While each core can access all node memory, access to memory "local" to the core's socket is much faster. For 12 OpenMP threads the placing is again optimal for the architecture and the efficiency increases again to around 85% but does not reach the level of smaller OpenMP thread counts. With 24 OpenMP threads the speed-up is slightly above 10 or 13 for the two versions of *NS3D*, resulting in an efficiency of around 40–55% only. For this configuration the same problem as for 8 OpenMP threads shows off, as half the used cores constantly need to access memory placed on the far side of a ccNUMA node.

If the results for the mixed MPI- and OpenMP-approach are assessed, the hardware-induced deficits can be bypassed and super-linear behavior can be achieved for increasing numbers of CPU cores used for the spanwise decomposition. This results in a maximum increase in efficiency of about 300% compared to the former code version or to the pure OpenMP decomposed case simulated with the new code (MPI=768). The significant increase in efficiency and speed-up which can be observed for an increasing number of total CPU cores used can most likely be traced back to more efficient cache use and MPI-communication. These make highly decomposed cases about 30% more efficient than less decomposed ones in the present study. Note that the good scaling behavior in spanwise direction now matches the scaling by domain decomposition in the x-y-plane, which has been investigated in the past, see [7].

6 Conclusions

In the present publication, a new version of our numerical code *NS3D* has been presented providing a tool for the direct numerical simulation of fundamental flow problems. A scaling study has been performed to investigate the parallel behavior of the new code on the Cray XC40, which illustrates a significant increase of code efficiency of up to 300% compared to the former code version. Since the new code is used for test cases exceeding the 80,000 core limit, a highly efficient use of the resources provided by the HLRS can be guaranteed. As a representative example, direct-numerical-simulation results are presented for compressible adverse-pressure-gradient turbulent boundary layers with three pressure gradient distributions. The simulations show Clauser-parameter distributions being constant at $\beta = 0.25$, 0.6 and 1.0 for long distances in the investigated domain, indicating the self-similarity of the resulting boundary layers in the streamwise direction. This similarity is the most important condition for the investigation of turbulent boundary layers influenced by canonical pressure gradient distributions and will be the foundation for ongoing investigations.

Acknowledgements The financial support by the Deutsche Forschungsgemeinschaft, DFG, under reference number RI680/31-1, SPP1881, SFB/TRR 40-A4, and the provision of computational resources on Cray XC40 by the Federal High Performance Computing Center Stuttgart (HLRS) under grant GCS Lamt (LAMTUR), ID 44026, are gratefully acknowledged.

References

1. A. Babucke, Direct numerical simulation of noise-generation mechanisms in the mixing layer of a jet. PhD thesis, University of Stuttgart, 2009
2. S.M.A. Bhuiyan, R.R. Adhami, J.F. Khan, Fast and adaptive bidimensional empirical mode decomposition using order-statistics filter based envelope estimation. EURASIP J. Adv. Signal Process. (2008). DOIurlhttps://doi.org/10.115/2008/728356
3. F.H. Clauser, The turbulent boundary layer, in *Advances in Applied Mechanics*, vol. 4 (Elsevier, 1956), pp .1–51
4. T. Colonius, S.K. Lele, P. Moin, Boundary conditions for direct computation of aerodynamic sound generation. AIAA J. **31**(9), 1574–1582 (1993)
5. W.K. George, L. Castillo, Boundary layers with pressure gradient: another look at the equilibrium boundary layer. Near Wall Turbulent Flows (1993), pp. 901–910
6. M. Keller, Numerical investigation of gaseous film and effusion cooling in supersonic boundary-layer-flows. PhD thesis, University of Stuttgart, 2016

7. M. Keller, M.J. Kloker, Direct numerical simulations of film cooling in a supersonic boundary-layer flow on massively-parallel supercomputers, in *Sustained Simulation Performance 2013* (Springer, 2013), pp. 107–128

8. M.A. Keller, M.J. Kloker, DNS of effusion cooling in a supersonic boundary-layer flow: influence of turbulence. AIAA-2013-2897 or AIAA paper 2013-2897 (2013)

9. M.A. Keller, M.J. Kloker, Effusion cooling and flow tripping in laminar supersonic boundary-layer flow. AIAA J. **53**(4), 902–919 (2014)

10. M.A. Keller, M.J. Kloker, Direct numerical simulation of foreign-gas film cooling in supersonic boundary-layer flow. AIAA J. **55**(1), 99–111 (2016)

11. Y. Kim, I.P. Castro, Z.T. Xie, Divergence-free turbulence inflow conditions for large-eddy simulations with incompressible flow solvers. Comput. Fluids **84**, 56–68 (2013)

12. V. Kitsios, A. Sekimoto, C. Atkinson, J.A. Sillero, G. Borrell, A.G. Gungor, J. Jiménez, J. Soria, Direct numerical simulation of a self-similar adverse pressure gradient turbulent boundary layer at the verge of separation. J. Fluid Mech. **829**, 392–419 (2017)

13. M. Klein, A. Sadiki, J. Janicka, A digital filter based generation of inflow data for spatially developing direct numerical or large eddy simulations. J. Comput. Phys. **186**, 652–665 (2003)

14. H.B.E. Kurz, M.J. Kloker, Receptivity of a swept-wing boundary layer to micron-sized discrete roughness elements. J. Fluid Mech. **755**, 62–82 (2014)

15. S.K. Lele, Compact finite difference schemes with spectral-like resolution. J. Comput. Phys. **103**, 16–42 (1992)

16. J. Linn, M.J. Kloker, Numerical investigations of film cooling. RESPACE—Key Technologies for Resuable Space Systems (ed. A. Gülhan) NNFM 98, pp. 151–169

17. J. Linn, M.J. Kloker, Effects of wall-temperature conditions on effusion cooling in a supersonic boundary layer. AIAA J. **49**(2), 299–307 (2011)

18. T.S. Lund, X. Wu, K.D. Squires, Generation of turbulent inflow data for spatially-developing boundary layer simulations. J. Comput. Phys. **140**(2), 233–258 (1998)

19. G.L. Mellor, D.M. Gibson, Equilibrium turbulent boundary layers. J. Fluid Mech. **24**(2), 225–253 (1966)

20. NN, MPI—a message-passing interface standard. Technical Report CS-94-230 (University of Tennessee, Knoxville, 1994)

21. A. Perry, I. Marusic, M. Jones, On the streamwise evolution of turbulent boundary layers in arbitrary pressure gradients. J. Fluid Mech. **461**, 61–91 (2002)

22. P.R. Spalart, Direct simulation of a turbulent boundary layer up to $Re_\theta = 1410$. J. Fluid Mech. **187**, 61–98 (1988)

23. S. Stolz, N. Adams, Large-eddy simulation of high-Reynolds-number supersonic boundary layers using the approximate deconvolution model and a rescaling and recycling technique. Phys. Fluids **15**(8), 2398–2412 (2003)

24. E. Touber, Unsteadiness in shock-wave/boundary-layer interactions. PhD thesis, University of Southampton, 2010

25. A.A. Townsend, *The Structure of Turbulent Shear Flow* (Cambridge University Press, 1956)

26. M.R. Visbal, D.V. Gaitonde, On the use of higher-order finite-difference schemes on curvilinear and deforming meshes. J. Comput. Phys. **181**, 155–185 (2002)

27. C. Wenzel, B. Selent, M.J. Kloker, U. Rist, DNS of compressible turbulent boundary layers and assessment of data/scaling-law quality. J. Fluid Mech. **842**, 428–468 (2018)

28. S. Xu, M.P. Martin, Assessment of inflow boundary conditions for compressible turbulent boundary layers. Phys. Fluids **16**(7), 2623–2639 (2004)

Towards a Direct Numerical Simulation of Primary Jet Breakup with Evaporation

Jonathan Reutzsch, Moritz Ertl, Martina Baggio, Adrian Seck
and Bernhard Weigand

Abstract The primary breakup of liquid jets plays an important role in fuel injection for combustion engines and gas turbines. Due to the ambient conditions the liquid also evaporates during breakup. Direct Numerical Simulations (DNS) are well suited for the analysis of this phenomena. As a first step towards an understanding of this problem, a DNS of an evaporating jet is carried out and a comparison with a non evaporating jet is presented. We use the in-house 3D Computational Fluid Dynamics (CFD) code Free Surface 3D (FS3D) to solve the incompressible Navier-Stokes equations. The Volume of Fluid (VOF) method is used in combinations with Piecewise Linear Interface Calculation (PLIC) to reconstruct a sharp interface. The energy equation is solved to obtain the phase change at the interface. We were able to conduct a purely Eulerian simulation of an evaporating jet during atomization. The morphology of the liquid jets and the vapour concentration are shown and analysed. The droplet size distribution shows the influence of evaporation, leading to smaller droplets. The presented results are in good accordance with our expectation as well as with other investigations in the literature. In addition, we present the results of an investigation into replacing pointers to fields in our code with arrays to improve computational performance. With this change we are able to increase the efficiency of the code and obtain a speed up of more than 40%.

1   Introduction

The breakup of liquid jets into droplet sprays are prevalent in many modern technical applications as well as in nature. They can be found in spray drying, spray painting and medical inhalers. One highly relevant application is the injection of liquid fuel into a combustion chamber of a gas turbine, a combustion engine or a rocket engine. Here the jet breakup is of importance not just due to the economic impact of its efficiency, but also due to the connected environmental impact. In these cases of

J. Reutzsch (✉) · M. Ertl · M. Baggio · A. Seck · B. Weigand
Institut für Thermodynamik der Luft- und Raumfahrt, Universität Stuttgart,
Pfaffenwaldring 31, 70569, Stuttgart, Germany
e-mail: jonathan.reutzsch@itlr.uni-stuttgart.de

© Springer Nature Switzerland AG 2019
W. E. Nagel et al. (eds.), *High Performance Computing in Science
and Engineering '18*, https://doi.org/10.1007/978-3-030-13325-2_15

fuel injection, the jet breakup is not only influenced by the fluid dynamics between the liquid and the gas, but also by the evaporation of the liquid due to the high temperatures involved in this process.

Experimental investigations by Hetsroni and Sokolov [11] are concerned with jets from an internally mixing nozzle for cases with and without evaporation. Their analysis of the velocities show that the jet with evaporation is narrower and that the vapour phase reduces the generation of turbulence. Investigations by Chen et al. [4] on an acetone jet from an air-blast nozzle with phase Doppler interferometry and laser induced florescence analysed anisotropy in the dispersion of droplets and the influence of different flows and droplet sizes on the evaporation rates. In addition to research on evaporating jets, there are also investigations of jets with combustion [37] or jets at supercritical conditions [1]. A good overview of experimental results can be found in the work of Smallwood and Goulder [31].

Numerical simulations can provide useful insights in addition to experiments, as they not only allow for investigations into internal mechanisms, which often prove hard to be obtained from experiments, but also they allow for any variations of material properties and boundary conditions. Due to the highly transient nature of primary jet breakup, Direct Numerical Simulations (DNS) with their high temporal and spatial resolutions are uniquely suited for simulating this breakup. Early DNS investigations by Miller and Bellan [16] look at the interactions of a flow laden with evaporating hydrocarbon droplets in an Euler-Lagrange formulation with a higher temperature gas stream in the opposite direction, focusing on the mixing layer. They investigate the growth of the mixing layer, the saturation of the gas streams and the resulting flow field and analyse the effects of these on the movement and positioning of the droplets and their evaporation. Zeng et al. [38] used a coupled Euler-Euler Level-Set method in combination with a surface regression model to simulate an evaporating jet with a very high Reynolds number. Furthermore, a Lagrange model is used for small particles. To the authors of the present report, this simulation seems unsuited to DNS due to its coarse resolution.

An overview of research on the simulations of reactive multiphase flows is given by Reveillon et al. [22]. They use Euler-Lagrange DNS in combination with a simple reaction rate model to investigate the combustion of a liquid jet. They use their numerical simulations to classify flame sprays and to investigate the influence of acoustic modulation and also present the works of other authors on those topics. Recent results by Barba and Picano [2] present a hybrid Eulerian-Lagrangian DNS of the primary breakup of an acetone jet in the atomization regime. They investigated the reasons for clustering of droplets as well as the influence of clustering on the evaporation.

While research into DNS of the primary jet breakup without evaporation generally uses a purely Eulerian approach for simulations, see Gorokhovski and Hermann [9], for simulations of jets with evaporation, generally, Lagrangian methods combined with a breakup model are used for the treatment of the evaporating droplets. As far as the authors are aware, no DNS of jet breakup exists, where the droplets are not simplified as Lagrangian particles. This however makes it very hard to properly investigate the influence of evaporation on the primary jet breakup. Since the authors of the present report are experienced in the purely Eulerian simulation of primary jet breakup [5, 6] as well as of evaporation [28] they decided to approach this issue.

This work is a first step towards a purely Eulerian simulation of an evaporating jet during atomization. The results will help to better understand the influence of evaporation on the primary breakup.

2 Formulation and Numerical Method

FS3D is a CFD code for multiphase flows which has been developed at ITLR (Institute of Aerospace Thermodynamics) in Stuttgart. *Direct Numerical Simulation* (DNS) is employed to solve the incompressible Navier Stokes equations, hence, no turbulence models are used. FS3D is based on the *Volume of Fluid* (VOF) method [12] and makes use of the *piecewise linear interface reconstruction* (PLIC) method [23]. The energy equation with temperature dependent thermo-physical properties is included and phase change processes, such as evaporation and ice growth, can be considered. Several recent studies in the last twenty years show the efficiency and ability to simulate highly dynamic processes with the code. These comprise dynamic processes, droplet deformation [25], droplet impact onto thin films [8], droplet collisions [27], droplet wall interactions [26], bubbles [36], evaporating droplets [30] as well as rigid particle interactions [20] and liquid jets [6, 29, 39].

The flow field is computed by solving the governing equations for mass, momentum and energy conservation which are given by

$$\rho_t + \nabla \cdot (\rho \mathbf{u}) = 0, \tag{1}$$

$$(\rho \mathbf{u})_t + \nabla \cdot [(\rho \mathbf{u}) \otimes \mathbf{u}] = \nabla \cdot (\mathbf{S} - \mathbf{I}p) + \rho \mathbf{g} + \mathbf{f}_\gamma, \tag{2}$$

$$(\rho c_p T)_t + \nabla \cdot (\rho c_p T \mathbf{u}) = \nabla \cdot (k \nabla T) + T_0 \left[(\rho c_p)_t + \nabla \cdot (\rho c_p \mathbf{u}) \right] + \Delta h_v \dot{m}'''. \tag{3}$$

Here ρ denotes the density, $\mathbf{u}$ the velocity vector, p the static pressure and $\mathbf{S}$ is the viscous stress tensor that is defined by $\mathbf{S} = \mu \left[\nabla \mathbf{u} + (\nabla \mathbf{u})^T \right]$ for Newtonian fluids with μ representing the dynamic viscosity; $\mathbf{g}$ describes the volume forces, such as gravity, and $\mathbf{f}_\gamma$ the body force which is used to model surface tension in the vicinity of the interface. Regarding the energy Eq. (3), c_p denotes the specific heat capacity at constant pressure, T the temperature, T_0 the reference temperature for zero enthalpy and k the heat conductivity. Δh_v represents the latent heat of the evaporating fluid and $\dot{m}'''$ is the volumetric mass source of vapour. The Navier-Stokes equations are discretised using a finite volume method and a Marker and Cell (MAC) grid [10], hence, velocities are stored on cell faces, scalars at cell centres, respectively. To identify different phases, in our case liquid and gas, an additional field variable f_1 is introduced based on the VOF method of Hirt and Nichols [12], defined as

Fig. 1 Schematic of scalar fields of both VOF-variables f_1 (black) and f_2 (red)

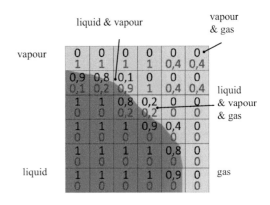

$$f_1(\mathbf{x}, t) = \begin{cases} 0 & \text{in the gaseous phase,} \\ ]0;\,1[& \text{at the interface,} \\ 1 & \text{in the liquid phase.} \end{cases}$$

The transport equation of the VOF-variable f_1 reads

$$(f_1)_t + \nabla \cdot (f_1 \mathbf{u}_\Gamma) = -\dot{m}'''/\rho_l \tag{4}$$

Due to evaporation of the liquid phase we introduce a second VOF-variable f_2, which describes the volume fraction of vapour. A schematic of both variables and the appearance of the different phases is depicted in Fig. 1. The advection of the volume fraction is described analogously to the f_1 transport by

$$(f_2)_t + \nabla \cdot (f_2 \mathbf{u}_{gp}) = \nabla \cdot (\mathfrak{D}_{bin} \nabla f_2) + \dot{m}'''/\rho_v . \tag{5}$$

The velocities in both transport equations $\mathbf{u}_\Gamma$ and $\mathbf{u}_{gp}$, which denote the advection velocity of the interface and the gaseous phase, respectively, differ due to the volume generation due to evaporation. Furthermore, ρ_l and ρ_v are the liquid and vapour density. The diffusion of vapour inside the gas is considered by the binary diffusion coefficient $\mathfrak{D}_{bin}$. Following the *one-field-formulation* the local fluid properties are calculated with the volume fractions, e.g. the density is determined as

$$\rho(\mathbf{x}, t) = \rho_l f_1(\mathbf{x}, t) + \rho_v f_2(\mathbf{x}, t) + (1 - f_1(\mathbf{x}, t) - f_2(\mathbf{x}, t))\rho_g . \tag{6}$$

The volumetric mass source of vapour $\dot{m}'''$ is present only in interfacial cells. It is obtained by multiplying the local area specific mass source of vapour $\dot{m}''$, which is based on the gradient of vapour mass fraction X_v, with the local interface density a_Γ. Following the derivation of Kays et al. [13] it can be calculated as

$$\dot{m}'' = \frac{\mathfrak{D}_{bin}\rho_{gp}}{1 - f_1} \nabla X_v \hat{\mathbf{n}}_\Gamma \tag{7}$$

The vapour pressure at the surface is assumed to be the saturation pressure and is estimated using the Wagner equation. For more details regarding this method the reader is referred to Schlottke and Weigand [30].

For the calculation of the fluxes FS3D makes use of the *piecewise linear interface reconstruction* (PLIC) method proposed by Rider and Kothe [23]. Therewith, the required sharp interface can be achieved in order to suppress numerical dissipation. Two advection methods are implemented in FS3D. An algorithm, based on Strang splitting [32], where three one-dimensional non-conservative transport equations in each direction are considered successively , and a fully three dimensional advection method, which uses a construction of a six-faced polyhedron for a more realistic approximation of the volume fluxes [21]. The surface tension can be computed using different models. The conservative continuous surface stress (CSS) model by Lafaurie et al. [14], the continuum surface force model (CSF) by Brackbill et al. [3], and the balanced force approach by Popinet [17] are implemented. The Poisson equation for pressure, which is a condition resulting from the incompressible formulation, is solved implicitly. To solve this, two methods are currently implemented: the first one solves the resulting set of linear equations with a multigrid solver, which uses a Red Black Gauss Seidel algorithm for smoothing and can be run in a V- or W-cycle scheme [24]. The second one uses the recently integrated software package UG4, which is a massively parallel geometric multigrid solver [33–35]. It was developed at the Goethe Center for Scientific Computing at the Goethe University in Frankfurt, and it was designed for the efficient solution of partial differential equations. For more details regarding the integration into FS3D the reader is referred to Ertl et al. [5]. Due to the high computational effort of DNS the code is fully parallelised using MPI and OpenMP. Up to now simulations with a maximum of eight billion computational cells have been conducted at the High Performance Computing Center Stuttgart (HLRS) and first tests with over eight billion cells have been run successfully. FS3D is well validated and has good performance on the Cray XC40 Hazel Hen supercomputer [5, 18].

3 Numerical Setup

All simulations are performed on a three-dimensional regular grid. All jet simulations are performed twice, first without evaporation and then with evaporation according to the method described in Sect. 2. An overview of the computational domain is shown in Fig. 2. It is identical for both kind of simulations. The nozzle inflow boundary condition, which is placed at the centre of the left side, has a diameter of $D_0 = 2.5 \cdot 10^{-3}$ m. The whole domain is of rectangular shape with a length $l = 20D$ and a quadratic base with $h = w = 8D$. The left side, except for the nozzle, is set as a no-slip wall, whereas all other sides have a continuous (Neumann, zero gradient) boundary condition. For the inlet velocity profile we choose a parabolic profile with a maximum velocity of $U_0 = 10.3$ m/s, a turbulent intensity of $Tu = 10\%$

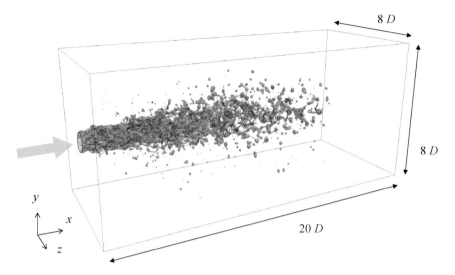

Fig. 2 Computational domain and coordinate system of jet simulation. No slip wall and inflow boundary condition on the left side. Orange arrow indicates the direction of injection.

and turbulent length scale of $L_t = D_0/8 = 3.125 \times 10^{-4}$ m. Gravitational acceleration has been neglected.

In x-direction the domain is discretized with 1536 cells and in y- and z-direction with 512 cells, respectively. In the y-z-plane we apply a grid refinement, which leads to a higher grid resolution in the centre around the jet and, therefore, the setup satisfies the necessary spatial resolution according to the Kolmogorov length scale for the considered Reynolds number and the Ohnesorge number. The dimensionless numbers are calculated as

$$Re = \frac{\rho_l U_0 D}{\mu_l} = 30{,}000\,, \tag{8}$$

$$Oh = \frac{\mu_0}{\sqrt{\rho_l \sigma D_0}} = 0.002\,. \tag{9}$$

This places the jet into the atomisation regime. Due to preceding simulations with evaporation and several validations discussed in Sect. 2, we choose water at 300 K as the liquid and dry air at the same temperature as the surrounding gas. Further investigations with higher temperature and, hence, higher evaporation rates are planned in the future. All required material properties are shown in Table 1. The diffusion coefficient is only defined for vapour and air, and diffusion into the liquid phase is neglected. For the simulation without evaporation only density, viscosity and surface tension are needed.

Table 1 Material properties

	Symbol	Unit	Water	Vapour	Air
Density	ρ	kg/m^3	996.2	0.722	1.162
Visocsity	μ	μPa s	853.5	18.94	9.735
Sp. heat capacity	c_p	J/(kg K)	4180	1869	1001
Heat conductivity	λ	W/(m K)	0.609	0.026	0.018
Surface tension	σ	mN/m		72.75	
Diffusion coefficient	$\mathfrak{D}_{bin}$	cm^2/s		0.232	
Enthalpy	Δh_v	kJ/kg		2256	

4 Results

In the following the results of the simulation are analysed, starting with an investigation of the morphology of the jet. Therefore, the structure, and the generation of ligaments and droplets, as well as their development and behaviour are discussed. The main focus is put on the comparison of the two simulations and the evaluation not only of occurring differences but also of potential similarities. For an easier differentiation we call the simulation without evaporation "*Case 1*" and the one with evaporation "*Case 2*", respectively. Furthermore, we use the following dimensionless time for our investigations

$$t^* = \frac{U_0 \, t}{D} \, . \tag{10}$$

Since the evaporation method was never applied before in a simulation with many droplets of different sizes at the same time or even never in a jet simulation, it is essential to get an initial rough impression of the behaviour of the evaporating jet. In Fig. 3 the simulation of the evaporating jet is shown at $t^* = 21.0$. Both VOF-variables are depicted, f_1 as a contour plot, showing the iso-surface through $f_1 = 0.5$, representing the liquid water jet surface and f_2 as a volume rendering indicating the concentration of vapour. Close to the nozzle exit (inflow boundary) the jet exhibits a cylindrical liquid core, with visible wave-like surface deformations. In stream wise direction the jet breaks up into a spray composed of ligaments and droplets. In the regions around this spray the vapour concentration shows peaks. The breakup of the liquid jet as well as the development of ligaments and droplets are in good accordance with existing investigations [6]. Likewise, the propagation of the vapour is as expected. In the central region of the spray, which due to the large amount of ligaments and droplets provides the highest amount of liquid surface, we get the highest amount of evaporation. Indeed, evaporation happens around all droplets, which can be well observed for separated droplets moving away from the jet. The physical behaviour exhibited in the simulation are in good accordance with the behaviour of evaporating jets described in literature [11].

Fig. 3 Simulation of an evaporating jet simulated with FS3D. The vapour field, which is represented by the VOF-variable f_2, is depicted as a rendered volume. The fluid is shown as a contour in blue $f_1 = 0.5$

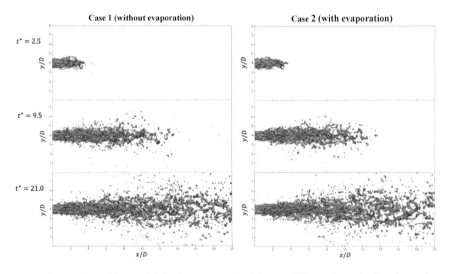

Fig. 4 Comparison of the spatial developement of both jets at different dimensionless times $t^* = 2.5$, $t^* = 9.5$ and $t^* = 21.0$. Both sequences show the contours of the f_1-field, on the left side without evaporation and on the right side with evaporation

The evolution of both jets after injection at times $t^* = 2.5, t^* = 9.5$ and $t^* = 21.0$ is shown in Fig. 4; on the left side the jet without evaporation (Case 1) and on the right side with evaporation (Case 2) are shown. Only the f_1-field is shown for both simulations, to reduce visual occlusion. At $t^* = 2.5$, after the injection, the jet core still has a cylindrical shape. The jet core is disturbed by the turbulence of the nozzle flow, as well as by the interaction with the ambient atmosphere, which causes three

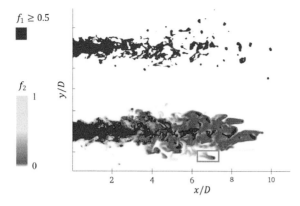

Fig. 5 Comparison of Case 1 (top) and Case 2 (bottom) at $t^* = 7.0$. Both jets show 2D slices at the centre of the z-direction. The liquid is represented by the f_1-variable with a value larger than 0.5. In the jet with evaporation the vapour is highlighted for the sake of visibility in a yellow-red-white colour scheme

dimensional waves to form on the jet surface. These waves are less developed for the simulation with evaporation. This can be seen especially well on the surface of the parabolic shaped core close to the nozzle exit. In addition, the breakup process in Case 1 seems more developed, showing a larger expansion of the jet and a higher number of detached droplets. This behaviour is in good accordance with literature, where i.e. Hetsroni and Sokolov [11] found, that evaporation damps the turbulence, thereby smoothing the disturbances.

Small droplets, which naturally form in both cases, can be observed to disappear in Case 2. This is because the droplets fully evaporate. This process becomes clear from Fig. 5. Here a 2D slice through the centre of the z-direction of the liquid is shown for both cases at $t^* = 7.0$, with the vapour concentration overlaid for Case 2. The simulation without evaporation, which is displayed at the top, differs particularly in regions further downstream and further in radial direction, compared to the one with evaporation. In Case 1 more droplets can be observed in these regions, whereas in Case 2 droplets have already evaporated and only swathes of vapour remain, indicated by the detached red areas of high vapour concentration.

These small developed vapour flow fields in the wake of single detached droplets are in good accordance with the literature, as similar phenomena have been investigated in the past. For instance Schlottke and Weigand [30] simulated (DNS) evaporating single droplets and investigated the vapour distribution in the zone around and especially behind a droplet in an airflow. A good example of these flow fields is highlighted in the green rectangle in Fig. 5.

The jets generally evolve in a similar way towards times $t^* = 14.2$ and $t^* = 21.0$, as depicted in Fig. 4. This relates especially to the deformation of the surface due to the interaction with the surrounding air. As a consequence, surface waves start to detach from the jet core and form ligaments. These then disintegrate further and break up into droplets. The atomisation of the jet can be seen clearly. This behaviour

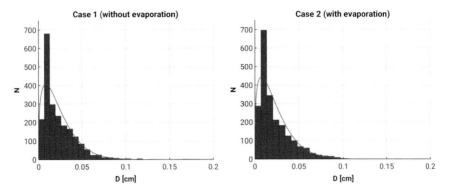

Fig. 6 Droplet size distribution for Case 1: no evaporation (left) and Case 2: with evaporation (right), with the Weibull distribution in red for time $t^* = 21.0$

is in good agreement with the classification of breakup regimes by Ohnesorge [15] as was to be expected from previous studies (e.g. [6]).

The size distribution for the droplets is shown in Fig. 6, for both cases, together with the Weibull distribution. To obtain this result the separated liquid components were identified in a post-processing step and their equivalent diameter was calculated from their liquid mass. The results are in accordance with our other observations, showing more droplets in the three smallest size categories for the evaporating jet. It is especially noticeable, that without evaporation a very small amount of droplets in the size categories 0.12 and 0.17 cm are visible, while in the case without evaporation none were observed. The peak of the Weibull curve is also shifted farther towards small droplet sizes for this case. The average droplet diameters have been calculated to be $d^{av}_{Case1} = 0.0233$ cm for Case 1 and as $d^{av}_{Case2} = 0.0229$ cm for Case 2. Thus, the evaporation leads to smaller droplets.

A calculation of the jet angle gives for Case 1 $\alpha_{Case1} = 26.90°$ and for Case 2 $\alpha_{Case2} = 26.54°$ for time $t^* = 21.0$. The jet angle was calculated from the distance of the centre of mass of the droplets to the centre of the nozzle exit (inflow boundary condition). The 5% most distant results were discarded as outliers. The influence of evaporation on the jet angle is neglectable in this case.

5   Computational Performance

5.1   Dynamic Memory Allocation: Pointer Versus Allocatable Arrays

The majority of FS3D data field are dynamically allocated. That is, their size is specified during execution and memory space is allocated and deallocated at run time. In the Fortran 95 standard, allocatable arrays or arrays with pointer attribute can

Table 2 Strong scaling setup

Problem size	512^3			
MPI-processes	2^3	4^3	8^3	16^3
Cells per process	256^3	128^3	64^3	32^3
Nodes	2	4	32	256
Processes per node	4	16	16	16

both be used for dynamic memory management. Allocatable arrays are associated unequivocally to a specific place in memory after their allocation, whereas pointers can refer to different memory spaces during execution. This feature of pointer variables is useful in numerous programming applications, such as abstract data types or linked lists. It has been reported by HLRS experts that the use of pointers can hinder good optimization by the compiler, because at compilation time the memory spaces referred to by the pointer variables are unclear. Therefore, it is recommended to replace pointers with allocatable arrays when possible.

For historical reasons, the great majority of FS3D data field have, however, been declared with the pointer attribute, even if they could have been replaced by allocatable arrays. With the hope of enhancing FS3D computational efficiency, we created a novel version of the code in which a great part of the pointer arrays were replaced by allocatable arrays.

The performance analysis of the new code version with allocatable arrays instead of pointers is carried out by simulating an oscillating droplet. The baseline case is an elongated droplet which is initialised as an ellipsoid with the semi-principal axis $a = b = 1.357$ and $c = 0.543$ mm at the centre of a cubic domain with an edge length of $x = y = z = 8$ mm. The fluid of the droplet is water at 293 K. We used a fixed computational Cartesian grid with 512^3 cells and varied the number of processors from 2^3 up to 16^3. Since we did not use hyperthreading, the number of processors corresponds to the number of processes. The details of the analysis setup are shown in Table 2. Here, we used only spatial domain composition with MPI parallelization. For information on FS3D performance with hybrid OpenMP and MPI operation see [19] and [7].

All the simulations were carried out for the same number of cycles and with a constant processor clock rate (2.5 GHz). From the average processor time, we estimated the number of cycles per hour (CPH), which in turn was used to calculate the strong scale efficiency (SSE). The latter is defined as:

$$SSE = \frac{CPH_N}{N \times CPH_1}.$$ (11)

As we don't have a case with just one processor, Eq. (11) changes to

$$SSE = \frac{CPH_N}{N/8 \times CPH_8}.$$ (12)

The number of cycles per hour for each simulation are shown in Table 3. As expected, the code version with allocatable arrays is generally faster (see Fig. 7,

Table 3 Number of cycles per hour for the simulated cases

Processors	CPH allocatable	CPH pointer
8	81.478	62.376
64	400.727	329.111
512	1488.482	1032.787
4096	396.458	480.496

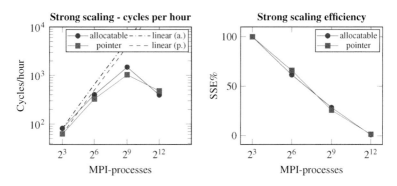

Fig. 7 Number of cycles per hour (on the left) and strong scaling efficiency (on the right) over the number of processes for the code versions with allocatable and pointer arrays. In the first graph, the linear speed up for both code versions is shown for comparison

left), with the only exception of the 4096 processor case. Here however, the MPI-communication takes about 96% of the entire simulation time (93% for the pointer version). We can infer therefore, that this latter case is not representative of the performance of allocatable over pointer arrays. From the right side of Fig. 7, it can be seen that the two code versions have a very similar trend for the strong scale efficiency. This indicates that the conversion from pointer to allocatable arrays had little influence on the MPI-communication. The larger difference was observed for the 512 processor case (64^3 cells per processor), where the code with allocatable arrays was about 44% faster.

6 Conclusions

We set up direct numerical simulations of the breakup of a liquid water jet at moderate temperatures and Reynolds numbers injected into an air atmosphere at the same temperature. We used the Volume of Fluid (VOF) method to capture the liquid phase and we solved the energy equation in combination with the introduction of a second volume fraction for the vapour phase for simulating evaporation of the jet. We did an additional simulation without evaporation for comparison.

We analysed the morphology of the liquid phase, as well as the distribution of the vapour concentration. We observed a damping effect of the evaporation on the turbulence disturbing in the jet. We also observed the evaporation of liquid, especially for droplets, due to their larger surface area. This result has been confirmed via calculating the droplet size distribution, showing more smaller droplets in the case of evaporation. These observations are in good accordance with results from literature analysing experimental and numerical investigations.

Due to the isothermal initial conditions and an ambient atmosphere, which is far from the boiling point, the evaporation rate was comparably small. Consequently, it doesn't affect the fluid dynamic of the jet and the atomisation as strongly as, for instance, flash boiling would do. Nevertheless, differences could already be observed and quantified.

The main objective of this study was to examine the feasibility and assess, if FS3D is capable of simulating a jet with evaporation. After a successful evaluation of this, further investigations are planned in the future, especially with higher temperatures and, therefore, with higher evaporation rates.

In addition, we replaced the large pointer arrays in our Fortran code with allocatable arrays. This measure led to an increase in serial performance, which effects also scale up to several thousand processors.

Acknowledgements The authors kindly acknowledge the *High Performance Computing Center Stuttgart* (HLRS) for support and supply of computational time on the Cray XC40 platform under the Grant No. FS3D/11142. In addition the authors kindly acknowledge the financial support by the Deutsche Forschungsgemeinschaft (DFG) for SFB-TRR75, GRK 2160/1 and WE2549/36-1.

References

1. S. Baab, C. Steinhausen, G.B.W. Lamanna, F.J. Förster, A quantitative speed of sound database for multi-component jet mixing at high pressure. Fuel (2018). https://doi.org/10.1016/j.fuel.2017.12.080 (In Press)
2. F.D. Barba, F. Picano, Clustering and entrainment effects on the evaporation of dilute droplets in a turbulent jet. Phys. Rev. Fluids **3**, 034304 (2018). https://doi.org/10.1103/PhysRevFluids.3.034304
3. J.U. Brackbill, D.B. Kothe, C. Zemach, A continuum method for modeling surface-tension. J. Comput. Phys. **100**(2), 335–354 (1992)
4. Y.C. Chen, S.H. Starner, A.R. Masri, A detailed experimental investigation of well-defined, turbulent evaporating spray jets of acetone. Int. J. Multiph. Flow **32**, 389–412 (2006)
5. M. Ertl, J. Reutzsch, A. Nägel, G. Wittum, B. Weigand, Towards the implementation of a new multigrid solver in the DNS code FS3D for simulations of shear-thinning jet break-up at higher reynolds numbers, in *High Performance Computing in Science and Engineering' 17* (Springer, 2017), pp. 269–287. https://doi.org/10.1007/978-3-319-68394-2
6. M. Ertl, B. Weigand, Analysis methods for direct numerical simulations of primary breakup of shear-thinning liquid jets. At. Sprays **27**(4), 303–317 (2017)
7. C. Galbiati, M. Ertl, S. Tonini, G.E. Cossali, B. Weigand, DNS investigation of the primary breakup in a conical swirled jet, in *High Performance Computing in Science and Engineering '15 Transactions of the High Performance Computing Center, Stuttgart (HLRS)* (2016), pp. 333–347
8. H. Gomaa, I. Stotz, M. Sievers, G. Lamanna, B. Weigand, Preliminary investigation on diesel droplet impact on oil wallfilms in diesel engines, in *ILASS-Europe 2011, 24th European Conference on Liquid Atomization and Spray Systems*, Estoril, Portugal, September 2011 (2011)

9. M. Gorokhovski, M. Herrmann, Modeling primary atomization. Annu. Reviiew Fluid Mech. **40**, 343–66 (2008)
10. F.H. Harlow, J.E. Welch, Numerical calculation of time-dependent viscous incompressible flow of fluid with free surface. Phys. Fluids **8**(12), 2182–2189 (1965)
11. G. Hetsroni, M. Sokolov, Distributino of mass, velocity, and intensity of turbulence in a two-phase turbulent jet. J. Appl. Mech. **38**, 315–327 (1971)
12. C.W. Hirt, B.D. Nichols, Volume of fluid (VOF) method for the dynamics of free boundaries. J. Comput. Phys. **39**(1), 201–225 (1981). https://doi.org/10.1016/0021-9991(81)90145-5
13. W. Kays, M. Crawford, B. Weigand, Convective heat and mass transfer-4th edn. (McGraw-Hill, 2004)
14. B. Lafaurie, C. Nardone, R. Scardovelli, S. Zaleski, G. Zanetti, Modelling merging and fragmentation in multiphase flows with SURFER. J. Comput. Phys. **113**(1), 134–147 (1994)
15. A.H. Lefebvre, *Atomization and Sprays* (Taylor & Francis Group, 1989)
16. R.S. Miller, J. Bellan, Direct numerical simulation of a confined three-dimensional gas mixing layer with one evaporating hydrocarbon-droplet-laden stream. J. Fluid Mech. **384**, 293–338 (1999)
17. S. Popinet, An accurate adaptive solver for surface-tension-driven interfacial flows. J. Comput. Phys. **228**(16), 5838–5866 (2009). https://doi.org/10.1016/j.jcp.2009.04.042
18. P. Rauschenberger, J. Schlottke, K. Eisenschmidt, B. Weigand, Direct numerical simulation of multiphase flow with rigid body motion in an Eulerian framework, in *ILASS—Europe 2011, 24th European Conference on Liquid Atomization and Spray Systems*, Estoril, Portugal (2011)
19. P. Rauschenberger, J. Schlottke, B. Weigand, A computation technique for rigid particle flows in an eulerian framework using the multiphase DNS code FS3D, in *High Performance Computing in Science and Engineering '11 Transactions of the High Performance Computing Center, Stuttgart (HLRS)* (2011). https://doi.org/10.1007/978-3-642-23869-7_23
20. P. Rauschenberger, B. Weigand, Direct numerical simulation of rigid bodies in multiphase flow within an Eulerian framework. J. Comput. Phys. **291**, 238–253 (2015). https://doi.org/10.1016/j.jcp.2015.03.023
21. M. Reitzle, C. Kieffer-Roth, H. Garcke, B. Weigand, A volume-of-fluid method for three-dimensional hexagonal solidification processes. J. Comput. Phys. **339**, 356–369 (2017). https://doi.org/10.1016/j.jcp.2017.03.001
22. J. Reveillon, C. Perat, Z. Bouali, Examples of the potential of dns for the understanding of reactive multiphase flows. Int. J. Spray Combust. Dyn. 3(1), 63–92 (2011)
23. W.J. Rider, D.B. Kothe, Reconstructing volume tracking. J. Comput. Phys. **141**(2), 112–152 (1998). https://doi.org/10.1006/jcph.1998.5906
24. M. Rieber, Numerische Modellierung der Dynamik freier Grenzflächen in Zweiphasenströmungen. Dissertation, Universität Stuttgart (2004)
25. M. Rieber, F. Graf, M. Hase, N. Roth, B. Weigand, Numerical simulation of moving spherical and strongly deformed droplets, in *Proceedings ILASS-Europe* (2000), pp. 1–6
26. N. Roth, J. Schlottke, J. Urban, B. Weigand, Simulations of droplet impact on cold wall without wetting. ILASS (2008), pp. 1–7
27. N. Roth, H. Gomaa, B. Weigand, Droplet collisions at high weber numbers: experiments and numerical simulations, in *Proceedings of DIPSI Workshop 2010 on Droplet Impact Phenomena & Spray Investigation*, Bergamo (2010)
28. S. Ruberto, J. Reutzsch, N. Roth, B. Weigand, A systematic experimental study on the evaporation rate of supercooled water droplets at subzero temperatures and varying relative humidity. Exp. Fluids **58**(5), 55 (2017). https://doi.org/10.1007/s00348-017-2339-5
29. W. Sander, B. Weigand, Direct numerical simulation and analysis of instability enhancing parameters in liquid sheets at moderate reynolds numbers. Phys. Fluids **20**(5), 053301 (2008)
30. J. Schlottke, B. Weigand, Direct numerical simulation of evaporating droplets. J. Comput. Phys. **227**(10), 5215–5237 (2008)
31. G. Smallwood, O. Goulder, Views on the structure of transient diesel sprays. At. Sprays **10**, 355–386 (2000)

32. G. Strang, On the construction and comparison of difference schemes. SIAM J. Numer. Anal. **5**(3), 506–517 (1968)
33. A. Vogel, A. Calotoiu, A. Nägel, S. Reiter, A. Strube, G. Wittum, F. Wolf, Automated performance modeling of the UG4 simulation framework, in *Software for Exascale Computing—SPPEXA 2013–2015, LNCSE*, ed. by H. Bungartz, P. Neumann, W.E. Nagel, vol. 113 (2016), pp. 467–481. https://doi.org/10.1007/978-3-319-40528-5_21
34. A. Vogel, A. Calotoiu, A. Strubem, S. Reiter, A. Nägel, F. Wolf, G. Wittum, 10,000 performance models per minute-scalability of the UG4 simulation framework, in *Euro-Par 2015*, ed. by J. Träff, S. Hunold, F. Versaci, vol. 9233 (2015), pp. 519–531. https://doi.org/10.1007/978-3-662-48096-0
35. A. Vogel, S. Reiter, M. Rupp, A. Nägel, G. Wittum, UG4: a novel flexible software system for simulating PDE based models on high performance computers. Comput. Vis. Sci. **16**(4), 165–179 (2013). https://doi.org/10.1007/s00791-014-0232-9
36. H. Weking, J. Schlottke, M. Boger, C.D. Munz, B. Weigand, DNS of rising bubbles using VOF and balanced force surface tension, in *High Performance Computing on Vector Systems* (Springer, Berlin, Heidelberg, New York, 2010)
37. J.F. Widmann, C. Presser, A benchmark experimental database for multiphase combustion model input and validation. Combust. Flame **129**(1), 47–86 (2002). https://doi.org/10.1016/S0010-2180(01)00374-1
38. P. Zeng, B. Binninger, N. Peters, M. Herrmann, Simulation of primary breakup for diesel spray with phase transition, in *Proceedings ICLASS 2009* (2009)
39. C. Zhu, M. Ertl, B. Weigand, Numerical investigation on the primary breakup of an inelastic non-newtonian liquid jet with inflow turbulence. Phys. Fluids **25**, 083102 (2013)

Direct Numerical Simulation of Turbulent Flow Past an Acoustic Cavity Resonator

Lewin Stein and Jorn Sesterhenn

Abstract The present project studies the sound pressure inside an acoustic resonant cavity driven by turbulent flow. With this numerical study as an interim result, the ultimate goal is to improve the industrial layout of resonant cavities, in general: For example to circumvent resonance conditions. The current rise of high-performance computing allows us to simulate the dynamics of the non-linear turbulence-acoustic interaction by the high-quality method of a Direct Numerical Simulation (DNS). For the first time, the three dimensional geometry investigated here can be studied numerically in full detail without simplification. So far numerical studies of Helmholtz resonators with resolved neck shape do not consider an inflowing turbulent boundary layer or do not resolve all system scales, but assume some form of turbulence model. To effectively run the DNS on a supercomputer, a multi-block parallelization method is newly implemented for complex geometries, which consist of multiple, different sized blocks. Both in strong and weak scaling test a previous single-block parallelization is outperformed. The optimal load of gridpoints per core is identified and a distinction between misleading and meaningful weak scaling tests is made.

1   Introduction

Whenever a compressible fluid flows along a cavity acoustic fluid-structure interaction plays a major role. Many examples can be listed in which this interaction between acoustics and the turbulent flow takes place. Noise silencers consisting of cavity arrays are installed in the intake and the exhaust ducts of combustion engines ranging from ships to airplanes. Beside silencing properties cavities may give rise to desirable tones of wind instruments like transverse flutes and organs or undesired "sunroof buffeting" of moving vehicles. In any case, there is a lack of understanding about the coupling between the acoustics and the turbulence and the associated interchange of acoustical and flow energy.

L. Stein (✉) · J. Sesterhenn
Institut für Strömungsmechanik und Technische Akustik,
Müller-Breslau-Str. 15, 10623 Berlin, Germany
e-mail: Lewin.Stein@tu-berlin.de

J. Sesterhenn
e-mail: Joern.Sesterhenn@tu-berlin.de

© Springer Nature Switzerland AG 2019
W. E. Nagel et al. (eds.), *High Performance Computing in Science and Engineering '18*, https://doi.org/10.1007/978-3-030-13325-2_16

259

This paper is organized as follows. Section 1.1 specifies the configuration of the present study. Section 2 deals with the computational aspects: subdivided into a parallelization Sect. 2.1 and a justification how the resources were used 2.2. In Sect. 3 the subject-specific results are presented. The arranged subsections are the DNS setup 3.1, the visualization of the flow 3.2, the evaluation of $k_x - \omega$—spectra 3.3, and the comparison with theoretical and experimental results 3.4. A conclusion is given in Sect. 4.

1.1   Present Study

The case under investigation consists of a turbulent flat plate flow on top (x is stream-wise) and a rectangular cavity below (negative y-direction). The cavity is joined via a rectangular opening/neck with the flat plate. All walls are defined by adiabatic, no-slip conditions. In spanwise z-direction, the boundary layer is periodically continued. The top side is open. Since the neck volume is much smaller than the volume of the cavity, the classical idealization of a Helmholtz resonator is effectual. A sketch of the geometry is provided in Fig. 1. The spatial dimensions match an experimental study by [1]. In the following, this allows comparisons. The characteristic parameters are a Reynolds number $Re_{\delta_{99}}$ approx. 4200 with respect to the boundary layer thickness δ_{99} at the upstream edge of the neck; and a low Mach number of ≈ 0.1, which corresponds to a free stream velocity of $u_0 \approx 40$ m/s. The constant Prandtl number is $Pr = 0.7$. Standard temperature and pressure (273 K, 100 kPa) are used.

2   Computational Aspects

Simulations were performed on our in-house code "Navier-Stokes Fortran", abbreviated below as NSF (not licensed). Since 2000 NSF is being continuously developed and validated [2–5]. NSF solves the compressible, three-dimensional Navier-Stokes equations in real space and time using the Finite Difference Method on a structured grid. The Navier-Stokes equations can be either formulated in a characteristic form [2] or in a skew-symmetric form [3]. In this project, we work with the skew-symmetric form. Despite the Finite Difference Method, the skew-symmetric form guarantees additional conservation properties in order to deal properly with acoustics and shocks. In particular, not only explicit entities like mass, momentum and the total energy are conserved but also implicit entities like the kinetic or internal energy (see Fig. 2).

The numerical differentiation is calculated with either implicit or explicit derivation schemes between 4th and 8th order with a spectral-like resolution of the wavenumbers [6]. For time integration both classical low-storage Runge-Kutta Methods of 4th order and exponential time-stepping using Krylov subspaces can be applied. The code utilizes the libraries ARPACK, LIBPACK, MPI, and OpenMPI.

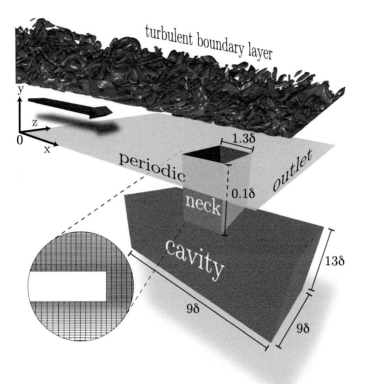

Fig. 1 Setup of the studied Helmholtz resonator under grazing turbulent flow. In red and blue isosurfaces of the vorticity are rendered. All units are normalized with the boundary layer thickness δ_{99}, defined at the upstream neck edge $\delta_{99,neck} \cong 1$cm. Near the opening, the grid is further refined. The origin of the coordinate system "0" is located at the leading $(x = 0)$, upper $(y = 0)$, front $(z = 0)$ flatplate corner

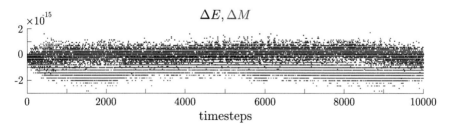

Fig. 2 Conservation down to machine double precision of mass (blue) and internal plus kinetic energy (black) for 10,000 steps

2.1 Parallelization

The parallelization (MPI) is achieved by a hybrid approach using a multi-block-code with a ghost layer synchronization as well as block internal decomposition into CPUs. Thereby the multi-block-code parallelization strategy was newly developed in the scope of this project.

Single-Block (SB) parallelization Before performing a derivative the single-block scheme redistributes the data between the cores such that stripes in the direction of the desired derivative are created (Fig. 3). By this, the derivatives can be done in the whole block at once without dealing with core-to-core interfaces. This is especially useful for compact (i.e. implicit) derivatives. Since units of well-structured data are sent in a predictable way, this procedure is surprisingly efficient. However, if the number of cores gets beyond 1000, the communication overhead becomes the bottleneck. Moreover, the computing space is restricted to a single Cartesian block, suitable for large academic research studies with a simple geometry.

Multi-Block (MB) parallelization The most straightforward procedure to parallelize the calculation of large and complex geometries is to split up the grid into manageable sub-pieces, called blocks here (Fig. 4). In order to exchange information between neighbored blocks, small overlapping regions are synchronized. The size of this strip of so-called ghost-points matches the width of the derivative stencil. This is especially simple for explicit derivatives, while for compact derivatives, the parallel solution creates extra effort. In contrast to the SB parallelization scheme the optimal number of effective gridpoints per core is reduced by the number of ghost points, but therefore the communication overhead almost grows

Fig. 3 Single-block parallelization scheme: a cube-shaped domain is reshaped according to the direction of the derivatives

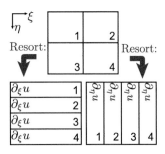

Fig. 4 Multi-block parallelization scheme: inter block communication between differently shaped cubes is accomplished by synchronized layers of overlapping ghost points

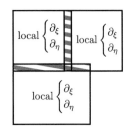

no worse than linear (see weak scaling in Fig. 6). Besides, the more efficient data communication, this method has a further advantage: more general multi-block grids can be created.

Thereby, the SB and the MB parallelization can be combined easily. This allows load balancing and adjustment to different hardware. Additionally, the code uses a compilation flag for auto-vectorization.

Scaling Test

To prove the suitability of our solver on a Tier-0/1 system we discuss now its scalability. All the following scaling tests were conducted on a Cray XC40 machine with a Haswell E5-2680 V3 processor (30 MB cache per core, 24 cores and 128 GB RAM per node). NSF's optimal load of gridpoints per core is $\approx 29^3$, our "sweet spot". The measurements of wall time per steps were averaged over 100 steps.

In the case of **strong scaling**, all values were rescaled in such a way, that the SB data point closest to the sweet spot exactly lies on the ideal linear curve (both $384 \times 384 \times 24$ points per 144 cores and $384 \times 384 \times 96$ points per 576 cores result in $\approx 29^3$ gridpoints per core). The used rescaling factors are denoted in the legend of Fig. 5 behind the @-sign. As shown in Fig. 5 the values below the sweet spot display a super-linear scaling, probably due to a cache overflow (too many gridpoints per core). Values above the sweet spot have no cache bottleneck anymore (few gridpoints per core) and most likely suffer from the communication overhead. To achieve comparability the MB scaling runs have the same number of physical gridpoints as the SB runs (see legend of Fig. 5). Since the effective computational domain is bigger due to a ghost layer of four gridpoints per dimension the sweet spot of the MB case is shifted to slightly higher values. Especially, this is visible for the larger case with a resolution of $384 \times 384 \times 96$. To include the loss of runtime because of an increased computational domain the rescaling factor of the SB case is used for the MB case too. Roughly speaking, both parallelization schemes display a similar strong scaling speedup.

Figure 6 shows **weak scaling**: In the case of weak scaling the gridpoints per core directly match our sweet spot $\approx 29^3$. Therefore, cache effects are minimized. The comparable effective computational domain in the MB case is $\approx (29-4)^3$ with a ghost layer of four points. Again for both parallelization schemes, the same rescaling factor is used. This is the wall time per step and per core in case of 24 cores with SB parallelization. While the SB method loses 50% efficiency between 24 and 9216 cores, the MB method stays close to a linear scaling. The present fluid solver always works on the sweet spot (weak scaling), and such makes use of this nearly ideal MB scaling.

Informative Value of Strong Scaling Tests

An issue rarely mentioned is the manipulability of strong scaling tests: With increasing core number the reduction of cache/RAM overflow can be equally balanced with an increase of the communication overhead. Such two problems cancel out each

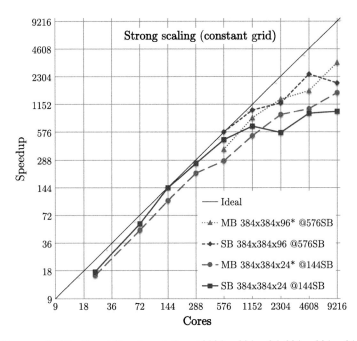

Fig. 5 Strong scalability. Two different resolutions of 384 × 384 × 96, 384 × 384 × 24 and both available parallelization strategies SB, MB are tested. * denotes the number of computational grid-points including virtual ghost points

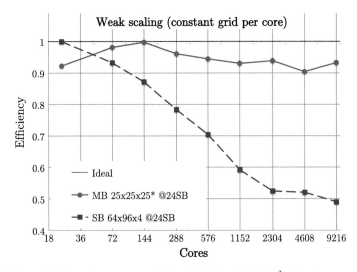

Fig. 6 Weak scalability at the sweet spot of $64 \times 96 \times 4 \approx (25 + 4)^3$ gridpoints per core. With more cores, the MB parallelization strategy has a much better efficiency than the SB scheme. * denotes the number of physical gridpoints excluding virtual ghost points

other and result in ideal scaling. An example is given in Fig. 7. The upper spuriously ideal curve is rescaled at 512 cores and calculated with 81^3 gridpoints per core, while the lower curve is rescaled at 32 cores and calculated with 32^3 gridpoints per core. In the second case, the problem size reflects the sweet spot of the program, with a minimized cache overflow. This case holds an actual informative value of the communication overhead above 32 cores (x-axis of Fig. 7) and of the cache overflow below 32 cores (super linear scaling). In the first spuriously ideal case the problem size is 16 times bigger and such implies heavy cache/RAM overflow superposed with communication overhead. The presented ideal scaling curve has no informative value.

As done here, it is recommended to invest some time to identify the machine dependent optimal load of gridpoints per core (sweet spot) of the program by starting with a low number of gridpoints per core first, which clearly excludes cache/RAM exhaustion effects.

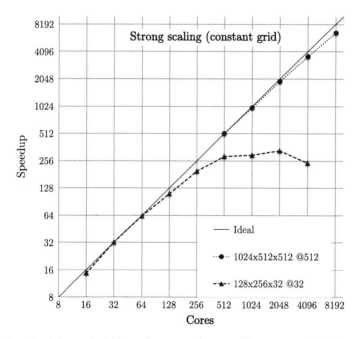

Fig. 7 Example of the manipulability of strong scaling tests. The upper spuriously ideal curve is normalized at 512 cores with $\approx 81^3$ gridpoints per core, causing cache/RAM overflow and communication overhead at once. The lower curve is calculated on the same supercomputer solving the same problem but is normalized at 32 cores with 32^3 gridpoints per core, mainly causing communication overhead

2.2 *How the Resources were Used*

Over the course of the project 5 Million CPU-h were utilized. Beyond the initial test-ing and validation period, several mayor consecutive production runs were conducted as described in Table 1.

To justify these numbers more detailed considerations follow now. Our geometry consists of four blocks. The blocks are depicted in Fig. 8. Their spatial resolution and core number are listed in Table 3. For proper statistical averaged information (including low frequencies) all four blocks require running five cycles. Thereby the statistical measures are calculated on-the-fly. The post-processing is executed locally at our institute cluster. More physical aspects are discussed in Sect. 3.1.

Table 1 Typical parameters of a production run

Wall time	24 h
Number of cores	40,000
Number of gridpoints	1.1×10^9
Number of time steps	32 000
Time step (no I/O)	1.9 s
Mean time step (incl. I/O and initialization)	2.7 s
Physical time step	3×10^{-7} s
Full dump (every 1000 steps, allows restart)	91 GB
Mini dump (every 50 steps)	330 MB
Total data written	3.1 TB
Total data read	91 GB
Size of NSF's executable	328 MB

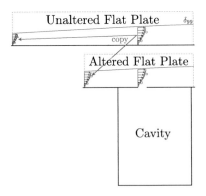

Fig. 8 Qualitative side view of the two parallel executed simulations in this project. The upper simulation is a turbulent flat plate flow. The lower simulation is a turbulent flat plate flow altered by a wall-mounted cavity. Walls are plotted as solid black. Nonreflective and inlet boundaries are drawn as dashed gray lines. MPI copied information within and between the simulations is denoted by red arrows

Table 2 Underlying physical and numerical parameters of the simulation

M	0.11	Is the mach number
N_{3D}	3	Is the spatial dimension
N_{vars}	5	Are the state variables density ρ, velocities u, v, w and pressure p
CFL	0.7	Is the set Courant-Friedrichs-Lewy number for the used RK4
W_{pi}	1.9 s	Is the walltime per rank / iteration step with on-the-fly data evaluation
g_{refine}	0.1	Is the grid refinement factor between the smallest Δx_{min} and the average grid spacing $\overline{\Delta x}$
S_{dp}	8 byte	Is the size of a double precision number

Table 3 The spatial resolution of the simulation. Physical issues ([+]notation) are discussed in [5]

Block	Range of grid spacing			Number of gridpoints			CPUs per dimension		
	Δx^+	Δy^+	Δz^+	n_x	n_y	n_z	c_x	c_y	c_z
Channel	1.7−2.4	1.6−5.9	3.7	1440	288	672	32	9	32
Neck	1.7	1.6	3.7	180	24	582	6	2	6
Cavity	1.7−2.4	1.6−6.0	3.7	912	912	576	48	19	16
Flat plate	1.7−2.4	1.6−5.9	3.7	2304	288	672	48	6	48

The computational requirements of the project (Table 1) can be deduced from the parameters given in Tables 2 and 3 using the relations.

$$N_{steps} = \frac{\text{x-gridpoints}}{g_{refine}} \frac{1 + M^{-1}}{\text{CFL}} \text{ cycles,} \tag{1}$$

$$\text{CPU-h} = \text{cores } N_{steps} W_{pi}[h], \tag{2}$$

$$\text{Field size} = \text{total-gridpoints } N_{vars} S_{dp}. \tag{3}$$

3 Subject-Specific Results

3.1 Setup of the Direct Numerical Simulation

The mesh of the Direct Numerical Simulation needs to resolve the smallest Kolmogorov length scale η. Thereby η depends on the viscous length scale δ_v like $\eta \approx 2\delta_v$. To calculate the viscous length scale the mean velocity profile needs to be known. A renowned analytic solution was formulated for example by Van-Driest. A more detailed discussion of the initial and boundary condition can be found in [5]. The validation of a sufficient resolution and a comparison with the analytical Van-Driest solution of a mean turbulent boundary layer profile is published in [5], too. Where physically permitted, mesh refinement (see Fig. 1) reduces the cost of the computation.

To avoid boundary reflections, characteristic non-reflective boundary conditions, as well as a sponge function in the wall near regions, are utilized. More details are provided in [5]. Here we just want to mention a utilized feature of our Multi-block code: In order to simulate a turbulent boundary layer flow grazing past a cavity, a realistic turbulent flat plate flow is a precondition. A consistent well-established turbulence generation procedure for flat plate flow is the Recycling-Rescaling method [7]. Besides our actual plate-cavity case, this method requires a second simulation of an isolated flat plate to generate unaltered physical turbulence. Within this second isolated turbulent boundary layer simulation the turbulence information is looped (recycled). More specifically the fluctuations are copied from a downstream location upwards to the inlet. Using the Multi-block code we simulate both cases in parallel. In this way, we can copy the turbulence information twice not only within the plate itself but also from the plate to the coupled plate-cavity case, too (as sketched in Fig. 8). All operations take place within the fast RAM only. All the time-dependent inflow information does not need to be saved and reread from the work disk. The application of the Recycling-Rescaling method in the main plate-cavity case directly would disturb the subsonic turbulent boundary layer.

Additionally, the secondary flat plate simulation serves as a perfect comparable reference case with identical inflow conditions: To compare the turbulent flow of a flat plate (without a cavity) with the modified flow over a wall-mounted cavity.

3.2 Visualization of the Turbulence-Acoustic Interaction at a Helmholtz Cavity

The fully developed flow field above and inside a Helmholtz cavity is shown in Figs. 9 and 10: instantaneous isosurfaces of the vorticity $|\nabla \times \mathbf{u}| = 3000$ Hz and pressure $p = 100,010$ Pa are rendered. The broadband noise of the boundary layer excites cavity modes, which are radiated as large radial structures out of the opening into the free semi-half space of the upper flat plate. The appearance of turbulence is restricted to the upper side of the wall, and the immediate downstream surrounding of the opening. For later modeling approaches, this allows assuming potential flow above the boundary layer and pure acoustical propagation inside the cavity. For a further investigation of the turbulent boundary layer thickness δ_{99} over the streamwise length, an interested reader may refer to [5].

3.3 Comparison of the $k_x - \omega$—Spectra of a Flat Plate Flow with and without a Helmholtz Cavity

In this subsection, we show how fluid structures and their acoustical properties are affected by the presence of a resonant cavity. To distinguish acoustical and fluid-related effects, a spectral representation is favorable, since it separates acoustic and

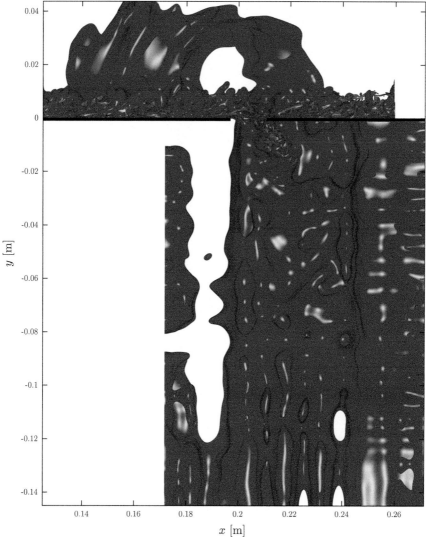

Fig. 9 xy side view of instantaneous isosurfaces are shown: the vorticity is red; the pressure is blue; walls are black. At the upper upstream and the upper and lower downstream edge of the cavity opening vortex separation is visible, accompanied by small pressure structures of the same size. This coupling of acoustics and flow at the opening is a crucial part of the set-up of acoustic models

flow by its distinctive propagation speeds ($c \approx 340$ m/s vs. $u \lesssim 40$ m/s), respectively. For sound prediction models, the frequency dependent sound pressure level is the typical quantity of interest, too. To convert the pressure fluctuations from the real $x - t-$space to the $k_x - \omega$—space (streamwise wavevector and angular frequency), we execute a Fourier transformation in time and space. More specifically, we calculate

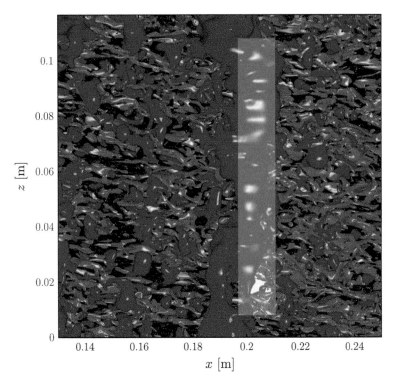

Fig. 10 xz top view of instantaneous isosurfaces are shown: the vorticity is red; the pressure is blue. The underlying opening is highlighted by a rectangle, while the wall is colored black

the power spectral density, which is defined as Fourier transformation of the time-averaged autocorrelation of pressure fluctuations p':

$$\Phi_{pp}(k_x, \omega) = \frac{1}{8\pi^3} \iint\limits_{-\infty}^{\infty} < p'(x,t)p'(x+r,t+\tau) > \ e^{-i(k_x r - \omega \tau)} dr d\tau \ . \quad (4)$$

In case of our discrete DNS data, the power spectral density can be approximated by [8]

$$\Phi_{pp}(k_x, \omega) \cong \frac{1}{NMf_{st}f_{sx}} \left| \sum_{m=1}^{M} \sum_{n=1}^{N} p'(x_m, t_n) e^{i\omega t_n} e^{i k_x x_m} \right|^2 , \quad (5)$$

with the sampling rate $f_{st} = N/T$ of the time period T with N time steps and the sampling rate $f_{sx} = M/L$ of the length L with M gridpoints, respectively. From the power spectral density Φ_{pp} the sound pressure level (SPL) spectrum follows directly:

$$SPL(k_x, \omega) = 10 \; \log_{10} \left[f_{bin} k_{bin} \frac{\phi(k_x, \omega)}{p_{ref}^2} \right] \mathrm{dB} \; , \qquad (6)$$

with the narrowband frequency bin $f_{bin} = 1/T = 50$ Hz, the wavevector spacing $k_{bin} = 1/L$ and the standard threshold of human hearing $p_{ref} = 2 \times 10^{-5} Pa$.

The resulting SPL $k_x - \omega$—spectra of Figs. 11 and 12 are averaged over all spanwise z locations for a constant height y above the bottom wall. All acoustic fluctuations are located around the frequency axis between the phase velocities $u_0 \pm c$, while most fluid related pressure fluctuations are centered around the convective ridge with a phase velocity below the free stream velocity u_0. Inside the convective ridge, most energy is stored, which is also reflected by the highest sound pressure levels.

The mounting of a cavity inside an initial undisturbed flat plate flow results in a secondary vortex sheet near the wall. This increases the SPL at smaller scales (larger k_x) and results in a broader distribution of the convective ridge (greater variance in $\pm k_x$). Beside the convective ridge, the most distinctive modifications induced by the cavity are the acoustic cavity modes itself, which are characterized by localized dots of maximal SPL near the frequency axis. An agreement between frequencies of these acoustic cavity modes and the analytical predictions based on the speed of sound and the spatial dimension exists. Rossiter modes are not distinctive in Figs. 11 and 12.

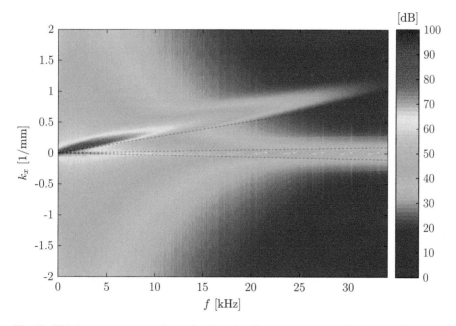

Fig. 11 SPL $k_x - \omega$—spectrum of turbulent flat plate flow at a constant wall distance of $y_+ = 13$ (Eq. 6). The characteristic phase velocities u_0 and $u_0 \pm c$ are marked by dashed black lines

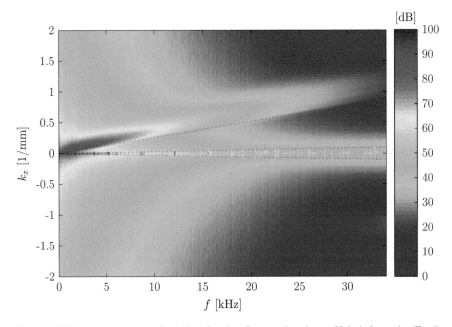

Fig. 12 SPL $k_x - \omega$-spectrum of turbulent flat plate flow grazing above a Helmholtz cavity (Eq. 6). The wall distance is $y_+ = 13$

3.4 Comparison of the Sound Pressure Level of the DNS with a Theoretical Solution and an Experimental Measurement

In this subsection, the results are compared to a semi-analytical solution by [9] and an experiment by [1].

Figure 13 shows the SPL of the turbulent flat plate flow directly at the wall ($y = 0$). The model of [9] agrees with the DNS results within ± 2 dB. Only the first three data-points vary more due to the non-periodic FFT input (short 20 ms time series, top-hat window).

In Fig. 14 the SPL of the Helmholtz cavity is evaluated at a microphone position in the no-flow region (cf. Sect. 3.2), which is centered at the bottom of the cavity. An agreement between DNS and the experiment is visible. The SPL error varies between ± 7 dB. The extrema frequencies of the experiment and the DNS results agree very well: The first peak at 300 Hz can be identified as Helmholtz base frequency (the operating frequency of a Helmholtz silencer). The second peak at 1300 Hz is the first vertical (y) cavity mode. Higher transversal (x, z) cavities modes begin with 1700 Hz. Typically, cavity silencers operate at subsonic and low-frequency conditions near the Helmholtz base frequency far below the cutoff frequency of higher cavity modes. Hence, Fig. 14 focuses on low frequencies (similar to [1]).

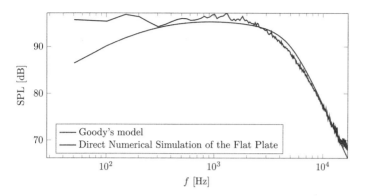

Fig. 13 Narrowband SPL spectrum ($f_{bin} = 50$ Hz). To calculate the SPL (ω) of the DNS a time series of the pressure fluctuations at the wall ($y = 0$) of the flat plate is evaluated with Eqs. 5 and 6 for $L_{x-plate} = 7.5\,\delta_{99,neck}$ (averaged over all k_x)

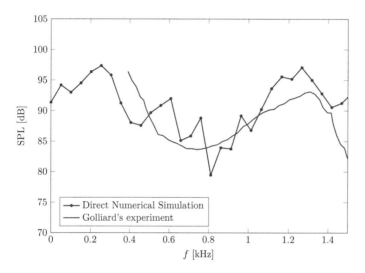

Fig. 14 Narrowband SPL spectrum ($f_{bin} = 50$ Hz). To calculate the SPL(ω) of the DNS a time series of the pressure fluctuations centered at the bottom of the Helmholtz cavity is evaluated with Eqs. 5 and 6 for a single spatial point

4 Conclusion

For the first time, a Direct Numerical Simulation of a Helmholtz cavity driven by a fully turbulent, grazing flow is conducted [5], using a newly developed multi-block-code. Based on detailed strong and weak scaling tests the performance of the new multi-block-code is demonstrated. In doing so, the optimal load of gridpoints per core is identified and a distinction between misleading and meaningful weak scaling tests is made.

An agreement between the results of the DNS, a semi-analytical solution (Fig. 13) and an experiment are shown (Fig. 14). Based on the present DNS dataset, the follow-up publication [10] presents a sound prediction model in detail including an interpretation of the physics. Instead of costly experimental parameter studies in an aeroacoustic wind tunnel, the new model shall simplify the design process of acoustic cavity resonators in turbulent flow.

In a nutshell, utilizing a Direct Numerical Simulation this work ultimately aims at a key challenge of the rapidly growing transport sector: the handling of cavity oscillations of planes, ships, and land vehicles.

References

1. J. Golliard, Noise of Helmholtz-resonator like cavities excited by a low Mach-number turbulent flow. Ph.D. thesis, Université de Poitiers (2002)
2. J. Sesterhenn, A characteristic-type formulation of the Navier-Stokes equations for high order upwind schemes. Computers and Fluids **30**, 37 (2001)
3. J. Reiss, *A Family of Energy Stable* (Skew-Symmetric Finite Difference Schemes on Collocated Grids. J. Sci, Comput, 2015)
4. S. Bengoechea, L. Stein, J. Reiss, J. Sesterhenn, *Numerical Investigation of Reactive and Non-reactive Richtmyer-Meshkov Instabilities*. Active Flow and Combustion Control (2014)
5. L. Stein, J. Reiss, J. Sesterhenn, *Numerical simulation of a resonant cavity: Acoustical response under grazing turbulent flow*. Notes on Numerical Fluid Mechanics and Multidisciplinary Design pp. 671–681 (2017)
6. S.K. Lele, Compact finite difference schemes with spectral-like resolution. J. Comput. Phys. **103**(1), 16 (1992)
7. S. Pirozzoli, M. Bernardini, Turbulence in supersonic boundary layers at moderate Reynolds number. J. Fluid Mech. **688**, 120 (2011)
8. S.M. Alessio, *Digital Signal Processing and Spectral Analysis for Scientists* (Springer, 2016)
9. M. Goody, Empirical Spectral Model of Surface Pressure Fluctuations. AIAA Journal **42**(9), 1788 (2004)
10. L. Stein, J. Sesterhenn, An acoustic model of a Helmholtz resonator under a grazing turbulent boundary layer. Acta Mech. **230** (2019). https://doi.org/10.1007/s00707-018-2354-5

High-Fidelity Direct Numerical Simulation of Supercritical Channel Flow Using Discontinuous Galerkin Spectral Element Method

Fabian Föll, Sandeep Pandey, Xu Chu, Claus-Dieter Munz, Eckart Laurien and Bernhard Weigand

Abstract Supercritical fluids are suggested as one of the potential candidates for the next generation nuclear reactor by Generation IV nuclear forum to improve the thermal efficiency. But, supercritical fluids suffer from the deteriorated heat transfer under certain conditions. This deteriorated heat transfer phenomenon is a result of a peculiar attenuation of turbulence within the flow. This peculiarity is difficult to predict by conventional turbulence modeling. Therefore, direct numerical simulations were used in the past employing the low-Mach assumption with a finite volume code. As a next step, we extend the discontinuous Galerkin spectral element method for direct numerical simulation of supercritical carbon dioxide. The higher-order of accuracy and fully compressible code improve the fidelity of the simulations. A computationally robust and efficient implementation of the equation of state was used which is based on adaptive mesh refinement. The objective of this report is to demonstrate the usage of the code in complex flow. Therefore, channel geometry is adopted and simulations were conducted at different Mach numbers to observe the effects of compressibility in the supercritical fluid regime. The isothermal boundary conditions were used at the walls of the channel. The mean profile of pressure,

F. Föll · C.-D. Munz
Institute of Aerodynamic and Gas dynamics, University of Stuttgart, Stuttgart, Germany
e-mail: foell@iag.uni-stuttgart.de

C.-D. Munz
e-mail: munz@iag.uni-stuttgart.de

S. Pandey (✉) · E. Laurien
Institute of Nuclear Technology and Energy Systems, University of Stuttgart,
Stuttgart, Germany
e-mail: sandeep.pandey@ike.uni-stuttgart.de

E. Laurien
e-mail: eckart.laurien@ike.uni-stuttgart.de

X. Chu · B. Weigand
Institute of Aerospace Thermodynamics, University of Stuttgart, Stuttgart, Germany
e-mail: xu.chu@itlr.uni-stuttgart.de

B. Weigand
e-mail: bernhard.weigand@itlr.uni-stuttgart.de

© Springer Nature Switzerland AG 2019
W. E. Nagel et al. (eds.), *High Performance Computing in Science and Engineering '18*, https://doi.org/10.1007/978-3-030-13325-2_17

density and temperature are drastically affected by the Mach number variation. In the end, scalability tests were conducted and code shows a very good parallel scalability up to 12,000 cores.

1 Introduction

The Supercritical Water-Cooled Reactor, also known as the High-Performance Light Water Reactor (HPLWR) in Europe, is one of the promising options among other Generation-IV International Forum's (GIF) reactors. However, the structural material may degrade under high temperature and extremely corrosive environment. Because, the safety of nuclear power plants cannot be compromised, and to circumvent this problem, supercritical carbon dioxide (sCO$_2$) is one of the alternatives with the 0 ozone depleting potential and 1 global warming potential. The lower critical pressure and temperature of carbon dioxide (p_c = 7.38 MPa, T_c = 304.25 K) as compared to water (p_c = 22.06 MPa, T_c = 674.09 K) provide an opportunity to generate electrical power in a reduced operating range. Therefore, it can operate in mid-temperature range (723–973 K), and improves the plant reliability. When a fluid is heated in the near-critical region then it undergoes a transition from liquid-like state to gas-like state (vice versa for cooling), and these states are separated by a coexistence line, commonly known as Widom line [3]. Along the Widom line, a sharp cross-over of different thermophysical properties takes place. Inspired by it, we have investigated the heat transfer and hydraulic characteristics of supercritical CO$_2$ by means of direct numerical simulation (DNS) using a finite volume based solver, OpenFOAM. Our findings suggest that when CO$_2$ flows upward and a uniform heat flux is applied to the wall of the tube then heat transfer is poor between the wall and bulk fluid and this results in a peak in the wall temperature [7]. A similar observation was reported during the heat rejection process of CO$_2$ in the near-critical region in the downward flow. Further investigation unveils that reduction in sweep and ejection events are the culprit for this peculiar phenomenon [20]. Further examination was conducted for horizontal flow of CO$_2$ with the heating to find out the effects of gravity [18]. In this type of configuration, thermal stratification was noticed in which low-density fluid accumulates at the top of the pipe [8]. These studies were not only limited to get fundamental insight but also by these numerical experiments [7, 9, 20], a comprehensive database was developed, which is being used at several places to improve the heat transfer models [21, 22] and development of new methods based on machine learning using deep neural networks [5, 6].

All these investigations were based on the low-Mach assumption which decouples the acoustic interaction from the compressibility effects. In order to study the effect of compressibility in the trans- and supercritical region, fully-compressible simulations are necessary. Kawai [16] conducted a direct numerical simulation of heated transcritical boundary layers with zero pressure gradients under supercritical conditions. The effects of compressibility were significant in trans-critical flow even at low Mach numbers. Recently, Sengupta et al. [23] performed a fully compressible

DNS at low-Mach numbers for channel flow with a hot and cold wall. For the same, they deployed a finite-volume code using the Peng-Robinson (PR) equation of state (EoS). They showed that compressibility effects can exist in the near-wall region at low Mach numbers when the pseudo-critical temperature transition occurs near the cold wall. Kim et al. [18] also performed DNS for channel flow under transcritical temperature conditions with a hot an cold wall at a Mach number of 0.26. In their simulations, a fully compressible and fully conservative approach was used and they tackle the numerical stability problem via systematic grid refinement.

All these fully-compressible DNS studies were conducted at low-Mach numbers with PR as EoS by using finite volume method. The PR-EoS is computationally less expensive but it deviates from the standard database such as RefProp by NIST [1]. Therefore, in this study, we made an attempt to extend a high-order code based on the discontinuous-Galerkin spectral element method (DGSEM) developed by the group of Prof. C.-D. Munz at IAG, University of Stuttgart, for supercritical DNS. This code has shown massively parallel scalability over 100,000 cores on HPC cluster at HLRS and it is another motivation for us to implement this code to our case. The EoS was derived from the standard database of CoolProp [4], which is an accurate and open-source database. To demonstrate our approach, a simple geometry of channel is chosen. Both the walls are maintained at a different temperature to account for the variable properties. The Mach number was varied in order to observe its effect on mean variables and instantaneous flow. The parallel-computation performance of present case is also evaluated in the end.

2 Computational Details

This section describes the governing equation, numerical method, equation of states, computational details along with the supporting theory for the analysis.

2.1 Governing Equations

Fully compressible Navier-Stokes equations were used to describe the flow in this DNS. In the following Eqs. 1–3 show the mass, momentum and energy conservation for this DNS, respectively;

$$\frac{\partial \rho}{\partial t} + \nabla \cdot (\rho \mathbf{u}) = 0 \tag{1}$$

$$\frac{\partial \rho \mathbf{u}}{\partial t} + \nabla \cdot [\rho \mathbf{u}\mathbf{u} + p\mathbf{I} - \tau] = \zeta \tag{2}$$

$$\frac{\partial e}{\partial t} + \nabla \cdot [(e + p)\mathbf{u}] = \nabla \cdot [\tau \cdot \mathbf{u} - \mathbf{q}] + \mathbf{u}\zeta \tag{3}$$

$$e = \varepsilon + \frac{1}{2}\rho\mathbf{u} \cdot \mathbf{u}; \quad \varepsilon = f(\rho, p); \quad p = f(\rho, \varepsilon) \tag{4}$$

In Eqs. 1–4, ρ is the density, $\mathbf{u}$ is the velocity vector, p is the pressure, e is the total energy per unit volume, ε represents the internal energy per unit volume. The viscous stress is denoted by τ and heat flux by $\mathbf{q}$, defined by Eqs. 5 and 6, respectively. A source term (ζ) was added to the streamwise momentum and correspondingly to the energy equation to drive the flow. This forcing term is analogous to the pressure gradient in pressure driven flows. The forcing term is applied only in the streamwise direction and remains constant throughout the domain. But, it varies in the time in such a way to keep the mass flux and bulk density constant. In Eq. 6, λ is the thermal conductivity.

$$\tau = \mu\left[[\nabla\mathbf{u} + (\nabla\mathbf{u})^T] - \frac{2}{3}(\nabla \cdot \mathbf{u})\mathbf{I}\right] \tag{5}$$

$$\mathbf{q} = -\lambda\nabla T \tag{6}$$

2.2 Numerical Method and Flow Domain

For solving the Navier-Stokes equations, we adopted a CFD solver based on the discontinuous-Galerkin spectral element method [15]. A density-based approach for the solution of the compressible Navier-Stokes equations is applied along with the equation of state described in the next Section. A brief description of the solver follows; A typical conservation equation is shown in Eq. 7. Here, U is the solution vector, F is the flux for viscous and heat conduction.

$$\mathbf{U_t} + \mathbf{F}(\mathbf{U}, \nabla\mathbf{U}) = 0 \tag{7}$$

During the computation, the whole domain is subdivided into non-overlapping hexahedral elements, followed by a mapping of each element onto the reference cube element, $E = [-1, 1]^3$ by a mapping $x(\xi)$, where $\xi = (\xi, \eta, \zeta)$ is the master element coordinate vector. Due to this mapping, Eq. 7 turns to:

$$J\mathbf{U_t} + \varphi(\mathbf{U}, \nabla_\xi\mathbf{U}) = 0 \tag{8}$$

where J is the Jacobian and ∇_ξ is the nabla operator for mapped space. As a next step, the solution and fluxes are approximated by a tensor-product basis of 1-D Lagrange interpolating polynomials; volume and surface integrals are replaced by Gauss-Legendre or Gauss-Legendre-Lobatto quadrature. The evaluation of the numerical fluxes at the element faces is done by formulating a Riemann problem. The time discretization was performed using an explicit fourth-stage Runge-Kutta scheme.

2.3 Equation of State

For the equation of state, we used an EoS library tool based on the adaptive tabulation using a quad-tree approach. In this method, we first build tables for EoS prior to the simulations. It relies on interpolation instead of iteration and uses L_2 projection method to calculate the polynomial coefficients [11]. The tool has the capability to derive these tables from typical EoS (e.g. Peng-Robinson, Redlich-Kwong, and ideal gas) and standard database (e.g. CoolProp [4] and RefProp [1]). Figure 1 shows a result from this implementation for carbon dioxide derived from CoolProp database for speed of sound and isochoric specific heat as a function of density and pressure. The grids are also shown in the figure to illustrate the adaptive mesh refinement capabilities of the library. The mesh is refined close to the saturation curve where thermophysical properties show drastic variations. But as one moves away from the critical region, cells become larger in size. The mesh refinement is governed by the error level and maximum steps for refinement. As a next step, these EoS tables are supplied to the solver before starting the simulation. During the simulation, an iteration of a specific quantity is replaced by an interpolation of the polynomial basis. The valid cell for the EoS interpolation is found by a grid traversal through the quad-tree until the lowest level is reached [14]. Compared with the analytical EoS (e.g. Peng-Robinson) and the full IAPWS-IF97, this EoS table achieves a high-accuracy combined with acceptable computational costs at the same time. A speedup of three orders of magnitude was reported by [11] as compared with the IAPWS-IF97 EoS.

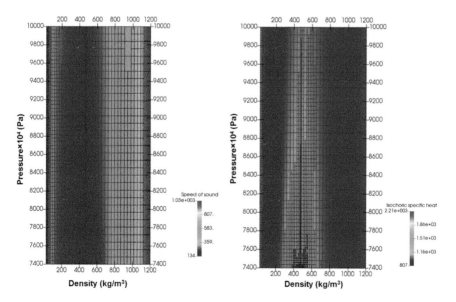

Fig. 1 Implementation of EoS library to CO_2 derived from CoolProp [4]

2.4 Computational Details

A plane channel flow configuration as depicted in Fig. 2 is chosen for this study. The channel has dimension $L_x \times L_y \times L_z = 4\pi H \times 2H \times 2\pi H$, with H as the half channel height. The channel half width is adjusted to match the Reynolds and Mach number. The periodic boundary conditions are employed in the homogeneous streamwise and spanwise directions. The flow in each case including the validation study was initialized with a laminar velocity profile superimposed with random fluctuations;

$$u_x = u_{x,max}(1 - y^2)(1 + \Upsilon), u_y = 0, u_z = 0 \qquad (9)$$

where $u_{x,max}$ is the maximum velocity in a laminar channel flow and it is fixed to 1.5 of the mean velocity; and Υ is the perturbation function. This perturbation function is defined as; $\Upsilon = A \times random - number$. The constant A can be increased or decreased to speed up the development of turbulent flow in the channel. It is set to 0.5 in the present study. For other variables (e.g. temperature, density), a uniform value was imposed in the flow-field. The computational domain consists of $N_x \times N_y \times N_z = 128 \times 56 \times 128$ with a polynomial degree of 3–5. It results in approximately 58–200 Millions degree of freedom for the domain. This mesh resolution is comparable to Sengupta et al. [23] (used 186 Mio.) and Ma et al. [19] (used 37 Mio.). The mesh resolution was chosen to resolve the Kolmogorov and Batchelor scales. We are aiming to simulate three distinct cases, which correspond to different bulk Mach numbers as shown in Table 1. The friction Reynolds number (Re_τ) is based on half channel height. The simulated cases are for CO_2 with variation in Mach number. These cases have a bulk Reynolds number of 2800 based on half channel height, bulk velocity and viscosity at reference condition at 8 MPa pressure and 340 K temperature. The wall temperature is fixed to 335 K for the bottom wall and 345 K for the top wall. With the increase in Mach number, wall-normal temperature profile will differ from incompressible flow and it shows the effect of non-linear property change.

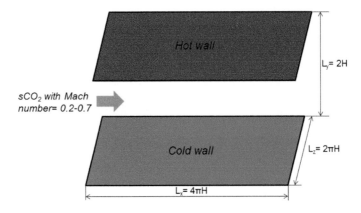

Fig. 2 Geometry for fully compressible turbulent channel flow

Table 1 A summary of simulation conditions; reference conditions are at $p = 8$ MPa, $T = 340$ K

Sr. No.	Case	Fluid	Ma	Re	Re_τ	T_{up} (K)	T_{bottom} (K)	ρ_b (kg/m^3)	p_b (MPa)
1	CM02R28	CO$_2$	0.2	2800	190				8.020
2	CM08R28	CO$_2$	0.5	2800	190	345	335	178	8.189
3	CM15R28	CO$_2$	0.7	2800	190				8.394

3 Results and Discussion

Direct numerical simulations were performed for CO_2 as well as ideal gas for validation. In this section, any generic quantity (say ϕ) is averaged in time as well as in streamwise and spanwise, and $\overline{\phi}$ shows the same while fluctuating part is shown as ϕ'.

3.1 Validation

The here used high-order code has been used to study liquid-vapor flow with phase transition and surface tension [13], two-phase flow [12] and shock-capturing [24]. Moreover, the code is extensively validated for compressible and incompressible cases including periodic-hill, NACA 0012 airfoil, spatially-developing supersonic turbulent boundary layer [2]. Additionally, a validation study which is relevant for our test case is provided in this section. Identical conditions have been used as of the original case. The low Mach number channel flow corresponding to case M02 in Table 2 is validated with the well known incompressible channel flow DNS data of Kim et al. [17] (acronym as KMM). Figure 3 shows the wall-normal profiles for the mean velocity and root mean square velocity fluctuations normalized with the bulk velocity (U_b). A fairly good agreement can be observed from Fig. 3 for the incompressible flow case. For high-Mach number cases, dynamic viscosity was calculated with the power law with an exponent of 0.7. The agreement for high-Mach numbers (case M15) is also good with the data of Coleman et al. [10] corresponding to case CKM15 (see Fig. 4).

Table 2 Physical and numerical parameter for validation study

Case	M	Re	Re_τ	$L_x \times L_y \times L_z$	$N_x \times N_y \times N_z$
KMM	0	2800	178	$4\pi \times 2h \times \frac{4\pi}{3}$	$192 \times 129 \times 160$
M02	0.2	2700	181	$4\pi \times 2h \times \frac{4\pi}{3}$	$256 \times 112 \times 112$
CKM15	1.5	3000	222	$4\pi \times 2h \times \frac{4\pi}{3}$	$144 \times 119 \times 80$
M15	1.5	3000	222	$4\pi \times 2h \times \frac{4\pi}{3}$	$256 \times 112 \times 112$

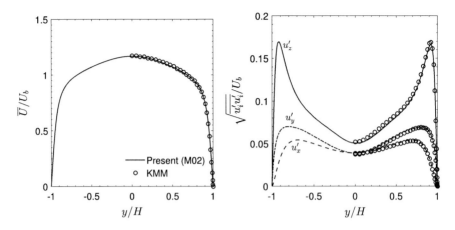

Fig. 3 Validation of the present compressible DNS code with Kim et al. [17] for low-Mach numbers; left: mean velocity profile, right: root mean square fluctuations

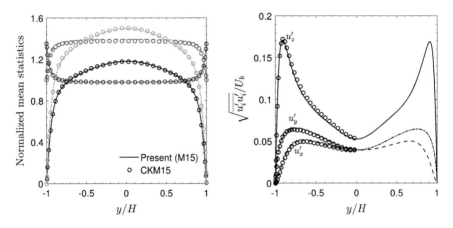

Fig. 4 Validation of the present compressible DNS code with Coleman et al. [10] for high-mach numbers; black: velocity, blue: density, red: temperature, green: mach number based on bulk velocity and speed of sound at the wall; left: mean profile, right: root mean square fluctuations

3.2 Mean Flow Parameters

Figure 5 illustrates the mean velocity, temperature, and density profiles for all three cases shown in Table 1. The mean velocity, mean density and mean pressure were normalized with the respective bulk values while the temperature is normalized by a reference temperature. The peak in mean velocity shifted upward with the increase in Mach number while there is no change in profile in the near wall region. The wall-normal mean temperature profiles vary significantly as depicted in Fig. 5b. At the low-Mach number, the profile is 'S' shaped which is typically observed in hot-cold channel flows (for instance in Sengupta et al. [23]). With the increase in Mach

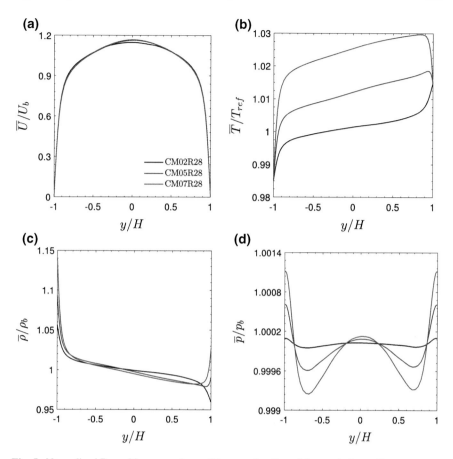

Fig. 5 Normalized Reynolds averaged quantities as a function of the vertical coordinate; **a** mean velocity, **b** mean temperature, **c** mean density and **d** mean pressure

number, the temperature profiles changes. The maximum value in temperature occurs near to the hot wall, but not exactly at the hot-wall.

The effects of Mach number can also be seen on mean density profiles in Fig. 5c. In these DNS, we opted for the same Reynolds number and the same bulk density (corresponding to the reference conditions of 8 MPa and 340 K) with the isothermal wall. With the increase in Mach number, the pressure increases to enforce a constant bulk density. Therefore, the values of density differ at $y = -H$ and $y = H$ (i.e. at the no-slip walls). Our procedure differs from that of Kim et al. [18] in which they kept the same bulk pressure while varied the bulk density for different cases. Figure 5d illustrates the variation of mean pressure along the y/H. The mean pressure is symmetric, unlike the other variables. As expected, pressure fluctuations increase with increasing Mach numbers. But, the fluctuations in mean pressure are insignificant quantitatively and it also supports the applicability of the low-Mach assumption in earlier investigations.

3.3 *Instantaneous Flow*

In this section, instantaneous flow fields are visualized in the YZ-plane for two extreme cases. Figure 6 demonstrates the contours of the instantaneous field. At the low Mach number, density varies by almost ±5% from the reference value while the temperature variations are within ±1.5%. A sharp change in both variables is observed in the near-wall region while the core region has almost the same magnitude. As soon as Mach number increases, this trend changes as can be observed in Fig. 7 for case Ma07Re2800. The fluctuations around the reference value also increase as compared to the low Mach number case. The change in temperature in the near-wall region is more pronounced. This leads of course to change in the density.

To further analyze the turbulence in near-wall region, we examined the instantaneous fluctuations of velocity for the high Mach number case (i.e. case Ma07Re2800). Figure 8 shows the velocity streaks in XZ-plane in the near-wall region (corresponding to $y^+ = 25$, based on the reference values). It is worth to mention here that we do not use semi-local scaling here because the variation in properties is not drastic in this range. Low and high-speed streaks are present near both walls. Both the streaks are comparatively longer near to the cold wall while streaks are shorter and thicker in near to the hot wall.

3.4 *Computational Performance*

The computational code used in this study is based on the DGSEM method. A computationally-efficient equation of state based on the adaptive quad-tree data structure is being used. The code is parallelized using OpenMPI and uses the HDF5

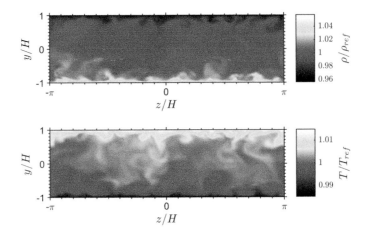

Fig. 6 Instantaneous contours; top: density and, bottom: temperature for case Ma02Re2800

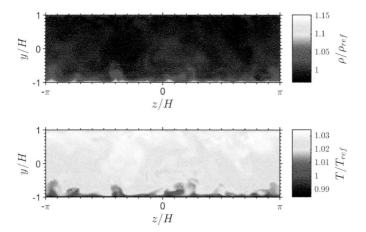

Fig. 7 Instantaneous contours; top: density, and bottom: temperature for case Ma07Re2800

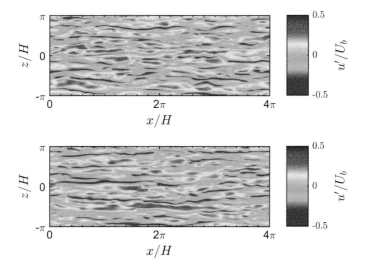

Fig. 8 Instantaneous contours of the streamwise velocity fluctuations at $y^+ = 25$; top: hot wall, and bottom: cold wall for case Ma07Re2800

library for I/O. It is programmed in Fortran-90. The domain decomposition is automatic based on the space-filling curves. In the past, this code has shown massively parallel scalability for ideal gas. But with our case, scaling behavior and calculation time may get adversely affected due to the inclusion of more sophisticated EoS and real gas effects in the supercritical region.

In this section, we present strong scaling characteristics of the code without multithreading. The solver has not shown any benefit of multithreading [2], therefore, it is not considered in this work. The simulations were performed on *Hazel Hen* located at the High-Performance Computing Center (HLRS) Stuttgart for a range of 96–12,288

Table 3 Computational parameter for parallel-scalability investigation

Case	N_x	N_y	N_z	#Elements	N	#DoF
1	32	14	32	14,336	4	1,792,000
2	128	56	128	917,504	2	24,772,608
3	128	56	128	917,504	5	198,180,864

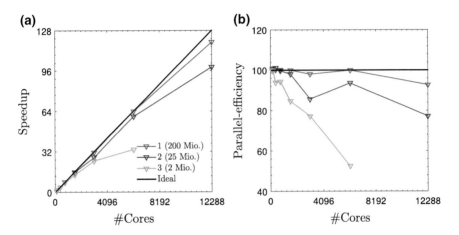

Fig. 9 Comparison of strong scaling of the present solver without multi-threading for varying degree of freedom; **a** speedup, **b** parallel-efficiency

physical cores. The first case with low Mach number (i.e. case Ma02Re2800) is used to characterize the scaling behavior of the code. For the same, two different domain viz. Low with $32 \times 14 \times 32$ elements and High with $128 \times 56 \times 128$ elements were selected with varying degrees of freedom as mentioned in Table 3. All cases use the same numerical setup except the number of elements and degree of polynomial (N) in the mesh. These cases are executed for 200 time steps and speedup is calculated only for the simulation time without considering the initialization and other I/O. We also calculate performance index ($PID = \frac{wall-clock-time \cdot \#cores}{\#DOF \cdot \#time-steps \cdot \#Runge\ Kutta-stages}$), which is a measure of the performance of different setup, and it decouples the of #DoF, #cores and #time-steps. Our calculation time per time step is increased by a factor of ≈ 3.8 compared with ideal gas due to the inclusion of the sophisticated EoS.

Figure 9 shows the speedup and parallel efficiency for the solver with our testcase. We see a very good scalability of the code at 200 Mio. degrees of freedom until 1024 nodes with a parallel efficiency of 90–100%. Interestingly, there is also a slightly super-linear speed-up at 192–384 nodes. The efficiency and speedup degrade as the degrees of freedom decrease. But for a realizable DNS of supercritical CO_2 which resolve the small turbulence scale, typical DoF is in the same order of case-1. The performance index (PID) has the least value for 200 Mio DoF case and it remains nearly constant over the range of employed nodes (refer to Fig. 10).

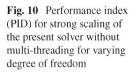

Fig. 10 Performance index (PID) for strong scaling of the present solver without multi-threading for varying degree of freedom

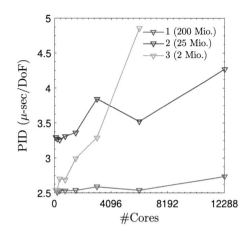

3.5 *Conclusion*

In this work, direct numerical simulations were conducted for a channel flow at supercritical pressure using a fully-compressible high-order computational code based on the discontinuous Galerkin spectral element method. The bulk Reynolds number was set to 2800 based on the channel half-width and Mach number was varied in the range of 0.2–0.7 with a temperature difference of 10 K. The degree of freedom was in between 58 and 200 Mio. to resolve the small-scale turbulence motions. For the equation of state, a computationally robust and highly accurate approach has been used based on the quadtree tabulation method. The validation with the ideal gas cases at both low and high Mach numbers warrants the accuracy of our implementation. We examined the three distinct cases with supercritical channel flow while varying the Mach number. The mean velocity profile is not affected much by the Mach number variations. On the other hand, mean density and mean temperature profiles change significantly along the wall-normal direction. The fluctuations of density and temperature also increase around the mean with the increase in the Mach number. The low and high speed streaks are comparatively longer near to the cold wall. In addition, we investigated the parallel scaling characteristics of our testcase on *Hazel Hen* supercomputer. We achieved an excellent speedup until 12,288 computation cores. As a next step, we aim to investigate the role of compressibility in the trans-critical region at higher Mach and Reynolds number with a degree of freedom in the range of 400–600 Mio.

Acknowledgements The authors are sincerely thankful to the High Performance Computing Center (HLRS) Stuttgart for providing access to *Hazel Hen* under project DNSTHTSC. FF gratefully acknowledges the Deutsche Forschungsgemeinschaft (DFG) for support through SFB-TRR 40. SP is grateful to the Forschungsinstitut für Kerntechnik und Energiewandlung (KE) e.V., Stuttgart, for the fellowship. XC and BW extend their gratitude to DFG for financial support in the framework of SFB-TRR 40 and SFB-TRR 75.

References

1. NIST chemistry webbook, in *NIST Standard Reference Database Number 69*, ed. by E. Lemmon, M. McLinden, D. Friend, P. Linstrom, W. Mallard (National Institute of Standards and Technology, Gaithersburg, 2011)
2. M. Atak, A. Beck, T. Bolemann, D. Flad, H. Frank, C.D. Munz, High fidelity scale-resolving computational fluid dynamics using the high order discontinuous Galerkin spectral element method. in *High Performance Computing in Science and Engineering*, vol 15, (Springer, 2016), pp. 511–530
3. D.T. Banuti, M. Raju, M. Ihme, Similarity law for widom lines and coexistence lines. Phys. Rev. E **95**, 052,120 (2017)
4. I.H. Bell, J. Wronski, S. Quoilin, V. Lemort, Pure and pseudo-pure fluid thermophysical property evaluation and the open-source thermophysical property library coolprop. Ind. Eng. Chem. Res. **53**(6), 2498–2508 (2014). https://doi.org/10.1021/ie4033999. URL http://pubs.acs.org/doi/abs/10.1021/ie4033999
5. W. Chang, X. Chu, A.F.B.S. Fareed, S. Pandey, J. Luo, B. Weigand, E. Laurien, Heat transfer prediction of supercritical water with artificial neural networks. Appl. Therm. Eng. **131**, 815–824 (2018)
6. X. Chu, W. Chang, S. Pandey, J. Luo, B. Weigand, E. Laurien, A computationally light data-driven approach for heat transfer and hydraulic characteristics modeling of supercritical fluids: from DNS to DNN. Int. J. Heat Mass Transf. **123**, 629–636 (2018)
7. X. Chu, E. Laurien, Direct numerical simulation of heated turbulent pipe flow at supercritical pressure. J. Nucl. Eng. Radiat. Sci. (2) (2016)
8. X. Chu, E. Laurien, Flow stratification of supercritical CO_2 in a heated horizontal pipe. J. Supercrit. Fluids **116**, 172–189 (2016)
9. X. Chu, E. Laurien, D.M. McEligot, Direct numerical simulation of strongly heated air flow in a vertical pipe. Int. J. Heat Mass Transf. **101**, 1163–1176 (2016)
10. G.N. Coleman, J. Kim, R.D. Moser, A numerical study of turbulent supersonic isothermal-wall channel flow. J. Fluid Mech. **305**, 159183 (1995)
11. M. Dumbser, U. Iben, C.D. Munz, Efficient implementation of high order unstructured WENO schemes for cavitating flows. Comput. Fluids **86**(Supplement C), 141–168 (2013)
12. S. Fechter, C.D. Munz, A discontinuous Galerkin-based sharp-interface method to simulate three-dimensional compressible two-phase flow. Int. J. Numer. Methods Fluids **78**(7), 413–435 (2015)
13. S. Fechter, C.D. Munz, C. Rohde, C. Zeiler, Approximate Riemann solver for compressible liquid vapor flow with phase transition and surface tension. Comput. Fluids (2017)
14. F.F.S. Föll, T. Hitz, C.D. Munz near critical jet simulations with a DG method and tabulated general equations of state, in *Fabian Föll at ICNMMF 2017* (Tokyo, Japan , 2017)
15. F. Hindenlang, G.J. Gassner, C. Altmann, A. Beck, M. Staudenmaier, C.D. Munz, Explicit discontinuous galerkin methods for unsteady problems. Comput. Fluids **61**, 86–93 (2012)
16. S. Kawai, H. Terashima, H. Negishi, A robust and accurate numerical method for transcritical turbulent flows at supercritical pressure with an arbitrary equation of state. J. Comput. Phys. **300**, 116–135 (2015)
17. J. Kim, P. Moin, R. Moser, Turbulence statistics in fully developed channel flow at low reynolds number. J. Fluid Mech. **177**, 133–166 (1987)
18. K. Kim, J.P. Hickey, C. Scalo, Pseudophase-change effects in turbulent channel flow under transcritical temperature conditions. ArXiv e-prints (2017)
19. P.C. Ma, X.I.A. Yang, M. Ihme, Structure of wall-bounded flows at transcritical conditions. Phys. Rev. Fluids **3**, 034,609 (2018). https://doi.org/10.1103/PhysRevFluids.3.034609. URL https://link.aps.org/doi/10.1103/PhysRevFluids.3.034609
20. S. Pandey, X. Chu, E. Laurien, Investigation of in-tube cooling of carbon dioxide at supercritical pressure by means of direct numerical simulation. Int. J. Heat Mass Transf. **114**, 944–957 (2017)
21. S. Pandey, E. Laurien, Heat transfer analysis at supercritical pressure using two layer theory. J. Supercrit. Fluids **109**, 80–86 (2016)

22. S. Pandey, E. Laurien, X. Chu, A modified convective heat transfer model for heated pipe flow of supercritical carbon dioxide. Int. J. Therm. Sci. **117**, 227–238 (2017)
23. U. Sengupta, H. Nemati, B.J. Boersma, R. Pecnik, Fully compressible low-mach number simulations of carbon-dioxide at supercritical pressures and trans-critical temperatures. Flow Turbul. Combust. **99**(3), 909–931 (2017)
24. M. Sonntag, C.D. Munz, Efficient parallelization of a shock capturing for discontinuous galerkin methods using finite volume sub-cells. J. Sci. Comput. **70**(3), 1262–1289 (2017)

Application and Development of the High Order Discontinuous Galerkin Spectral Element Method for Compressible Multiscale Flows

Andrea Beck, Thomas Bolemann, David Flad, Nico Krais, Jonas Zeifang and Claus-Dieter Munz

Abstract This paper summarizes our progress in the application and feature development of a high-order discontinuous Galerkin (DG) method for scale resolving fluid dynamics simulations on the Cray XC40 Hazel Hen cluster at HLRS. We present the extension to Chimera grid techniques which allow efficient computations on flexible meshes, and discuss data-based development of subgrid closure terms through machine learning algorithms. We also show the first results of the simulation of a challenging, high Reynolds number supersonic dual nozzle flow.

1 Introduction

The efficient and accurate numerical simulation of multiscale problems arising in fluid mechanics and aeroacoustics is one of the key computational challenges in the next decades [23]. While the time-averaged simulation methods like the Reynolds averaged Navier Stokes equations (RANS) approach are used extensively and

A. Beck · T. Bolemann · D. Flad · N. Krais (✉) · J. Zeifang · C.-D. Munz
Institute of Aerodynamics and Gasdynamics, University of Stuttgart,
Pfaffenwaldring 21, 70569, Stuttgart, Germany
e-mail: krais@iag.uni-stuttgart.de

A. Beck
e-mail: beck@iag.uni-stuttgart.de

T. Bolemann
e-mail: bolemann@iag.uni-stuttgart.de

D. Flad
e-mail: flad@iag.uni-stuttgart.de

J. Zeifang
e-mail: zeifang@iag.uni-stuttgart.de

C.-D. Munz
e-mail: munz@iag.uni-stuttgart.de

© Springer Nature Switzerland AG 2019
W. E. Nagel et al. (eds.), *High Performance Computing in Science
and Engineering '18*, https://doi.org/10.1007/978-3-030-13325-2_18

routinely also in industrial applications, the step towards scale-resolving simulations (LES and DNS) requires the development and design of numerical schemes, pre- and postprocessings tools and overall novel frameworks suitable for HPC and massive data.

The research group of Prof. Munz is active in the development and application of numerical solvers for LES and DNS for basic research and engineering applications. The FLEXI framework,[1] based on a high order discontinuous Galerkin spectral element method (DGSEM), has been established as a highly scalable and efficient open source solver for the compressible Navier Stokes equations. A number of successful HPC computations with FLEXI have been published [1, 5, 7, 10, 13], however, to increase its range of applicability, further developments and features are necessary.

In this work, we report on three research projects initiated during the last computing period. For high order numerical methods, high grid quality in complex domains is of critical importance to ensure a consistent geometry discretization. However, grid generation methods that meet these requirements are not matured yet. To ameliorate this problem and to allow greater flexibility, in particular for mesh movement, we have implemented a Chimera grid method in FLEXI. Details on the theoretical considerations, the implementation and scaling results can be found in Sect. 2. Another goal from the last research period was to develop optimized subgrid closures for LES. In Sect. 3, we present results for closure terms derived from DNS data through artificial neural networks. The basic idea behind this approach is to learn the functional relationship between DNS data and its LES terms without a priori assumptions through an optimization approach. Finally, we want to pave the way for complex LES at both high Reynolds numbers and strongly compressible flows for future applications. In Sect. 4, we investigate a challenging supersonic dual nozzle flow serving as a benchmark case for the European TILDA project.

2 Explicit Discontinuous Galerkin Chimera Overset Method

The Chimera overset method describes a technique to discretize a computational domain using several independent but overlapping grids. Originally developed to overcome restrictions of structured meshes in designing grids for complex geometries, the method is also useful to facilitate the process of unstructured grid generation and to accommodate for large movements in computations with dynamic boundaries.

During the last grand period, our solver FLEXI was extended to support Chimera meshes. Since our code employs explicit time stepping and thus requires the fast computation of many small time steps compared to implicit algorithms that are mainly used for Chimera methods, special emphasis has to be put on the computational

---

[1] www.flexi-project.org.

performance of the method. We report here the implementation of the method in the context of explicit discontinuous Galerkin methods and some preliminary performance results obtained on the Cray XC 40 system.

2.1 Chimera Method for Discontinuous Galerkin Schemes

The Chimera method relies on the discretization of the computational domain with not a single, contiguous mesh, but several meshes that do overlap to a certain extend. The boundaries of these meshes that are not part of the physical boundaries of the considered problem form artificial boundaries and their treatment is the main ingredient of the Chimera method. Conventional chimera methods used e.g. for Finite-Volume methods rely on the introduction of fringe cells next to the artificial boundaries. The current state of the flow in these fringe cells must then be determined using some sort of interpolation algorithm from one or several donor cells of the other considered meshes. Once the state of the fringe cells is know, the usual stencil can be used in the inner cells to advance the solution in time.

The introduction of fringe cells can lead to several problems, especially if employing higher order methods. To be able to use the inner stencil, possibly several layers of fringes cells have to be used, leading to a great amount of necessary overlap, which implies a restriction in flexibility and a severe amount of overhead. However, if one uses the discontinuous Galerkin method, the natural way of coupling different domains is by using the surface fluxes which depend on the conservative state and their gradients from the inside and the outside of an element. This means only the surface solution from the outside needs to be interpolated and no values from the volume, reducing the number of interpolation points and disposing of the need for fringe cells.

We are following Galbraith et al. [11] in implementing the method. Assume we want to compute the surface flux of the surface Γ_1 of the blue mesh on the left side of Fig. 1. First, an adjustable number of Gaussian quadrature points is created on the surface. Since the DG method is implemented after transforming the equations to a reference element, the coordinates ξ of these points in the reference space of the blue mesh are known. The inner values can be computed simply by interpolating the continuous polynomial solution inside of the blue element to the surface. To get the outer values, we need to interpolate the solutions from one or multiple of the cells in the red mesh—in this case from elements Ω_1 and Ω_2. Unfortunately, we can not directly evaluate the solution in the red cells, since we do not know the reference coordinates in the red mesh. Instead, we need to compute the physical coordinates $\mathbf{x}$ of the quadrature points by evaluating the given mapping from the reference space to physical space in the blue mesh and then search for the reference coordinates in the red cells. This is done by inverting the mapping from reference to physical space in the red cells using Newton's method. When the reference coordinates in the red cells are known, the outside solution can be interpolated and the surface flux calculated.

2.2   Validation and Performance Results

As a testcase for validation and performance analysis we choose the laminar flow around a sphere at $Re = 300$, which allows us to compare the results to reference solutions [16] obtained on a single grid. In the future, we will also employ tests using a dynamic version of the Chimera method to calculate the flow around a rotating sphere, which will significantly increase the complexity of the algorithm, since the positions of the interfaces are changing constantly.

The computational domain consists of a regular cylindrical background mesh and body fitted spherical mesh around the embedded sphere. The area occupied by the spherical mesh is cut out from the background grid except for a very small overlap region as can be seen on the right side of Fig. 1. This highlights that the DG Chimera method does not need to employ fringe cells, in fact coinciding mesh boundaries are sufficient in this context.

At the given Reynolds number, the flow is still laminar but unsteady and vortices are shed into a planar-symmetric wake. Figure 2 shows a visualization of the vortices at a specific time in the periodic process. Just behind the sphere, a vortex is forming, initially still attached to the sphere, which will be shed later and convected downstream, where it induces a counter-rotating vortex. The results are in very good agreement to a reference solution obtained on a single grid and show that the Chimera interface close to the sphere does not influence the development of the vortices or introduce spurious errors.

Special care is taken to avoid significant performance impact of the additional operations caused by the Chimera method, both in parallel and on core level. For instance, the number of calls to Newton's method is reduced by using bounding boxes for each element to identify those cells that possibly contain a given physical coordinate. Likewise, to reduce communication each MPI region calculates a bounding

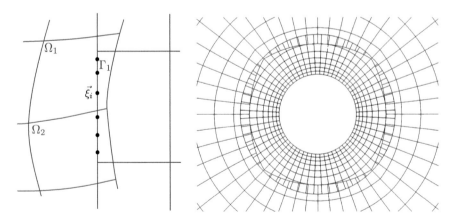

Fig. 1 *Left*: Schematic view of two overlapping grids. Positions in reference space $\vec{\xi}$ of Gauss quadrature points on surface Γ_1 in the blue mesh marked in black. *Right*: Closeup of overlapping grid used in the computation of the flow around a sphere

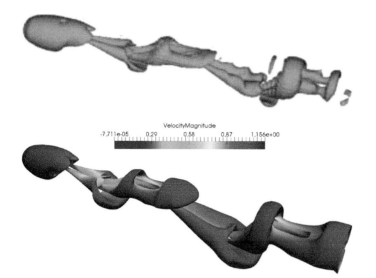

Fig. 2 Isosurface of $\lambda_2 = 0$ colorized by the velocity magnitude for flow around a sphere at $Re = 300$. *Top*: Reference solution obtained on a single grid taken from [16]. *Bottom*: Solution obtained using the discontinuous Galerkin Chimera method

box, which allows us to only communicate data necessary for the coordinate search to those MPI ranks that have a chance of actually including these points. Some preliminary performance results are reported here that are part of an ongoing analysis of the algorithm on the Cray XC 40 system.

Strong scaling of the algorithm has been tested using a spherical inner mesh consisting of 34,914 elements and a background mesh of 177,672 cells. We consider a sixth order computation, which leads to a problem size of about 46 million degrees of freedom. The simulations are run for a number of different configurations, ranging from 24 processors on a single node to 3072 processors on 128 nodes. All runs are repeated several times in order to accomodate for fluctuating performance on the high performance system. Results for the performance index PID are shown in Fig. 3. The PID is defined as the time it takes to advance a single degree of freedom on one core by one stage in the Runge-Kutta time intergration scheme:

$$PID = \frac{\text{wall-clock-time} \cdot \#\text{cores}}{\#\text{DOF} \cdot \#\text{time steps} \cdot \#\text{RK-stages}}. \tag{1}$$

To get reference performance data without the influence of the Chimera interface, we first replace the interface by an inflow boundary condition. In Fig. 3, these results are shown in black. One can see that the performance is nearly constant across the considered range of loads, showing the excellent parallel performance of the baseline DG scheme. When the chimera interface is activated, this picture changes at first. The blue crosses represent results obtained with the Chimera boundaries

Fig. 3 Performance index
(PID) over load per
processor for strong scaling
test of Chimera algorithm.
Black crosses indicate
reference solutions obtained
without Chimera interface.
Blue crosses symbolize
Chimera calculations
without load balancing, red
crosses with load balancing

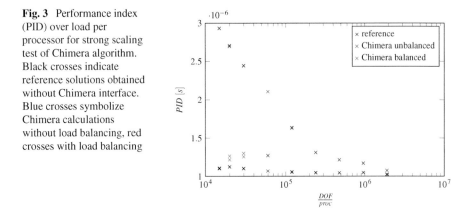

Fig. 3 Performance index (PID) over load per processor for strong scaling test of Chimera algorithm. Black crosses indicate reference solutions obtained without Chimera interface. Blue crosses symbolize Chimera calculations without load balancing, red crosses with load balancing

and show a strong increase of computational time with an increasing number of processors used in the simulation. This behaviour is to be expected and is a result of the imbalance introduced by the Chimera method, since only elements directly at the interface are affected by the additional work of interpolating solutions to the boundary. To make efficient use of the available resources, it is thus imperative to employ a load balancing method. In this case, we measure the computational time excluding communication on each processor and try to redistribute load evenly by shifting one or several elements across the MPI ranks. This iterative process leads to a great increase in performance after just one or two steps in the scheme, nearly eliminating the overhead introduced by the imbalance. First results are given in Fig. 3, where the red crosses show the performance of simulations with activated load balancing. The load balancing method prevents the increase of computational time with increasing processor numbers very efficiently and leads to a reasonable performance loss compared to the reference solution, especially when considering the overhead introduced by the load balancing method itself.

We have shown that we can construct a Chimera algorithm that does not massively influence the performance of the explicit DG scheme, if one takes care to avoid the imbalance inherently introduced by the interface. The next step will be to investigate the performance in combination with moving boundaries and to analyze different methods to redistribute the loads evenly.

3   Data-Optimized Subgrid Closures

The development of closure models for Large Eddy Simulation can be traced roughly along two lines of research: Based on ab initio physical considerations, explicit closure models are generated, which aim to mimic or approximate the unknown sugbrid terms. Contrary to this method, implicit model generation focuses on manipulating the discretization error terms in a way that either physical considerations are met or—as a minimum requirement—the resulting scheme remains stable in an underresolved setting. Both approaches however share the commonality that they try to model the unknown functional relationship between the grid resolved quantities $\bar{U}$ and the unknown closure terms. Data-based approaches can provide an alternative to this method, in that the functional relationship can be *learned* from data via machine learning algorithms like Artificial Neural Networks (ANN). ANNs with a certain complexity can be shown to be universal function approximators, given that sufficient training data is available to tune their internal parameters [3, 8].

We have initiated a research project in which we try to approximate the closure models for LES based on an approximation via these ANNs. The project is an extension and formalization of the ideas presented in [12], with the novel focus on incorporating structural information in the learning process. In a first step described in the following sections, the training data for the models is generated using our flow solver FLEXI on Hazel Hen. It will become obvious that a number of large scale simulations are necessary to generate sufficient training data. We have trained first, rather small models on a NVIDIA K40c Tesla GPU, which already give promising results presented here. Training more sophisticated networks requires more data runs of FLEXI on Hazel Hen and more powerful GPUs like the NVIDIA P100 cluster at HLRS.

3.1 Problem Definition

The governing equations for the large scale (or grid scale) solution of the incompressible Navier-Stokes equations are given by

$$\frac{\partial \bar{u}_i}{\partial t} + \frac{\partial \left(\bar{u}_i \, \bar{u}_j \right)}{\partial x_j} + \frac{\partial \bar{p}}{\partial x_i} = \frac{1}{Re} \frac{\partial^2 \bar{u}_i}{\partial x_j x_j} + \frac{\partial \tau_{ij}}{\partial x_j}$$
$$\frac{\partial \bar{u}_i}{\partial x_i} = 0, \tag{2}$$

where the operation $\overline{(.)}$ denotes an explicit or implicit filtering. Regardless of the filter shape, the residual or subgrid stress tensor τ_{ij} introduced by this operation for a grid converged, explicitly filtered LES is given by

$$\tau_{ij}^{DNS}(u, \overline{(.)}) = \overline{u_i \, u_j} - \overline{u}_i \overline{u}_j \tag{3}$$

As obvious from Eq. 3, the second term in τ depends on the unfiltered velocity field and it thus not available from the resolved data. In the following, we refer to the analytical subgrid stresses from Eq. 3 as τ_{ij}^{DNS} to distinguish it from the modelled or approximate $\tau_{ij}^{LES}(\bar{u}, \Upsilon)$ (here, Υ denotes model parameters, e.g. the Smagorinsky constant [24]). Several closure strategies exist for finding an approximation for this term, among them eddy viscosity approaches, scale similarity models [2] and deconvolution methods [15]. In our data-driven approach, no a priori assumptions or physical considerations are necessary except for the postulation shared among all models that $\tau_{ij}^{LES}(\overline{u}, \Upsilon) \approx \tau_{ij}^{DNS}(u, \overline{(.)})$. We thus aim to construct a model that relates grid quantities $\overline{u}$ to τ_{ij}^{DNS} via deep learning with ANNs. Since the terms $\overline{u_i}\,\overline{u_j}$ are known, we will only construct a model for the remaining term $\overline{u_i u_j}$ to which we will refer to as τ_{ij}^{ANN}.

3.2 Learning Framework and Strategy

With the definitions in place, we can now describe the general strategy for learning τ_{ij}^{ANN}. Figure 4 summarizes the different elements of the learning framework. Starting from the upper left, we use the CFD solver FLEXI to generate high resolution DNS data for the most general form of turbulence, the decaying homogeneous isotropic turbulence (DHIT) case [9]. Since the initial velocity and pressure fields for this test cases are randomly chosen but follow a specified spectral distribution of kinetic energy [22], each initialization leads then to a deterministic flow field. Thus, using only a single DNS run to train the model provides too little variance in the data for generalization and would prevent the model from working for different realizations of this flow. Thus, we require a large number of distinct DHIT DNS runs, but with different initial fields to ensure that the training data is meaningful. With the DNS data generated, we can now define the LES quantities $\overline{U}$ and exact closure terms τ_{ij}^{DNS} by filtering the DNS data. These quantities define the inputs and output vectors for the ANN training, in which τ_{ij}^{ANN} is to be learned from the grid data $\overline{U}$. We train the ANN on this data in a *supervised learning* method, and find an approximation to the exact closure term from the data. In doing so, the DG discretization of FLEXI gives us two data stencil choices: Firstly, we can learn in a point-to-point fashion, where input and output quantities are chosen from a single point in the computational

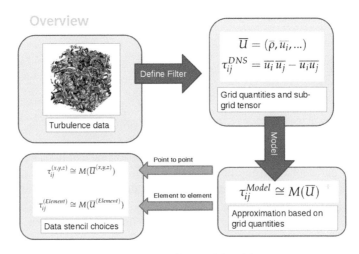

Fig. 4 Overview of machine learning strategy for subgrid models

domain. Secondly, we can choose an element-wise approach, in which we correlate the input and output data in a 3D region of the flow field, i.e. in a whole DG element. This second approach has the advantage that it includes spatial information and neighboring relationships in the learning, and can thus be expected to give a better approximation of turbulent structures.

3.3 Details on Data Generation

ANNs are very data-hungry in the sense that the more complex the network, the larger the number of training samples needs to be to adjust the weights optimally. For our training on DNS data, this creates the requirement for a large number of independent realizations of the flow, i.e. for a large number of distinct DNS runs and subsequent postprocessing of the data. To this end, we have conducted 25 DHIT computations with 134 Mio. spatial DOF per run (computed on a 64^3 element grid with 512 DOF per element). All runs were conducted with the compressible Navier-Stokes solver FLEXI at $Ma = 0.1$ and $Re_\lambda \approx 170$, which has been shown to give excellent scaling results on Hazel Hen for these types of problems [6]. A typical timestep for the explicit time integration was 5.5×10^{-5}, leading to about 54,000 time steps from $T = 0$ to $T = 3.0T^*$. The PID was $1.3 \times 10^{-6} s/DOF$ on 8200 cores of Hazel Hen, leading to about 15,000 CPUh/run.

Figure 5 (left) shows the temporal evolution for the kinetic energy for randomly selected runs. We found that the self-similar decay region is approximated well by $t^{-2.2}$, which is in good agreement with the predicted exponent for low Re DHITs [4]. The right plot in Fig. 5 shows the temporal evolution of the kinetic energy spectrum

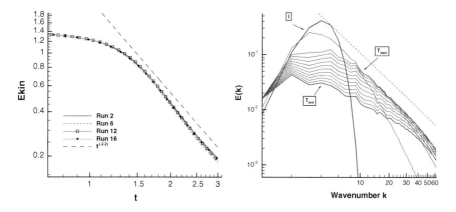

Fig. 5 *Left*: Temporal evolution of kinetic energy for 4 selected DHIT runs; *Right*: Temporal evolution of the kinetic energy spectrum of run 11 from $T = 0$ to $T = 2.0T^*$; I: initial spectrum, T_{Start}: start of data collection $T = 1.0T^*$, T_{End}: end of data collection $T = 2.0T^*$, dashed line: $k^{-\frac{5}{3}}$

of a single run, from the initial state to the end of data sampling at $T = 2.0T^*$. For training, we decided to store the DNS solution at intervals of $\Delta t = 0.1T^*$, i.e. 11 samples per run. We found this interval to be a good compromise between a high sampling rate per run (more data, but stronger correlation between data sets, which slows down learning) and a lower sampling rate per run, which would have required more DHIT runs.

After all runs were completed, the stored data was transfered to an 8^3 element grid with a polynomial approximation of degree $N = 5$ per L_2 projection and used to compute the training pairs $(\bar{U}, \tau_{ij}^{DNS})$. This configuration was chosen as it represents a typical LES setup. In a preliminary study, we found that choosing smooth filter shapes instead of the local projector did not significantly affect the training success.

3.4 Artificial Neural Networks

Artificial Neural Networks can be seen as general function approximators consisting of compound linear and non-linear functions [3, 8]. In their most basic form, they consist of so-called linear neurons, which compute the linear combination of their inputs plus a bias term and then pass the results through a non-linear activation function [19]. These neurons can be arranged in layers, which themselves can be interconnected in various manners. The simplest architectural form of an ANN is the Multilayer Perceptron, in which all neurons in two adjacent layers are connected to each other [20]. These networks are versatile, but have a high number of weights which can make training difficult. Convolutional Neural Networks (CNNs) are more advanced designs, as their weights are organized in convolutional kernels which pass over the data [17, 18]. These weights are thus "shared" by the data and can be used more easily to extract structural features of the input (Fig. 6). To make training of deep networks with more than 100 layers possible, convolutional layers are equipped with identity shortcuts (Residual Neural Networks, RNNs [14]).

Initially, with random weights and biases, the network is in an untrained state. Its representation is improved by training, i.e. by comparing the network prediction Y for a known input $\hat{X}$ to the true output $\hat{Y}$. In our case, $\bar{U}$ corresponds to the known inputs, and τ_{ij}^{DNS} to the expected outputs. The prediction error or loss is measured in a suitable norm, and the loss is differentiated w.r.t. all weights in the network to obtain the gradient vector for an optimization algorithm. This optimizer then employs a variant of the gradient descent method to find the optimal weights and thereby the optimal network.

In our investigations, we examined various ANN architectures, among them MLP and CNN/RNN networks. The MLP were trained in a point-to-point and cell-to-cell fashion, the CNN/RNN only for cell data. All networks were implemented in TensorFlow 1.6 on a NVIDIA Tesla K40c GPU. Training process and generalization capabilities of the networks were checked by observing training and validation loss and cross correlation histories, where the latter quantity is defined as

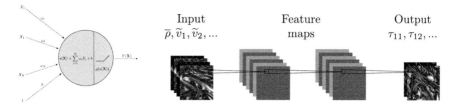

Fig. 6 *Left*: A single artificial neuron, consisting of a linear mapping of the inputs X to $a(\mathbf{X})$ with the weights ω_i and bias b following by a non-linear activation $Y(\mathbf{X}) = g(a(\mathbf{X}))$ *Right*: Schematic of a convolutional neural network design with two convolutional layers with 6 feature maps each

$$\mathscr{CC}(a,b) = \frac{cov(a,b)}{var(a)var(b)} = \frac{\sum_i (a_i - \bar{a})(b_i - \bar{b})}{\sqrt{\sum_i (a_i - \bar{a})^2}\sqrt{\sum_i (b_i - \bar{b})^2}}. \tag{4}$$

3.5 Results

In this section, we present selected results of the training process. All networks investigated were designed with 4 hidden layers, but varying number of neurons/feature maps per layer. The achieved cross correlation coefficients are summarized in Table 1. For the point-to-point MLP architectures, training time was very fast, but the achievable correlation was limited to about 75%. For the element-to-element setup, the number of neurons in the input and output layers and thus in the network itself is considerably larger. With this architecture, the achievable cross correlation on the training data was almost perfect, while the results on the validation set were much weaker (Fig. 7). This is a typical overfitting scenario, i.e. the network did not generalize to unknown data. To overcome this scenario, we switched to CNN and RNN networks. For both architectures, we were able to achieve results of over 90% on both training and test data [21]. It should be noted that in particular for the RNN, further training with larger networks and more DHIT data would have likely brought an additional improvement (Fig. 8).

Table 1 Training results for different networks. The number of neurons per layer (for MLPs), of feature maps per layer (for CNN) and of residual network blocks is given in the first column

Network, neuron/feature maps	DOF	$\mathscr{CC}_{train}(\%)$	$\mathscr{CC}_{val}(\%)$	Time (h)
Point to point, MLP, 64	13.000	75.7	76.0	0.13
Point to point, MLP, 512	800.000	70.6	75.6	0.15
Elementwise, MLP, 512	2.400.000	89.0	53.6	8.6
Elementwise, MLP, 2048	19.200.000	97.8	30.2	22.9
CNN, 32	95.000	91.7	91.5	53.3
RNN, 8	680.000	93.6	93.8	169

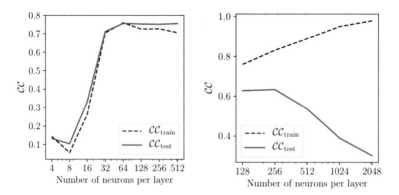

Fig. 7 Cross correlation for MLP networks as a function of neurons in the hidden layers [21], *Left*: point-to-point fitting, 4 hidden layers. The training and validation cross correlations are in good agreement, so no overfitting and good generalization can be expected. *Right*: element-to-element fitting, 4 hidden layers. The network is able to fit the training data very well, but fails on the validation set. This is a typical overfitting case

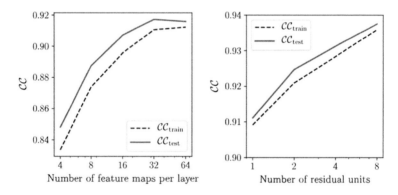

Fig. 8 Cross correlation for CNN networks [21], *Left*: CNN, *Right*: RNN

Figure 9 gives a visual comparison of the exact and predicted τ_{11} components in the computational domain for a network with 84% correlation. The overall shape of the stress structures and their magnitudes are well predicted by the ANN. In the future, we will investigate how to construct an actual LES model from the learned closure terms. It is likely that merely replacing the closure terms in Eq. 2 by the ANN output will not automatically result in a stable closure model. In a first step, a combination of an eddy viscosity model with the ANN term (akin to a mixed model in LES) will thus be investigated. This approach should among others answers the question if the effort of training ANN models to very high accuracy (both in terms of computing time and data generation) is justified.

In summary, this project is a first and promising step towards data-based LES closure models. The approach relies solely on DNS data collected from runs on Hazel

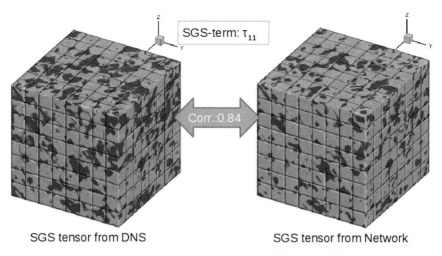

Fig. 9 τ_{11} component in the computational domain. *Left:* τ_{11}^{DNS} *Right:* τ_{11}^{ANN}

Hen, and offline training on this data on our local GPU. The training process was successful, but further improvements are still likely through two factors: Significantly more DNS data would not only increase the training accuracy, but would also allow to train more general SGS models, for example for boundary layer or compressible flows. Access to more suitable GPUs for machine learning would allow the creation and training of more sophisticated networks. To this end, a test account on the Laki cluster at HLRS with access to the NVIDIA P100 GPUs has been applied for.

4 Bypass Noise-Suppressing Supersonic Nozzle

While DG codes are known to perform well for low and medium Reynolds numbers and flows with moderate Mach number, Large Eddy Simulations at high Reynolds numbers, involving strong compressibility effects are still rare. We want to assess the capabilities of our numerical framework using the example of a supersonic dual nozzle configuration.

The simulation setup for the bypass nozzle is based on a test case [27] from the European TILDA project, with experimental results being available. The nozzle emits cold air from a plenum chamber into an Eiffel chamber. A schematic of the dual nozzle setup and the computational mesh are given in Fig. 10. We subsequently denote by the subscripts 1 and 2 the quantities in the nozzle core and the nozzle bypass, with dimensionless lengths being defined relative to the reference length $D_{a2} = 83.74$ mm, the bypass diameter. The nozzle is characterized by high Reynolds number of $Re_1 = 0.96 \times 10^6$ and $Re_2 = 2.87 \times 10^6$, with the core being in a subsonic ($M_1 = 0.91$) and the bypass in a supersonic ($M_2 = 1.14$) regime. The

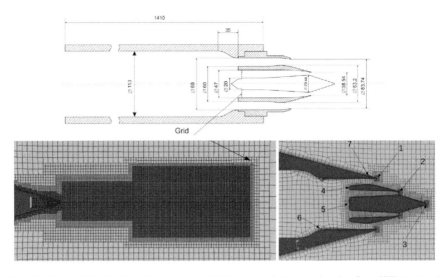

Fig. 10 Setup of the dual nozzle geometry with bypass and plenum chamber from [27] (top) and computational domain with non-conforming refinement regions (bottom)

flow is driven by a pressure gradient with a stagnation pressure of $p_0 = 199.1$ kPa in the settling chamber and $p_c = 88.4$ kPa in the Eiffel chamber, with the respective stagnation temperatures being $T_0 = 269$ K and $T_c = 273$ K. Since the nozzle is a highly demanding test case, prior to conducting fully three-dimensional simulations, which will follow in a later project, we first investigate the feasibility and resolution requirements for a 2D slice through the domain. Due to the nozzle being axisymmetric this is expected to yield results comparable to the 3D version, though effects due to turbulence, being inherently three dimensional, cannot be recovered. Since there was no information about the backward facing cone, the rear of the nozzle has been assumed as being planar for simplicity.

The simulation has been conducted at a polynomial degree of $N = 3$, where the mesh comprises of 39,459 cells, yielding at total of 6.3×10^5 DOFs. To simplify the mesh generation, non-conforming interfaces are used, allowing for very local refinements in critical regions [26].

The presence of shocks, occurring especially at the bypass, leads to severe instabilities in the unmodified DG scheme. Thus a hybrid discontinuous Galerkin / Finite-Volume scheme is used, as outlined in [25]. In this scheme critical cells are identified by a troubled cell indicator and the cell is switched to an Finite-Volume representation, with one subcell per degree of freedom (DOF) of the DG solution. During the switching process, the DG solution is converted to a set of integral mean values. The Finite-Volume cells can predominantly be found close to the nozzle boundaries and in the wake, where the largest instabilities and density gradients are present.

Figure 12 gives an impression of the shock structure in the experiment and simulation by a Schlieren visualization. While the location of the shock structures in

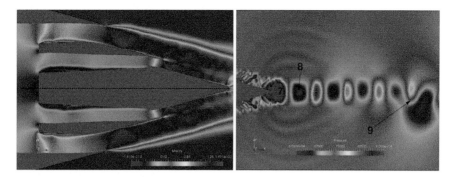

Fig. 11 Distribution of Mach number within the nozzle (left) and the development of shock diamonds in the free-stream (right)

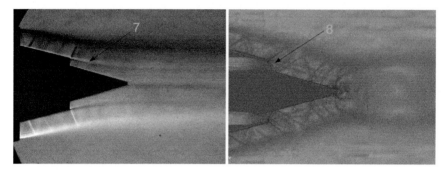

Fig. 12 Comparison of the Schlieren visualizations of the nozzle. Experimental results from [27] (left), simulation results (right)

the simulation is in good agreement to the experiment, their angles are noticeably steeper, especially close to the nozzle exit. This can be explained by a significantly higher Mach number at these positions, as shown in Fig. 11. Here, we can see that the Mach number at the throat is well above the values measured in the experiment, and most notably the flow being supersonic in the core section. Downstream of the nozzle we can observe the development and collapse of shock diamonds in the free-stream, with strong acoustic waves being emitted from the shear layer in front of the nozzle tip.

The higher Mach numbers observed in the simulation may be caused by various reasons. The most probable one is the lack of the cone at the rear of the nozzle, causing the flow to stagnate there. Thus a large separation region develops at the backward facing leading edges, which causes artificial blocking effects. Furthermore the lack of turbulence, otherwise serving as a dissipative mechanism, can lead to an increase in the flow velocity in the nozzle throat. Due to the major simplification of the test case such inaccuracies were to be expected. However, the investigations have shown no major problems for follow-up 3D simulations of this case.

5 Summary and Outlook

In this paper, we have shown the current development both in the capabilities of our simulation framework FLEXI as well as in the pursuit of optimal modelling approaches for large-eddy simulations. By employing machine learning algorithms trained on high quality DNS data, we can approximate the perfect LES closure terms. The large amount of data necessary to train neural networks requires efficient simulation tools run on high performance computing systems.

In order to facilitate the application of our framework to more complex geometries and to allow for large motions of the boundaries, a Chimera scheme was implemented. By careful design of the underlying parallelization strategy, we can assure that the additional work introduced by the artificial boundaries does not negatively influence the excellent scaling properties of the baseline DG scheme.

Building on the extension of the DG scheme with shock capturing capabilities developed in the previous project period, we now present a first challenging application of supersonic high Reynolds number dual nozzle.

Acknowledgements The research presented in this paper was supported in parts by the Deutsche Forschungsgemeinschaft (DFG), the Boysen Stiftung and the Simtech Cluster of Excellence PN5-21. We truly appreciate the ongoing kind support by HLRS and Cray in Stuttgart.

References

1. M. Atak, A. Beck, T. Bolemann, D. Flad, H. Frank, F. Hindenlang, C.-D. Munz, Discontinuous Galerkin for high performance computational fluid dynamics. in *High Performance Computing in Science and Engineering '14*. (Springer International Publishing, 2015), pp. 499–518
2. J. Bardina, J. Ferziger, W. Reynolds, Improved subgrid-scale models for large-eddy simulation. In *13th Fluid and Plasma Dynamics Conference* (1980), p. 1357
3. A.R. Barron, Universal approximation bounds for superpositions of a sigmoidal function. IEEE Trans. Inf. Theory **39**(3), 930–945 (1993)
4. G. Batchelor, A. Townsend, Decay of isotropic turbulence in the initial period. Proc. R. Soc. Lond. Math. Phys. Eng. Sci. **193**(1035), 539–558 (1948)
5. A. Beck, T. Bolemann, D. Flad, H. Frank, G. Gassner, F. Hindenlang, C.-D. Munz, High-order discontinuous Galerkin spectral element methods for transitional and turbulent flow simulations. Int. J. Numer. Methods Fluids **76**(8), 522–548 (2014)
6. A. Beck, T. Bolemann, D. Flad, H. Frank, N. Krais, K. Kukuschkin, M. Sonntag, C.-D. Munz, Application and development of the high order discontinuous Galerkin spectral element method for compressible multiscale flows. in *High Performance Computing in Science and Engineering '17*. (Springer, 2018), pp. 387–407
7. A. Beck, D. Flad, C. Tonhäuser, G. Gassner, C.-D. Munz, On the influence of polynomial de-aliasing on subgrid scale models. Flow Turbul. Combust. 1–37 (2016)
8. G. Cybenko, Approximation by superpositions of a sigmoidal function. Math. Control Signals Syst. **2**(4), 303–314 (1989)
9. D. Flad, A. Beck, C.-D. Munz, Simulation of underresolved turbulent flows by adaptive filtering using the high order discontinuous Galerkin spectral element method. J. Comput. Phys. **313**, 1–12 (2016)

10. D. Flad, G. Gassner, On the use of kinetic energy preserving DG-schemes for large eddy simulation. J. Comput. Phys. **350**, 782–795 (2017)
11. M. Galbraith, J. Benek, P. Orkwis, M. Turner, A discontinuous Galerkin chimera scheme. Comput. Fluids **98**, 27–53 (2014)
12. M. Gamahara, Y. Hattori, Searching for turbulence models by artificial neural network. Phys. Rev. Fluids **2**(5), 054604 (2017)
13. G. Gassner, A. Beck, On the accuracy of high-order discretizations for underresolved turbulence simulations. Theor. Comput. Fluid Dyn. **27**(3–4), 221–237 (2013)
14. K. He, X. Zhang, S. Ren, J. Sun, Deep residual learning for image recognition. in *Proceedings of the IEEE Conference on Computer Vision and Pattern Recognition* (2016), pp. 770–778
15. S. Hickel, N. Adams, A. Domaradzki, An adaptive local deconvolution method for implicit LES. J. Comput. Phys. **213**(1), 413–436 (2006)
16. D. Kim, H. Choi, Laminar flow past a sphere rotating in the streamwise direction. J. Fluid Mech. **461**, 365–386 (2002)
17. A. Krizhevsky, I. Sutskever, and G. Hinton, ImageNet classification with deep convolutional neural networks. in F. Pereira, C. Burges, L. Bottou, and K. Weinberger, (Eds.), *Advances in Neural Information Processing Systems 25*, (Curran Associates, Inc., 2012), pp. 1097–1105
18. Y. LeCun, B. Boser, J. Denker, D. Henderson, R. Howard, W. Hubbard, L. Jackel, Handwritten digit recognition with a back-propagation network. In *Advances in Neural Information Processing Systems* (1990), pp. 396–404
19. W. McCulloch, W. Pitts, A logical calculus of the ideas immanent in nervous activity. Bull. Math. Biophys. **5**(4), 115–133 (1943)
20. M. Minsky, S. Papert, Perceptron (expanded edition) (1969)
21. S. Reuschen, *Development of artificial neural networks to determine the closure terms in large eddy simulations* (University of Stuttgart, Thesis, 2018)
22. R. Rogallo, *Numerical Experiments in Homogeneous Turbulence*, vol. 81315. (National Aeronautics and Space Administration) (1981)
23. J. Slotnick, A. Khodadoust, J. Alonso, D. Darmofal, W. Gropp, E. Lurie, D. Mavriplis, CFD vision 2030 study: a path to revolutionary computational aerosciences. Technical Report (NASA Langley Research Center; Hampton, VA, United States, 2014)
24. J. Smagorinsky, General circulation experiments with the primitive equations: I. the basic experiment. Month. Weather Rev. **91**(3), 99–164 (1963)
25. M. Sonntag, C.-D. Munz, Efficient parallelization of a shock capturing for discontinuous Galerkin methods using finite volume sub-cells. J. Sci. Comput. **70**(3), 1262–1289 (2017)
26. M. Spraul, *Numerical simulation of a dual nozzle with the discontinuous Galerkin spectral element method*. Thesis, University of Stuttgart, 2018
27. V. Zapryagaev, Test case p4. bypass noise-suppressing nozzle test case. Test case description, ITAM, Novosibirsk

Application of the SPH Method to Predict Primary Breakup in Complex Geometries

G. Chaussonnet, T. Dauch, S. Braun, M. Keller, J. Kaden, C. Schwitzke, T. Jakobs, R. Koch and H. -J. Bauer

Abstract Understanding the process of primary breakup of liquids of air-assisted atomization systems is of major importance for the optimization of spray nozzles. Experimental investigations face various limitations. Conventional numerical methods for the simulation of the process are subject of various shortcomings. In the present paper the "Smoothed Particle Hydrodynamics" (SPH)-method and its advantages over conventional methods are presented. The suitability of the method is demonstrated by analyzing two different air-assisted atomization systems. First, a methodology for the investigation of the two-phase flow in fuel spray nozzles for aero-engine combustors is introduced and predictions of the flow in a typical nozzle geometry are presented. Second, numerical simulations of the atomization of a viscous slurry, which is used in gasifiers for biofuel production, are presented and compared to experimental results. Highly valuable results are retrieved, which will improve the fundamental understanding of primary breakup and will enable to optimize nozzle geometries.

1 Introduction

Aiming at the reduction of pollutant emissions from air traffic, academia and industry both invest in research to investigate processes causing the formation of pollutants of aero-engines. One aspect influencing the formation of pollutants is the quality of

G. Chaussonnet (✉) · T. Dauch · S. Braun · M. Keller · J. Kaden · C. Schwitzke ·
R. Koch · H.-J. Bauer
Institut für Thermische Strömungsmaschinen KIT, Kaiserstraße 12,
76131 Karlsruhe, Germany
e-mail: geoffroy.chaussonnet@kit.edu

T. Dauch
e-mail: thilo.dauch@kit.edu

T. Jakobs
Institüt für Technische Chemie, Herrmann-von-Helmholtz-Platz 1,
76344 Eggenstein-Leopoldshafen, Germany

© Springer Nature Switzerland AG 2019
W. E. Nagel et al. (eds.), *High Performance Computing in Science and Engineering '18*, https://doi.org/10.1007/978-3-030-13325-2_19

the combustion, which directly depends on the injected fuel spray and its placement inside the combustion chamber.

Because of limited optical access and challenging thermodynamic conditions, experimental studies are costly and cannot provide detailed information about the breakup of the fuel in the close vicinity of fuel injectors. Hence, more and more numerical investigations are employed for analyzing the two-phase flow inside the combustor.

At the "Institut für Thermische Strömungsmaschinen" (ITS) a numerical code based on the Lagrangian "Smoothed Particle Hydrodynamics (SPH)"-method has been developed [9, 14]. The objective is to predict the liquid fuel breakup in the close vicinity of the fuel spray nozzle. Conventional grid based methods exhibit a variety of inherent shortcomings, which can be overcome by a fully Lagrangian approach. In recent publications the potential of the code in terms of two-phase flow predictions has been demonstrated successfully [2, 6, 13].

Current state-of-the art methods for combustor design do not take into account the details of primary breakup. Correlations based on empirical studies are employed to impose droplet initial conditions for subsequent Euler-Lagrangian CFD predictions of the reacting flow. These correlations must be tuned to each individual setup and need to be calibrated. Most of the correlations are valid only for low pressures and temperatures. A detailed simulation of primary breakup can overcome these shortcomings.

The aim of this paper is to demonstrate how SPH predictions can be used to investigate the details of primary breakup in two industrial configurations. The first is a nozzle typically utilized in aero-engines at realistic operating conditions. The second type of configuration is a twin-fluid nozzle which is used for atomizing a highly viscous slurry. In the context of renewable energies, the gasification process allows to transform organic waste into liquid fuel [5]. The gasification occurs in a pressurized reactor where the reactant, a non-Newtonian slurry of high viscosity, is injected as a spray. The quality of the spray has a strong influence on the overall efficiency of the process [11]. Hence, the design of the nozzle is of primary importance. Gasification takes place at high pressure and temperature (p = 80 bar and T ≈ 1500 °C), which limits the instrumentation of the gasifier. Hence, the numerical simulation is well suited for analyzing the breakup process occurring at these operating conditions.

The paper is organized as follows. The numerical method is presented in Sect. 2, then HPC characteristics of the code are addressed in Sect. 3. The two configurations are investigated in Sects. 4 and 5, followed by the conclusion.

2 Numerical Method

The *Smoothed Particle Hydrodynamics* (SPH) method was originally developed in the context of astrophysics [7, 15]. Nowadays, SPH is employed for the analysis of different problems in science and engineering [4, 6, 13]. In contrast to conventional grid based CFD methods, SPH is a fully Lagrangian method and does not require a computational mesh. The flow domain is discretized by means of moving elements, referred to as *particles*. Each element carries the physical quantities of interest and is advected inside the domain at the fluid velocity. The Lagrangian aspect of the particles enables to avoid the discretization of the non-linear convective acceleration. In the context of multiphase flows, phase interfaces are inherently advected. Once particles are initialized as liquid, they remain liquid. Thus, the contour of the ligaments is uniquely defined by the location of the liquid particles. Consequently, no interface reconstruction is required. As primary breakup is entirely resolved, no breakup criterion is required. Recently, the capabilities of the SPH method was demonstrated by Koch et al. [14].

In the SPH mehod, the physical quantities and their gradient at a particle location (a) are determined by the interpolation over the values of its neighbors (b) [17]:

$$f(r_a) = \sum_{b \in \Omega_a} V_b \, f(r_b) \, W(r_b - r_a, h) \tag{1}$$

where V_b is the volume of the adjacent particles. The term W is a weighting function (the *kernel*). It depends on the inter particle distance $r_b - r_a$ and a characteristic length scale h called the *smoothing length*. The kernel promotes the influence of closer neighbors as illustrated in 2D in Fig. 1 (*top*). When the neighbors are located outside the *sphere of influence* of the central particle (Fig. 1 *bottom*), they are not taken into account. Hence, the kernel has a compact support. In the present studies, two isothermal weakly-compressible non-miscible fluids are considered. In the Navier-Stokes equations the source terms in the momentum equation are the pressure gradient, the viscous term proportional to the Laplacian of the velocity, the surface tension force and the gravity. No evaporation is taken into account. These equations are formulated in a Lagrangian frame of reference and solved using the SPH method. The interested reader is referred to [4] for further details about the modeling.

Fig. 1 Top part: Surface of a 2-D kernel. Bottom part: Particle distribution and illustration of the sphere of influence

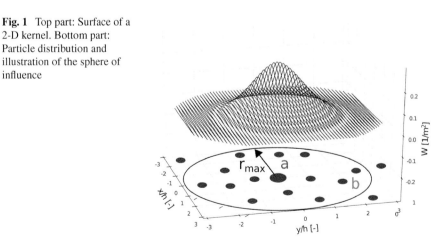

3 HPC Characteristics of the Code

The code used in this present studies is `super_sph`. It was developed at the *Institut für Thermische Strömungsmaschinen*.

It is implemented in C++ and relies on MPI for the parallel processing. As pointed out previously, the SPH method is a particle method where particles are influenced by their neighbours. Hence, the neighbours of each particles are required at each time step. This is done by a list-search algorithm on a Cartesian background grid which shows a complexity of $O(n)$.

The parallelization is performed by domain-partitioning. In typical simulations, all particles are evenly distributed in space, which greatly facilitates the partitioning of the domain. In this case, partitions of equal volume lead to partitions of equal number of particles. The inter-partition communications are achieved by transferring so-called *halo-particles* located close to the interfaces of the subdomains to the neighbouring processors. This is done by a non-blocking communication. In an earlier version of the code, the strong scalability was approximately 0.77 on 4000 cores, with an ideal behaviour of 1. However, it was observed that the node-level performance was not optimal. Several optimizations were performed, especially the data management was changed from an array-of-structure to a structure-of-array architecture, which showed a more cache-friendly behavior. Due to these modifications, the code provides a weaker scalability, but it allows to simulate a larger physical time at the same computational resources than in its old version. Most importantly, the code demonstrated superior performance compared to a commercial code and to an open-source code at the same computational configuration. For 1.6 millions particles on 400 cores with an inter particle spacing of 5 μm, `super_sph` demonstrated the capability of computing 15 μs physical time per invested CPUh versus 2 and <0.5 μs/CPUh for a commerical code and OpenFOAM, respectively. For more quantitative details, the reader is referred to [1].

The code allows two types of I/O methods. In the first method, each processor will write all the data into one HDF5 file. This has the advantage to avoid parallel writing into the same file. However, to post-process the final results (e.g. with Paraview), it is necessary to recompose all data into one single file. Depending on the number of particles and exported time steps, this can lead to massive files of several Terabytes. The second strategy is to use parallel writing. All processors write one block of data in the same individual file per exported quantity and per exported time step. Performance tests were conducted on this workflow and the simulations showed similar performances as compared to the first strategy.

4 Application to an Aeroengine Airblast Atomizer

The main objective of the first study is the analysis of the multiphase flow in the vicinity of the prefilming edge of an airblast atomizer as used in jet engines. In order to assess the eligibility of cheaper 2D predictions in comparison to more expensive 3D predictions, differences between results of simulations in 2D and 3D are discussed. The focus is on the interaction of the inner air flow with the fuel in the fuel supply duct, the formation of ligaments and their disintegration.

4.1 Fuel Spray Nozzle Geometry

The fuel spray nozzle investigated in the present study features three annular, coaxial air ducts referred to as inner air, outer air and dome air according to the setup presented by Mansour and Benjamin [16]. All air flows exhibit a swirl component because of swirl generators located inside each duct as depicted in Fig. 2 on the left. Due to the high angular momentum of all air flows, high centrifugal forces occur, which are forcing the flow in radial direction. As the ratio of angular and axial momentum is large, it comes to vortex breakdown. A stagnation point (SP) is formed, in the vicinity of the center axis as indicated in Fig. 2 on the right hand side.

The main objective of the dome air flow is to shape the recirculation zone inside the combustor. In between the inner and the outer air duct, fuel is supplied through an annular, coaxial slot across the whole perimeter. The fuel is released into the shear zone between inner and outer air. The shear stress imposed by the air will act against the surface tension, and finally the fuel will be disintegrated.

Local recirculation zones are to be expected downstream the bluff edges of each shroud as indicated in Fig. 2 on the right. The flows of inner and outer air form a stream tube. Hence, no mass transfer takes place across these boundaries. The boundaries are indicated as dashed lines in Fig. 2 on the right.

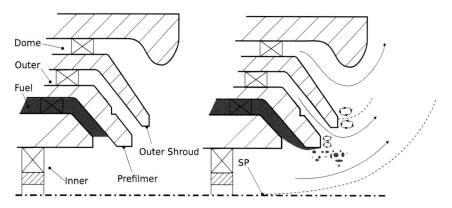

Fig. 2 Fuel spray nozzle—LEFT: Geometry—RIGHT: Local air flow

4.2 Computational Setup

In 3D the numerical domain is a three dimensional sector based on the axisymmetric
cross-section as depicted in Fig. 3. There are two inlet boundaries for air and one
inlet boundary for fuel. The velocity profiles for both air flows were extracted from
grid-based CFD predictions and imposed as boundary conditions. In Fig. 4 the shape
of the velocity profiles imposed at each inlet is presented. The axial velocity profile
of the fuel flow at the inlet is based on a correlation for laminar flows. At the fuel
inlet, the circumferential velocity profile recovers a free vortex in the center. Close to
the walls the profile is modified by a linear distribution in order to realize the no-slip
condition right on the wall. The pressure gradient is set to zero at all inlet boundaries.
At the outlet cross-section of the domain, the pressure is set constant and the velocity
gradient in normal direction is imposed as zero. In order to capture the influence of
vortex breakdown, the boundaries of the stream tube containing the inner and outer
air are extracted from grid based one-phase flow predictions and imposed as slip walls
in the SPH setup as depicted in Fig. 3 on the left. The thermodynamic conditions of
the underlying operating point are listed in Table 1. Surface tension coefficients are

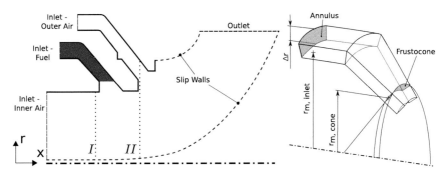

Fig. 3 Computational domain—LEFT: Cross-section—RIGHT: Frustocone

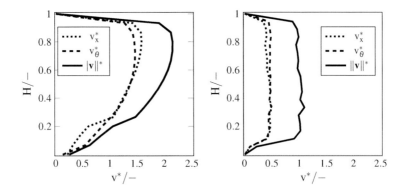

Fig. 4 Inlet boundary conditions—LEFT: Inner—RIGHT: Outer

Table 1 Thermodynamic conditions and fluid properties

		Air	Fuel
T	K	649	293.15
p	MPa	1.213	1.213
ρ	kgm^{-3}	6.51	770
μ	10^{-3} Pa s	0.03255	1.56
σ	10^{-3} Nm^{-1}	–	27.5

set for the liquid to recover static contact angles of 60°. As depicted in Fig. 3, the cross-section of the fuel duct at the inlet constitutes an annulus. Further downstream, the duct is deflected in radial direction and forced to smaller radii. Whenever the main direction of the flow exhibits a radial component, the cross-section constitutes a frustoconical surface. The area of the frustoconical surface scales with the radius. The same effect applies to the outer air duct.

The presented setup enables the computation of the local multi-phase flow in the vicinity of the atomizer at a high spatial resolution of 10 μm. The 3D sector covers 5° in circumferential direction. We are aware of the fact, that in order to resolve details of the flow in circumferential direction, the size of the segment is too small. However, swirl effects can inherently be captured. Furthermore, the sector recovers the inherently three dimensional shape of the flow cross-section along the flow path as illustrated in Fig. 3 (right). Aiming at lower computational costs, predictions based on two-dimensional domains are desired. However, 2D predictions do not recover all three dimensional aspects of fuel atomization. For the present it implies case a planar extension in lateral direction, which does not capture the frustoconical shape of the flow cross-sections as presented before. In 2D the circumferential velocity at the inlet of each duct is used as input to a 2D swirl model representing centrifugal forces [6].

All the other boundary conditions were identical. A discretization study in 2D up to a resolution of 2.5 μm has been conducted, demonstrating that a spatial resolution of at least 10 μm is required for the present geometry. Throughout the project the usual average runtime on ForHLR-II was 3 days using 400–600 cores for 2D cases (up to 43,200 CPUh). 3D computations required approximately 9 days using 800 cores (172,800 CPUh).

4.3 Computational Results

In Fig. 5 the velocity profiles at position II (3) obtained from 2D and 3D predictions are presented. In contrast to the inner air flow, the differences of the velocity magnitude in the outer air flow are significant. In 2D there is no contribution of the circumferential velocity to the magnitude. Furthermore, the flow is accelerated because of the decrease of the flow cross-section as illustrated in Fig. 2. In 2D this effect cannot be captured. Hence, only a deflection of the flow, but no acceleration

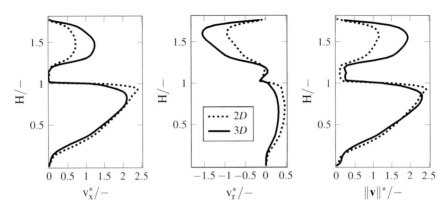

Fig. 5 Velocity profiles: 2D versus 3D at Pos. II

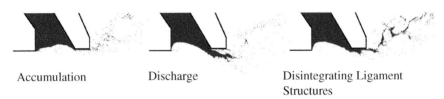

Accumulation   Discharge   Disintegrating Ligament Structures

Fig. 6 2D SPH particle data: Periodic filling and discharge

takes place resulting in a much smaller magnitude of the velocity. The nozzle configuration triggers a periodic filling and discharge effect at the exit region of the fuel duct. In Fig. 6 three snapshots presenting the different phases of the periodic discharge are depicted. In the beginning, fuel is accumulated at the exit region of the fuel duct. Once the cavity is filled up with fuel, a ligament is formed reaching out into the inner air duct. The momentum of the inner air flow pushes the ligament against the prefilming surface. The ligament is released into the shear zone and finally breaks up into smaller ligaments. Once the ligament is extracted, the fill level of the cavity reaches its minimum. Then for a period of time fuel is again accumulated in the cavity, before the formation of a new ligament starts. In 2D (Fig. 6) and 3D (Fig. 7) the same periodic effect can be observed.

5 Application to an Twin-Fluid Atomizer

5.1 Reference Experiment and Numerical Setup

The reference experiment consists of a twin-fluid external mixing nozzle as depicted in Fig. 8 (left) that discharges in a pressurized cavity [12, 18]. The liquid is injected at low velocity as a cylindrical jet, and the gas is injected in a coaxial high-speed

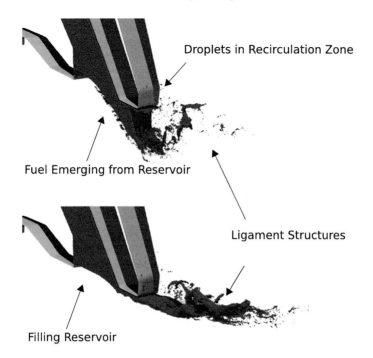

Fig. 7 3D Reconstructed phase interface—TOP: Snapshot of discharge—BOTTOM: Disintegrating structures after discharge

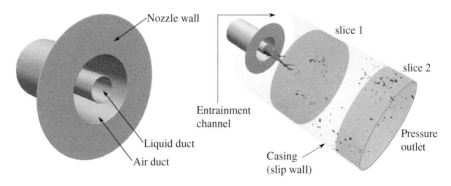

Fig. 8 Numerical domain superimposed with the virtual slices

annular stream enclosing the liquid. The liquid duct diameter D_l, the height H_g of the annular gas duct and the separator thickness e_s are 2, 1.6 and 0.1 mm, respectively.

For research studies the slurry was replaced by a mix of Glycerol and water. Its characteristics are a density ρ_l of 1233 kg/m^3, a constant dynamic viscosity μ_l of 0.2 Pa s, and a surface tension σ of 63.6 N/m. The vessel is pressurized to 11 bar and the temperature is kept to 20 °C, resulting into a density ρ_g of 13.25 kg/m^3 and a dynamic viscosity μ_g of 18.61 µPa s. The mean gas velocity is 58 m/s.

The aerodynamic Reynolds number $Re_g = 2H_g U_g / \nu_g$ and the Weber We $= \rho_g D_l U_{rel}^2 / \sigma$ number are 137,000 and 1375, respectively. The gas-to-liquid mass flow rate ratio GLR and the gas-to-liquid momentum ratio M are equal to 4.4 and 49, respectively. The numerical domain is depicted in Fig. 8. It contains the nozzle and a fraction of the reactor, covering an axial distance downstream of 30 mm from the nozzle exit and a radius R_c of 8.7 mm. The trajectory of the spray droplets might be influenced by this confined domain, but the fundamental process of atomization will not be affected. The length of the gas channel is 6 H_g and that of the liquid channel is only D_l long due to the low Reynolds number Re_l. A turbulent and a laminar profile are imposed at the gas and liquid channel inlet, respectively. The wall of the outer crown of the nozzle is limited to 3 mm. On the walls of the nozzle, a no-slip boundary condition is imposed. An additional annular gas inlet of 3 mm channel height is added on the outer radius of the cavity in order to take into account the entrainment of air by the coaxial jet (see Entrainment channel in Fig. 8 on the right). The entrainment velocity is set to 14.84 m/s. It ensures an appropriate entrainment rate up to the outlet of the cavity [10]. On the side of the cavity a wall without friction is imposed. Finally, at the outlet a constant pressure of 11 bar is set. Figure 8 shows the slices where the spray characteristics will be analyzed.

At the initiation of the simulation, the cavity is filled with gas particles and the liquid jet is initialized as a 5 mm long cylinder. The initial inter-particle spacing is equal to 33.33 μm, which results into 208 million particles. The simulation was run on 2000 cores for a physical time of 45 ms.

With a mean time step Δt of 28 ns, the simulation is run over 1.6 million time iterations. This amounts to a total computational cost of 1.2 millions CPU-hours. In total 1553 time steps were saved, corresponding to a sampling frequency of 34.5 kHz. The data of each saved time step has a total size of 11.69 GB.

5.2 Results and Analysis

The results of the simulation will be qualitatively compared to the high-speed images of the experiment in Fig. 9. The simulation captures well the flapping motion of the jet that occurs intermittently in the experiment, and which is characterized by a shape of a hook. The agreement is good, both in terms of the length scales as well as the temporal evolution of the hook shape.

The volume drop size distribution of the spray was calculated for the two slices as presented in Fig. 8. The results are shown in Fig. 10, superimposed with the SMD as dashed vertical line. The dotted line corresponds to best fit of a Rosin-Rammler function. The characteristics of the spray evolve from a SMD of 234 μm and a maximum diameter of 1.4 mm within slice 1, to a SMD of 144 μm and a maximum diameter of 0.9 mm within slice 2. It is observed that the shape of the PDF resolved by the numerical simulation is truncated at the lower diameters range, which compromises the determination of the peak value of the PDF. This is due to the too coarse spatial resolution, which requires to discard the small diameters from 0 to 66 μm. Neverthe-

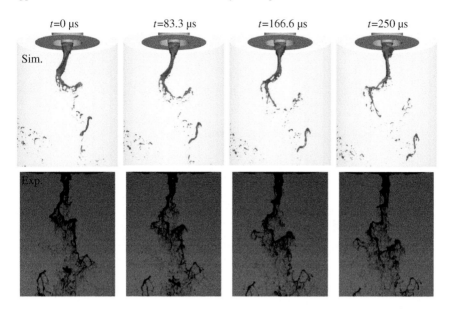

Fig. 9 Time series of the simulation and the experiment

Fig. 10 Volume PDF close to nozzle exit (grey) and outlet (black). Vertical lines indicate SMD and dashed lines Rosin-Rammler fits of VPDFs

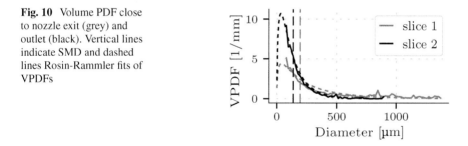

less, the fitting of the data with the Rosin-Rammler distribution in Fig. 10 shows that the resolved scales of the droplet size can be represented by this type of distribution, which is a promising result. In addition, the fitting allows to illustrate the shape of the distribution, and especially, the increase of the peak on slice 2, which demonstrates that secondary breakup occurs leading to the homogeneization of the spray.

Due to the Lagrangian nature of the SPH method, it is possible to build the tree of fragmentation from the set of the particle clusters. The tree of fragmentation is defined here as a chained object that contains all the relationships between fragmented droplets, as illustrated in Fig. 11. This tree of fragmentation enables to extract various quantities to be presented subsequently. By projecting all breakup events on a background grid, it is possible to define the *breakup activity* N_ϕ, which is defined here as the number of breakup event per volume and per unit time. Figure 12 shows the axial and radial profiles of the breakup activity N_ϕ. The axial evolution of N_ϕ shows that the maximum of the breakup activity occurs at $z \approx 11$ mm, which corresponds to the

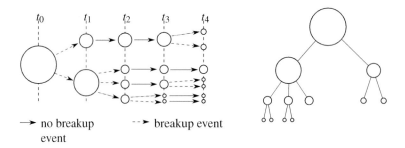

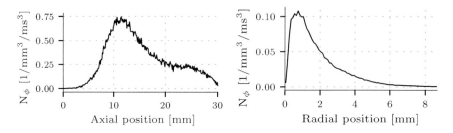

Fig. 11 Time evolution of the breakup of a single droplet (left) and its corresponding tree of fragmentation (right)

Fig. 12 Axial (left) and radial (right) profiles of the number of breakup events per volume and time unit

mean intact length of the liquid jet. At the outlet of the domain, N_ϕ goes to zero, suggesting that the spray is in a stable form, i.e. its drop size distribution will not change anymore. With regards to industrial applications, this is a valuable information for predicting where the spray has reached its steady drop size distribution. The radial profile of N_ϕ shows that the most active region is at $0.5 < r < 1$ mm, which is the location of the prolongation of the liquid duct. This is an unexpected result because one might expect that most of the breakup events would occur within the shear layer of the air stream, initially located a $r > 1$ mm. In the following, the distribution of $q = d_c/d_m$, the ratio of the child droplet diameter d_c to the mother droplet diameter d_m is investigated. The statistical distribution of q is presented in Fig. 13 in terms of Probability Density Function (PDF) and Cumulative Density Function (CDF). The values have been determined for the whole domain. The dashed line corresponds to $q = 1$ and marks the limit between breakup and coalescence. According to the CDF, 83.4% of the events are of breakup type.

On the PDF, the two different phenomena can be discriminated by two different slopes in the log-log plot depicted in Fig. 12 on the right hand side.

A maximum in the PDF is located at $q \approx 3 \times 10^{-4}$. This very low value of q indicates a breakup by peeling off large liquid structures. It cannot be stated if this maximum corresponds to physics or if it is a numerical artifact due to the limited spatial resolution. Large values of q (>5) represent a droplet impacting on a much large liquid lump, and could be labeled *impaction* instead of *coalescence*.

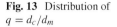

Fig. 13 Distribution of $q = d_c/d_m$

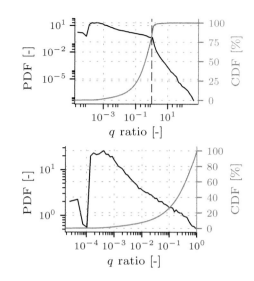

Fig. 14 Fragmentation intensity spectrum

The same analysis is applied to the statistical set limited by $q < 1$ and plotted in Fig. 14. It is labeled *fragmentation intensity spectrum* by Gorokhovski and Saveliev [8]. The region $q < 10^{-4}$ is not discussed because of the lack of confidence in the spatial resolution. For $q > 10^{-3}$, the PDF is piecewise linear. This linear part of the log-log plot corresponds to a fragmentation spectrum as a power law, which is typical for a fragmentation spectrum [3].

6 Conclusion

Both predictions of primary atomization have demonstrated that the SPH approach is an appropriate method for predicting the physics of liquid air-assisted atomization of different industrial applications. The predictions of the flow in a typical air-blast nozzle for aero-engines revealed that temporal fluctuations and flow instabilities can be studied using SPH simulations in 3D.

In the second case the predictions of the flow in a gasifier of a biofuel plant demonstrated that the SPH method does capture the whole breakup cascade, from primary instabilities to the spray characteristics. Finally, the `super_sph` code developed at ITS has shown strong HPC capabilities, that make it a serious candidate for simulations of larger scales.

Acknowledgements The authors would like to thank the Helmholtz Association of German Research Centres (HGF) for funding (Grant No. 34.14.02), and Rolls-Royce Deutschland Ltd. & Co KG for the outstanding cooperation. This work was performed on the computational resource ForHLR Phase I and II funded by the Ministry of Science, Research and the Arts Baden-Württemberg and DFG ("Deutsche Forschungsgemeinschaft").

References

1. S. Braun, R. Koch, H.J. Bauer, Smoothed particle hydrodynamics for numerical predictions of primary atomization, (Springer 2016), pp. 321–336. https://doi.org/10.1007/978-3-319-47066-5_22
2. S. Braun, M. Krug, L. Wieth, C. Hfler, R. Koch, H.J. Bauer, simulation of primary atomization: assessment of the smoothed particle hydrodynamics (SPH) method, in *13th Triennial International Conference on Liquid Atomization and Spray Systems (ICLASS)*, (Tainan, Taiwan, 2015)
3. W.K. Brown, A theory of sequential fragmentation and its astronomical applications. J. Astrophys. Astron. **10**(1), 89–112 (1989)
4. G. Chaussonnet, R. Koch, H.J. Bauer, A. Sänger, T. Jakobs, T. Kolb, Smoothed particle hydrodynamics simulation of an air-assisted atomizer operating at high pressure: influence of non-newtonian effects. J. Fluids Eng. **140**(6), 061,301 (2018)
5. N. Dahmen, E. Dinjus, T. Kolb, U. Arnold, H. Leibold, R. Stahl, State of the art of the bioliq process for synthetic biofuels production. Environ. Prog. Sustain. Energy **31**(2), 176–181 (2012). https://doi.org/10.1002/ep.10624
6. T. Dauch, S. Braun, L. Wieth, G. Chaussonnet, M. Keller, R. Koch, H.J. Bauer, Computation of liquid fuel atomization and mixing by means of the SPH method: application to a jet engine fuel nozzle, in *Proceedings of ASME Turbo Expo 2016: Turbine Technical Conference and Exposition*, (Seoul, South Korea, 2016). https://doi.org/10.1115/GT2016-56023
7. R. Gingold, J. Monaghan, Smoothed particle hydrodynamics-theory and application to non-spherical stars. Monthly Not. R. Astron. Soc. **181**, 375–389 (1977)
8. M. Gorokhovski, V. Saveliev, Statistical universalities in fragmentation under scaling symmetry with a constant frequency of fragmentation. J. Phys. D: Appl. Phys. **41**(8), 085,405 (2008)
9. C. Höfler, S. Braun, R. Koch, H.J. Bauer, Modeling spray formation in gas turbines-a new meshless approach. J. Eng. Gas Turbines Power **135**(1), 011,503011,503 (2012). https://doi.org/10.1115/1.4007378
10. H.J. Hussein, S.P. Capp, W.K. George, Velocity measurements in a high-Reynolds-number, momentum-conserving, axisymmetric, turbulent jet. J. Fluid Mech. **258**, 31–75 (1994)
11. T. Jakobs, N. Djordjevic, S. Fleck, M. Mancini, R. Weber, T. Kolb, Gasification of high viscous slurry R&D on atomization and numerical simulation. Appl. Energy **93**, 449–456 (2012)
12. T. Jakobs, N. Djordjevic, A. Sanger, N. Zarzalis, T. Kolb, Influence of reactor pressure on twin-fluid atomization: basic investigations on burner design for high-pressure entrained flow gasifier. At. Sprays **25**(12), (2015)
13. M.C. Keller, S. Braun, L. Wieth, G. Chaussonnet, T. Dauch, R. Koch, C. Hfler, H.J. Bauer, Numerical modeling of oil-jet lubrication for spur gears using smoothed particle hydrodynamics, in *Proceedings of the 11th International SPHERIC Workshop*, pp. 69–76 (2016)
14. R. Koch, S. Braun, L. Wieth, G. Chaussonnet, T. Dauch, H.J. Bauer, Prediction of primary atomization using smoothed particle hydrodynamics. Eur. J. Mech. B/Fluids **61, Part 2**, 271–278 (2017). https://doi.org/10.1016/j.euromechflu.2016.10.007
15. L. Lucy, A numerical approach to the testing of the fission hypothesis. Astron. J. **82**, 1013–1024 (1977)
16. A. Mansour, M. Benjamin, Pure airblast nozzle. U.S. Patent No. 6,622,488 B2 (2003)
17. J.J. Monaghan, Smoothed particle hydrodynamics. Rep. Prog. Phys. **68**, 1703–1759 (2005)
18. A. Sänger, T. Jakobs, N. Djordjevic, T. Kolb, Experimental investigation on the influence of ambient pressure on twin-fluid atomization ofliquids with various viscosities, in *Proceedings of the Triennal International Conference on Liquid Atomization and Spray System (ILASS)* (2015)

Enhancement and Application of the Flow Solver FLOWer

Felix Frey, Johannes Herb, Johannes Letzgus, Pascal Weihing, Manuel Keßler and Ewald Krämer

Abstract Recent enhancements and applications of the flow solver FLOWer are presented in this paper. A locally formulated laminar-turbulent transition model is implemented and used for simulations of a steady and pitching finite wing. Corresponding experimental results show good agreement on transition points. In CFD, a separation bubble is observed that triggers laminar-turbulent transition. Furthermore, the paper focuses on CFD simulations conducted on Airbus Helicopters' compound helicopter RACER and the underlying setup, including flexible main rotor blades, engine boundary conditions and deformable flaps. Finally, interaction phenomena and interesting flow characteristics are described for hovering and cruise conditions.

1 Introduction

The progressing development in high-performance computing (HPC) enables the broad application of flow simulations. Being a capable and affordable means of investigating aerodynamic phenomena, Computational Fluid Dynamics (CFD) is also widely used on helicopters. At the Helicopter Group of the Institute of Aerodynamics and Gas Dynamics (IAG) of the University of Stuttgart, it is not only used on rather fundamental topics of helicopter aerodynamics, but also on the investigation of the

F. Frey (✉) · J. Herb · J. Letzgus · P. Weihing · M. Keßler · E. Krämer
Institute of Aerodynamics and Gas Dynamics (IAG), University of Stuttgart,
Stuttgart, Germany
e-mail: frey@iag.uni-stuttgart.de

J. Letzgus
e-mail: letzgus@iag.uni-stuttgart.de

P. Weihing
e-mail: weihing@iag.uni-stuttgart.de

M. Keßler
e-mail: kessler@iag.uni-stuttgart.de

E. Krämer
e-mail: kraemer@iag.uni-stuttgart.de

© Springer Nature Switzerland AG 2019
W. E. Nagel et al. (eds.), *High Performance Computing in Science
and Engineering '18*, https://doi.org/10.1007/978-3-030-13325-2_20

aerodynamic performance of actual helicopter models, both during their design phase and their operational phase.

In this paper, numerical simulations for both of these applications are presented. Whereas the investigation of laminar-turbulent transition is part of the department's projects on basic research on helicopter aerodynamics, the simulations on the compound helicopter RACER are part of a cooperation with Airbus Helicopters, focused on the de-risking prior to the demonstrator's first flight.

All of the simulations presented below were carried out on the HLRS system Cray XC40 (Hazel Hen) within the research project HELISIM.

2 Numerical Method

2.1 Flow Solver FLOWer

At IAG's Helicopter Group, CFD simulations are conducted with the block-structured finite volume flow solver FLOWer [19]. Since the code's original development by the German Aerospace Center (DLR), significant enhancements and extensions have been made by IAG. The additional features not only comprise several fifth- and sixth-order spatial weighted essentially nonoscillatory (WENO) schemes, hybrid mesh capability [9], a new coupling library for fluid-structure interaction and state of the art detached-eddy simulation (DES) methods [24], but also extensive optimizations for high-performance computing [10, 12, 21]. Most recently, an Adaptive Mesh Refinement (AMR) functionality was implemented [17], allowing for the definition of flow criteria triggering an automated mesh refinement in areas where these are met.

For the present studies, an implicit dual time-stepping method of second order for time integration and both second-order Jameson Schmidt Turkel (JST) [8] and fifth-order WENO as spatial scheme is used. A three-level multigrid method is applied in order to achieve convergence acceleration.

2.2 Implementation of the γ-Re_θ Laminar-Turbulent Transition Model

Most URANS and detached-eddy simulations assume the flow to be fully turbulent, although it is known that both fixed-wing aircraft wings and helicopter rotor blades feature laminar-turbulent boundary-layer transition. While the effect of transition is rather small on lift, it significantly alters the profile drag and might influence flow separation [22]. Thus, a profound understanding of the phenomenon as well as a method to model it in CFD is of great importance. Most commonly, the semi-empirical e^N envelope method [3] is used, which is based on the integration of boundary-layer quantities along streamlines starting from the stagnation point, which requires numerous of non-local operations. Unfortunately, this requires a certain block-infrastructure of the computational domain, whereby the whole boundary layer of an airfoil section has to be handled by a single process [7]. This restriction severely hinders parallelization and is not in accordance with the fact that high-fidelity grids apply hundreds of thousands of grid cells to resolve the boundary layer, but only up to 30,000 grid cells per block are an optimum with the FLOWer code [12]. A possible remedy is the modern local-correlation based γ-Re_θ transition model [11] that introduces two additional transport equations, but is formulated solely using locally available quantities and is therefore best suited for HPC. The model accounts for natural transition by Tollmien-Schlichting instabilities, laminar separation bubbles and bypass transition. The basic part for the determination of transition onset is the comparison of an empirically known critical momentum thickness Reynolds number with a locally estimated one from the flow solution. It was recently implemented in the FLOWer code by Weihing [23] and is used in this work to simulate laminar-turbulent transition on a finite wing, see Sect. 3.

3 Laminar-Turbulent Transition on a Finite Wing

3.1 Experimental Setup and Flow Conditions

The experimental study [13, 14] was conducted in the Cross Wind Simulation Facility (SWG) at DLR Göttingen. A finite wing in the shape of a rotor blade tip model was mounted in a test section of the closed-return wind tunnel with a cross section of 2.4 m $\times$ 1.6 m. The wing has a chord length c of 0.27 m and a span b of 1.62 m. The wing has a positive linear twist, it uses a DSA-9A airfoil and has a parabolic tip. 21 unsteady pressure transducers of type Kulite XCQ-093 were used to obtain surface pressure data in three spanwise sections, at $y/b = 0.49$, 0.68 and 0.86, with y being the local coordinate in spanwise direction (see Fig. 3). Among others, a static case, with a wing-root angle of attack of $\alpha = 5°$, and a dynamic case, in which the wing was sinusoidally pitched with $\alpha = 5° \pm 6°$ and a pitching frequency of 3.2 Hz, were investigated. The detection of transition points in the experiment is based on the fact that there is a slight unsteady movement of the transition point that leads to measurable pressure fluctuations. By analyzing the standard deviation of the time- or phase-averaged pressure signal in the static or dynamic case, respectively, a transition point can be determined [6]. The free stream velocity was 55 m/s, leading to a Reynolds number of about 900,000 and a Mach number of 0.16. In the test section of the wind tunnel a turbulence intensity of 0.2% was measured.

3.2 Computational Setup

The numerical setup used in this work is based on an investigation by Letzgus et al. [12] and includes the rigid finite wing which is positioned in the middle of the wind tunnel. The boundary layers of the wind tunnel walls are not resolved, thus an inviscid boundary condition is applied there. At in- and outlet a far field boundary condition is used. The wing grid is of C-H type, has about 21 million grid cells and is embedded into a Cartesian background grid with about 11 million grid cells using an overset technique. URANS simulations are carried out with a physical time step size of 4.75×10^{-4} s representing $1/10$ of a convective time unit, which is the time it takes the flow to convect from leading to trailing edge. The γ-Re_θ model is used for transition modeling, while for turbulence modelling Menter-SST is applied. The computational domain is split into 1902 block that were allocated to 1632 cores on Hazel Hen. Due to the two additional transport equations of the transition model, the computation time of one physical time step increased by about 20% compared to a fully turbulent simulation, but there is room for improvement in the further development of the implementation.

3.3 Results: Static Case with $\alpha = 5°$

In the first test case, the wing-root angle of attack was static with $\alpha = 5°$. Overall, the computational pressure distributions at the spanwise station $y/b = 0.68$ using fully turbulent flow (dashed brown line) and considering transition (solid blue lines) are very similar and agree very well with the experiment—see Fig. 1. The error bars indicate the standard deviations of the pressure sensors and are multiplied by a factor of 10 for the purpose of illustration. The significant increase of the standard deviation of the pressure sensor at $x/c = 0.1$ on the suction side indicates laminar-turbulent transition around this location in the experiment, as discussed in Sect. 3.1.

Using the transition model, it was found that surface pressure is unsteady and oscillates around $x/c = 0.1$ on the suction side. Therefore, results of many physical time steps (light blue lines) are plotted in Fig. 1 as well as the mean (dark blue line). This mean pressure distribution exhibits a short plateau of almost constant pressure at $x/c = 0.1$, which is a strong indication of flow separation and, more precisely, a laminar separation bubble. In fact, Fig. 2, which shows instantaneous streamlines in a spanwise cut at $y/b = 0.68$, reveals a narrow separation bubble at the leading part of the upper side of the wing. The top detail in Fig. 2 also shows the contours of the intermittency γ that is supposed to be zero in fully laminar and one in fully turbulent flow. Around the maximum wall-normal height of the separation bubble a rapid increase in γ is noticeable, which proves the transition model to operate reasonably. In general, it is not unusual that a laminar boundary layer facing an adverse pressure gradient separates first, then transitions to turbulent flow and reattaches [18]. With the present experimental setup, a laminar separation bubble is believed to occur at

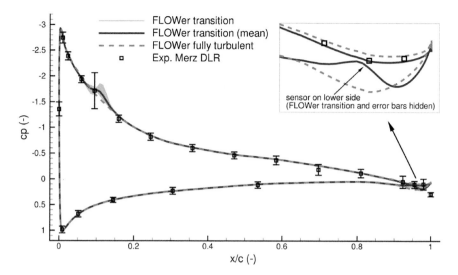

Fig. 1 Pressure distributions of FLOWer computations and experiment at $y/b = 0.68$. The standard deviation indicated by the error bars is multiplied by a factor of 10 for the purpose of illustration (static case), experimental data: [13]

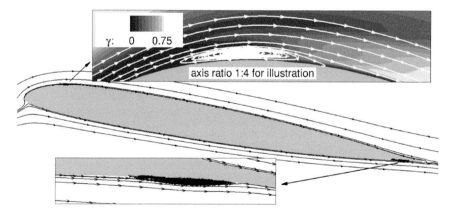

Fig. 2 Instantaneous streamlines at $y/b = 0.68$ indicate narrow separation bubbles around $x/c = 0.1$ on the suction side and $x/c = 0.95$ on the pressure side (static case)

higher angles of attack [14], but at $\alpha = 5°$ there is so far no indication that a separation bubble is present as well [5]. Therefore, further investigations need to clarify whether the phenomenon of flow separation during laminar-turbulent transition is a numerical artifact.

Concerning the lower side of the wing, the flow is expected to stay laminar for a long distance. There are too few pressure sensors at the rear part of the lower side of the wing to determine a transition point in the experiment, however, the surface pressure at the highlighted position of the transitional computation agrees

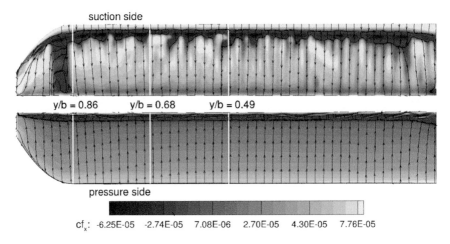

Fig. 3 Instantaneous surface streamlines and contours of the skin friction coefficient cf_x indicate flow separation on the suction and pressure side (static case)

significantly better with the experiment then the fully turbulent computation—see Fig. 1 again. As Fig. 2 shows there also is a small—supposably laminar—separation bubble.

So far, only one spanwise station at $y/b = 0.68$ was investigated. However, the findings also apply to other spanwise stations. The instantaneous surface streamlines and contours of the skin friction coefficient cf_x in Fig. 3 indicate the separation bubbles on the suction and pressure side of the wing and prove that there is little variation over the span.

In CFD simulations, one indication of laminar-turbulent transition is a sudden increase in skin friction. However in the present case, flow separation is believed to be caused by a laminar separation bubble, therefore there is no straightforward method to determine a transition point, especially since laminar-turbulent transition is a gradual process and does not occur at a fixed position [2]. In this work, two approaches were carried out to determine a transition point on the suction side in case of a separation bubble: Firstly, the transition point was defined by the downstream edge of the plateau gained from a time-averaged pressure distribution (denoted cp), as proposed by O'Meara and Mueller [18]. Secondly, in accordance with Alam and Sandham [1] the transition point is defined by the point of maximum negative (mean) skin friction (denoted cf_x). With both methods, the transition point is expected to be located near the maximum wall-normal height of the separation bubble. Table 1 lists the thereby determined points of transition at three spanwise stations. Overall, both methods yield quite similar results that agree very well with the measurements. In detail, the transition point determined by the pressure distribution agrees slightly better and the deviation to the experimental results increases with the wing span, but stays below two percent of the chord length.

Table 1 Comparison of the transition point on the suction side between CFD with two detection methods and experiment at three spanwise stations (static case)

y/b	0.49	0.68	0.86
FLOWer, based on cp	0.12	0.11	0.15
FLOWer, based on cf_x	0.13	0.12	0.16
Exp. Merz DLR [13]	0.11	0.10	0.14

3.4   Results: Dynamic Case with $\alpha = 5° \pm 6°$

In this case, the wing was sinusoidally pitched around its quarter-chord point with $\alpha = 5° \pm 6°$ and a pitching frequency of 3.2 Hz. As in the static case, laminar-turbulent transitions occurs in the form of separation bubbles. However, simple time-averaging of the pressure or skin friction distribution is no longer possible, since the flow conditions constantly change as the wing pitches up and down. Therefore, a time window containing several physical time steps was defined and used for averaging. It showed that considering 10 to 15 time steps per averaging window yields mean values that reasonably enable the determination of a transition point. Of course, the window size depends on the time step size, the current phase of the pitching cycle and leaves room for optimization.

In Fig. 4 the transition points on the suction side during one pitching cycle are compared between CFD results and the experiment, again at three spanwise stations. During the upstroke, the transition points move upstream towards the leading edge. Accordingly, they move downstream again during the downstroke, however, there is a hysteresis. These characteristics are reproduced very well with the transitional computation. Again, both numerical methods to determine the transition point yield similar results. Compared to the static case, the deviation to the measurements has slightly increased, especially at the outermost station at $y/b = 0.86$. Since the dynamic case is more challenging both in the experiment and the simulation, and introduces new uncertainties, e.g. with the averaging process, the results obtained are quite satisfactory.

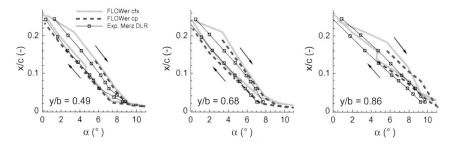

Fig. 4 Comparison of the movement of the laminar-turbulent transition point on the suction side during one pitching cycle between CFD with two detection methods and experiment at three spanwise stations (dynamic case), experimental data: [13]

4 Aerodynamic Simulations for Airbus Helicopters' Compound Helicopter RACER

In order to close the gap between the operational capabilities of rotorcraft and fixed-wing aircraft, various efforts have been made in the last decades in developing promising new concepts. After none of these could be well-established for a long time, various helicopter manufacturers increasingly have been focusing on the development of new high-speed rotorcraft configurations in recent years.

Airbus Helicopters (AH) took up the compound approach by combining a helicopter with wings and lateral rotors in the form of the X^3 demonstrator. The supplementary wings provided the rotorcraft with additional lift, allowing for the unloading of the main rotor. This, in turn, significantly mitigated the detrimental aerodynamic effects on the main rotor blades in fast horizontal flight which usually limit the flight envelope. The lateral rotors contributed to the total thrust and therefore helped extend this limit towards higher horizontal flight velocities.

This was successfully realized in the X^3's configuration, leading to a new—unofficial—flight speed record. After proving the high-speed capabilities of the compound configuration, AH decided to develop a more production-oriented demonstrator. Within the Fast Rotorcraft section of the European Union's Clean Sky 2 research programme, RACER (Rapid and Cost-Efficient Rotorcraft) is designed in co-operation with multiple international partners from industry and research institutes. As one of them, IAG supports the aerodynamic and aeroacoustic analysis of the trimmed, complete compound rotorcraft within the project CA^{3}TCH (Coupled Aerodynamic-Aeroacoustic Analysis of a Trimmed Compound Helicopter). This essential contribution towards the de-risking of a broad spectrum of flight cases and topics prior to the demonstrator's first flight is achieved by conducting advanced high fidelity simulations.

The findings of these simulations on aerodynamics topics are contained in the following section, in contrast to recent publications on compound helicopters, which concentrated on design [15], optimization [20, 25] or flight mechanics topics [4]. Further information and more specific results of the conducted research can be found in [16].

4.1 Computational Setup

The CFD model underlying the flow simulations in FLOWer consists of 101 structured meshes representing individual parts of the helicopter's geometry. These are connected via the Chimera technique (see Fig. 5 for topology) and embedded into a Cartesian off-body mesh which is automatically generated with IAG's in-house tool Backgrid and enables refinement towards the helicopter surface with the help of hanging grid nodes.

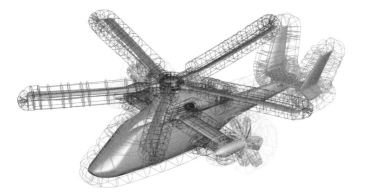

Fig. 5 Topology of RACER body meshes, different structures connected via Chimera technique

In order to account for specific flow conditions and their respective wake convection, the off-body mesh is individually refined in areas of particular interest. Whereas for the studies at hand, these areas have been defined manually prior to the simulations, and the off-body mesh has been generated accordingly, the newly implemented AMR functionality will be applied for future simulations.

Due to these individual refinements, the overall number of cells slightly varies between the examined flight cases. A standard setup consists of roughly 150 mio. cells in total, among which 50 mio. cells are situated in the off-body mesh. The latter's dimensions are set to 17.3 main rotor diameters in all spatial directions for the simulation of hovering conditions in order to allow for the proper development of a stream tube without a significant impact of the far field boundary conditions. For the flight cases with substantial flight velocities and consequently a prescribed convection speed at the boundaries, however, this extent can be significantly reduced to five main rotor diameters.

4.1.1 Main Rotor Deformation

Due to the importance of correctly represented main rotor aeroelasticity on the general quality of simulation results, the main rotor is treated as elastic structures. The deformations on discrete rotor blade positions are calculated by AH's in-house comprehensive tool HOST and transferred to FLOWer. Based on these offsets and with the help of linear interpolation in between, the blades' surface mesh is deformed (see Fig. 6 for blade deformation in cruise flight). The surrounding volume grid cells are treated with a Radial Basis Function (RBF) method to ensure their deformation accordingly. The blades' pitching, lagging and flapping motions are realized via spherical junctions between blade roots and rotor hub, on which the respective movements can be prescribed around corresponding hinges. The transition between these spherical movements and the deformed blades is then assured by applying additional deformations of the blade roots' geometry.

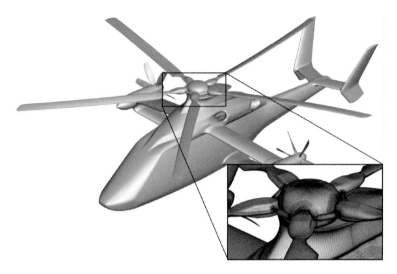

Fig. 6 Main rotor deformation in cruise flight and representation of rotor hub geometry

In order to ensure a most realistic representation of the rotor hub, also the inter-blade dampers are included in the CFD model (also see Fig. 6). As they experience a change in damper length due to relative blade root motions, this is compensated by further RBF deformations.

For the investigations at hand, the lateral rotors have been treated as rigid structures due to their small radius and the expected small deformations. For future simulations, however, this is currently improved to a procedure similar to the main rotor blades.

4.1.2 Engine Integration

The importance of an adequate representation of the helicopter's engines is twofold: On one hand, the effect of the flow characteristics around the helicopter on the inlet is of particular interest, as the inflow quality impacts the engines' performance. On the other hand, the large flow velocities and the increased temperatures of the exhaust should not be neglected when analysing the aerodynamic and thermal loads on the helicopter's surfaces. For this reason, the engines are included in the CFD setup with the help of respective boundary conditions inside the engine inlet ducts and the exhaust nozzles. While allowing to ignore the highly complex flow in the engines' core and the combustion itself, this still enables the definition of required mass flux and thermodynamic quantities on the boundary surfaces.

The outlet is divided into two different sections: While the core comprises the actual exhaust with its mass flux being coupled every time step to the inlet's mass flux for mass conservation, an additional cooling flow surrounds it. Similar to further cooling flows emerging from the lateral rotors' spinners, this is not coupled to any inflow, which, however, is acceptable due to the relatively small mass flux.

4.1.3 Flaps Integration

RACER's configuration with various aerodynamic surfaces on boxwing and H-tail allows it to be equipped with numerous flaps in order to modify the individual lift characteristics. Hereby, not only rudders and elevators on the tail but also flaps on the wings contribute to the degrees of freedom (DOFs) in helicopter trim.

A wing flap deflection affects the lift share between wings and main rotor, whereas the rudders and elevators allow for additional DOFs regarding the rotorcraft's yaw and pitch attitude. As all these aspects are vital for the helicopter's flight mechanic and aerodynamic behaviour, they also have to be taken into account within the CFD simulations. For this reason, the respective control surfaces on wings and H-tail are treated with the same deformation functionality as the main rotor, enabling deflections according to the desired trim state prescribed by HOST, representing the pilot input.

4.1.4 Computational Performance

With the well-implemented advanced hybrid parallelization in FLOWer demonstrated in [12], the CFD setup used for the analysis of RACER with a total number of roughly 150 mio. cells can be divided in a way that allows for each core to handle a maximum of 20,000 cells. Most simulations were therefore conducted on roughly 7,000 processors of Hazel Hen, which resulted in a computational time of approximately three hours for 360 CFD time steps. Hence, up to eight trim iterations of one main rotor revolution each could be realized within 24 hours with a time resolution of 1° main rotor azimuth per time step. The time required for the CA tool to calculate updated control data (e.g. deformation data and control angles) on a local machine and to broadcast it to Hazel Hen between each of the CFD trim cycles can be neglected as it was performed within minutes.

Consequently, most of the flight cases that were examined required roughly one day of computational time in order to achieve a converged trim state. For flight conditions with small flow convection, however, the number of time steps per trim iteration had to be increased in order to reach a fully developed flow field. Furthermore, with the occurrence of significant flow interactions, an extended load averaging period was required for a more stable trim process.

4.2 Results

Within the course of the CA³TCH project, approximately 40 different flight cases have been examined to date. Due to their antipodal characteristics, the hovering condition as well as the cruise flight at a horizontal velocity of 220 kts are of particular interest. The respective flow fields and vortex systems are displayed in Fig. 7 with the help of the λ_2 criterion.

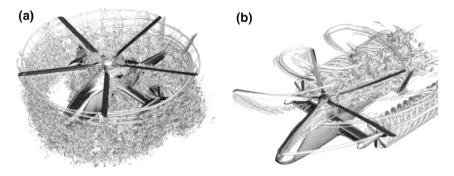

Fig. 7 λ_2 flow field visualization of RACER under different flight conditions [16]. **a** Hover. **b** Cruise

In hover, several flow characteristics and interaction phenomena can be clearly determined from the highly resolved flow field: (1) The lateral rotors are fully positioned within the main rotor downwash and therefore do not interact with main rotor tip vortices. (2) The tail surfaces are not impinged on by the main rotor downwash as the slipstream passes the tail boom slightly in front of them. Consequently, no significant empennage loads have to be expected in hover. (3) Due to a combination of lateral rotor thrust and main rotor downwash, the left lateral rotor's wake rapidly convects back- and downwards. (4) Different characteristics of the wake convection can be observed on the right lateral rotor, which operates in reverse thrust conditions. (5) The flow around fuselage and wings shows massive flow separation on their bottom sides.

The cruise flight, in contrast, is dominated by the high convection velocity and the resulting large advance ratio, nicely visible through the significant distance between the single main rotor blade tip vortices and their relatively horizontal convection. As a result, the main rotor downwash does not impinge on any other component of RACER, but shows several interesting characteristics in itself. When the retreating blades enter or exit the large reverse flow area, so-called reverse flow edge vortices emerge. Additionally, the blade-on-blade interaction occurs here when the blade is impinged on by its own tip vortex near the root area. Finally, downstream of the rotor head, the convecting wake leads to another blade interaction. Similar to the lateral rotors, however, the rotor head's wake does not impinge on the empennage but passes in between and besides the fins, respectively.

The effect of the described flow phenomena on aerodynamic interactions between RACER's main rotor, wings, lateral rotors and empennage has been thoroughly analysed for hover and cruise and is described in detail in [16].

5 Conclusions

The modern γ-Re_θ laminar-turbulent transition model was implemented in the flow solver FLOWer, since its local formulation is best suited for HPC. The transition points of both the simulation of a steady and a pitching finite wing in a wind tunnel agreed very good with experimental results. However, in the computations, transition occurred in form of a laminar separation bubble, which was not recorded in the experiment and thus needs to be further investigated.

A CFD setup was established for Airbus Helicopters' compound helicopter RACER, incorporating flexible main rotor blades, respective boundary conditions representing engine flows and deformable flaps on wings and control surfaces. This setup was used for high-fidelity simulations of hovering and cruise conditions. The former showed significant interaction phenomena between the helicopter's components, whereas for the latter, interesting flow characteristics could be observed on the main rotor due to the large advance ratio.

Acknowledgements The provided supercomputing time and technical support of the High Performance Computing Center Stuttgart (HLRS) of the University of Stuttgart within the HELISIM project is gratefully acknowledged. Parts of the research presented in this study were performed in cooperation of the Institute of Aerodynamics and Gas Dynamics of the University of Stuttgart and the Airbus Helicopters division of Airbus SE within the European research framework Clean Sky 2 under Grant 686530. We would like to express our thanks to the European Union for providing us the resources to realize this project. The authors would like to thank Airbus Helicopters Germany and France for the esteemed cooperation within this project and beyond. Furthermore, the investigation is based on the long-standing cooperation with the German Aerospace Center (DLR) making us their CFD code FLOWer available for advancements and research purpose, which we would like to thank for.

References

1. M. Alam, N.D. Sandham, Direct numerical simulation of 'short' laminar separation bubbles with turbulent reattachment. J. Fluid Mech. **410**, 1–28 (2000). https://doi.org/10.1017/S0022112099008976
2. M. Brendel, T.J. Mueller, Boundary-layer measurements on an airfoil at low Reynolds numbers. J. Aircr. **25**(7), 612–617 (1988). https://doi.org/10.2514/3.45631
3. M. Drela, M.B. Giles, Viscous-inviscid analysis of transonic and low Reynolds number airfoils. AIAA J. **25**(10), 1347–1355 (1987). https://doi.org/10.2514/3.9789
4. K. Ferguson, V. Khromov, Maneuverability assessment of a compound helicopter. J. Am. Helicopter Soc. **61**(1), 1–15 (2016). https://doi.org/10.4050/jahs.61.012008
5. A.D. Gardner, C.B. Merz, C.C. Wolf, Effect of sweep on a pitching finite wing. in *American Helicopter Society 74th Annual Forum* (2018)
6. A.D. Gardner, K. Richter, Boundary layer transition determination for periodic and static flows using phase-averaged pressure data. Exp. Fluids **56**(6), 119 (2015). https://doi.org/10.1007/s00348-015-1992-9
7. C.C. Heister, A. Klein, E. Krämer, RANS-*Based Laminar-Turbulent Transition Prediction for Airfoil and Rotary Wing Applications Using Semi-empirical Criteria*, (Springer, Berlin, Heidelberg, Germany, 2003) pp. 313–320. https://doi.org/10.1007/978-3-642-35680-3_38

8. A. Jameson, W. Schmidt, E. Turkel, Numerical solution of the Euler equations by finite volume methods using Runge Kutta time stepping schemes. in *14th Fluid and Plasma Dynamics Conference* (1981). https://doi.org/10.2514/6.1981-1259

9. U. Kowarsch, T. Hofmann, M. Keßler, E. Krämer, Adding hybrid mesh capability to a CFD-solver for helicopter flows. (Springer International Publishing, Cham, Switzerland, 2017), pp. 461–471. https://doi.org/10.1007/978-3-319-47066-5_31

10. P. Kranzinger, U. Kowarsch, M. Schuff, M. Keßler, E. Krämer, Advances in parallelization and high-fidelity simulation of helicopter phenomena. (Springer International Publishing, Cham, Switzerland, 2016), pp. 479–494. https://doi.org/10.1007/978-3-319-24633-8_31

11. R.B. Langtry, F.R. Menter, Correlation-based transition modeling for unstructured parallelized computational fluid dynamics codes. AIAA J. **47**(12), 2894–2906 (2009). https://doi.org/10.2514/1.42362

12. J. Letzgus, L. Dürrwächter, U. Schäferlein, M. Keßler, E. Krämer, Optimization and HPC-applications of the flow solver FLOWer. (Springer International Publishing, Cham, Switzerland 2018), pp. 305–322. https://doi.org/10.1007/978-3-319-68394-2

13. C.B. Merz, Der dreidimensionale dynamische Strömungsabriss an einer schwingenden Rotorblattspitze. Ph.D. thesis, Gottfried Wilhelm Leibniz Universität Hannover (2016)

14. C.B. Merz, C.C. Wolf, K. Richter, K. Kaufmann, A. Mielke, M. Raffel, Spanwise differences in static and dynamic stall on a pitching rotor blade tip model. J. Am. Helicopter Soc. **62**(1), 1–11 (2017). https://doi.org/10.4050/JAHS.62.012002

15. A.M. Moodie, H. Yeo, Design of a cruise-efficient compound helicopter. J. Am. Helicopter Soc. **57**(3), 1–11 (2012). https://doi.org/10.4050/jahs.57.032004

16. C. Öhrle, F. Frey, J. Thiemeier, M. Keßler, E. Krämer, Coupled and trimmed aerodynamic and aeroacoustic simulations for airbus helicopters' compound helicopter RACER. in *American Helicopter Society Technical Conference on Aeromechanics Design for Transformative Vertical Flight* (2018)

17. C. Öhrle, U. Schäferlein, M. Keßler, E. Krämer, Higher-order simulations of a compound helicopter using adaptive mesh refinement. in *American Helicopter Society 74th Annual Forum* (2018)

18. M. O'Meara, T.J. Mueller, Laminar separation bubble characteristics on an airfoil at low Reynolds numbers. AIAA J. **25**(8), 1033–1041 (1987). https://doi.org/10.2514/3.9739

19. J. Raddatz, J.K. Fassbender, *Block Structured Navier-Stokes Solver FLOWer.* (Springer, Berlin/Heidelberg, Germany 2005), pp. 27–44. https://doi.org/10.1007/3-540-32382-1_2

20. O. Rand, V. Khromov, Compound helicopter: insight and optimization. J. Am. Helicopter Soc. **60**(1), 1–12 (2015). https://doi.org/10.4050/jahs.60.012001

21. C. Stanger, B. Kutz, U. Kowarsch, E.R. Busch, M. Keßler, E. Krämer, *Enhancement and Applications of a Structural URANS Solver.* (Springer International Publishing, Cham, Switzerland 2015), pp. 433–446. https://doi.org/10.1007/978-3-319-10810-0_29

22. M.R. Visbal, D.J. Garmann, Analysis of dynamic stall on a pitching airfoil using high-fidelity large-eddy simulations. AIAA J. **56**(1), 46–63 (2017). https://doi.org/10.2514/1.J056108

23. P. Weihing, *Transport Equation based Transition Modeling in FLOWer* (University of Stuttgart, User guide. Institute of Aerodynamics and Gas Dynamics, 2017)

24. P. Weihing, J. Letzgus, G. Bangga, T. Lutz, E. Krämer, Hybrid RANS/LES capabilities of the flow solver FLOWer—application to flow around wind turbines. in *6th Symposium on Hybrid RANS-LES Methods* (2016)

25. H. Yeo, W. Johnson, Optimum design of a compound helicopter. in *American Helicopter Society International Meeting on Advanced Rotorcraft Technology and Lift Saving Activities* (2006)

CFD Simulations in Complex Nuclear Containment Configurations

A. Mansour, C. Kaltenbach and E. Laurien

Abstract Two-phase flows with water droplets have a significant influence on the thermal-hydraulic behaviour within Pressurized Water Reactors PWR. Such flows take place in the form of spray cooling, inter alia, in French nuclear reactors. In the case of a leak in the primary circuit of a PWR, hot steam will be released in the containment, which results in a pressure and temperature increase. The spray system ensures the reduction of the containment pressure and may, thus, help to avoid nuclear incidents. This work presents firstly a CFD simulation of spray cooling as well as aerosol particle washout by means of a spray system in a real size nuclear containment. The parallel performance of the simulations within a two-room model containment called $THAI^+$ is also investigated. Due to the large size and geometric complexity of this configuration, numerical grids with high refinement levels have to be generated to get accurate simulation results. For this reason, a good scalable CFD code is indispensable in order to achieve accurate simulations in realistic compulational times. Since another aim of this work is to define guidelines for the optimum use of computational resources, the scalability tests will be performed on four different grid styles with six mesh resolutions up to $39-40 \times 10^6$ elements.

1 Introduction

In the case of a leak in the primary circuit in a Pressurized Water Reactor PWR, steam is released into the containment atmosphere. This may result in a pressure and temperature increase, which can affect the integrity of the reactor. Moreover, also radioactive core particles can be distributed in the containment. Cold water spray

A. Mansour (✉) · C. Kaltenbach · E. Laurien
Institute of Nuclear Technology and Energy Systems, University of Stuttgart,
Pfaffenwaldring 31, 70569 Stuttgart, Germany
e-mail: abdennaceur.mansour@ike.uni-stuttgart.de

C. Kaltenbach
e-mail: christian.kaltenbach@ike.uni-stuttgart.de

E. Laurien
e-mail: eckart.laurien@ike.uni-stuttgart.de

© Springer Nature Switzerland AG 2019
W. E. Nagel et al. (eds.), *High Performance Computing in Science
and Engineering '18*, https://doi.org/10.1007/978-3-030-13325-2_21

systems, which are mounted in the upper part of the containment, are an effective method to reduce the stress on the containment building (reduction of pressure and temperature) and to prevent the spread of radioactive particles. Such spray systems are an important part in the reactor safety concept, e.g. in the European Pressurized Water Reactor (EPR), see [1]. For a better understanding of these flow mechanisms in containment applications, the methods of Computational Fluid Dynamics CFD have been recently used in the nuclear reactor safety [2, 3]. CFD methods may present a wide application range and a better accuracy than the traditional Lumped Parameter LP codes used in the reactor safety analysis. CFD methods, nevertheless, require relatively large computing time since very fine grids are to be generated in order to have a good accuracy. For this reason, investigations on the parallel computing of CFD codes are significant in order to achieve reliable calculation results in realistic computional times.

The aim of this work is to test the applicability of two-phase CFD simulations with spray in a complex real size nuclear reactor containment. For this, models with heat and mass transfer for spray cooling and aerosol particle washout are implemented using the CFD code ANSYS CFX 16.1. After the integration of a nozzle system in the containment, the physical mechanisms of spray cooling and particle washout are presented in the frame of a generic incident scenario within a simplified PWR building. Furthermore, the parallel performance of the CFD simulations in the multi-room model containment $THAI^+$ will be investigated. To this end, four different grid styles with refinement levels up to $39-40 \times 10^6$ elements are generated. The results of the investigations on parallel performance may present guidelines for an optimum use of computational resources in dependence on the grid style and mesh resolution.

2   Investigation of Spraycooling and Aerosol Particle Washout in a Real Size Nuclear Containment with CFD Methods

2.1   Purpose of the Simulation

This section presents the extrapolation of thermohydraulic modells dealing with spraycooling and aerosol particle washout by spray to a generic PWR containment (KONVOI). Both modells are developed by Kaltenbach and Laurien [4, 5] and they are separatly validated with THAI experiments. Investigations in model containments like THAI or THAI$^+$ are first steps towards complete three dimensional simulations in real size nuclear power plant containments. This section describes the cooling effect of water injected in a heated containment atmosphere in combination with particle washout by water spray in a real size containment.

Fig. 1 Section through KONVOI containment, primary circuit and crane is marked in red

2.2   Real Size Containment: Geometrical Modell and Mesh

Nuclear power plants according to the KONVOI design comprise the complete primary circuit and the heat exchanger for the secondary circuit in the containment. Its made of steel (thickness of 28 mm) with 56 m in diameter. To ensure the integrity of the steel containment (e.g. in case of a plane crash), it is surrounded by concrete walls with a thickness of 1.8 m. The basic geometrical modell in this work is a KONVOI containment, which was first investigated with CFD methods by Zhang and Laurien [6]. Figure 1 shows a centered section through the KONVOI containment with the primary circuit and crane marked in red and all the concrete in grey.

A generic spray system is added to the containment to show flow phenomena directed to spray injection. Altogether, 25 spray nozzles with an outlet diameter of 8 mm are introduced in rings in a height of 45 m. Figure 2 presents the mesh of the modified KONVOI containment with spray nozzles. The flow volume is meshed unstructured with tetrahedral volume elements (13.8 mio.) and a strong refinement towards the spray nozzles. Concrete structures represented by walls, primary circuit etc. are neglected.

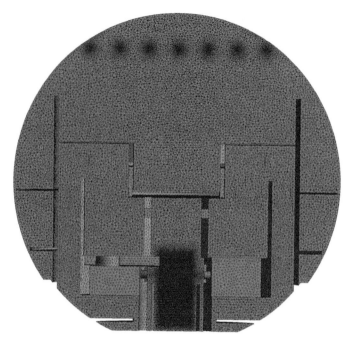

Fig. 2 Schematic mesh with refinement towards the spray nozzle system

2.3 Accident Scenario

During the accident scenario, a breach in the primary circuit is considered. Water coolant evaporates and heats up the containment atmosphere in conjunction with a pressure increase. While coolant is leaking out of the breach, also radioactive partikels are distributed to the containment atmosphere. The particles arise from the damaged core inventory. The current accident scenario assumes a higher pressure and gas temperature in the containment and moreover a particle distribution. Boundary conditions consist of a mixture of two validation experiments (THAI HD-31-SE [7] and THAI AW 4 [8]) used to validate the developed physical modells for spraycooling and aerosol particle washout in the model containment THAI. Table 1 summarizes

Table 1 Initial and boundary conditions in the simulation

Variable	Value	
Gas temperature	90	(°C)
Droplet temperature	20	(°C)
CsI mass	0.7	(g_{CsI}/m^3)
Absolute pressure	1.5	(bar)
Nozzle water mass flow (each nozzle)	1	(kg/s)
Const. droplet diameter d_L	830	(μm)
Const. particle diameter d_P	1.76	(μm)

the initial and boundary conditions. At the beginning of the simulation transient, a constant distribution of aerosol particles and gas temperature in the whole containment is assumed. The scenario neglects the continuous production of hot steam and particles. Only the initially assumed gas and particle configuration is washed out respectively and also cooled down. Spray is injected constant with a nozzle water mass flow rate of 1 kg/s. The simulation runs for 600 s.

2.4   Numerical Setup

The simulation is based on an Euler-Euler two-fluid approach [9]. Due to the high complexity of the geometry, the simulation is done with two separate velocity fields (gas/particles and water droplets). Spray droplets and particles are modeled monodisperse during the numerical investigation. A transient simuation is done with a constant timestep of 0.05 s. Boundaries which comprise the flow volume are assumed adiabatic. Gas temperature reduction inside the volume is only possible through spray injection. Water accumulation in the sump region is neglected to reduce the computational effort.

2.5   Results

Figure 3 presents the gas temperature distribution in the containment in a two dimensional section at two different timesteps (t = 150 s and t = 600 s). Spray is activated at the beginning of the transient and cools down the gas present in the spray area. As time goes by, the cooled gas is distributed to other regions in the containment by convection, see Fig. 3a. Beside the gas cooling in the spray region, there is still uncooled gas mass, especially above the spray nozzle system and near the reactor pressure vessel, see Fig. 3b. Complex wall structures in the containment lead to an uneffective convection flow in this area.

Figure 4 presents the quantitative behaviour of gas temperature and absolute pressure at MP1, MP2 and MP3. The position of all measure points are obvious in Fig. 3a. MP1 represents the region above the spray nozzle system, MP2 is placed in the spray center and MP3 is covering peripheral region of the spray. Spray cooling is most effective in the center (MP2) and in the peripheral area (MP3) of the spray. MP2 shows a strong gas temperature increase at 100 s. This increase is due to the movement of uncooled gas volume during the formation of the flow field. Later timesteps show a similar behaviour with a fluctuating temperature. A completly different behaviour is considered at MP1. No gas temperature decrease is obvious up to 100 s. This can be explained by the fact, that during the flow formation no gas movement is obvious in this area. Temperature reduction in this area is only possible with convection of cooled gas. Figure 4d shows the pressure reduction due to a constant cooling process.

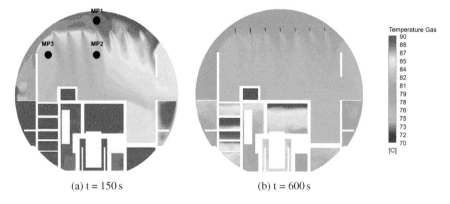

(a) t = 150 s   (b) t = 600 s

Fig. 3 Gas temperature at different time steps

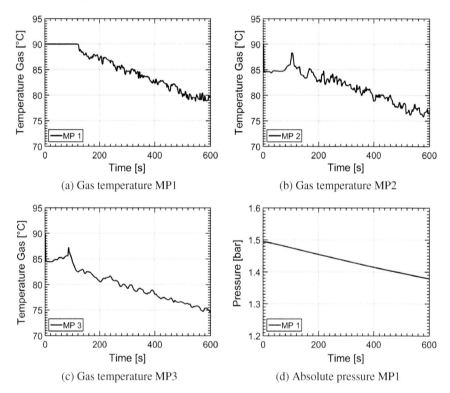

(a) Gas temperature MP1   (b) Gas temperature MP2

(c) Gas temperature MP3   (d) Absolute pressure MP1

Fig. 4 Gas temperature and absolute pressure at MP1, MP2 and MP3

Figure 5 shows the distribution of CsI particles in a two dimensional section at two different timesteps (t = 150 s and t = 600 s). The CsI content is normalized from zero (no particles available) to one (maximal particle content). In the same way like

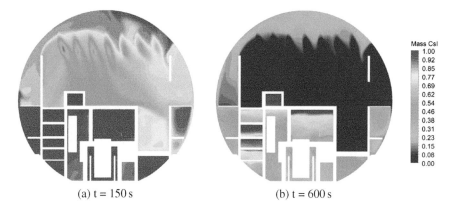

(a) t = 150 s   (b) t = 600 s

Fig. 5 Normalized CsI mass at different time steps

spray cooling, the most effective particle washout is the spray region due to direct contact between droplets and particles. Most of the particles can be removed with a spray activation time of 600 s.

3 Parallel Computing

3.1 $THAI^+$ Test Facility

$THAI^+$ is a test facility, which has been constructed by Becker Technologies in Eschborn, Germany [10]. It results from the extension of the original THAI vessel (Thermal-hydraulics, Hydrogen, Aerosols and Iodine) into a new vessel called PAD (Parallel Attachable Drum). THAI vessel has a cylindrical shape form with a height of 9.2 m and a diameter of 3.2 while the new PAD vessel is characterized by a height of 9.8 m and a diameter of 1.6 m. DN 500 pipelines connect the vessels with each other, which enables the flow circulation between them.

THAI vessel contains also some internal installations, such as an inner cylinder, condensate trays and gutters which mainly serve to collect the condensed liquid in case of condensation, see Fig. 6. Due to the geometric complexity of THAI vessel, this latter corresponds to the plant room of a pressurized water reactor PWR which presents many components and rooms. In contrast, the PAD vessel stands for the operating room of the PWR. Thus, the $THAI^+$ test facility enables thermal-hydraulic investigations in a complex multi-room geometry and represents a first step to simulations in a PWR configuration.

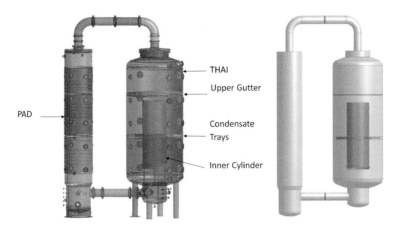

Fig. 6 CAD form of $THAI^+$ with its different components (left). Simplified configuration for CFD calculations (right)

3.2 Scalability Tests on Different Grid Styles

To measure the parallel performance of the calculations, the speedup and efficiency will be considerd. These are defined by

$$Sp = \frac{t_{ref}}{t_p} \quad , \tag{1}$$

$$Efficiency = \frac{t_{ref}}{p \cdot t_p} = \frac{Sp}{p} \quad . \tag{2}$$

In Eqs. 1 and 2, t_{ref} is the wall clock time on the reference core number and t_p stands for the wall clock time on p cores.

For grid partitioning in ANSYS CFX, a node-based partitioning method is used according to the Multilevel Graph Partitioning Software MeTiS which is the default partitioning algorithm [11]. The scalability investigations are performed on the CRAY XC-40 Hazel Hen of the High Performance Computing Center Stuttgart HLRS. Hazel Hen is a supercomputer which contains around 7712 computing nodes with 24 processing cores and 128 GB memory each.

In the frame of this study, part simulations of a natural convection flow in $THAI^+$ are considered, in which a total simulation time of t = 150 s and a time step $\Delta t = 0.25$ s are set, see [12]. Since a main aim of this work is to define a kind of Best Practice Guidelines for the optimal use of computational resources in dependence on the grid resolution, the scalability tests have been carried out on four grid styles [13]. The first grid is composed of pure hexahedral elements while the second one is composed of pure tetrahedral elements. The two other grids are hybrid grids which

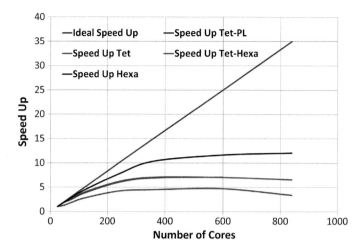

Fig. 7 Speedups on four grid styles with a grid resolution of about 1.2×10^6 elements

are a combination of tetrahedral and hexahedral elements as well as tetrahedral and prism elements. For each of these grid styles, three refinement levels are considered. The coarse grids 5 with 1.2×10^6 elements, the mid-fine grids 3 with 7.1×10^6 elements and the fine grids 1 with $39-40 \times 10^6$ elements.

Figures 7, 8, 9, 10, 11, 12, 13, 14 and 15 report the results of the scalability study. For the coarse grid 5 with 1.2×10^6 elements, the speed up curves show a very similar tendency on all grid styles. The maximum values are achieved on more than 840 cores for the hexahedral grid and on almost the same core number of about 600 cores for all other grid styles. However, the hexahedral grid shows the best parallel efficiency and the highest maximum speed up of more than 12, but also higher computational times. For instance, the simulations on the hexahedral grid 5 required a computational time of 7.18 h on 24 cores. This is almost twice as big as the time needed on the other grid styles since this varies between 3.34 and 3.7 h.

For the same grids 5, the pure tetrahedral grid is characterized by the lowest efficiency as well as the lowest maximum speed up which is equal to 4.69. In contrast, the hybrid grids Tet-PL and Tet-Hexa show a better efficiency with maximum speed up values of approximately 7.11 and 7.08. On the other hand, very similar computational times are achieved on the Tet, Tet-PL and Tet-Hexa grids.

For the mid-fine grid 3 with 7.1×10^6 elements, almost the same speed up behaviors as on the coarse grids 5 are observed. The pure hexahedral grid Hexa is here also characterized by the best efficiency showing a maximum speed up of 5.55 on 960 cores, while maximum values of about 4.26, 4.13 and 3.82 are detected on the Tet-Hexa, Tet-Pl and Tet grids. Nevertheless, the CFD calculations on the hexahedral grid led to higher computing times than on the three other mesh styles, on which similar times are measured.

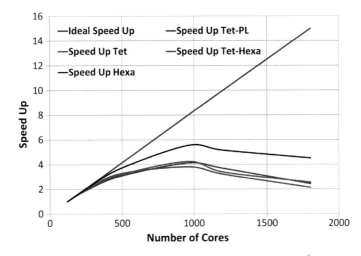

Fig. 8 Speedups on four grid styles with a grid resolution of about 7.1×10^6 elements

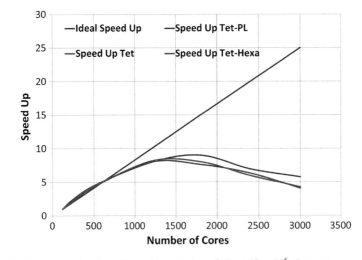

Fig. 9 Speedups on grid styles with a grid resolution of about 40×10^6 elements

In Figs. 9, 12 and 15, the results of the scalability study on the finest grids 1 with $39–40 \times 10^6$ elements are also reported. Due to quality issues, no simulations could be carried out on the fine hexahedral grid 1. The comparison of the other grid styles shows that, as for grids 5 and 3, no significant differences can be seen for the grids Tet-PL, Tet and Tet-Hexa since very similar parallel efficiencies and computing times are detected.

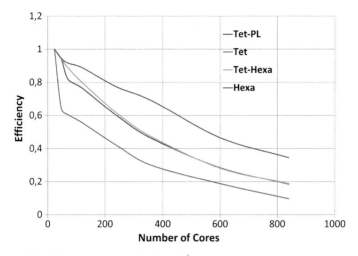

Fig. 10 Parallel efficiency on grids with 1.2×10^6 elements

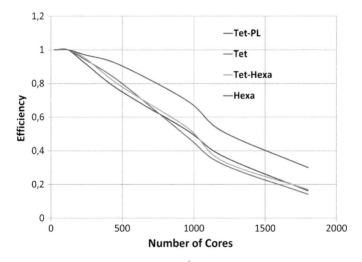

Fig. 11 Parallel efficiency on grids with 7.1×10^6 elements

Based on these results, it can be concluded that, apart from the mesh resolution, the CFD calculations on the hexahedral grids are characterized by the highest parallel efficiency. However, they also require the largest computating times. Calculations on the tetrahedral and hybrid grids show similar parallel efficiencies, which are lower than the pure hexahedral grid efficiency. However, simulations on the tetrahedral and hybrid grids led to less computational times than on the hexahedral grids.

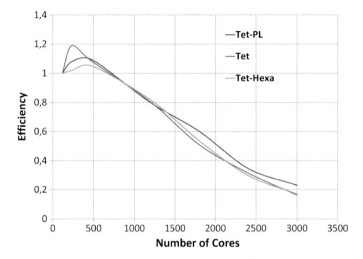

Fig. 12 Parallel efficiency on grids with approximately 40×10^6 elements

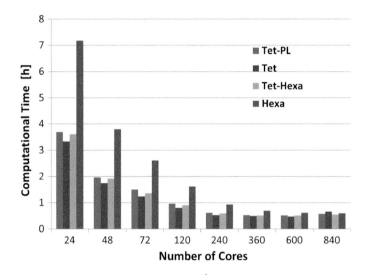

Fig. 13 Computational times on grids with 1.2×10^6 elements

The results of these scalability investigations may present guidelines for the optimum use of computational resources in CFD calculations in dependence on the grid style and mesh resolution.

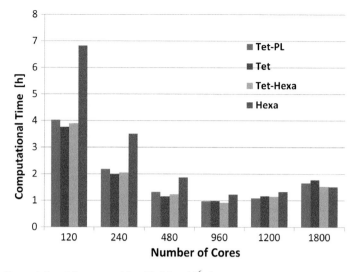

Fig. 14 Computational times on grids with 7.1×10^6 elements

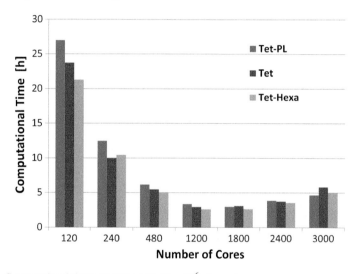

Fig. 15 Computational times on grids with 40×10^6 elements

4 Conclusion

In this work, the applicability of two phase CFD simulations with spray water droplets in a real size generic pressurized water reactor was tested. To this end, models for spray cooling and aerosol particle washout mechanisms were developed and implemented in the CFD software ANSYS CFX. The results showed that such calculations are qualitatively possible in this highly complex geometry. Furthermore, the parallel efficiency of CFD simulations on different grid styles were investigated. The results of these investigations may present guidelines for a better use of the computational resources in CFD applications.

Acknowledgements This work is financially funded by the German Federal Ministry for Economic Affairs and Energy (BMWi) on the basis of the project number 1501493.
The THAI project was carried out on behalf of the Federal Ministry for Economic Affairs and Energy (project no. 1501455) on the basis of a decision by the German Bundestag.

References

1. G. Kessler, A. Veser, F.-H. Schlüter, W. Raskob, C. Landman, J. Päsler-Sauer, *The Risks of Nuclear Energy Technology* (Springer, Berlin, GER, 2014)
2. M. Babić, I. Kljenak, B. Mavko (ed.), Simulation of Atmosphere Mixing and Stratification in the THAI Experimental Facility with a CFD Code, in *Nuclear Energy for New Europe*, Bled, Slovenia, 5–8 September 2005
3. M. Houkema, N.B. Siccama, J.L.A. Nijeholt, E. Komen, Validation of the CFX4 CFD code for containment thermal-hydraulics. Nucl. Eng. Des. **238**, 590–599 (2008)
4. C. Kaltenbach, E. Laurien, CFD simulation of spray cooling in the model containment THAI. Nucl. Eng. Des. **328**, 359–371 (2018)
5. C. Kaltenbach, E. Laurien, CFD simulation of aerosol particle removal by water spray in the model containment THAI. J. Aerosol Sci. **120**, 62–81 (2018)
6. J. Zhang, E. Laurien, Numerical simulation of flow with volume condensation in presence of non-condensable gases in a PWR containment, in Proceedings of International Congress on Advances in Nuclear Power Plants (ICAPP), Charlotte, NC, USA, 6–9 April 2014
7. S. Gupta, G. Langer, M. Colombet, Hydrogen combustion during spray operation: Tests HD-30, HD-31 and HD-32, OECD-NEA THAI-2-Project. Technical Report 1501420-TR-HD-30-32. (Eschborn, GER, 2014)
8. E. Schmidt, M. Colombet, M. Freitag, S. Gupta, B. von Laufenberg, Aerosol spray test with CsI aerosols. Technical Report No. 1501455 (Eschborn, GER, 2015)
9. M. Ishii, K. Mishima, Two-fluid model and hydrodynamic constitutive relations. Nucl. Eng. Des. **82**, 107–126 (1984)
10. M. Freitag, E. Schmidt, S. Gupta, M. Colombet, A. Khnel, B. Von Laufenberg, Technical report: commissioning test of the two vessel configuration test TH.-27. Reposrt No. 1501455 FB/TR TH27 (Becker Technologies GmbH, Eschborn, Germany, 2016)
11. G. Karypis, V. Kumar, A fast and high quality multilevel scheme for partitioning irregular graphs. SIAM J. Sci. Comput. (1998)
12. A. Mansour, E. Laurien, Numerical error analysis for three-dimensional CFD simulations in the two-room model containment THAI+: grid convergence index, wall treatment error and scalability tests. Nucl. Eng. Des. **326**, 220–233 (2018)
13. A. Mansour, E. Laurien, Uncertainty quantification techniques, grid convergence and computational efforts on hexahedral, tetrahedral and hybrid mesh styles for three-dimensional unsteady CFD simulations in a complex multi-room configuration. Int. J. Eng. Comput. (under review, 2018)

Cavitation Simulations of a Tip Leakage Vortex for a NACA0009 Hydrofoil and a Francis Turbine at Stable Full Load Operating Point

Jonas Wack and Stefan Riedelbauch

Abstract In the first part of the study, the ability of homogeneous and inhomogeneous two-phase modeling approach is investigated using a NACA0009 hydrofoil with a cavitating tip leakage vortex. The results indicate that the inhomogeneous model does not increase accuracy of the simulation results. This can be explained by the small diameter of the cavitation bubbles, which results in a strong coupling between liquid and vapor phase. Based on the results for the test case, homogeneous model is applied for two-phase simulations of a Francis turbine at a stable full load operating point. The mesh study demonstrates the need for fine meshes with a size of approximately 50 million elements. While the effect of the cavitation constants can be neglected, the geometry of the runner nut and enabling curvature correction in the SST turbulence model has a relevant impact on simulation accuracy. The use of a transient rotor stator interface is challenging for computational performance. Applying multipass partitioning method increases the speedup by around 10%. Further performance improvement can be achieved with the expert parameter *parallel optimization level*. All in all, a parallelization on 1200 cores is reasonable.

1 Introduction

In recent years, hydro power plants are more and more operated at off-design operating points due to the integration of other renewable energies into the electrical grid. This causes secondary flow phenomena that often result in cavitation. In hydro turbines, different forms of cavitation can appear at the same time. The presence of different forms of cavitation makes it challenging to choose a suitable multiphase modeling approach. Two-phase simulations on a Francis turbine at deep part load conditions with the widely used homogeneous model showed shortcomings to accurately reproduce the regions of cavitation compared to experiments [22]. This raises the question whether the more advanced inhomogeneous model, also called

J. Wack (✉) · S. Riedelbauch
Institute of Fluid Mechanics and Hydraulic Machinery, Pfaffenwaldring 10,
70569 Stuttgart, Germany
e-mail: jonas.wack@ihs.uni-stuttgart.de

© Springer Nature Switzerland AG 2019
W. E. Nagel et al. (eds.), *High Performance Computing in Science and Engineering '18*, https://doi.org/10.1007/978-3-030-13325-2_22

two-fluid model, can increase simulation accuracy. In the first part of this study, the two multi-phase modeling approaches are investigated on a NACA0009 hydrofoil that is facing a cavitating tip leakage vortex and results are compared to experimental data from Dreyer et al. [6, 7]. Even though tip vortex cavitation occurs only in Kaplan and not in Francis turbines, this test case is selected due to the appearance of vortex cavitation and the coexistence of cloud cavitation at the leading edge.

In the second part of the report, cavitation simulations are performed on a full load operating point of a Francis turbine with discharge 30% increased compared to best efficiency point. For the chosen pressure level a stable cavitating vortex rope forms in the draft tube cone. The study serves as a basis to select a suitable simulation setup at stable conditions that can be applied to investigations of an unstable operating point.

2 Numerical Methods

The commercial code Ansys CFX version 18.1 is used for CFD simulations. Two-phase simulations are performed except for the Francis turbine mesh study.

2.1 Two-Phase Modeling

For cavitation simulations, the homogeneous and inhomogeneous models are used as two-phase modeling approach. The more complex inhomogeneous model comprises of one set of governing equations (mass and momentum conservation) per phase that results in a significantly higher computational effort compared to single phase simulations. This modeling approach can handle separate velocity fields for the different phases, which results in interfacial forces like drag. For simulations with the inhomogeneous model, the particle model is applied for interfacial transfer. The minimum volume fraction for area density is set to 10^{-7}.

When relative motion between the phases is neglected, liquid and vapor phase can be treated as one common mixture. That is the assumption of the homogeneous model, where only one set of governing equations and a transport equation for the volume fraction α has to be solved for the mixture phase. Due to the reduced number of equations that needs to be solved, the computational effort is significantly lower compared to the inhomogeneous model. More detailed information about homogeneous and inhomogeneous model can be found in literature (e.g. [9, 24]).

The phase change is considered by a cavitation model. In the current study, the Zwart model [25] is applied. It contains model constants F_e for evaporation and F_c for condensation. A common approach is to calibrate the model constants on a simplified test case, as a calibration for an advanced application is not feasible. While Zwart et al. [25] suggest the use of $F_e = 50$ and $F_c = 0.01$, Morgut et al. [16] determined $F_e = 300$ and $F_c = 0.03$ for sheet cavitation on a NACA66(mod) hydrofoil and successfully investigated the flow in a Kaplan turbine and a marine propeller [17]. In this study, both pairs of model constants are applied to investigate the impact.

2.2 Turbulence Modeling

For the NACA0009 test case, the SBES model [15] is chosen, that can operate in scale resolving simulation (SRS) mode. The idea of SRS is to have a hybrid RANS-LES model that uses a RANS model in the wall boundary layers and resolves large eddies away from the wall by applying an LES model [13]. It has been shown that the SBES model can accurately predict the velocity distribution at the strong swirling flow conditions of the test case [23]. For the Francis turbine, RANS simulations are performed with the widely used SST turbulence model [12]. It is less restrictive in the time step size and is thus selected for general investigations at this operating point. In one simulation, curvature correction is applied, due to the fact that eddy-viscosity models have problems in correctly predicting strongly swirling flows and system rotation. With curvature correction the turbulence production term is modified to account for these effects [20].

2.3 Discretization

A second order backward Euler scheme is applied for temporal discretization. The selection of spatial discretization depends on the turbulence model. For RANS simulations (Francis turbine) a high resolution scheme is prescribed and for the SBES model (NACA0009 test case) a bounded central differencing scheme is used. This takes into account that LES simulations need schemes with low numerical dissipation [14].

3 Test Case NACA0009

The NACA0009 test case is selected to investigate the suitability of the homogeneous and inhomogeneous two-phase modeling approach. A variety of experiments have been carried out in the high speed cavitation tunnel of the EPFL for this hydrofoil at different gap widths and incidence angles [6, 7].

3.1 Computational Setup

The hydrofoil has an initial chord $c_0 = 0.11$ m, a span of 0.15 m and a maximum thickness $h = 9.9$ mm. It is cut off to a chord of $c = 0.1$ m. To reduce cavitation in the clearance, the foil tip on the pressure side is rounded with a radius of 1 mm [6], which is taken into account in the simulation model. However, a small radius of 0.1 mm remains on the suction side of the blade for the simulations, which can be explained by a facilitated meshing.

The investigated setup has an incidence angle of $10°$ and a normalized tip clearance $\tau = 1$. A block profile with a velocity $w_\infty = 9.78$ m/s and a turbulence intensity of 5% is set at the inlet. This is equivalent to the measurement with $w_\infty = 10$ m/s and considers the boundary layer close to the wall [6]. At the outlet, the static pressure is set to have an inlet pressure $p_\infty = 1$ bar. The inlet boundary condition is set 3.5 chords upstream of the hydrofoil and the outlet 12 chords downstream. To reduce the number of cells, the simulation domain is split up into three subdomains with the finest mesh extending from 1.5 chords upstream to 4 chords downstream of the wing to resolve the tip leakage vortex. A mesh study [23] resulted in a grid of 46M cells and a wall resolution with $y^+ = 1$, which is used for this study. The selected time step corresponds to an RMS Courant number of 0.35.

3.2 Numerical Results

Two configurations of cavitation model constants are examined for the homogeneous model. For the inhomogeneous model different treatment of the dispersed vapor phase is investigated. Two simulations are performed with a monodisperse bubble diameter of the vapor phase of 0.01 mm and 0.1 mm, respectively. These values are selected based on investigations by Maeda et al. [11] who determined that the bubble radius in a cavitation cloud is in the range of some micro meters up to around 100 μm. Furthermore, one simulation is carried out with a polydispersed treatment based on the homogeneous MUSIG approach with ten size groups, a maximum bubble diameter of 1 mm and a minimum bubble diameter of 0.001 mm. The upper limit of 1 mm is selected as in the test case other forms of cavitation than cloud cavitation occur. For all simulations performed with the inhomogeneous model, the drag force is considered while other interfacial forces like lift force are neglected. The drag coefficient is set constant to a value of $C_D = 0.44$.

A side view on the cavitation regions is shown in Fig. 1. The simulation results are averaged over 20,000 time steps and the cavitation volume is visualized by two different thresholds of the vapor volume fraction α_v: 0.5 and 0.05. Three different forms of cavitation can be observed for the investigated NACA0009 hydrofoil. Cloud cavitation is present on the suction side of the leading edge due to the high incidence angle. Furthermore, a cavitating tip separation vortex and a cavitating tip leakage vortex are forming caused by the flow through the gap.

A comparison of the simulations with the homogeneous model (see Fig. 1b, c) indicates that the averaged cavitation volume for all forms of cavitation is slightly bigger with the cavitation model constants from Morgut. Both, the vapor volume of cloud cavitation and the tip separation vortex seem to be slightly overestimated compared to the experiment. The simulation results with the inhomogeneous model (see Fig. 1d–f) show a slight increase in cavitation volume for cloud cavitation and the tip leakage vortex going from bubble diameter 0.01 mm to 0.1 mm to a poly-dispersed treatment. A comparison of Fig. 1c, f shows similar cavitation behavior for tip separation and tip leakage vortex and only a small difference for the vapor

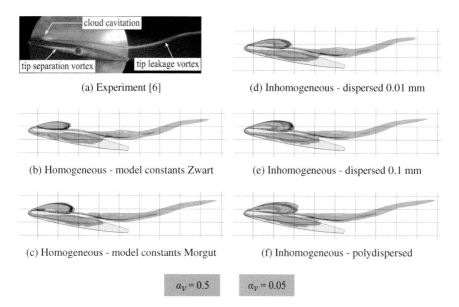

(a) Experiment [6]

(b) Homogeneous - model constants Zwart

(c) Homogeneous - model constants Morgut

(d) Inhomogeneous - dispersed 0.01 mm

(e) Inhomogeneous - dispersed 0.1 mm

(f) Inhomogeneous - polydispersed

$\alpha_v = 0.5$ $\alpha_v = 0.05$

Fig. 1 Visualization of the cavitation volume with two isosurfaces—$\alpha_v = 0.5$ and $\alpha_v = 0.05$—of the time-averaged vapor volume fraction

volume for cloud cavitation. This suggests that with optimized model constants of the cavitation model, similar results can be achieved with the homogeneous model as with the inhomogeneous model with a polydispersed treatment of the vapor phase. This result can be explained by the Stokes number St:

$$St = \frac{\tau_V}{\tau_F} = \frac{\frac{\rho_v D^2}{18 \mu_l}}{\tau_F} \tag{1}$$

The symbol τ_V represents the momentum response time and is a function of density of the vapor phase ρ_v, bubble diameter D and dynamic viscosity of the liquid phase μ_l. For small bubble diameters ($St \ll 1$), the vapor bubble responds fast to changes in the flow field, that happen within a characteristic time scale τ_F. Consequently, a velocity equilibrium between the two phases can be assumed, as the bubble has enough time to adapt to changes in the velocity of the liquid phase [4] and the inhomogeneous model does not have any benefit for small Stokes numbers. However, it has to be highlighted that for big bubble diameters the two-fluid model might be beneficial and is therefore often used for other applications like bubble column simulations.

All in all, it can be stated that simulation results show a reasonable agreement with the experiments. Deviations can be observed for the tip separation vortex in the region close to the leading edge. There, the cavitation volume is overestimated. Furthermore, the simulated vapor volume is overestimated in the region where the tip leakage vortex unites with the tip separation vortex. It can be stated that the differences between the investigated simulation setups are relatively small. For the investigated case, it is more essential to have a fine mesh resolution and a suitable turbulence model to resolve the pressure minimum in the vortex core of the tip leakage vortex [23].

4 Francis Turbine

After investigating different two-phase modeling approaches on the test case, a stable full load operating point at cavitating conditions is examined on a Francis turbine at model scale. Extensive measurements on this Francis turbine have been performed within the European Union project HYPERBOLE (see e.g. [8, 19]). At certain full load conditions with low σ-level, it comes to auto-oscillations in this turbine. Experimental investigations indicate that this unstable behavior is caused by different inflow condition into the draft tube due to cavitation on the runner blades. However, Müller [18] highlights that this physical mechanism still needs to be proved or refuted. Numerical simulations of the full load instability showed deviations in the prediction of head and torque as well as shortcomings to correctly reproduce the unstable behavior [5, 21]. The goal of this investigation is thus to increase the accuracy of the simulation results for a stable full load operating point that serves as a basis for analyzes at unstable conditions.

4.1 Computational Setup

The simulation domain is displayed in Fig. 2. It ranges from the spiral case inlet to the draft tube outlet. A transient rotor stator approach is applied at the interfaces between stationary parts and the rotating runner. At the inlet boundary condition a constant mass flow is prescribed and at the outlet a constant pressure is set according to the measurements. For all simulations, a time step is prescribed, which corre-

Fig. 2 Simulation domain Francis turbine

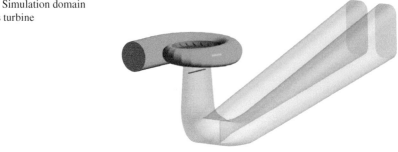

sponds to 0.96° of runner revolution. In a time step study it has been identified that further decreasing the time step size does not increase accuracy for RANS simulations of this operating point. Based on the results from the NACA0009 test case, all simulations for the Francis turbine are performed with the homogeneous two-phase modeling approach. The investigated operating point has the dimensionless quantities $Q_{ED} = 0.260$, $n_{ED} = 0.288$ and $\sigma = 0.37$. Two different geometries of the runner nut are simulated. The configuration without hole has initially been used. However, as deviations of the geometry were observed compared to the experimental setup, a configuration with hole is also investigated to examine the impact.

In the scope of this study, the results of four different simulation setups are compared. They differ in the geometry of the runner nut and the cavitation constants. Furthermore, one simulation is performed with curvature correction for the turbulence model. The setups are listed in Table 1.

4.2 Preliminary Investigations

A mesh study is performed with steady state single phase simulations. The procedure is to start with a basic mesh for the four subdomains spiral case (SC), stay and guide vanes (SVGV), runner (RU) and draft tube (DT) with an averaged y^+-value of approximately 50. Starting with the SC, the mesh is refined subdomain by subdomain until an acceptable mesh convergence is obtained. The resolution of the boundary layer is kept constant within the first step. Head, torque, minimum pressure and volume of cells below vapor pressure are selected as quantities to evaluate the level

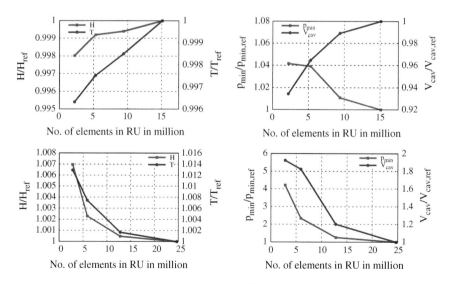

Fig. 3 Mesh study for subdomains SVGV (top) and RU (bottom). Left: Head and torque. Right: Minimum pressure and volume of cells below vapor pressure

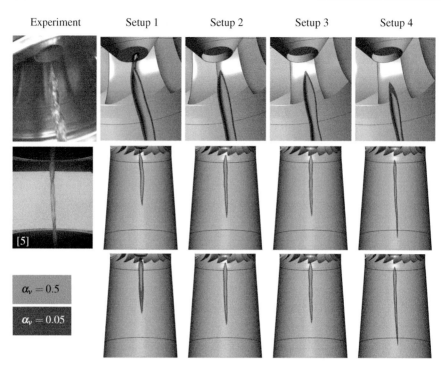

Experiment	Setup 1	Setup 2	Setup 3	Setup 4

Fig. 4 Visualization of the cavitation volume for different simulation setups with two isosurfaces of the vapor volume fraction. Top: Instantaneous snapshot—zoom to the runner nut. Middle: Instantaneous snapshot. Bottom: Averaged results

Table 1 Simulation setups

Setup	Runner nut	Cavitation constants	Curvature correction
1	No hole	Zwart	no
2	Hole	Zwart	no
3	Hole	Morgut	no
4	Hole	Zwart	yes

of convergence. For all quantities the values from the finest mesh of the investigated subdomain are used. The results for the subdomains SVGV and RU are presented in Fig. 3.

In the subdomain SVGV, four steps of mesh refinement are compared. With growing number of elements head and torque are slightly increasing. Nevertheless, the deviation between the finest and coarsest mesh is below 0.2% for head and 0.5% for torque. For minimum pressure and volume of cells below vapor pressure, only minor differences can be observed but a tendency towards mesh independent results can be stated. Even though torque in the SVGV mesh study seems not compeletly converged, the mesh with 9.4 M elements is selected, as the main objective of this study is not the exact forecast of integral quantities but to accurately predict cavitation behavior.

A more clear mesh convergence can be observed in the runner. Here, the mesh with 12.8 M elements is within 0.2% accuracy compared to the finest mesh and is thus selected for further investigations. For minimum pressure and volume of cells below vapor pressure it can be seen that the refinement of the runner is much more important to accurately resolve cavitation in the draft tube compared to the mesh at stay and guide vanes.

For spiral case two meshes (0.7 M and 1.1 M) and for draft tube three meshes (5 M, 13 M and 26 M) are investigated. In both subdomains the refinement does not lead to any significant change in the four analyzed quantities. Thus, for both subdomains the coarsest mesh is selected for further studies. In the mesh study for the NACA0009 test case [23], it was observed that the pressure minimum in the vortex core needs to be well resolved. This is the case with the 5 M mesh in the draft tube subdomain as at least 9 cells in radial direction capture the region where the pressure is falling below vapor pressure. At this point it has to be highlighted that the mesh study is performed for RANS simulations. For hybrid RANS-LES simulations a refinement of the draft tube mesh might increase accuracy.

In the second step, the cell distribution in the far field is kept constant and the resolution of the boundary layer is increased to $y^+ = 1$. This leads to a significant improvement when comparing head and torque with the measurement results. For head, underestimation is 3.8% instead of 6.1% with $y^+ = 50$. For torque, the deviation to the experiment reduces from approximately 7.3% to 4.7%. Consequently, the use of wall functions introduces a relevant error and the resolution of the boundary layer plays an important role, which has already been reported by Krappel et al. [10] on a different Francis turbine.

4.3 Numerical Results

In Fig. 4, the cavitating vortex rope in the draft tube cone is visualized by isosurfaces of the vapor volume for the different simulation setups. A comparison between setup 1 and 2 indicates that the geometry of the runner nut has a relevant impact on the simulation results. For setup 1, the cavitation volume attaches to the runner nut and the vortex rope develops an asymmetric shape. Furthermore, the length of the cavitation region is noticeably underestimated compared to the experiment. In setup 2, the hole leads to a symmetric vortex rope which is not connected to the runner nut. This is in better agreement to the measurements. However, the length of the cavitating vortex rope is still underestimated.

The adjustment of the cavitation constants in setup 3 does not have a significant impact on the cavitation volume of the vortex rope. Both, shape and size are very similar compared to setup 2. Contrary to this, the results of setup 4 indicate that curvature correction has a substantial impact on the results. It can be observed that the distance of cavitating vortex rope to the runner nut is larger than in setup 2. A

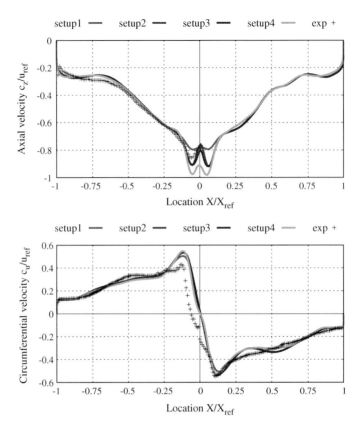

Fig. 5 Averaged axial and circumferential velocity distribution along line in the draft tube cone 0.39 times the runner outlet diameter below the runner. Experiments performed at EPFL and published by Decaix et al. [5]

comparison to experimental results shows that in this region curvature correction has a negative effect on the simulation accuracy. Nevertheless, the general shape of the cavitating vortex rope is thinner and extends further downstream which is in better agreement with the measurements. For setup 4, visualization of instantaneous and averaged volume fraction looks similar due to the symmetric shape of the vortex rope caused by the runner nut geometry. The same can be stated for setup 2 and 3.

Particle image velocimetry (PIV) measurements are available in the draft tube cone. Averaged results of axial and circumferential velocity distribution along a line in the draft tube cone (see Fig. 2), which is located 0.39 times the runner outlet diameter below the runner, is displayed in Fig. 5. Both velocities are nondimensionalized with the rotational speed of the runner at the outer diameter of the outlet. All setups have in common that there is a deviation of the axial velocity close to the draft tube wall compared to the measurement. The main differences between the setups can be found in the center of the draft tube cone where cavitation occurs.

Table 2 Deviation of head and torque to experiment

Setup	$\frac{H-H_{exp}}{H_{exp}}(\%)$	$\frac{T-T_{exp}}{T_{exp}}(\%)$
1	+0.9	+3.0
2	+0.8	+3.0
3	+0.8	+3.0
4	+0.2	+2.5

For setup 1, the results show that the maximum axial velocity is underestimated close to the center of the draft tube. Furthermore, the velocity deficit in the center caused by the occurrence of cavitation is not met. This can be explained by the asymmetric shape of the vortex rope, which is caused by the runner nut geometry. The asymmetry leads to the fact that the cavitation region does not remain in the center of the draft tube. Consequently, the velocity deficit evens out for this setup, which is not in agreement with the measurements. The peak of the circumferential velocity is lower for setup 1 compared to the other simulations. This corresponds to a reduced swirl in the region of the vortex rope and coincides with the observations that for setup 1 the shortest vapor region forms.

Setups 2 and 3 have only minor differences in axial and circumferential velocity distribution. The highest difference for these setups can be observed in the region of velocity deficit caused by cavitation. This deficit is best reproduced by these setups. However, it has to be mentioned that experimental results have to be taken with caution in the cavitation region, as cavitation has a negative impact on measurement accuracy. All in all, it can be stated that the cavitation constants have a negligible impact for this operating point.

Applying curvature correction (setup 4) has a relevant impact on the velocity distribution. The plateau shape of the axial velocity around $X/X_{ref} = -0.75$ is better met, but still shows deviations to the experiment. In the center of the draft tube cone, the maximum axial velocity is overestimated but the velocity deficit due to cavitation is met. A significant difference to the other setups can be observed for the circumferential velocity around $X/X_{ref} = 0.5$. There, the simulation with curvature correction better agrees with the measurement but still shows significant deviations. For the circumferential velocity, it has to be highlighted that the deviations to experimental results in the section $-0.1 < X/X_{ref} < 0.1$ have to be taken with caution. The reason is the reduced measurement accuracy due to cavitation.

The deviation of head and torque to the measurement for the different setups is listed in Table 2. Setup 1, 2 and 3 have similar results for head and torque. The overestimation is around 1% for head and 3% for torque. With curvature correction, a better agreement can be achieved. The results for head are close to the experiment and torque is overestimated by 2.5%. All in all, the unsteady simulations are in better agreement for head and torque compared to the steady state investigations in the mesh study (see Sect. 4.2). Furthermore, head and torque as well as velocity components are better met compared to unsteady simulations with similar setup and a coarser mesh resolution (see Decaix et al. [5] for comparison). This leads to the conclusion that the mesh resolution is essential for accurate results.

5 Computational Resources

All numerical simulations have been carried out on the CRAY XC40 (Hazel Hen) at the High Performance Computing Center Stuttgart. The Cray XC40 Hazel Hen architecture includes 7712 compute nodes with Intel Haswell processors with 12 cores, 128 GB memory and Cray Aries interconnect. A strong scaling test is carried out for the unsteady turbine simulation using the standard cavitation constant and the SST turbulence model without curvature correction. The mesh consists of approximately 50 M elements. The corresponding speedup curves are presented in Fig. 6 with the results normalized to the simulation with 480 cores and CFX version 19.0.

First, the different versions are compared (see Fig. 6 top, left). From version 17.2 to 18.1 a slight performance improvement of around 3% can be observed. Versions 18.1 and 19.0 have a similar performance behavior. Nevertheless, for version 19.0 an improvement in parallel performance is proclaimed by increasing the expert parameter *parallel optimization level* [3]. This parameter applies pre-set optimizations that, however, may have the drawback to reduce robustness [2]. Indeed, this parameter increases parallel performance as it can be seen in Fig. 6 top, right. Especially for a high number of cores the performance can be increased up to 23% by setting the *parallel optimization level* to 2. A comparison of the curves shows that with

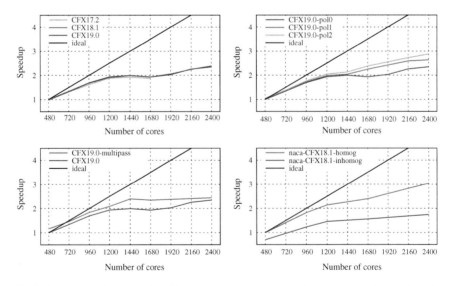

Fig. 6 Speedup test for unsteady turbine and test case simulations for mesh sizes of approximately 50M elements

increasing values for *parallel optimization level* performance improves. For simulations with 1200 cores, the parallel performance improves by 3% when setting the *parallel optimization level* to 1 and by 6% for a value of 2.

Unsteady turbine simulations often suffer in parallel performance due to the interface between rotating runner and steady simulation domains. Multipass partitioning uses circumferentially-banded partitions adjacent to each rotor stator interface. This has the advantage that interface nodes stay in the same partition when the domains slide relative to each other [1]. The performance curves for simulations with and without multipass partitioning are displayed in Fig. 6 bottom, left. It can be observed that enabling the multipass partitioning leads to an increased parallel performance of approximately 10%.

The results show that a sufficient performance can be obtained up to 1200 cores, which corresponds to approximately 40,000 cells per core. For the unsteady turbine simulations about 80 runner revolutions should be carried out. With the selected time step and the 50M mesh, this results in approximately 40 computation days on 1200 cores. Caused by the high utilization of the high performance cluster, the pure computation time can increase up to a factor of two due to queuing time.

In addition, a strong scaling test is carried out for the test case (see Fig. 6 bottom, right). Again sufficient performance can be received up to 1200 cores. A comparison of the different two-phase modeling approaches shows that the computational effort of the inhomogeneous model is around 1.5 times higher compared to the homogeneous model.

6 Conclusion and Outlook

For accurate simulation results the choice of an appropriate two-phase modeling approach is essential. The presented simulation results for a NACA0009 hydrofoil at cavitating conditions with the occurrence of a tip leakage vortex show that the homogeneous model has the same level of accuracy as the inhomogeneous model. This can be explained by the small vapor bubbles that have a small response time to changes in the liquid flow field. Thus, the assumption of a velocity equilibrium made by the homogeneous model is satisfied. For the NACA0009 test case, a fine mesh resolution that can resolve the pressure minimum in the tip leakage vortex is more important as the pressure level affects the mass transfer for cavitation.

A mesh study for the Francis turbine results in a mesh of approximately 50M elements. Furthermore, the wall resolution of the boundary layer has a significant impact on the prediction of head and torque. The simulation results indicate that the geometry of the runner nut has a relevant impact on cavitation volume and shape. Variations in the cavitation constants lead to negligible changes in the simulation results. The application of curvature correction has a significant effect on the solution. With curvature correction the cavitation region close to the runner nut is underestimated but compared to the other setups, the length of the cavitating vortex rope is predicted more accurately.

The results for the Francis turbine indicate that the turbulence model has a relevant effect on the accuracy of the simulation results. Consequently, the hybrid RANS-LES SBES model will be applied for further investigations. After that the findings from the stable operating point will be transferred to an unstable full load operating point with the goal to examine the physical mechanism behind the auto-oscillations.

Acknowledgements The authors would like to thank the HLRS Stuttgart for support and supply of computation time. Furthermore, the authors would like to thank the EPFL for providing the experimental data.

References

1. Ansys, Release 19.0: CFX-Pre User's Guide (2018)
2. Ansys, Release 19.0: CFX-Solver Modeling Guide (2018)
3. Ansys, Release 19.0: Release Notes (2018)
4. C.T. Crowe, J.D. Schwarzkopf, M. Sommerfeld, Y. Tsuji, *Multiphase Flows with Droplets and Particles* (CRC Press, 2011)
5. J. Decaix, A. Müller, A. Favrel, F. Avellan, C. Münch, URANS models for the simulation of full load pressure surge in Francis turbines validated by particle image velocimetry. J. Fluids Eng. **139**, 121103 (2017)
6. M. Dreyer, J. Decaix, C. Münch-Alligné, M. Farhat, Mind the gap: a new insight into the tip leakage vortex using stereo-PIV. Exp. Fluids **55**, 1849 (2014)
7. Dreyer, M.: Mind the gap: tip leakage vortex dynamics and cavitation in axial turbines. Ph.D. thesis, École Polytechnique Fédérale de Lausanne (2015)
8. A. Favrel, A. Müller, C. Landry, K. Yamamoto, F. Avellan, LDV survey of cavitation and resonance effect on the precessing vortex rope dynamics in the draft tube of Francis turbines. Exp. Fluids **57**, 168 (2016)
9. M. Ishii, T. Hibiki, *Thermo-Fluid Dynamics of Two-Phase Flow* (Springer, 2010)
10. T. Krappel, A. Ruprecht, S. Riedelbauch, Turbulence resolving flow simulations of a Francis turbine with a commercial CFD code, in *High Performance Computing in Science and Engineering '15* (Springer, 2015), pp. 421–433
11. M. Maeda, H. Yamaguchi, H. Kato, Laser holography measurement of bubble population in cavitation cloud on a foil section, in *Proceedings of the 1st Joint ASME/JSME Fluids Engineering Conference* (Portland, USA, 1991)
12. F.R. Menter, Two-equation eddy-viscosity turbulence models for engineering applications. AIAA J. **32**, 1598–1605 (1994)
13. F.R. Menter, J. Schütze, M. Gritskevich, Global vs. zonal approaches in hybrid RANS-LES turbulencemodelling, in *Progress in Hybrid RANS-LES Modelling* (Springer, 2012), pp. 15–28
14. F.R. Menter, Best practice: scale-resolving simulations in ANSYS CFD—Version 2.0. ANSYS Germany GmbH (2015)
15. F.R. Menter, Stress-blended Eddy simulation (SBES)–a new paradigm in hybrid RANS-LES modeling, in *Proceedings of the 6th HRLM Symposium* (Strasbourg, France, 2016)
16. M. Morgut, E. Nobile, I. Biluš, Comparison of mass transfer models for the numerical prediction of sheet cavitation around a hydrofoil. Int. J. Multiph. Flow **37**, 620–626 (2011)
17. M. Morgut, D. Jošt, E. Nobile, A. Škerlavaj, Numerical predictions of the turbulent cavitating flow around a marine propeller and an axial turbine. J. Phys.: Conf. Ser. **656**, 012066 (2015)
18. A. Müller, Physical mechanisms governing self-excited pressure oscillations in Francis turbines. Ph.D. Thesis, École Polytechnique Fédérale de Lausanne (2014)
19. A. Müller, A. Favrel, C. Landry, K. Yamamoto, F. Avellan, On the physical mechanisms governing self-excited pressure surge in Francis turbines, in *IOP Conference Series: Earth and Environmental Science*, vol. 22 (2014), p. 032034

20. P.E. Smirnov, F.R. Menter, Sensitization of the SST turbulence model to rotation and curvature by applying the spalart-shur correction term. J. Turbomach. **131**, 041010 (2009)
21. J. Wack, S. Riedelbauch, Two-phase simulations of the full load surge in Francis turbines, in *IOP Conference Series: Earth and Environmental Science*, vol. 49 (2016), p. 092004
22. J. Wack, S. Riedelbauch, K. Yamamoto, F. Avellan, Two-phase flow simulations of the interblade vortices in a Francis turbine, in *Proceedings of the 9th International Conference on Multiphase Flow* (Florence, Italy, 2016)
23. J. Wack, S. Riedelbauch, Application of homogeneous and inhomogeneous two-phase models to a cavitating tip leakage vortex on a NACA0009 hydrofoil, in *Proceedings of the 10th International Symposium on Cavitation* (Baltimore, USA, 2018)
24. G.H. Yeoh, J. Tu, *Computational Techniques for Multiphase Flows* (Elsevier, 2009)
25. P. Zwart, A. Gerber, T. Belamri, A two-phase flow model for predicting cavitation dynamics, in *Proceedings of the 5th International Conference on Multiphase Flow, Paper No. 152* (Yokohama, Japan, 2004)

Part V
Transport and Climate

The simulations in the category "Transport and Climate" are representative of the numerical research performed at Karlsruhe Institute of Technology and at University Hohenheim. In this granting period there have been four projects running on HazelHen (HLRS) and eight on ForHLR I/II (SCC), consuming nearly 70 million core-hours in total. All of the projects run on these computing systems were based upon three numerical model families (COSMO, WRF and ICON). The present collection features results from two climate science studies employing the "Weather Research and Forecasting" (WRF) model which has been developed under the leadership of NCAR.

The authors of the first study (Warrach-Sagi et al.) have targeted regional climate projections for Germany for a period until the end of the 21st century, employing spatial resolutions down to 12 km grid size. In another series of simulations the grid is locally refined in order to allow for the representation of convection events (instead of parameterizing them), underlining the benefit of supercomputing capabilities in climate research. This project specifically highlights the necessity of proper uncertainty quantification strategies in future climate research.

The second project, authored by Schwitalla et al., focuses on seasonal forecasts of weather extremes in a broad latitude belt. Here the authors likewise explore the influence of convection permitting resolutions (with horizontal grid widths of the order of a few kilometers), leading to simulations with more than 10^9 grid nodes covering $\mathcal{O}(10^6)$ time steps. They conclude that, although some difficulties in obtaining fine-grained reference data exists, the increased resolution leads to a significant advance in predictive quality of the model.

Markus Uhlmann
Institute for Hydromechanics, Karlsruhe Institute of Technology
Kaiserstr. 12, 76131 Karlsruhe, Germany
e-mail: markus.uhlmann@kit.edu

Climate Change Studies for Germany and Europe Using High Resolution WRF Simulations

Kirsten Warrach-Sagi, Viktoria Mohr, Josipa Milovac, Thomas Schwitalla and Volker Wulfmeyer

Abstract During the 21st century, not only the global mean surface temperature but also the amount and strength of weather extremes will increase. This is mainly due to the continuous emission of greenhouse gases by human activities. Since warming over many land areas is larger than over the oceans, climate conditions will be regionally highly variable. In order to make regional projections of the climate in the future and to evaluate the ability of models to represent such phenomena, multi-model ensembles of regional climate simulations are required. This is e.g. realized in the World Climate Research Program initiated COordinated Regional climate Downscaling EXperiment CORDEX. The objective of WRFCLIM at HLRS is to contribute to these multi-model ensembles with simulations based on the Weather and Research Forecasting (WRF) model. Five simulations were carried out within the framework of the BMBF funded Project ReKliEs-De (Regionale Klimasimulationen Ensemble für Deutschland) for the simulation time period 1958–2100 at 12 km resolution. These simulations represent the first WRF climate projections which have been realized on the HLRS supercomputer yet. The results demonstrate that the ensemble members provided by the WRF provide an excellent contribution to the spread of the ReKliEs-De ensemble results. This gives a much better confidence not only in the estimation of the evolution of medians of atmospheric variables due to climate change but also in the estimation of extreme values such as hot and ice days. The corresponding climate indices provided by ReKliEs-De underline the importance for society and economy to mitigate climate change. Furthermore, within the CORDEX Flagship Pilot Study framework simulations are carried out between 1999 and 2014 at 15 km

K. Warrach-Sagi (✉) · V. Mohr · J. Milovac · T. Schwitalla · V. Wulfmeyer
Institute of Physics and Meteorology, Garbenstrasse 30, 70599 Stuttgart, Germany
e-mail: kirsten.warrach-sagi@uni-hohenheim.de

V. Mohr
e-mail: viktoria.mohr@uni-hohenheim.de

J. Milovac
e-mail: j.milovac@uni-hohenheim.de

T. Schwitalla
e-mail: thomas.schwitalla@uni-hohenheim.de

V. Wulfmeyer
e-mail: volker.wulfmeyer@uni-hohenheim.de

© Springer Nature Switzerland AG 2019
W. E. Nagel et al. (eds.), *High Performance Computing in Science and Engineering '18*, https://doi.org/10.1007/978-3-030-13325-2_23

resolution further downscaling to the convection permitting (CP) grid on 3 km in central Europe to assess mainly diurnal cycles and high impact weather. The results and the performance analyses on the CP scale will be fundamental for to the set up and improvement of the next generation climate simulations. Therefore, HLRS and UHOH are prepared to contribute to a series of national and international projects concerning seasonal and climate simulations. The results highlight the necessity of CP simulations for next generation earth system models, and multi-model ensembles to assess climate change, as the latter provide indispensable uncertainty measures for climate change signals and extremes for Germany.

1   Introduction

Anthropogenic emissions of CO_2 and other greenhouse gases will provoke significant changes of the climate from the next decades to centuries. Substantial consequences for society, economy, and agriculture can be expected demonstrating the need of mitigation and adaptation. Although climate change is a global phenomenon, it will reveal its impacts on the regional scale. Mitigation, adaptation, and decicion making require robust climate data on high resolution in space and time. Therefore the World Climate Research Program initiated the COordinated Regional climate Downscaling EXperiment CORDEX [8] to numerically downscale the coarse resolution global coupled ocean-atmosphere general circulation models (GCMs) (grid cells > 120 km) analyzed in the IPCC (Intergovernmental Panel on Climate Change) report 2013 (http://www.ipcc.ch) with regional climate models (RCMs) to 50 km and 12 km resolution, respectively. These climate projections are typically run until 2100 and are following the so-called Representative Concentration Pathways (RCPs) for greenhouse gas emission scenarios.

Climate projections require an ensemble of simulations to assess not only the climate change signals, extreme values, and their uncertainty. Furthermore, the RCMs require an evaluation simulation to separate their biases from the biases induced by the GCMs in climate projections. Therefore, the regional climate models (RCMs) had to be evaluated against observations. This was achieved by downscaling the ERA-Interim [7] reanalysis data from 1989 to 2009. The European results (EURO-CORDEX, https://www.euro-cordex.net) were published e.g. by [34] and [14, 15] for the Weather Research and Forecasting (WRF) model [29] simulations of the University of Hohenheim at HLRS, and by [21, 32] and [20] for the whole EURO-CORDEX Ensemble. Afterwards, the GCM climate simulations are downscaled with the RCMs for the historical period until 2005 with observed greenhouse gas concentrations and the projections from 2006 to 2100 assuming the RCPs' greenhouse gas concentrations (e.g. [16]).

In order to densify the CORDEX ensemble and to increase the confidence in the estimation of extreme values such as hot days, the BMBF-funded (Bundesministerium für Bildung und Forschung) project ReKliEs-De (Regionale Klimaprojektionen Ensemble für Deutschland) was established. By the end of the REKliEs-De project in December 2017, 22 simulations for the RCP8.5 scenario and 6 additional simulations for the RCP2.6 were completed to complement the EURO-CORDEX ensemble. Five of these simulations were conducted by the University of Hohenheim with WRF at HLRS. This is the largest RCM projection ensemble until 2100 at the high resolution of 12 km, which has ever been produced for a specific region of this planet yet. With a total of 37 simulations of the RCP8.5 scenario and 14 simulations of the RCP2.6 scenarios, it is now possible to produce robust climate change information on high spatial resolution for Germany. Within the framework of ReKliEs-De, special climate indices were calculated and analyzed in detail with a focus on Germany and its main river catchment areas [12]; http://reklies.hlnug.de/).

For many applications, information about climate indices on the daily and monthly scales at 12 km is provided with a high level of confidence. However, some climate change signals are required at higher spatial and temporal resolution, e.g., the simulation of the diurnal cycle of precipitation suffers from the still not available convection permitting (CP) resolution, for instance, in case of high impact weather in summer. Previous work already demonstrated that further downscaling to a higher resolution on the CP scale (grid increments of 3 km or less) is not only beneficial for improved weather forecasting (e.g. [1]) but also for regional climate simulations [34]. At this resolution, the simulation of land-surface heterogeneities and orography is improved and the convection parameterization can be omitted. This view is supported by recent climate studies on the CP scale over the Alpine region [28] and over southern UK [18]. However, it is expected that the model uncertainties at these scales, where local interactions including land-atmosphere feedback play a more prominent role, will have even stronger impact on the final results than at the coarser RCMs applied e.g. in EURO-CORDEX. A multi-model ensemble has been set-up as a CORDEX flagship pilot study (FPS) to investigate present and future convective processes and related extremes over Europe and the Mediterranean [6]. We participate in this ensemble and the ReKliEs-project with WRF simulations in the frame of the WRFCLIM project at HLRS.

2 WRF Simulations at HLRS

For the regional climate simulations at 12 km resolution for Europe, the open source WRF model version 3.6.1 (http://www2.mmm.ucar.edu/wrf/users/) was applied with 5400 cores in hybrid mode (MPI+OpenMP) to achieve the best speed-up for our problem. The scalability of the WRF model is depending on the problem size (see [33] for more technical details). For the CP CORDEX-FPS test cases, WRF version 3.8.1 was applied with 240 cores to ensure all participating WRF configurations are applied in Europe with as little technical differences as possible. The CP climate evaluation simulation is now running with 1920 cores for central Europe including the southern half of Germany.

2.1 Climate Projections

Within the ReKliEs-De project future climate projections were carried out, applying four different GCMs with two different RCP scenarios as boundary forcing for the WRF model [24, 33]. The historical scenario of the GCMs covers the period from 1958 to 2005. The simulations of the historical period are forced by observed atmospheric composition changes of anthropogenic and natural sources. The projections carried out with the RCP scenarios covering the period from 2006 to 2100, represent mitigation scenarios that assume policy actions will be taken into account to achieve certain emission targets [30]. The numbers of the RCPs give a rough estimate of the range in the change of the radiative forcing due to projected GHG emissions by the year 2100 relative to the pre-industrial values. The RCP8.5 scenario is the so-called business-as-usual scenario which would lead to an approximated rise of the global temperature of around 4 °C.

The RCP2.6 is the so called climate-protection scenario to keep the global temperature rise beneath 2 °C. All regional climate models were applied to the EURO-CORDEX domain specified by CORDEX (see http://www.euro-cordex.net/). The analyses of ReKliEs-De is limited to Germany and its main river catchments (Fig. 1).

The WRF simulations for ReKliEs-De were started in 2016 at HLRS [33] and completed in October 2017. The computational resources from Juli 2017 to June 2018 are listed in Table 1. WRF was applied with the land surface model NOAH [3, 4], the Morrison two-moment microphysics scheme [25], the Yonsei University

Fig. 1 Model terrain height (m) in the model domain for 12 km horizontal resolution. The ReKliEs-De area is highlighted with a black contour line

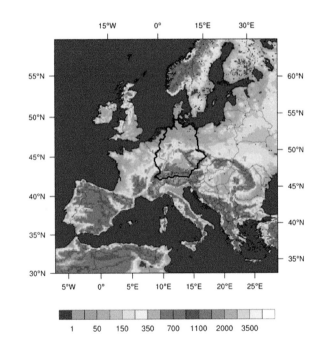

Table 1 Computing resources of WRFCLIM from July 2017 to June 2018 on Cray XC40

Simulations	Node-h	HPSS
Climate projections	174,000	Computing resources of WRFCLIM from July 2017 to June 2018 on Cray XC40
CP climate simulations	121,000	150 TB

(YSU) planetary boundary layer (PBL) scheme [11], the Kain-Fritsch-Eta convection scheme [17] and the radiation parameterization scheme CAM for longwave and shortwave radiation [5].

The raw model output led to an output size of 1.6 PB stored at HPSS and post-processed to obtain the EURO-CORDEX format required to archive and publish the daily and monthly data on the archive system of ESGF (Earth System Grid Federation, https://esgf-data.dkrz.de/projects/esgf-dkrz/). The postprocessing and analyses of the diurnal cycles of variables like precipitation is ongoing and step by step the raw data will be deleted from the HPSS. In Sect. 3 highlights of this treasure of climate simulations are reported.

These activities required the development of respective scripts, processing steps, and infrastructures which are now available at HLRS. Therefore, we can state that we realized a worldwide known hub on earth system modeling and climate projections in southern Germany which is a well requested partner for upcoming national and international projects and contributes significantly to one of the most important problems of human kind in this century: this is climate change.

2.2 Convection Permitting Climate Simulations

Since WRF version 3.8.1 it is possible to run WRF also with the more sophisticated NOAHMP [27] land surface model with OpenMP and hybrid mode respectively. Based on our previous sensitivity studies with respect to the representation of land-atmosphere feedbacks and the PBL moisture profiles [22] at CP scale, for these simulations the MYNN PBL scheme [26] with NOAHMP was chosen. The RRTMG scheme was used for longwave and shortwave radiation [13], Thompson microphysics scheme [31], GRIMS (Global/Regional Integrated Modelling System) scheme for shallow convection, and Grell-Freitas ensemble scheme [9] for deep convection parametrization in the outer domain. Prior to CP climate model simulations within a CORDEX FPS for Europe, an experiment was designed in order to illustrate advantages and possible problems of this very high resolution RCMs in a multi-model ensemble [6]. For the experiment three case studies with strong precipitation events each with different background forcing, have been chosen to be conducted in so called two modes:

1. with weather like initialization (WL)
2. with the climate mode initialization (CM).

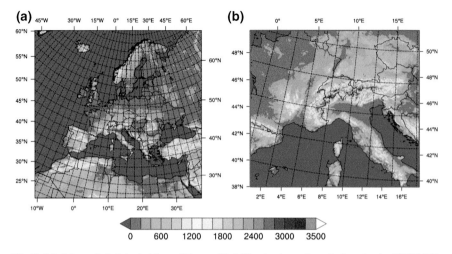

Fig. 2 Model terrain height (m) for **a** 12 km and **b** 2.8 km horizontal resolutions for the CORDEX-FPS convection permitting climate simulations

We performed the WRF simulations for CORDEX-FPS at HLRS. All the simulations were forced with ERA-Interim reanalysis data, and carried out over the EURO-CORDEX domain at 12 km, and further downscaled to 2.8 km covering central Europe (Fig. 2). WL simulations were initiated shortly prior to the event and carried out for 5–7 days depending on the case study. CM simulations were initiated with a longer time-frame prior to the event continuing for mostly ~1 month.

Even though the model is able to run in hybrid mode (MPI+OpenMP), it was decided that all the ensemble members run with different WRF configurations by the participating groups run the model with 240 cores. Therefore this was also applied for these case studies at HLRS. Within the 24 hour wall-time ~1 month was simulated. The simulations were completed in autumn 2017 and analyzed as part of the multi-model ensemble [6].

Since February 2018 such an evaluation run from 2000 to 2014 is ongoing, with 1 year of spin-up (1999). By the end of April 2018, 3 years are simulated. The model setting is similar to the setting used for the test cases, only the coarser domain is changed to a horizontal resolution of 15 km due to the results obtained from the test cases. Further the simulations are run at hybrid mode with 1920 cores. Table 2 summarizes the simulation details.

Table 2 Simulation details for the CORDEX-FPS convection permitting climate simulations

Forcing Data (GCM)	Scenario and target grid	Sim. period	Grid resolution	No. of grid cells	Nr. CPUs with openMPI	Sim. time and output size	Simulation status
ERA-Interim	Reanalysis dx = 0.0275°	20 – 27 June 2009	0.11° 0.0275°	445 × 433 × 50 489 × 441 × 50	240	~6 h ~240 MB	Completed
ERA-Interim	Reanalysis dx = 0.0275°	23 – 28 October 2012	0.11° 0.0275°	445 × 433 × 50 489 × 441 × 50	240	~4 h ~180 MB	Completed
ERA-Interim	Reanalysis dx = 0.0275°	2 – 7 November 2014	0.11° 0.0275°	445 × 433 × 50 489 × 441 × 50	240	~4 h ~180 MB	Completed
ERA-Interim	Reanalysis dx = 0.0275°	1 June – 1 July 2009	0.11° 0.0275°	445 × 433 × 50 489 × 441 × 50	240	~25 h ~1 TB	Completed
ERA-Interim	Reanalysis dx = 0.0275°	1 October – 1 November 2012	0.11° 0.0275°	445 × 433 × 50 489 × 441 × 50	240	~26 h ~1 TB	Completed
ERA-Interim	Reanalysis dx = 0.0275°	1 October – 7 November 2014	0.11° 0.0275°	445 × 433 × 50 489 × 441 × 50	240	~31 h ~1 TB	Completed
ERA-Interim	Reanalysis dx = 0.0275°	1.1. 1999 – 31.12. 2014	0.13755° 0.0275°	360 × 351 × 50 491 × 441 × 50	1920	~1350 h ~220 TB	Ongoing

3 Results

3.1 Climate Projections

In the business-as-usual scenario (RCP 8.5) the results reveal for Germany an increase of the annual mean temperature by 4 °C (ensemble bandwidth of 2.8–5 °C) at the end of this century in comparison with the end of the 20th century (Fig. 3). This implies a significant increase of heat days (maximum temperature above 30 °C) and decrease of frost days (minimum temperature below 0 °C) (Fig. 3). In Germany winter precipitation will increase and due to the warming this will be rather rain than snow. South western Germany will become drier in summer ([23], http://reklies.hlnug.de/). In case of the climate protection scenario (RCP2.6) this increase would only be 1 °C (bandwidth of 0.7–2.4 °C) with a corresponding reduction of the number of extreme events. The strong difference in the evolution of mean temperature is demonstrated in Fig. 3. Obviously, the RCP8.5 must be considered as very dangerous for the society whereas the RCP2.6 scenario will lead to an increase of temperature which falls in the goals of the Paris Agreement in 2015. Therefore, the ReKliEs-De results do not only contain information for adaptation in different areas such as hydrology and agriculture but have also political implications for decision makers.

Within ReKliEs-De, 24 climate indices were calculated and data and graphics were made publically available (see http://reklies.hlnug.de/). Two of these indices are displayed exemplary in Figs. 4 and 5. Ice days are defined as number of days where maximum temperature stays below 0 °C. From 1960 to 2100, ice days decrease from

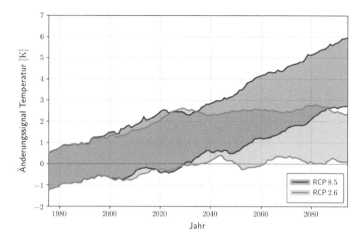

Fig. 3 Temporal development of the annual mean temperature change in the scenarios RCP8.5 (red) and RCP2.6 (blue) calculated for the ReKliEs-De domain from the ensemble members run for both scenarios. *Source* http://reklies.hlnug.de/

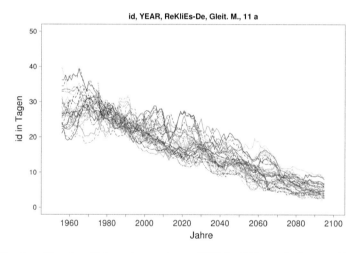

Fig. 4 Number of ice days id (max. temperature $> 0\,°C$) for different single ensemble members of RCP8.5 scenario for the ReKliEs-De domain with 11 year running mean. *Source* http://reklies.hlnug.de/

around 30 to less than 10 ice days per year. Although, the single simulations members revealing a large spread from 20 to 40 days in the 1960s towards the end of the 21st century, all members project to have less than 10 ice days per year. Figure 5 shows the decrease in summer precipitation for the climate protection scenario RCP2.6 and the business-as-usual scenario RCP8.5. The decrease of precipitation in the RCP8.5 scenario in most catchments is larger than 15%.

The climate indices reveal details about expected climate change that will impact our society, agriculture, health and economy at 12 km horizontal resolution and due

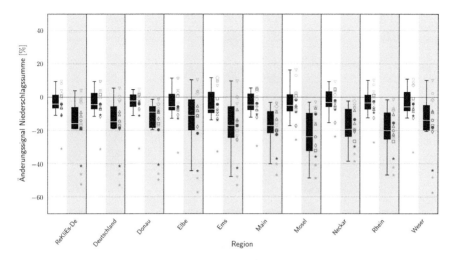

Fig. 5 Change in summer (JJA) precipitation (%) between 1971–2000 and 2071–2100 in case of RCP2.6 (white) and RCP8.5 (gray) for the ReKliEs-De domain, Germany and the larger river catchments within the ReKliEs-De domain. *Source* http://reklies.hlnug.de/startseite

to the large ensemble of model simulations with a high confidence. The data is now publically available and used by impact researchers and public authorities to assess climate change impact. Further they urge the need to follow the climate protection scenario [19].

3.2 Convection Permitting Climate Simulations

Some climate change signals are required at higher spatial and temporal resolution, e.g., the simulation of the diurnal cycle of precipitation and extreme events require convection permitting (CP) resolution. The CP CORDEX-FPS test cases were analyzed by [6] for precipitation onset, intensity, and the location of maxima for different regions in comparison with available observations. First case (C1) referred to an Intense Observation Period 16 (IOP16) during HyMeX measurement campaign in September and November 2012. Heavy precipitation event was observed from 23 to 28 October 2012 in southern France and NW Italy. Case two (C2) is a convective orographic precipitation event in Austria with weak but persistent large-scale forcing. Strong precipitation events were observed from 22 to 25 June 2009. Case 3 (C3) is the Foehn case observed in the period between 3 and 7 November 2014, when the slow eastward evolution of the trough caused persistent precipitation over the Alps with daily precipitation locally exceeding several hundreds of mm and reaching maximum of around 500 mm.

The ensemble comprised 18–21 members, depending on the test case. For the C1 case both WL and CM ensemble means underestimate the observed accumulated precipitation for the indicated period (Fig. 6a1–a3). The underestimation is more

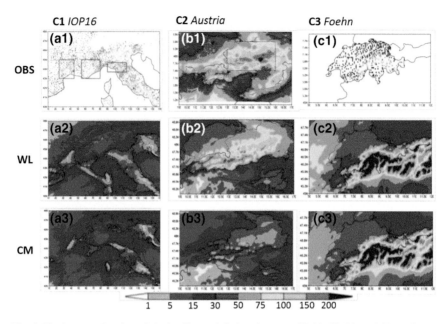

Fig. 6 Total accumulated precipitation [in mm] during the event C1 (a), C2 (b), and C3(c). Observations are given in the top row (1), the WL multi-model ensemble mean in the second row (2), and the CM multi-model ensemble mean (3). From [6]

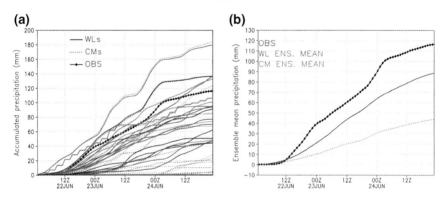

Fig. 7 Time series of accumulated precipitation (in mm on the y-axis) during the C2 event: **a** time series of the accumulated precipitation for each model and observations, **b** Time series of WL (blue) and CM (red) multi-model ensemble means, in comparison with observations (black) over the area of interest. From [6]

pronounced for the CM simulations, but the location of the maxima is well represented with both ensemble means. In C3 case, both the location of the precipitation maxima and the total accumulated precipitation amount is well represented with WL and CM ensemble means (Fig. 6c1, b2, b3). The most interesting case is the case C2. The CM ensemble mean strongly underestimates the observed precipitation (Figs. 6b3, 7a, b). The spread of the both ensembles is large, and most of the

CM simulations completely missed the event with accumulated precipitation close to 0 mm (Fig. 7a). The difference in behavior between the two ensembles is evident in Fig. 7b, where the CM ensemble mean shows an underestimation around the 60% of the correspondent observed curve. A detailed investigation of these test cases is still undergoing, but few factors appear to be responsible for the results of the most interesting C2 case: (1) problematic can be that the event is too close to the domain boundaries, and the involved modelling groups are using varying sponge layer depths and nesting strategies; (2) a relatively weak background synoptic state of the event would decrease the large scale forcing compared to local ones, and thus increase diversity across models [6].

4 Conclusion

The RCM simulations within ReKliEs-De revealed the necessity of a large number of ensemble members at high resolution for Europe to assess climate change at the regional scale. Considerably more confidence and robustness in the expected climate change signals were achieved by the evaluation of the large ensemble. Also, the required ensemble member number to receive robust information was derived in ReKliEs-De. Some climate change signals like mean temperature change can be assessed with a limited number of ensemble members, but some climate indices like tropical nights will require even more than the studied 37 model simulations to receive robust signals. The simulations from 1958 to 2100 with a model time step of 60 s and a 3 hourly model output frequency with WRF were facilitated at the CRAY XE6 and XC40 from summer 2016 to autumn 2017 (e.g. [2]. Each simulation needed 3600 hours walltime on 5400 cores. The resulting 400 TB raw output required postprocessing, archiving and publication on the ESGF archive system. This is still ongoing. The ReKliEs-De project demonstrated the importance of the HLRS infrastructure to enable the required five WRF climate projections within the project timeframe. This was only achieved by running the simulations in parallel, i.e. 27,000 core hours per job.

The results from ReKliEs-De highlight the importance of assessing climate change signals with a multi-model ensemble, WRF being one of them. The results of ReKliEs-De which are also published on http://reklies.hlnug.de/startseite, indicate a general increase of weather extremes in Germany. This fact is of great importance especially in agriculture, since large problems will evolve in the cultivation of different field plants. The development and timing of the sowing and harvesting of such plants are strongly dependent on optimum temperature ranges. The preliminary results from CORDEX-FPS test cases imply that for the long-term simulations which are at the moment ongoing, we can expect varying ranges of responses depending on types of the convective events, as well as to the specification of the model deployed [6]. In CP-RCM it has been seen that the precipitation pattern are moving with the local upper level flow, which is not the case in coarser RCM simulations where the precipitation remains locked at the mountain tops [10]. This has been also seen in these test cases. Such moving precipitation is a more realistic representation, which can be related primary to the convection that is resolved in CP-RCM, and not parameterised as is the case in coarser resolution models. Also such models have more detailed representation of land surface heterogeneity, which may strongly affect representation of precipitation distribution through land-atmosphere feedback processes. Further steps will include more thorough analysis of the impact of such processes on cloud formation and precipitation using these the CP-RCM simulations and various land-atmosphere feedback metrics. The currently running ensemble of CP climate simulations of the CORDEX-FPS will be evaluated at the end of this year before this CORDEX-FPS starts in its second phase of CP climate projections in 2019.

Acknowledgements This work is part of the ReKliEs-De project funded by the BMBF (Federal Ministry for Education and Research), and all the CORDEX-FPS simulations are funded by DFG within the frame of Research Unit on regional climate change FOR1695. We are grateful to Emanuela Pichelli for sharing visualization of the CORDEX-FPS results. We are thankful for the support from the staff of the DKRZ (Deutsches Klimarechenzentrum) for being able to access GCM data. Computational Resources for the model simulations on the CRAY XE6 and XC40 within WRFCLIM were kindly provided by HLRS, we thank staff for their great support.

References

1. H.S. Bauer, T. Weusthoff, M. Dorninger, V. Wulfmeyer, T. Schwitalla, T. Gorgas, M. Arpagaus, K. Warrach-Sagi, Predictive skill of a subset of models participating in D-PHASE in the COPS region. Q. J. R. Meteorol. Soc. **137**(S1), 287–305 (2011)
2. T. Bönisch, M. Resch, T. Schwitalla, M. Meinke, V. Wulfmeyer, K. Warrach-Sagi, Hazel Hen leading HPC technology and its impact on science in Germany and Europe **64** (2017)
3. F. Chen, J. Dudhia, Coupling an advanced land surface hydrology model with the Penn State NCAR MM5 modeling system. Part I: model implementation and sensitivity. Mon. Weather. Rev. **129**(4), 569–585 (2001)
4. F. Chen, J. Dudhia, Coupling an advanced land surface hydrology model with the Penn State NCAR MM5 modeling system. Part II: preliminary model validation. Mon. Weather. Rev **129**(4), 587–604 (2001)
5. W.D. Collins, P.J. Rasch, B.A. Boville, J.J. Hack, J.R. McCaa, D.L. Williamson, M. Zhang, Description of the NCAR community atmosphere model (CAM 3.0). Technical Report (2004)
6. E. Coppola, S. Sobolowski, E. Pichelli, F. Raffaele, B. Ahrens, I. Anders, N. Ban, S. Bastin, M. Belda, D. Belusic, A. Caldas-Alvarez, R.M. Cardoso, A. S Davolio, Dobler, J. Fernandez, L.F. Borrell, Q. Fumiere, F. Giorgi, K. Goergen, I. Guettler, T. Halenka, D. Heinzeller, O. Hodnebrog, D. Jacob, S. Kartsios, E. Katragkou, E. Kendon, S. Khodayar, H. Kunstmann, S. Knist, A. Lavin, P. Lind, T. Lorenz, D. Maraun, L. Marelle, E. van Meijgaard, J. Milovac, G. Myhre, H.-J. Panitz, M. Piazza, M. Raffa, T. Raub, B. Rockel, C. Schär, K. Sieck, P.M.M. Soares, S. Somot, L. Srnec, P. Stocchi, M. Tölle, H. Truhetz, R. Vautard, H. de Vries, K. Warrach-Sagi, A first-of-its-kind multi-model convection permitting ensemble for investigating convective phenomena over Europe and the Mediterranean. Clim. Dyn. (2018). https://doi.org/10.1007/s00382-018-4521-8
7. D.P. Dee, S.M. Uppala, A.J. Simmons, P. Berrisford, P. Poli, S. Kobayashi, U. Andrae, M.A. Balmaseda, G. Balsamo, P. Bauer, P. Bechtold, A.C.M. Beljaars, L. van de Berg, J. Bidlot, N. Bormann, C. Delsol, R. Dragani, M. Fuentes, A.J. Geer, L. Haimberger, S.B. Healy, H. Hersbach, E.V. Hlm, L. Isaksen, P. Kllberg, M. Khler, M. Matricardi, A.P. McNally, B.M.

Monge-Sanz, J.J. Morcrette, B.K. Park, C. Peubey, P. de Rosnay, C. Tavolato, J.N. Thpaut, F. Vitart, The ERA-Interim reanalysis: configuration and performance of the data assimilation system. Q. J. R. Meteorol. Soc. **137**(656), 553–597 (2011). https://doi.org/10.1002/qj.828

8. F. Giorgi, C. Jones, G.R. Asrar, Addressing climate information needs at the regional level: the CORDEX framework. WMO Bull. (2009)

9. G.A. Grell, S.R. Freitas, A scale and aerosol aware stochastic convective parameterization for weather and air quality modeling. Atmos. Chem. Phys. **14**(10), 5233–5250 (2014). https://doi. org/10.5194/acp-14-5233-2014

10. G.A. Grell, L. Schade, R. Knoche, A. Pfeiffer, J. Egger, Nonhydrostatic climate simulations of precipitation over complex terrain. J. Geophys. Res. Atmos. **105**(D24), 29595–29608 (2000). https://doi.org/10.1029/2000JD900445

11. S.Y. Hong, Y. Noh, J. Dudhia, A new vertical diffusion package with an explicit treatment of entrainment processes. Mon. Weather. Rev. **134**(9), 2318–2341 (2006). https://doi.org/10. 1175/MWR3199.1

12. H. Hübener, P. Hoffmann, K. Keuler, S. Pfeifer, H. Ramthun, A. Spekat, C. Steger, K. Warrach-Sagi, Deriving user-informed climate information from climate model ensemble results. Adv. Sci. Res. **14**, 261–269 (2017). https://doi.org/10.5194/asr-14-261-2017

13. M.J. Iacono, J.S. Delamere, E.J. Mlawer, M.W. Shephard, S.A. Clough, W.D. Collins, Radiative forcing by long-lived greenhouse gases: calculations with the AER radiative transfer models, J. Geophys. Res. Atmos. **113**(D13). https://doi.org/10.1029/2008JD009944

14. M. Ivanov, K. Warrach-Sagi, V. Wulfmeyer, Field significance of performance measures in the context of regional climate model evaluation. Part 1: temperature. Theor. Appl. Climatol. **132**(1), 219–237 (2018). https://doi.org/10.1007/s00704-017-2100-2

15. M. Ivanov, K. Warrach-Sagi, V. Wulfmeyer, Field significance of performance measures in the context of regional climate model evaluation. Part 2: precipitation. Theor. Appl. Climatol. **132**(1), 239–261 (2018). https://doi.org/10.1007/s00704-017-2077-x

16. D. Jacob, J. Petersen, B. Eggert, A. Alias, O.B. Christensen, L.M. Bouwer, A. Braun, A. Colette, M. Déqué, G. Georgievski, E. Georgopoulou, A. Gobiet, L. Menut, G. Nikulin, A. Haensler, N. Hempelmann, C. Jones, K. Keuler, S. Kovats, N. Kröner, S. Kotlarski, A. Kriegsmann, E. Martin, E. van Meijgaard, C. Moseley, S. Pfeifer, S. Preuschmann, C. Radermacher, K. Radtke, D. Rechid, M. Rounsevell, P. Samuelsson, S. Somot, J.F. Soussana, C. Teichmann, R. Valentini, R. Vautard, B. Weber, P. Yiou, EURO-CORDEX: new high-resolution climate change projections for European impact research. Reg. Nal Environ. Chang. **14**(2), 563–578 (2014). https://doi.org/10.1007/s10113-013-0499-2

17. J.S. Kain, The Kain Fritsch Convective Parameterization: an update. J. Appl. Meteorol. **43**(1), 170–181 (2004). https://doi.org/10.1175/1520-0450(2004)043<0170:TKCPAU>i2.0.CO;2

18. E.J. Kendon, N.M. Roberts, C.A. Senior, M.J. Roberts, Realism of rainfall in a very high-resolution regional climate model. J. Clim. **25**(17), 5791–5806 (2012). https://doi.org/10.1175/JCLI-D-11-00562.1

19. K. Keuler, H. Huebener, K. Blow, C. Menz, C. Steger, K. Warrach-Sagi, Climate change alternatives for Central Europe. Geophys. Res. Abstr. **20**(EGU2018-12), 882 (2018)

20. S. Knist, K. Goergen, E. Buonomo, O.B. Christensen, A. Colette, R.M. Cardoso, R. Fealy, J. Fernndez, M. Garca-Dez, D. Jacob, S. Kartsios, E. Katragkou, K. Keuler, S. Mayer, E. Meijgaard, G. Nikulin, P.M.M. Soares, S. Sobolowski, G. Szepszo, C. Teichmann, R. Vautard, K. Warrach-Sagi, V. Wulfmeyer, C. Simmer, Land-atmosphere coupling in EURO-CORDEX evaluation experiments. J. Geophys. Res. Atmos. **122**(1), 79–103 (2016). https://doi.org/10. 1002/2016JD025476

21. S. Kotlarski, K. Keuler, O.B. Christensen, A. Colette, M. Déqué, A. Gobiet, K. Goergen, D. Jacob, D. Lüthi, E. van Meijgaard, G. Nikulin, C. Schär, C. Teichmann, R. Vautard, K. Warrach-Sagi, V. Wulfmeyer, Regional climate modeling on European scales: a joint standard evaluation of the EURO-CORDEX RCM ensemble. Geosci. Model. Dev. **7**(4), 1297–1333 (2014). https:// doi.org/10.5194/gmd-7-1297-2014

22. J. Milovac, K. Warrach-Sagi, A. Behrendt, F. Spth, J. Ingwersen, V. Wulfmeyer, Investigation of PBL schemes combining the WRF model simulations with scanning water vapor differential

absorption lidar measurements. J. Geophys. Res. Atmos. **121**(2), 624–649 (2016). https://doi.org/10.1002/2015JD023927

23. V. Mohr, K. Blow, P. Hoffmann, H. Hbener, K. Keuler, C. Menz, K. Radtke, H. Ramthun, A. Spekat, C. Steger, V. Wulfmeyer, K. Warrach-Sagi, Weather extremes in an ensemble of downscaled CMIP5 simulations for Germany from 1971–2000. Geophys. Res. Abstr. **20**(EGU2018-13), 490 (2018)

24. V. Mohr, K. Warrach-Sagi, T. Schwitalla, H.S. Bauer, V. Wulfmeyer, High-resolution climate projections using the WRF model on the HLRS, in *High Performance Computing in Science and Engineering '16*, ed. by W. Nagel, D. Kroener, M. Resch (Springer International Publishing, 2016), pp. 577–587. https://doi.org/10.1007/978-3-319-47066-5

25. H. Morrison, G. Thompson, V. Tatarskii, Impact of cloud microphysics on the development of trailing stratiform precipitation in a simulated squall line: comparison of one and two-moment schemes. Mon. Weather. Rev. **137**, 991–1007 (2009)

26. M. Nakanashi, H. Niino, Development of an improved turbulence closure model for the atmospheric boundary layer. J. Meteorol. Soc. Japan. Ser. II **87**(5), 895–912 (2009). https://doi.org/10.2151/jmsj.87.895

27. G.Y. Niu, Z.L. Yang, K.E. Mitchell, F. Chen, M.B. Ek, M. Barlage, A. Kumar, K. Manning, D. Niyogi, E. Rosero, M. Tewari, Y. Xia, The community Noah land surface model with multiparameterization options (Noah-MP): 1. Model description and evaluation with local-scale measurements. J. Geophys. Res. Atmos. **116**(D12). https://doi.org/10.1029/2010JD015139

28. A.F. Prein, W. Langhans, G. Fosser, A. Ferrone, N. Ban, K. Goergen, M. Keller, M. Tölle, O. Gutjahr, F. Feser, E. Brisson, S. Kollet, J. Schmidli, N.P.M. Lipzig, R. Leung, A review on regional convection-permitting climate modeling: demonstrations, prospects, and challenges. Rev. Geophys. **53**(2), 323–361 (2015). https://doi.org/10.1002/2014RG000475

29. W.C. Skamarock, J.B. Klemp, J. Dudhia, D. Gill, D.O. Barker, M.G. Duda, W. Wang, J.G. Powers, A description of the advanced research WRF version 3. NCAR Technical Note TN-475+STR, NCAR, Boulder, CO (2008). http://www.mmm.ucar.edu/wrf/users/docs/arw_v3.pdf

30. K.E. Taylor, R.J. Stouffer, G.A. Meehl, An overview of CMIP5 and the experiment design. Bull. Am. Meteorol. Soc. **93**(4), 485–498 (2012). https://doi.org/10.1175/BAMS-D-11-00094.1

31. G. Thompson, P.R. Field, R.M. Rasmussen, W.D. Hall, Explicit forecasts of winter precipitation using an improved bulk microphysics scheme. Part II: implementation of a new snow parameterization. Mon. Weather. Rev. **136**(12), 5095–5115, https://doi.org/10.1175/2008MWR2387.1

32. R. Vautard, A. Gobiet, D. Jacob, M. Belda, A. Colette, M. Déqué, J. Fernández, M. García-Díez, K. Goergen, I. Güttler, T. Halenka, T. Karacostas, E. Katragkou, K. Keuler, S. Kotlarski, S. Mayer, E. van Meijgaard, G. Nikulin, M. Patarčić, J. Scinocca, S. Sobolowski, M. Suklitsch, C. Teichmann, K. Warrach-Sagi, V. Wulfmeyer, P. Yiou, The simulation of European heat waves from an ensemble of regional climate models within the EURO-CORDEX project. Clim. Dyn. **41**(9), 2555–2575 (2013). https://doi.org/10.1007/s00382-013-1714-z

33. K. Warrach-Sagi, V. Mohr, V. Wulfmeyer, High resolution WRF simulations for climate change studies in Germany, in *High Performance Computing in Science and Engineering '17*, ed. by W. Nagel, D. Kroener, M. Resch (Springer International Publishing, 2017), pp. 431–440 (2017). https://doi.org/10.1007/978-3-319-68394-2_25

34. K. Warrach-Sagi, T. Schwitalla, V. Wulfmeyer, H.S. Bauer, Evaluation of a climate simulation in Europe based on the WRF-NOAH model system: precipitation in Germany. Clim. Dynam. **41**(3–4), 755–774 (2013). https://doi.org/10.1007/S00382-013-1727-7

Seasonal Simulation of Weather Extremes (WRFXXXL)

Thomas Schwitalla, Volker Wulfmeyer and Kirsten Warrach-Sagi

Abstract To date, seasonal forecasts are often performed by applying horizontal resolutions of 75–150 km due to lack of computational resources and associated operational constraints. As this resolution is too coarse to represent fine scale structures impacting the large scale circulation, a convection permitting (CP) resolution of less than 4 km horizontal resolution is required. Most of the simulations are carried out as limited area model (LAMs) and thus require boundary conditions at all four domain boundaries. In this study, the Weather Research and Forecasting (WRF) model is applied in a latitude belt set-up in order to avoid the application of zonal boundaries. The horizontal resolution is 0.03° and 0.45° spanning a belt between 65°N and 57°S encompassing 12000 * 4060 * 57 grid boxes for the high-resolution domain. The simulations are driven by ECMWF analysis and high resolution SST data from multiple sources. The simulation period was February 1st–July 1st, 2015 which was a strong El Niño period.

1 Introduction

To date, climate simulations and long-term weather forecasts are extremely challenging. This is due to the complex interaction between large-scale and small-scale meteorological processes such as the monsoon circulation and mesoscale circulations induced by complex terrain and heterogeneous land surface properties. Furthermore, e.g. the African continent is strongly affected by teleconnections triggered by El Niño Southern Oscillation (ENSO) events. During an El Niño period, e.g. South East Africa and Australia can suffer from extreme aridness while countries like Kenya and Ethiopia can be influenced by additional precipitation during extreme events.

T. Schwitalla (✉) · V. Wulfmeyer · K. Warrach-Sagi
Institute of Physics and Meteorology, Garbenstrasse 30, 70599 Stuttgart, Germany
e-mail: thomas.schwitalla@uni-hohenheim.de

V. Wulfmeyer
e-mail: volker.wulfmeyer@uni-hohenheim.de

K. Warrach-Sagi
e-mail: kirsten.warrach-sagi@uni-hohenheim.de

© Springer Nature Switzerland AG 2019
W. E. Nagel et al. (eds.), *High Performance Computing in Science and Engineering '18*, https://doi.org/10.1007/978-3-030-13325-2_24

Thus, in principle, phenomena like ENSO, can influence weather pattern over the entire globe, and can maintain a predictability of extreme events of up to several months. Consequently, it should be possible to predict extreme events statistics up to the seasonal scale for the protection of the environment and humankind.

Unfortunately, still to date, it was not possible to exploit this potential of predictability in seasonal simulations and weather forecasts. Particularly, current model systems are inadequate to simulate droughts and extreme precipitation with acceptable predictive skill. This is due to the limited grid resolution of current global general circulation models (GCMs), poor model physics, particularly with respect to the simulation of the atmospheric boundary layer (ABL), the land-ABL interaction, clouds, and precipitation, as well as an inaccurate initialization of the state of the ocean-land-atmosphere system. For instance, current climate projections performed in the CORDEX project (www.cordex.org) for the African continent are performed with a horizontal resolution of 50 km, which is not sufficient to resolve mesoscale and orographically-induced circulations.

To enhance the prediction of climate change and extreme events, it is necessary to perform simulations on the convection permitting (CP) scale of the order of 1–3 km. References [4, 6, 8, 9] demonstrated that this resolution is leading to a considerable improvement of the predictive skill with respect to the simulation of surface temperatures, clouds, and precipitation. E.g., [2] stated that as long as a convection parametrization is applied, one cannot expect a major improvement in the spatial and temporal patterns of precipitation.

The commonly applied limited area models (LAM) are dependent on the lateral boundary forcing, usually provided by coarser scale global general circulation models (GCM). As these models often apply different physics schemes compared to LAMs, this can lead to considerable spin-up effects at the domain boundaries [1, e.g.]. To avoid these inconsistencies, latitude-belt simulations are performed with the Weather Research and Forecasting (WRF) model [11] where the external forcing is only required at the northern and southern domain boundaries. The experiment described here is a successive simulation based on the experience of [9].

2 Domain Setup

The applied model domain is shown in Fig. 1. The domain compasses 12000 ∗ 4060 cells for the CP resolution and 800 ∗ 271 cells for the coarse resolution simulation. Both experiments utilize 57 model levels and share the same physics packages apart from the convection parametrization. Compared to the previous set up of [9], the cloud microphysics now changed to the double-moment scheme of Thompson [12]. Further details about the experimental set up can be found in the intermediate report of 2017 [10].

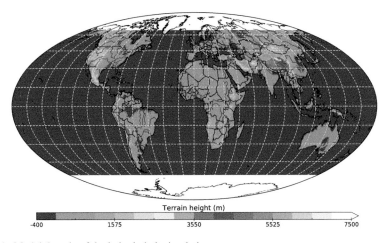

Fig. 1 Model domain of the latitude-belt simulation

3 Technical Aspects

The technical set up remains the same as shown in the intermediate report of [10]. During the last 12 months, approx. 641000 Node-h have been used. The simulation revealed some deficiencies of the Parallel NetCDF libraries. It appears that in case more than 10000 MPI tasks are applied, a limitation of either PNetCDF itself or the WRF I/O mechanism becomes evident. In this case the I/O rate is limited to about 7–8 GB/s. This issue is currently under investigation together with HLRS and Cray. The total amount of data stored in the HPSS tape library is about 250 TB. Out of this data volume, 150 TB are required for restart files. In case the user wants to repeat a certain time period with a high temporal resolution these files are necessary as they contain a memory imaging. In case the full WRF output.

Table 1 shows the scaling of the WRF model code with respect to a basic usage of 370 nodes when the high-resolution domain is applied. I/O is not considered in this evaluation.

Table 1 Runtime performance of the WRF model code in case the high-resolution domain is applied. The code was compiled using the Intel version 16 environment

# nodes	# Threads/node	# MPI tasks	Elapsed time (s)	Speedup	Theoretical speedup
4096	4	24576	0.21	11	11.1
4096	6	16384	0.22	10.5	11.1
700	6	2800	1.25	1.84	1.89
600	1	14400	1.6	1.44	1.62
380	2	4560	2.1	1.1	1.03
370	1	8888	2.3	–	–

4   Current Status and Preliminary Results

Both five month simulations for the 0.03° and 0.45° resolution have been successfully completed during Mai 2017. The additional limited are model (LAM) simulation proposed in the 2017 report will not be performed. This would lead to preference of the LAM simulation results as this type of simulation is driven by a coarser scale reanalysis (observations). Especially on longer time scales at very high resolution, we cannot rule out that the boundaries would have a major impact on the simulation results.

As the initial conditions were provided by the ECMWF global model, featuring a different description of applied model physics and soil properties, a soil moisture spin-up period is required. Figure 2 shows an area averaged time series of the soil moisture in 4 different layers for two different regions.

It is reasonable to determine the spin-up period by the behavior of the lowermost soil level (level 4 at 2 m depth) because the deep soil shows a kind of long-term memory. From the two timeseries shown in Fig. 2, it is clearly seen that e.g. over South East Asia, the adjustment takes about four weeks to reach an "equilibrium state" while in case of a dry soil over Australia the adjustment is accelerated. The

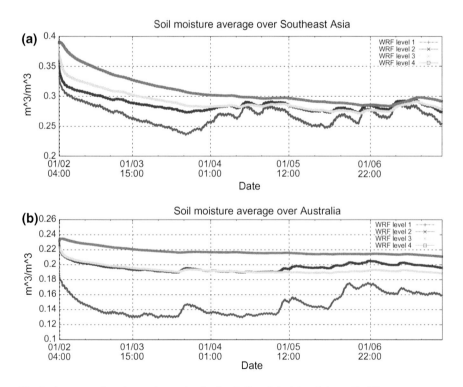

Fig. 2 Average soil moisture time series for South East Asia (**a**) and Australia (**b**)

variations in the uppermost soil layer over South East Asia are related to strong precipitation events while the very low soil moisture over Australia is due to a strong drying of the sand in the Australian desert.

For a seasonal forecasting system, it is also important, to accurately simulate the large scale pressure patterns. Figure 3 shows the mean sea level pressure averaged over the 5 month simulation period. The ECMWF analysis shows a typical situation with a pronounced Azores High pressure ridge and the heat low over the Arabian Peninsula and the Indian subcontinent. The CP resolution simulation shows a similar structure as the ECMWF analysis, although with more extreme values in case of the low pressure regions. The coarse resolution simulation strongly overestimates intensity of the high pressure ares and does hardly show any heat low over the previously called regions.

As precipitation and its extremes is one of the most important variables e.g. for agriculture and food security, precipitation data will be statistically evaluated against observations from the Global Precipitation Mission [5]. This data set provides 30 min precipitation estimates around the globe between 60°N and 60°S since 2014. Additionally, the simulations are verified against ECMWF operational analysis data. This data set provides the most accurate description of the large scale circulation as the ECMWF analysis merges global observations with a sophisticated data assimilation scheme [3]. Due to the large domain and its high resolution, it is a challenge to find suitable (gridded) observations so as not to smooth the information content from the high resolution.

Figure 4 shows the accumulated precipitation during the 5 month simulation period. The high precipitation amounts with more than 2500 mm in the Innertropical Convergence Zone (ITCZ) can be clearly identified. The coarse resolution simulation shows a northward shift of this particular precipitation pattern over the Pacific, associated with a strong overestimation of precipitation. The convection permitting simulation shows a more accurate representation of this precipitation pattern. In the Central Atlantic, the coarse resolution simulation shows an erroneous behavior with far too much precipitation while the CP simulation considerably underestimates the amount of precipitation. The overestimation appears to be the result of the applied convection parametrization which has deficiencies in representing convection in case the wind speeds and wind shear is weak. A similar behavior is observed in the Indian Ocean. It can be said that the CP resolution is beneficial in order to enhance the precipitation forecast quality.

As also the distributions of the 2-m temperatures are an important factor, Fig. 5 shows frequency histograms for the 00 Z, 06 Z, 12 Z and 18 Z time steps over the whole domain. It is seen that the shape agrees well with the ECMWF analysis, apart from the 0° region which is related to the different shape of the terrain. As the terrain is heavily smoothed in the NCP simulation, it is evident that higher temperatures are simulated whereas the CP terrain is a lot closer to the real orography.

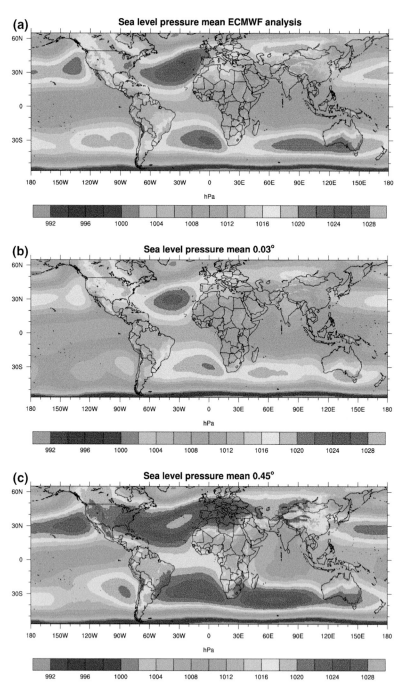

Fig. 3 Mean sea level pressure during the simulation period of: **a** ECMWF analysis, **b** CP simulation, and **c** coarse resolution

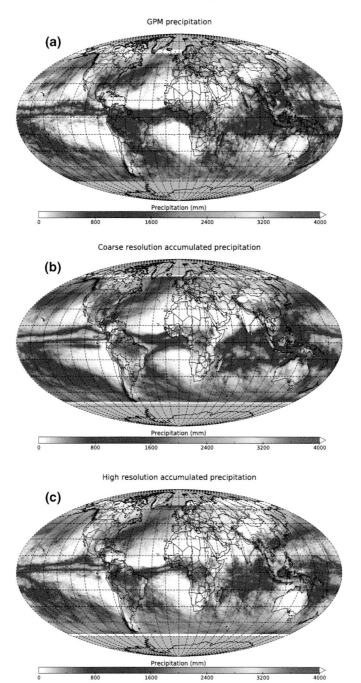

Fig. 4 Accumulated precipitation from the GPM data set (**a**), the coarse resolution simulation (**b**) and the convection permitting resolution (**c**). Grey areas denote missing values

Temperature distribution

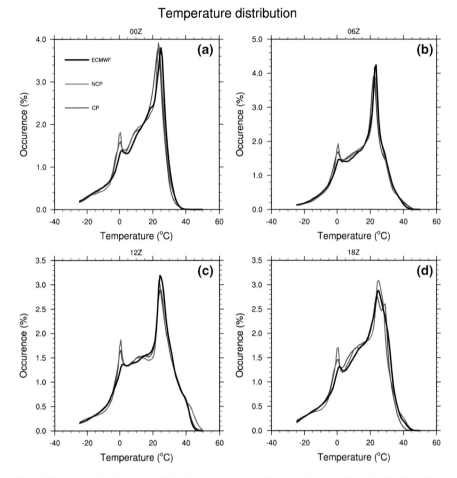

Fig. 5 Frequency distributions of the 2-m temperature of the whole model domain. The black line shows the ECMWF analysis (considered as a reference), the red line shows the coarser resolution (NCP), and the blue line indicates the CP resolution

5 Concluding Remarks

A convection permitting latitude belt simulation has been set up as a successor of a latitude belt simulation performed in 2014 [9]. The domain covers an area between 57°S and 65°N. The simulations were performed during an El Niñjo year and were started on February 1s, 2015 and continued until 1st July, 2015. Preliminary results show a beneficial impact on the precipitation, temperature and large scale circulation pattern forecast. it is interesting to note that the CP resolution apparently shows a higher variability in terms of precipitation and temperature as compared to the observations. Currently it is difficult to investigate the reasons for this behavior as usually

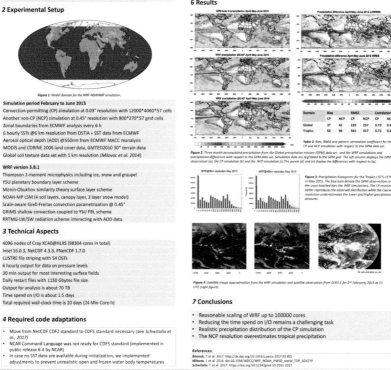

Fig. 6 Summary of the current status within the WRFXXXL project. This poster has been presented at the European Geosciences Union General Assembly in April 2018 in Vienna

the available data sets for this large domain often have a much coarser resolution than the CP simulation. As the computational resources are further increasing, this permits to run an ensemble of such large domains at this particular high resolution to further substantiate the improved model performance and provide potential users with an uncertainty of the forecast [7].

The attached poster (Fig. 6) provides more information on the current status and the quality of the seasonal forecast.

Currently, a manuscript is in preparation for submission to Nature.

Acknowledgements The required computation time for the WRF simulations was kindly provided by HLRS. We also highly appreciate the valuable technical support from HLRS and Cray.

References

1. N. Becker, U. Ulbrich, R. Klein, Systematic large-scale secondary circulations in a regional climate model. Geophys. Res. Lett. **42**(10), 4142–4149 (2015). https://doi.org/10.1002/2015GL063955
2. P.A. Dirmeyer, B.A. Cash, J.L. Kinter, T. Jung, L. Marx, M. Satoh, C. Stan, H. Tomita, P. Towers, N. Wedi, D. Achuthavarier, J.M. Adams, E.L. Altshuler, B. Huang, E.K. Jin, J. Manganello, Simulating the diurnal cycle of rainfall in global climate models: resolution versus parameterization. Clim. Dyn. **39**(1), 399–418 (2012). https://doi.org/10.1007/s00382-011-1127-9
3. ECMWF, *Part III: Dynamics and numerics*. IFS Documentation, ECMWF. Operational implementation, 12 May 2015
4. G. Fosser, S. Khodayar, P. Berg, Benefit of convection permitting climate model simulations in the representation of convective precipitation. Clim. Dyn. **44**(1), 45–60 (2015). https://doi.org/10.1007/s00382-014-2242-1
5. D. Huffman, D. Bolvin, D. Braithwaite, K. Hsu, R. Joyce, P. Xie, Integrated multi-satellite retrievals for GPM(imerg), version 4.4. NASA's precipitation processing center (2014). ftp://arthurhou.pps.eosdis.nasa.gov/gpmdata/
6. Y. Miyamoto, Y. Kajikawa, R. Yoshida, T. Yamaura, H. Yashiro, H. Tomita, Deep moist atmospheric convection in a subkilometer global simulation. Geophys. Res. Lett. **40**(18), 4922–4926 (2013), https://doi.org/10.1002/grl.50944, https://agupubs.onlinelibrary.wiley.com/doi/abs/10.1002/grl.50944
7. V.Mohr, T. Schwitalla, V.Wulfmeyer, K.Warrach-Sagi, High resolution climate projects using the WRF model on the HLRS, in *Sustained Simulation Performance 2016*, ed. by M.M. Resh, W. Bez, E. Focht, N. Patel, H. Kobayashi (Springer International Publishing, Cham, 2016), pp. 173–184. ISBN 978-3-319-46735-1
8. T. Sato, H. Miura, M. Satoh, Y.N. Takayabu, Y. Wang, Diurnal cycle of precipitation in the tropics simulated in a global cloud-resolving model. J.Clim. **22**(18), 4809–4826 (2009). https://doi.org/10.1175/2009JCLI2890.1
9. T. Schwitalla, H.S. Bauer, V. Wulfmeyer, K. Warrach-Sagi, Continuous high-resolution midlatitude-belt simulations for july-august 2013 with WRF. Geosci. Model. Dev. **10**(5), 2031–2055 (2017). https://doi.org/10.5194/gmd-10-2031-2017
10. T. Schwitalla, V. Wulfmeyer, K. Warrach-Sagi, Seasonal simulation of weather extremes, in *High Performance Computing in Science and Engineering '17*, ed. by W.E. Nagel, D.H. Kröner, M.M. Resch (Springer International Publishing, Cham, 2018), pp. 441–451. ISBN 978-3-319-68394-2
11. W.C. Skamarock, J.B. Klemp, J. Dudhia, D. Gill, D.O. Barker, M.G. Duda, W. Wang, J.G. Powers, A description of the advanced research WRF version 3. NCAR Technical Note TN-

475+STR, NCAR (Boulder, CO, 2008). http://www.mmm.ucar.edu/wrf/users/docs/arw_v3. pdf

12. G. Thompson, P.R. Field, R.M. Rasmussen, W.D. Hall, Explicit forecasts of winter precipitation using an improved bulk microphysics scheme. Part ii: implementation of a new snow parameterization. Mon. Weather. Rev. **136**(12): 5095–5115 (2008) https://doi.org/10.1175/2008MWR2387.1

Part VI
Computer Science

This year, again, two reports on projects labelled as "informatics" made it into the HLRS report volume and, thus, got an own chapter. By chance, both deal with molecular dynamics simulations.

The contribution *PetaFLOP Molecular Dynamics for Engineering Applications* by Philipp Neumann et al. reports on recent progress in the project GCS-MDDC using the software *ls1 mardyn*. This simulation software is a joint interdisciplinary undertaking of several groups to allow for highly scalable molecular dynamics simulations, with a particular focus on processes occurring in chemical engineering applications. Examples for those are nucleation, droplet formation, or droplet coalescence. *ls1 mardyn* was twice used for world-record molecular dynamics simulations with respect to the number of particles involved. In this paper, the authors present results on several OpenMP schemes for optimum node-level performance as well as on non-blocking communication and communication hiding for global collective operations.

The second report *Load Balancing and Spatial Adaptivity for Coarse-Grained Molecular Dynamics Applications by Dirk Pflüger et al.* is on research done within the *LAMD* project. There, the consortium works on the coupling of an AMR Lattice Boltzmann implementation with particle dynamics, which can be of particular interest for scale-bridging in multi-physics applications. The software framework used comprises *ESPResSo* and *p4est*. The focus in the paper presented here is on load balancing for short-range molecular dynamics as well as on spatial adaptivity on the Lattice Boltzmann side. Numerical experiments are presented for a cabin air filter simulation.

Hans-Joachim Bungartz
Institut für Informatik, Technische Universität München
Boltzmannstrasse 3, 85748 Garching, Germany
e-mail: bungartz@in.tum.de

PetaFLOP Molecular Dynamics
for Engineering Applications

Philipp Neumann, Nikola Tchipev, Steffen Seckler, Matthias Heinen, Jadran Vrabec and Hans-Joachim Bungartz

Abstract Molecular dynamics (MD) simulations enable the investigation of multicomponent and multiphase processes relevant to engineering applications, such as droplet coalescence or bubble formation. These scenarios require the simulation of ensembles containing a large number of molecules. We present recent advances within the MD framework *ls1 mardyn* which is being developed with particular regard to this class of problems. We discuss several OpenMP schemes that deliver optimal performance at node-level. We have further introduced nonblocking communication and communication hiding for global collective operations. Together with revised data structures and vectorization, these improvements unleash PetaFLOP performance and enable multi-trillion atom simulations on the HLRS supercomputer Hazel Hen. We further present preliminary results achieved for droplet coalescence scenarios at a smaller scale.

P. Neumann (✉)
Department of Informatics, Universität Hamburg, Bundesstr. 45a, 20146 Hamburg, Germany
e-mail: philipp.neumann@uni-hamburg.de

N. Tchipev · S. Seckler · H.-J. Bungartz
Department of Informatics, Technical University of Munich, Boltzmannstr. 3,
85748 Garching, Germany
e-mail: tchipev@in.tum.de

S. Seckler
e-mail: seckler@in.tum.de

H.-J. Bungartz
e-mail: bungartz@in.tum.de

M. Heinen · J. Vrabec
Thermodynamics and Process Engineering, Technical University of Berlin,
Ernst-Reuter-Platz 1, 10587 Berlin, Germany
e-mail: heinen@tu-berlin.de

J. Vrabec
e-mail: vrabec@tu-berlin.de

© Springer Nature Switzerland AG 2019
W. E. Nagel et al. (eds.), *High Performance Computing in Science
and Engineering '18*, https://doi.org/10.1007/978-3-030-13325-2_25

397

1   Introduction

Molecular dynamics (MD) simulations have become a valuable tool for energy applications. They enable the sampling of equations of state [10], the investigation of droplet coalescence [16] and bubble formation [11] at the nanoscale, or the investigation of interfacial flows [12]. Many of the aforementioned scenarios rely on the simulation of very large molecular systems; the single molecules may, fortunately, have rather simplistic shapes and sizes. A typical example is given by a mixture of acetone (that can be modeled by four Lennard-Jones (LJ) sites, a dipole and a quadrupole) and nitrogen (two Lennard-Jones sites with a quadrupole) which is frequently used to model fuel injection-like conditions in thermodynamic laboratories. In a long-term interdisciplinary effort of computer scientists and mechanical engineers, the MD framework *ls1 mardyn* has evolved over the last decade to investigate such large systems of small molecules [13]. *ls1 mardyn* has been used in various studies [20] and has been continuously extended to optimally exploit current HPC architectures [5, 17, 19].

In the following, we detail recent developments within the framework to achieve optimal performance at node and multi-node level. After introducing the actual problem setting of short-range molecular dynamics, related work and the original implementation of *ls1 mardyn* in Sect. 2, we provide an overview of OpenMP schemes that are available in *ls1 mardyn* in Sect. 3. Improvements to MPI communication are introduced in Sect. 4. We report performance results for extreme-scale experiments in Sect. 5 and preliminary results on droplet coalescence in Sect. 6. We close with a summary and an outlook to future work in Sect. 7. Scalability and performance results have recently been published in [18].

2   Short-Range Molecular Dynamics

2.1   Theory

In short-range MD, Newton's equations of motion are numerically solved [15]. In the following, considerations are restricted to small molecules. Due to their negligible conformational changes, molecules mainly undergo translational or rotational motion; both are included in the equations of motion and are solved simultaneously in *ls1 mardyn* using a leapfrog time integrator, without the need for iterative procedures (such as the SHAKE algorithm) to handle geometric constraints [15].

Molecules interact via force fields. In short-range MD, arising forces are only explicitly accounted for if the distance between two considered molecules is below a specified cut-off radius r_c. There are basically two variants to efficiently implement the cut-off condition: linked cells and Verlet lists [15]. Both methods turn the actual molecule-molecule interaction complexity from $O(N^2)$ to $O(N)$. *ls1 mardyn* makes use of the linked cell approach: a Cartesian grid with cell sizes $\geq r_c$ is introduced and covers the computational domain. The molecules are sorted into the cells. Molecular interactions only need to be considered for molecules that reside within the same cell or in neighboring cells.

Various interaction potentials have been proposed to model molecular behavior. The simulations reported in this contribution make use of the LJ potential [15]

$$U(r_{ij}) = 4\varepsilon \left(\left(\frac{\sigma}{r_{ij}} \right)^{12} - \left(\frac{\sigma}{r_{ij}} \right)^{6} \right), \tag{1}$$

with molecule-dependent parameters σ and ε and the distance of molecules i and j given by r_{ij}, as well as the electrostatic quadrupole potential for the interaction of two point-quadrupoles [7]. The force calculation is typically by far the most expensive step in MD and often contributes $\geq 90\%$ to the overall compute time.

2.2 Related Work

HPC and Related MD Implementaations

Various packages efficiently and flexibly implement (short-range) molecular dynamics algorithms, with the most popular ones given by Gromacs,[1] LAMMPS[2] and NAMD.[3] Gromacs leverages particularly GPUs but also supports OpenMP and large-scale MPI parallelism, and it also exploits SIMD instructions via a new particle cluster-based Verlet list method [1, 14]. A LAMMPS-based short-range MD implementation for host-accelerator systems is reported in [2] with speedups for LJ scenarios of 3–4. A pre-search process to improve neighbor list performance at SIMD level and an OpenMP slicing scheme are presented in [9, 21]. The arising domain slices, however, need to be thick enough, to actually boost performance at shared-memory level. This restricts the applicability of the method to rather large (sub-)domains per process.

ls1 mardyn

An approach to efficient vectorization built on top of the linked cell data structure within *ls1 mardyn* is presented for single- [4] and multi-site[4] molecules [3]. This method, combined with a memory-efficient storage, compression and data management scheme [6], allowed for a four-trillion atom simulation in 2013 on the supercomputer SuperMUC, phase 1 [5]. A multi-dimensional, OpenMP-based coloring approach that operates on the linked cells is provided in [19]. The method has been evaluated on both Intel Xeon and Intel Xeon Phi architectures and exhibits good scalability up to the hyperthreading regime. *ls1 mardyn* further supports load balancing. It

---

[1] www.gromacs.org.

[2] www.lammps.org.

[3] http://www.ks.uiuc.edu/Research/namd/.

[4] Molecules that consist of several interaction sites, e.g. two LJ sites.

uses k-d trees for this purpose. Recently, this approach has been employed to balance computational load on heterogeneous architectures [17]. A detailed overview of the original release of *ls1 mardyn* is provided in [13]. Various applications from process and energy engineering, including several case studies that exploit *ls1 mardyn*, are discussed in [20].

3   OpenMP Schemes

ls1 mardyn has recently been extended by two novel OpenMP parallelization schemes for the force calculation, complementing a preceding implementation. All three schemes are depicted in Fig. 1. The schemes **c08** and **c04**, depicted in Fig. 1a, c, employ coloring to the linked cells and thus enable parallelism at shared-memory level, avoiding race conditions when accumulating all force contributions per molecule. The scheme **c08** comes with finer granularity at cell level, but at the cost of more read/write operations per molecule. The scheme **sli** basically arranges all cells in a linear order and cuts this linearized cell accumulation into as many slices as threads are available. The scheme works particularly well for large domains since synchronization between the threads, achieved by a locking mechanism, is minimal in this case. Details on the schemes can be found in [18, 19].

We evaluated the schemes in single-site Lennard-Jones experiments on Hazel Hen, cf. Fig. 2, with 47×10^6 molecules, cut-off radius $r_c/\sigma = 3.5$ and density $\rho\sigma^3 = 0.78$. Performance is given in million molecule updates per second (MMUPS) per thread in this figure. All schemes exhibit good scalability at node level, showing some advantages of **sli** over **c04** over **c08**. This, however, is strongly scenario-specific. For example, we encountered better performance of **c04** in a two-phase vapor-liquid scenario (that is, a scenario with high load imbalances because one region is densely packed with molecules (liquid phase) and another region where molecules are sparsely present (vapor phase)).

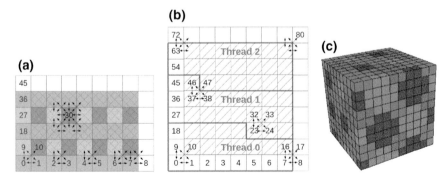

Fig. 1 OpenMP schemes integrated in *ls1 mardyn*: **a c08** coloring, **b sli**, **c c04** coloring

Fig. 2 Strong scaling on one node of Hazel Hen. HT denotes two threads pinned on the same core running `sli`. Turbomode was switched off

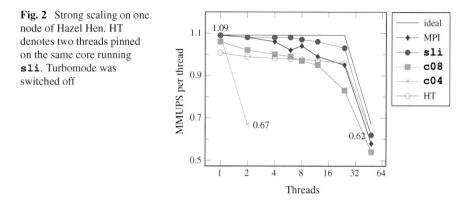

4   Communication Hiding

MD processes such as droplet coalescence operate on long time scales, yielding the requirement of (hundreds of) millions of time steps that have to be executed. Due to the sequential processing of the time steps and with the time scale (nonlinearly) increasing with the number of molecules, the size of the overall MD system becomes limited and improving "time-to-solution" is of utmost importance. As a consequence, codes such as *ls1 mardyn* have to feature optimal strong scalability. Therefore, we have improved MPI communication in *ls1 mardyn*.

Communication Between Neighboring Subdomains

ls1 mardyn applies a domain decomposition—regular Cartesian domain decomposition or decomposition based on k-d trees [17]—to split the MD domain and distribute computational load. Molecules are sent from one subdomain to another (a) if they have propagated to the other subdomain within a time step or (b) if they are located close to the other subdomain and are thus required for local force calculations. We have reduced the size of the messages that are exchanged between subdomains taking into account (a) and (b): if a molecule is required due to (b), we only send the molecular data that is required for the force calculation (i.e., the molecule position data). In case (a), we send the entire molecule data set to the other subdomain. All required molecular data of type (a) and (b) are collected in one buffer and sent in one message to a neighboring subdomain. Different communication schemes are supported and have been investigated in ls1 mardyn, including fully asynchronous data transfer, which is supported for all spatial decomposition schemes, as well as schemes that make a trade-off between the number of communication steps and the number of communication neighbors. The load balancing has been enhanced to support almost-arbitrary heterogeneous loads by using in-situ time measurements.

Nonblocking Collectives

Collectives are required to gather statistical properties from the MD system, such as temperature or energy values. These properties vary rather slowly over time. We exploit this "feature" and have incorporated MPI-3 based nonblocking collectives into *ls1 mardyn*, which assemble statistical properties by relying on data from a preceding time step. This allows to hide global reduction operations behind actual MD time step evaluations.

5 Scalability and Performance at Large Scale

We investigated both weak and strong scalability on Hazel Hen in various scenarios, cf. Fig. 3. The molecular ensemble consisted of single-site LJ molecules at a liquid density $\rho\sigma^3 = 0.78$, specifying a cut-off radius $r_c/\sigma = 3.5$.

In the weak scaling scenario, we were able to simulate up to 2.1×10^{13} particles on 7168 nodes using 1 MPI rank per node and 48 OpenMP threads at node-level; the latter setting yields minimum memory requirements compared to other hybrid MPI/OMP configurations. On 7168 nodes, the scenario would correspond to a cube with a width of $11.8\,\mu$m, referring to physical units for liquid xenon ($\sigma = 3.9450$ Å).

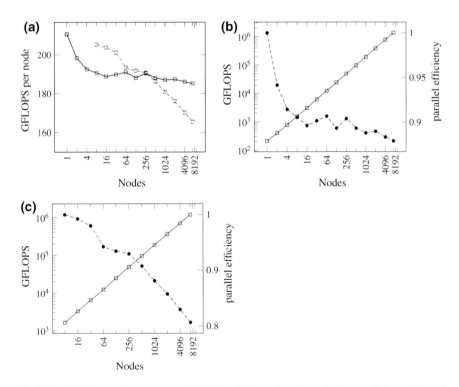

Fig. 3 Scalability experiments on up to 7168 Hazel Hen nodes. **a** GFLOPS per node in weak (solid line) and strong scaling (dashed line). **b** Total GFLOPS (solid line) and parallel efficiency (dashed line) in weak scaling. **c** Total GFLOPS and parallel efficiency in strong scaling

The weak scaling efficiency was measured to be $\geq 88\%$ (cf. Fig. 3b). A performance of 1.33 PFLOPS on 7168 nodes was achieved (9% of single precision peak of Hazel Hen), corresponding to 189×10^9 molecule updates per second.

Strong scalability was investigated using a system of 23.85 billion molecules. We achieved a performance of 1.18 PFLOPS and a parallel efficiency of 81% on 7168 nodes. In this case, we used 6 MPI ranks and 8 OpenMP threads per MPI rank which had been shown to provide slightly superior performance over the 1MPI/48OMP setting in previous tests. The simulation rate arose as 178×10^9 molecule updates per second.

6 Droplet Coalescence

Enabled by a new sampling module for recording the temporal development of a two-dimensional density field $\rho(z, r)$, preliminary results of two coalescing argon droplets were obtained from molecular dynamics simulations with *ls1 mardyn*, cf. Figs. 4 and 5. The first application area of this sampling module were rotationally

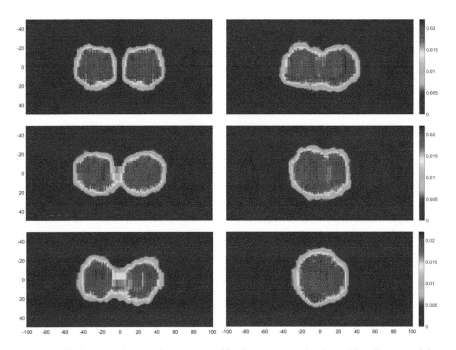

Fig. 4 Preliminary work on coalescence, specifically two argon droplets with a diameter of $d = 5$ nm and a number of $N \approx 10^3$ particles in equilibrium with their vapor at a temperature of $T = 110$ K. From top left to bottom right, the sequence shows a time period of $\Delta t \approx 0.1$ ns in total. The snapshots exhibit the two-dimensional density field $\rho(z, r)$, sampled by a new module of *ls1 mardyn* as depicted by a colormap

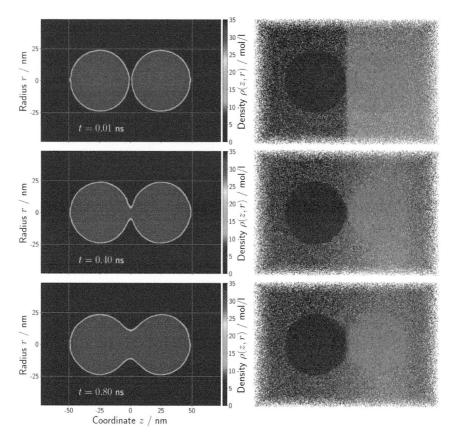

Fig. 5 Preliminary work on the subject of coalescence, specifically two argon droplets with a diameter of $d = 50$ nm containing a number of $N \approx 10^6$ particles in equilibrium with their vapor at a temperature of $T = 110$ K. The left column shows two-dimensional density fields, sampled by a new module of *ls1 mardyn* as depicted by a colormap. The right column shows snapshots of the visualization by the cross-platform visualization framework *MegaMol* [8], considering every single particle of the system ($\approx 10^6$ per droplet). The colors red and green were selected to be able to observe the diffusive mixing of particles that initially were located either in or around the left or right droplet. The sequence shows a time period of $\Delta t = 0.4$ ns

symmetrical systems around the z axis. In the first test scenario, two droplets with a diameter of $d = 5$ nm containing a number of particles $N \approx 10^3$ were centered along the z axis with a mutual distance of $\delta = 1$ nm, within and in thermodynamic equilibrium with a vapor phase. The centers of mass of the droplets initially had a relative velocity of zero. The workflow to setup such a scenario is to first separately equilibrate the two phases (liquid and vapor) in a cubic volume and subsequently cut the droplets out of the liquid phase and place them into the vapor phase, to invoke a physical configuration. The evolution of the coalescence process is depicted in Fig. 4, showing a time period of $\Delta t \approx 0.1$ ns in total. Due to the short distance between the droplets of only 1 nm, the particles located at their surfaces are mutually attracted by the opposite interface. Consequently, a liquid bridge forms within a fraction of a nanosecond. This bridge rapidly grows, driven by the surface tension. At the time instance when the thickness of the bridge is almost the same as the droplet diameter, the outer boundaries of the droplets on the left and right sides start to propagate towards the center of the connected droplets, forming a single larger one. The pictures in Fig. 4 clearly show that because of the small size of the droplets, their shape, represented by the interface, is strongly affected by fluctuations. In the field of nanotechnology, scientists might by interested in the dynamics of such small droplets. In this case molecular dynamics obviously would be the appropriate method to study these effects. The main objective of this study, however, is rather advanced. A major advantage of molecular dynamics is its sound physical background. Assumptions or approximations are limited to the molecular model, covering the force fields and hence the interactions between molecules. The system under investigation does not need any further assumptions to set up boundary conditions, e.g. at a vapor-liquid interface, like it is usually done in computational fluid dynamics (CFD). Hence, molecular dynamics can favorably serve as a benchmark for methods dealing with systems on more aggregated scales, like CFD, to validate their assumptions or chosen boundary conditions.

For studying the dynamics of coalescence processes of macroscopic droplets, and in this course to compare the results of atomistic simulations to those of continuum approaches, significantly larger droplets have to be considered. They have to be at least large enough to not be dominated by the strong fluctuations on the nano-scale. Therefore, a second scenario with by one order of magnitude larger droplets was created, i.e. with a droplet diameter of $d = 50$ nm. Since the volume and therefore also the number of particles of the droplet is proportional to the droplet diameter to the power of three, $N \propto d^3$, the larger droplets consisted of $N \approx 10^6$ particles. The results of this study are shown in Fig. 5. Obviously, the evolution of the interface's shape during the coalescence process can be visually followed very clearly, and therefore allows for quantitative comparison with other methods. These investigations are planned to be continued by varying different parameters of the molecular model, but also by variation of the thermodynamic state point, e.g. the system temperature T. Additionally, the calculations will be repeated with a further increase of the droplet size, to be able to extrapolate the results to macroscopic droplets.

7 Summary and Outlook

We have provided results from large-scale MD simulations leveraging the performance of basically the entire Hazel Hen supercomputer, achieving PetaFLOP performance for systems with billions up to 20 trillions of molecules. In the largest of our runs, we however encountered some issues in terms of performance loss by up to 50% and failing nodes. The cause of these performance decreases is under current investigation. Considering three different OpenMP schemes, we observed that one or the other scheme may be favorable in a simulation scenario. This confirms the need for auto-tuning at OpenMP level which is part of the work plan of our current GCS-MDDC compute project at HLRS. Preliminary work on the subject of droplet coalescence showed that to meet the target of being able to compare the results of the atomistic simulations to those of methods operating on macroscopic scales, e.g. CFD, utilizing continuum approaches, the droplets have to be at least large enough so that the dynamics of the coalescence process is not dominated by the strong fluctuations on the nano-scale. Future work will focus on integrating auto-tuning into the simulation and address even larger droplet coalescence scenarios.

Acknowledgements The presented work was carried out in the scope of the Large-Scale Project *Extreme-Scale Molecular Dynamics Simulation of Droplet Coalescence*, acronym GCS-MDDC, of the Gauss Centre for Supercomputing; the authors thank Bernd Krischok, Dr.-Ing. Martin Bernreuther and Prof. Dr. Michael Resch for their support throughout the project. Financial support by the Federal Ministry of Education and Research, project *Task-based load balancing and auto-tuning in particle simulations* (*TaLPas*), grant numbers 01IH16008A/B/E, is acknowledged.

References

1. M. Abraham, T. Murtola, R. Schulz, S. Páll, J. Smith, B. Hess, E. Lindahl, GROMACS: High performance molecular simulations through multi-level parallelism from laptops to supercomputers. SoftwareX **1–2**, 19–25 (2015)
2. W. Brown, P. Wang, S. Plimpton, A. Tharrington, Implementing molecular dynamics on hybrid high performance computers—short range forces. Comput. Phys. Commun. **182**(4), 898–911 (2011)
3. W. Eckhardt, Efficient HPC implementations for large-scale molecular simulation in process engineering. Dissertation, Dr. Hut, Munich, 2014
4. W. Eckhardt, A. Heinecke, An efficient vectorization of linked-cell particle simulations, in *ACM International Conference on Computing Frontiers*, ACM, New York, NY, USA, pp. 241–243 (2012)
5. W. Eckhardt, A. Heinecke, R. Bader, M. Brehm, N. Hammer, H. Huber, H.G. Kleinhenz, J. Vrabec, H. Hasse, M. Horsch, M. Bernreuther, C. Glass, C. Niethammer, A. Bode, H.J. Bungartz, 591 TFLOPS multi-trillion particles simulation on superMUC (Springer, Berlin, Heidelberg, 2013), pp. 1–12
6. W. Eckhardt, T. Neckel, Memory-efficient implementation of a rigid-body molecular dynamics simulation, in *Proceedings of the 11th International Symposium on Parallel and Distributed Computing—ISPDC 2012* (Munich, 2012). IEEE, pp. 103–110
7. C. Gray, K. Gubbins, Theory of Molecular Fluids: I: Fundamentals, in *International Series of Monogr*, vol.1 (Oxford University Press, Oxford, 1984)
8. S. Grottel, M. Krone, C. Müller, G. Reina, T. Ertl, MegaMol—a prototyping framework for particle-based visualization. IEEE Trans. Vis. Comput. Graph. **21**(2), 201–214 (2015)
9. C. Hu, X. Wang, J. Li, X. He, S. Li, Y. Feng, S. Yang, H. Bai, Kernel optimization for short-range molecular dynamics. Comput. Phys. Commun. **211**, 31–40 (2017)
10. A. Köster, T. Jiang, G. Rutkai, C. Glass, J. Vrabec, Automatized determination of fundamental equations of state based on molecular simulations in the cloud. Fluid Phase Equilibria **425**, 84–92 (2016)
11. K. Langenbach, M. Heilig, M. Horsch, H. Hasse, Study of homogeneous bubble nucleation in liquid carbon dioxide by a hybrid approach combining molecular dynamics simulation and density gradient theory. J. Chem. Phys. **148**(124702) (2018)
12. G. Nagayama, P. Cheng, Effects of interface wettability on microscale flow by molecular dynamics simulation. Int. J. Heat Mass Transf. **47**, 501–513 (2004)

13. C. Niethammer, S. Becker, M. Bernreuther, M. Buchholz, W. Eckhardt, A. Heinecke, S. Werth, H.J. Bungartz, C. Glass, H. Hasse, J. Vrabec, M. Horsch, ls1 mardyn: the massively parallel molecular dynamics code for large systems. J. Chem. Theory Comput. **10**(10), 4455–4464 (2014)
14. S. Páll, B. Hess, A flexible algorithm for calculating pair interactions on SIMD architectures. Comput. Phys. Commun. **184**(12), 2641–2650 (2013)
15. D. Rapaport, *The Art of Molecular Dynamics Simulation* (Cambridge University Press, Cambridge, 2004)
16. L. Rekvig, D. Frenkel, Molecular simulations of droplet coalescence in oil/water/surfactant systems. J. Chem. Phys. **127**(134701) (2007)
17. S. Seckler, N. Tchipev, H.J. Bungartz, P. Neumann, Load balancing for molecular dynamics simulations on heterogeneous architectures, in *2016 IEEE 23rd International Conference on High Performance Computing (HiPC)*, pp. 101–110 (2016)
18. N. Tchipev, S. Seckler, M. Heinen, J. Vrabec, F. Gratl, M. Horsch, M. Bernreuther, C. Glass, C. Niethammer, N. Hammer, B. Krischok, M. Resch, D. Kranzlmüller, H. Hasse, H.J. Bungartz, P. Neumann, TeTRiS: twenty trillion-atom simulation. Int. J. High Perform. Comput. Appl. (2019), https://doi.org/10.1177%2F1094342018819741
19. N. Tchipev, A. Wafai, C. Glass, W. Eckhardt, A. Heinecke, H.J. Bungartz, P. Neumann, Optimized Force Calculation in Molecular Dynamics Simulations for the Intel Xeon Phi (Springer International Publishing, Cham, 2015), pp. 774–785)
20. J. Vrabec, M. Bernreuther, H.J. Bungartz, W.L. Chen, W. Cordes, R. Fingerhut, C. Glass, J. Gmehling, R. Hamburger, M. Heilig, M. Heinen, M. Horsch, C.M. Hsieh, M. Hülsmann, P. Jäger, P. Klein, S. Knauer, T. Köddermann, A. Köster, K. Langenbach, S.T. Lin, P. Neumann, J. Rarey, D. Reith, G. Rutkai, M. Schappals, M. Schenk, A. Schedemann, M. Schönherr, S. Seckler, S. Stephan, K. Stöbener, N. Tchipev, A. Wafai, S. Werth, H. Hasse, Skasim—Scalable hpc software for molecular simulation in the chemical industry. Chem. Ing. Tech. **90**(3), 295–306 (2018)
21. X. Wang, J. Li, J. Wang, X. He, N. Nie, Kernel Optimization on Short-Range Potentials Computations in Molecular Dynamics Simulations (Springer, Singapore, 2016), pp. 269–281

Load-Balancing and Spatial Adaptivity for Coarse-Grained Molecular Dynamics Applications

Steffen Hirschmann, Michael Lahnert, Carolin Schober, Malte Brunn, Miriam Mehl and Dirk Pflüger

Abstract We present our approach for a scalable implementation of coupled soft matter simulations for inhomogeneous applications based on the simulation package ESPResSo and an extended version of the adaptive grid framework `p4est`. Our main contribution in this paper is the development and implementation of a joint partitioning of two or more distinct octree-based grids based on the concept of a *finest common tree*. This concept guarantees that, on all grids, the same process is responsible for each point in space and, thus, avoids communication of data in overlapping volumes handled in different partitions. We achieve up to 85% parallel efficiency in a weak scaling setting. Our proposed algorithms take only a small fraction of the overall runtime of grid adaption.

1 Introduction

Many scientifically relevant systems are composed of several different subsystems, each representing a different physical mechanism that is governed by different sets of equations. In case of particles subjected to a background flow, we need to model the fluid, the interaction of the particles, and the interaction between particles and fluid. Interactions between particles may include long-range electrostatic interactions, bonding, and electrokinetic influences by an ionic fluid. To model the complete system, each individual component interacts with all the other subsystems, thus, forming a challenging coupled problem across multiple time and length scales.

The underlying simulation package of our work is ESPResSo [2, 18], the extensible simulation package for research on soft matter systems. ESPResSo is a flexible multi-physics simulation software for soft matter simulations. They can have up to four different, distinct physics components that are all implemented and coupled with each other: Short-range MD, long-range MD, hydrodynamics, and electrokinetics.

S. Hirschmann (✉) · M. Lahnert · C. Schober · M. Brunn · M. Mehl · D. Pflüger
Institute for Parallel and Distributed Systems, University of Stuttgart, Universitätsstr. 38, 70569 Stuttgart, Germany
e-mail: steffen.hirschmann@ipvs.uni-stuttgart.de

© Springer Nature Switzerland AG 2019
W. E. Nagel et al. (eds.), *High Performance Computing in Science and Engineering '18*, https://doi.org/10.1007/978-3-030-13325-2_26

A common approach to reduce the computational cost and, thus, runtime of grid-based simulations is to use tree-structured adaptive grids for the grid-based part of our simulations. In this context, space-filling curves (SFC) are a well-known way for grid traversal and linearization, see, e.g., [6, 28]. Partitioning a simulation domain with a space-filling curve is usually done by leveraging the linearization defined by the SFC and performing a so-called chain-on-chain partitioning [22], i.e., basically cutting the pieces of the linearized list of grid cells into equal pieces. This approach is found to be efficient in terms of runtime and memory consumption and of good quality compared to other methods [4, 20]. A ready-to-use implementation of parallel algorithms for adaptive mesh refinement and dynamic load-balancing along a space-filling curve is provided by the `p4est` framework [5, 14]. It provides scalable algorithms for grids composed of several octrees (forest-of-octrees) based on Morton ordering [21] (Z-curve).

We provide dynamic spatial adaptivity for hydrodynamics using the lattice-Boltzmann method as well as a way to dynamically load-balance MD simulations in ESPResSo. This is based on grids implemented in an extended version of the `p4est` library [16, 17]. As the basis for the current work, we implemented each of the four subsystems to use a `p4est`-based grid: Load-balanced short-range MD [12] as well as spatially adaptive and load-balanced long-range MD [27], the lattice-Boltzmann method (LBM) [16, 17], and electrokinetics [27]. Coupling a grid-based system and a particle system with spatially adaptive grids on the one hand and dynamic load-balancing on the other hand is a challenging problem. If the respective partitions of both subsystems are not congruent, very high communication cost is generated due to the need to exchange data not only at partition surfaces but in overlapping volumes. Therefore, we propose a method for jointly partitioning two octree grids.

Our algorithms support an arbitrary number of `p4est` instances (subsystems). The only requirement is that they must share the same "macroscopic" setup, i.e., the number and mutual arrangement of trees must be the same for all instances. In this paper, we focus on the coupled simulation of short-range molecular dynamics (non-charged particles) with a background flow calculated with the lattice-Boltzmann method. These systems serve as a template for future extensions.

2   Cabin Air Filter Simulation: An Application Example

One of our applications of interest using a fully coupled system as described above is the simulation of particulate flow through cabin air filters. The aim is to model the separation behavior of highly electrostatically charged dust particles at the surface of complex filter media structures. The interaction of dust particles due to electrostatics and other inter-particle forces has been neglected in simulation approaches so far. Laboratory experiments, however, suggest these interactions are likely to play a decisive role in the particle collection process of the fibrous structures. In order to investigate the agglomeration rate of particles, we can focus on the inflow area of the filter, as illustrated in Fig. 1. Since the greatest influence of particle-particle

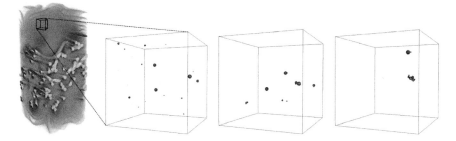

Fig. 1 Simulation setup and agglomeration formation over time

interactions is assumed to be in this area, the development of particle size distributions is investigated there omitting the filter structure. Figure 1 shows the simulation setup as well as the formation of agglomerates for particle whose sizes and charges have been chosen at random. All necessary steps for the parallel simulation of this scenario with the default version of ESPResSo have been completed [25].

For a more detailed analysis, the particle charge as well as the size distribution have been adapted to more realistic values based on published data from literature [9, 10, 15]. From these data, several samples were drawn randomly for fixed domain sizes until the target concentration of $75 \frac{mg}{m^3}$ (laboratory testing conditions) was reached.

We measured the runtime of 100 fluid time steps in a strong scaling setup. The MD time step is 1000 times smaller than the fluid time step. All simulations were performed with 16 processes per node in order to facilitate the equal distribution of LB cells to the subdomains. Note that a time step consists of updating the individual physical subsystems (short-range MD, long-range MD, and hydrodynamics every 1000th time step) as well as a coupling step.

To achieve realistic simulations in a reasonable simulation time, certain parameters have been altered: 1. number of particles (stochastically drawn distributions), 2. domain sizes, 3. resolution for LB grid, 4. boundary conditions for LB flow field (constant initial flow field vs. fix inflow and outflow velocities). As shown in Figs. 2 and 3, there is an optimal number of nodes in terms of runtime for each simulation setup. The optimal number of nodes depends on the ratio of the respective LB and MD efforts.

The resulting particle weight/size distribution, also plotted in Fig. 3, shows the importance of taking the particle-particle interactions into account and justifies the use of the cost intensive fully coupled system. In the case of larger particles, other collection mechanisms become effective. The increasingly effective inertial effects are presumably responsible for improved measurements in filtration efficiency. The latter statement, however, has to be verified with larger simulation setups as well as the additional simulation of the particulate air flow through the filter medium. In order to keep the computational effort for these extended scenarios feasible, we need efficient algorithms for partitioning and dynamic load-balancing as well as adaptive mesh refinement (to accurately resolve the regions around the filter structure and the particles).

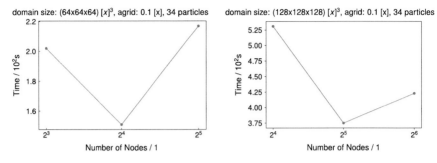

Fig. 2 Strong scaling measurement of 100 fluid time steps for different domain sizes. A single time step consists of updating the individual subsystems (short-range MD, long-range MD, and hydrodynamics every 1000th time step) as well as a coupling step

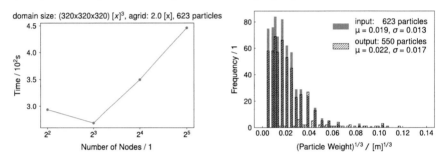

Fig. 3 Runtime of 100 fluid time steps for a larger domain size in a strong scaling scenario (left). Again, a single time step consists of updating the individual subsystems (short-range MD, long-range MD, and hydrodynamics every 1000th time step) as well as a coupling step. The particle weight distributions on the right indicate the importance of particle-particle interactions for different rates. The mean values μ and the respective standard deviations σ demonstrate the shift towards larger/heavier particles

3   Soft Matter Simulations Using ESPResSo

To understand the challenges and the implementation of coupled load-balancing, we briefly sketch the different subsystems of ESPResSo and their current spatially adaptive and/or load-balanced implementation in this section.

3.1   Molecular Dynamics

Molecular dynamics (MD) traces the trajectories of particles. Accordingly, these particles' positions and velocities represent the quantities of interest. Particles interact through inter- and intra-molecular force fields. The trajectories are given by solving Newton's translational and rotational equations of motion.

In this work, we focus on short-range intra-molecular interactions described by a pair-potential U. Intra-molecular forces are derived from U via

$$\mathbf{F}(\mathbf{x}_i) = -\nabla \sum_{j \neq i} U(\mathbf{x}_i, \mathbf{x}_j),$$

where $\mathbf{F}(\mathbf{x}_i)$ is the force acting on particle i at position $\mathbf{x}_i$. U is classified as short-range if and only if

$$\int_{\|\mathbf{x}\| > r_c} U(\mathbf{x}) d\mathbf{x} \xrightarrow{r_c \to \infty} 0.$$

In this case, linear complexity of the force calculations is achieved by cutting off U at a constant radius r_c, i.e., neglecting interactions with particles at distances larger than r_c. Given such a cut-off radius, the Linked-Cell method [13] can be used which discretizes the domain with a Cartesian grid with mesh width r_c. Particles are assigned to cells of this grid such that potential interaction partners of each particle can be detected with $\mathcal{O}(1)$ cost by searching in neighboring grid cells only.

In ESPResSo the Linked-Cell method is implemented via a so-called cell system [18]. In parallel simulations, this grid is subdivided evenly among the respective ranks. This subdivision is strictly Cartesian, i.e., partitions are rectangular subvolumes. In [12] we introduce a new cell system in a minimally invasive way which is able to leverage arbitrary domain decompositions. This in particular requires new communication algorithms as there is no strict 'upper-lower-left-right-front-back' relation between neighboring partitions anymore, and as the number of neighbor partitions becomes variable. For short-range MD, we do not port the algorithm itself to `p4est` but rely on the partitioning and communication functionality of `p4est` to generate information to fill ESPResSo's internal data-structures. This implementation supports dynamic re-partitioning of the grid and the associated numerical payload (particle data). As we left the internal algorithms untouched, the grid of our implemented Linked-Cell method remains regular.

3.2 Hydrodynamics

The molecular ensemble simulated by ESPResSo can be subjected to a background flow. This background flow is realized using a thermalized D3Q19 implementation of the lattice-Boltzmann method (LBM) as described in [7, 8, 24]. The LBM simulates fluid-flow based on probability densities of virtual particles with discrete velocities associated to the centers of grid cells. Each time step consists of two steps: 1. *collision*, a cell-local interaction of probability densities driving those densities towards their local equilibrium, and 2. *streaming*, propagating these effects to the respective neighbor cells according to the chosen stencil.

Extending the LBM to adaptive grids according to [23] requires that the time step is proportional to the local spatial resolution. Thus, grid adaptivity in space automatically induces local time step adaptivity, which is desirable from an application point of view as it reduces the computational costs of coarser grid regions further. The scheme transmits information between cells of different discretization levels by embedding virtual cells in coarse quadrants neighboring finer regions. Simply spoken, this allows to stream between fine cells and coarser neighbors through their virtual children, whereas streaming between coarse cells and their equally coarse neighbors happens at the coarse level. In this sense, streaming can be decomposed into level-wise steps, each of which only visits cells of the same refinement level. To transfer data from the finer virtual cells to their parent cells and vice versa, virtual cells are populated by evenly distributing the fluid mass among them, and, conversely, the fluid mass of virtual cells is agglomerated within the real quadrant. This scheme is used by multiple applications, e.g. [19, 26]. We have implemented it in [16, 17].

3.3 Coupling MD and LBM

ESPResSo uses a bidirectional, frictional coupling between particles and fluid adapted from [1, 7]. The force acting on a particle i that stems from the fluid is given as

$$\mathbf{f}_{fl,i} = -\zeta \left(\mathbf{v}_i - \mathbf{u}(\mathbf{x}_i, t) \right) + \mathbf{f}_{st,i},$$

where $\mathbf{v}_i$ is the particle velocity, $\mathbf{u}(\mathbf{x}_i, t)$ the interpolated fluid velocity at the particle's position and ζ the friction coefficient. An additional stochastic force $\mathbf{f}_{st,i}$ is added to each particle. The force $-\mathbf{f}_{fl,i}$ is transferred back to the fluid cell particle i resides in. As regions around particles are areas of interest, we refine the fluid grid around particles to the maximum refinement level. This also improves the accuracy of the coupling.

4   Dynamic Adaptivity and Load-Balancing

As described above, the coupling between the different physical subsystems is realized through a bi-directional force-coupling. Particles in the short-range MD reside in a regularly discretized grid—technically realized as a `p4est` octree—while the fluid uses a dynamically adaptive grid. In the following, we describe our common tree partitioning approach for the different grids. In a first part, however, we shortly introduce the underlying mapping between particles and fluid cells that define the technical coupling requirements.

4.1 Mapping Data Between Grids

To pass information between both grids, the regular Linked-Cell grid and the adaptive fluid grid, the coupling method needs to find the corresponding fluid cell (in the adaptive grid) for a given particle position. The respective algorithm consists of two steps: 1. We calculate the cell containing the particle in a (virtual) grid that is regularly

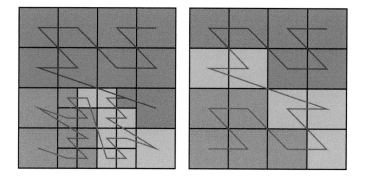

Fig. 4 Two differently discretized grids which are both partitioned independently based on their number of cells among three processes. This leads to a volume-to-volume mapping problem

refined according to the finest grid level present in the adaptive fluid mesh; 2. we calculate the Morton index of the respective cell, i.e., its position in a linearized ordering of all grid cells of this regular grid along the Morton curve; 3. we perform a binary search in the list of real cells of the adaptive fluid grid to determine the correct fluid cell (either the calculated cell of the virtual regular grid or a father/grandfather/…cell). If possible, we use intrinsics for calculating Morton indices, which includes interleaving bits of integers corresponding to a three-dimensional cell index (PDEP, parallel bit deposit on x86 Bit Manipulation Instruction Set 2).

If each of the two involved grids were partitioned (load-balanced) independently, a particle might be handled by different processes in the Linked-Cell partitioning than in the fluid partitioning, c.f. Fig. 4. When exchanging information between grids in such a partitioning scheme, this inevitably results in communication not only at partition surfaces, but also in overlapping subvolumes of the computational domain between processes. To avoid this kind of volume-volume mapping, we partition both grids jointly in such a way that volume overlap over partitions assigned to different processes is avoided and, as a result, process-local coupling between MD and LBM is ensured. Of course, this comes at the cost of not partitioning the involved grids optimally in terms of their own respective objective function (minimizing the time to solution).

4.2 Finest Common Tree and Joint Partitioning

We base the joint partitioning on the *finest common tree* (FCT) which is effectively an intersection of both grids. This means that an octant in the finest common tree is refined if it is refined in both grids, see Fig. 5. Each of the two input grids is given by a corresponding octree. The finest common tree is an "intersection" of both input-octrees in terms of the nodes of the tree. A node in the finest common tree exists if and only if it exists in both input-octrees.

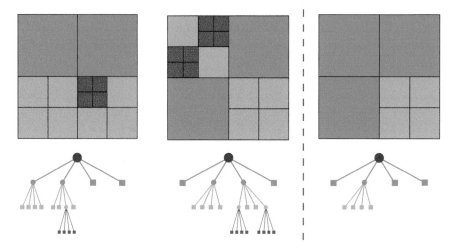

Fig. 5 Two arbitrary, adaptive, two-dimensional grids and their finest common tree on the right. Colors indicate the level of a quadrant. The finest common tree consists of all quadrants which have an equivalent quadrant in both grids or a refined version of it. The tree view below the grid illustrates why we speak of an "intersection": A node in the corresponding tree exists if and only if it exists in the two original grids (image from [3])

The intersection between two grids is well-defined if both grids use the same space-filling curve and the same macro-structure, i.e., the number and position of trees in the forest of octrees. We realize this by using the same `p4est_connectivity` structure for both grids, the regularly discretized grid for short-range MD and the dynamically adaptive grid for the background flow. The number of trees in the forest of octrees is chosen to be minimal given a required mesh width [12].

We calculate the FCT as shown in Algorithm 1. As an initial guess for the FCT, we copy one of the trees. On the resulting initial FCT guess, we call `p4est`'s coarsening function (line 9). This performs a depth-first post-order traversal of the FCT.[1] For every set of children of a tree node it calls a user-defined callback function (which is defined for the FCT generation in lines 12 ff.). The callback function looks for overlapping quadrants in the other tree by linearly traversing its quadrant list. This is valid due to the equal order in both trees: If we have two quadrants with indices i and j in tree FCT with $i < j$, also the corresponding overlapping quadrants' ids i' and j' in the other tree fulfill $i' \leq j'$. It is important to note that the post-order traversal induces a non-monotony in the quadrant search: If we have a set of sibling quadrants $0, 1, \ldots, 7$ and one of the siblings $1, \ldots, 7$ has been coarsened before, the index of $q[0]$ (line 13, 26) will be smaller than in the last call. We detect this case and reduce the index (i') of the concurrently traversed tree (line 20–22).

Furthermore, the algorithm assumes that the partitioning of both trees is aligned. Although this can be done manually [3], our approach is different: After creating

---

[1]The callback is called for child cells before it is called on a cell itself.

two p4est instances with the same connectivity as mentioned above on the same number of processes, the partitioning of the initial (coarse and regular) grid is aligned once. Since it is the sole purpose of our joint partitioning scheme to conserve this property, from there on we simply never let the partitionings diverge.

Algorithm 1 Finest common tree creation via P4EST_COARSEN. The coarsen call-back finds a corresponding quadrant in the other p4est and marks a quadrant for coarsening if its level is higher (more refined). While P4EST_COARSEN traverses one tree, the callback uses a global state to traverse the second tree concurrently. For the sake of brevity we omit pointers. The function VIRT_MORTON_IDX_OF_QUAD implements the mentioned virtual Morton index calculation.

1: $old\_tree \leftarrow \perp$ ▷ Global state for concurrent tree traversal
2: $qid \leftarrow 0$
3: $old\_morton\_idx \leftarrow 0$

4: **function** CREATE_FCT(p4est_t t_1, p4est_t t_2)
5:     $fct \leftarrow$ P4EST_COPY(t_1)
6:     $fct.user\_pointer \leftarrow t_2$
7:     $old\_tree \leftarrow \perp$
8:     $old\_morton\_idx \leftarrow 0$
9:     P4EST_COARSEN(fct, COARSEN_CALLBACK)
10:    **return** fct
11: **end function**

12: **function** COARSEN_CALLBACK(p4est_t t_1, int $tree\_no$, p4est_quadrant_t $qs[]$)
13:     $q \leftarrow qs[0]$
14:     $t_2 \leftarrow t_1.user\_pointer$
15:     $tree \leftarrow$ P4EST_TREE_ARRAY_INDEX($t_2.trees$, $tree\_no$)
16:     **if** $tree \neq old\_tree$ **then**   ▷ Reset state if prior call had different tree
17:        $old\_tree \leftarrow tree$
18:        $qid \leftarrow 0$
19:     **end if**
20:     **if** VIRT_MORTON_IDX_OF_QUAD(q) $< old\_morton\_idx$ **then**   ▷ Handle recursive calls
21:        $qid \leftarrow \max\{0, qid - (\text{P4EST\_CHILDREN} - 1)\}$
22:     **end if**
23:     $old\_morton\_idx \leftarrow$ VIRT_MORTON_IDX_OF_QUAD(q)
24:     **while** $qid < tree.quadrants.elem\_count$ **do**
25:        $p \leftarrow$ P4EST_QUADRANT_ARRAY_INDEX($tree.quadrants$, qid)
26:        **if** P4EST_QUADRANT_OVERLAPS(p, q) **then**
27:           **return** $p.level < q.level$
28:        **end if**
29:        $qid \leftarrow qid + 1$
30:     **end while**
31: **end function**

4.3   Joint Partitioning

Instead of partitioning both grid separately, we partition the finest common tree and transfer the resulting partitioning back to both input grids. This results in a joint partitioning in which a unique process is responsible for every point in space on both grids, see Fig. 6.

Based on the FCT, we implement the joint partitioning scheme as shown in Algorithm 2. Given two trees, it constructs the finest common tree. Then it linearly combines given repartitioning weights per quadrant for the input trees to obtain repartitioning weights for the FCT and performs the repartitioning of the FCT using chain-on-chain partitioning [22]. After that, it determines the partitioning of quadrants for the input trees such that these adhere the partitioning boundaries of the finest common tree. Finally, it asks p4est to repartition the trees using the newly determined boundaries.

Algorithm 2 Joint repartitioning function. Calculates the FCT and weights for its quads, then repartitions it. Afterwards, this partitioning is transferred into partitionings for t_1 and t_2, respectively. The inputs of this function are the p4est instances trees t_1 and t_2 as well as weights for their quadrants returned by the REPART_WEIGHTS and the factors α_1 and α_2 for the linear combination of them. The function for performing the chain-on-chain partitioning (DETERMINE_PARTITIONS) is not shown here, refer to [11, 22] for details.

1: **function** PARTITION_JOINTLY($\alpha_1, t_1, \alpha_2, t_2$)
2:   $fct \leftarrow$ CREATE_FCT(t_1, t_2)
3:   **for all** $(s_f, s_1, s_2) \in fct.trees \times t_1.trees \times t_2.trees$ **do**   ▷ Calculate weights for FCT
4:     **for all** $q \in s_f.quadrants$ **do**
5:       $Q_1 \leftarrow \{p \in s_1 : p \text{ overlaps } q\}$
6:       $Q_2 \leftarrow \{p \in s_2 : p \text{ overlaps } q\}$
7:       $w_1 \leftarrow \sum_{p \in Q_1}$ REPART_WEIGHTS(t_1, p)
8:       $w_2 \leftarrow \sum_{p \in Q_2}$ REPART_WEIGHTS(t_1, p)
9:       $W[q] \leftarrow \alpha_1 w_1 + \alpha_2 w_2$ ▷ Combined repart weight for FCT quad
10:      $N_1[q] \leftarrow |Q_1|$ ▷ Number of quads in t_1 for given FCT quad
11:      $N_2[q] \leftarrow |Q_2|$
12:    **end for**
13: **end for**
14: $\mathcal{P} \leftarrow$ DETERMINE_PARTITIONS(W)
15: **for all** $(r, P) \in \mathcal{P}$ **do**   ▷ Calculate the number of quadrants per proc. r for t_1, t_2
16:    $quads_1[r] \leftarrow \sum_{q \in P} N_1[q]$
17:    $quads_2[r] \leftarrow \sum_{q \in P} N_2[q]$
18: **end for**
19: P4EST_PARTITION_GIVEN($t_1, quads_1$)
20: P4EST_PARTITION_GIVEN($t_2, quads_2$)
21: **end function**

Note that the finest common tree and our combined partitioning method does not require one of the two involved grids to be regularly discretized. As one can see in

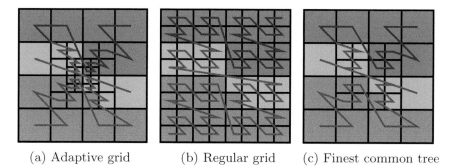

(a) Adaptive grid   (b) Regular grid   (c) Finest common tree

Fig. 6 Intersecting both grids leads to the resulting grid on the right. This grid is partitioned (here: based on the number of cells per process on three different processes as indicated by the colors). The obtained partition boundaries are projected back to the original grids

the algorithms and in Fig. 5, it works for arbitrary octree grids. Also note that the method is not constrained to two grids. It generalizes to an arbitrary number of grids. However, the penalty with respect to the quality of the partitioning is assumed to get worse if more grids are involved.

5 Results

We evaluate our approach of a joint partitioning for MD simulations with background flow in a weak scaling experiment. The setup is as follows: We consider a rectangular domain of size $2a \times a \times a$ with a given unit length a. The left half of this domain is filled with 24,000 particles interacting via Lennard-Jones potentials. Initially, the particles are randomly distributed across the domain (uniform probability distribution) and we prescribe an initial flow in x-direction. Particles and flow interact as described in Sect. 3.3. Particles are transported in flow direction, which induces a dynamically changing and non-uniform particle distribution. The lattice-Boltzmann grid is dynamically refined around the particles, meaning that the refinement is also changed if the particles move. We keep the difference between minimum and maximum refinement level constant at three. Figure 7 shows the setup and visualizes the effect of a joint partitioning with the FCT: The LBM grid is distributed in quadratic chunks of several quadrants. For visualization purposes, we deliberately reduced the FCT level.

We perform the weak scaling experiment on "Hazel Hen" at the High Performance Computing Center Stuttgart (HLRS). We start with 24 processes (1 node) and, then, scale the domain volume and the number of particles successively up by a factor of 8. Simultaneously, we increment the minimum and maximum refinement level of the LBM grid by 1. We present two scaling studies: 1. *case 3–6*: going from levels 3 (min) and 6 (max) on 24 cores over 4 (min) and 7 (max) on 192 cores to 5 (min) and

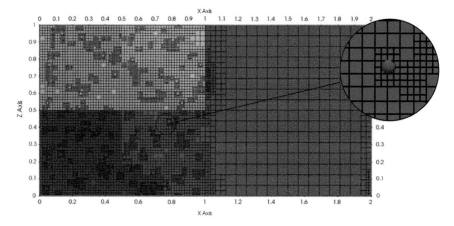

Fig. 7 Setup similar to the one we use in the experiment: For visualization purposes it has been reduced significantly. This figure shows the x-z plane at $y = 0$ with particles on the left and the LBM grid. Colors indicate subdomains. The LBM grid is refined around particles (minimum level: 4, maximum level: 7)

8 (max) on 1536 cores; 2. *case 4–7*: going from from levels 4 (min) and 7 (max) on 24 cores up to 6 (min) and 9 (max) on 1536 cores. In each experiment, we perform 160 time steps and average over three distinct runs. Every 16 time steps, we perform a grid adaptation and a joint repartitioning. We choose the number of particles as weights for MD cells and a uniform weighting for LBM cells. These weights are linearly combined, both with a factor of one.

The resulting runtimes can be found in Fig. 8. The individual runtimes of LBM and MD can naturally vary since we randomly generate particles. The results show a good weak scaling. The total runtime is largely dominated by the runtime of the LBM,

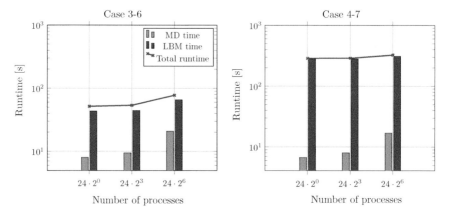

Fig. 8 Runtimes of the two test cases *case 3–6* (left) and *case 4–7* (right) in a weak scaling experiment. The x-axis shows the number of processes (24, 192, 1536) and the y-axis the runtime. Both plots also list the individual times for MD (green) and LBM (blue)

so the combined scaling reflects LBM scaling. Overall, we loose 50% performance from 24 to 1536 processes in *case 3–6* and only 13.5% performance in *case 4–7*. These losses are basically induced by the non-perfect scaling of the subcomponents.

To assess the quality of load-balancing, we inspect the imbalance I between all processes based on the maximum and average of the runtimes rs as: $I = \max\{rs\}/\mathrm{avg}\{rs\}$. For all cases, the imbalance of our jointly partitioned simulations is between 1.0 and 1.16 for the individual components (MD and LBM). We assess the runtime of Algorithms 1 and 2 in the context of grid adaption. This includes dynamic refinement around particles, repartitioning, payload migration (particles and LBM populations) and the necessary adaptions of ESPResSo's internal data structures. In this context, the runtime of Algorithms 1 and 2 for the largest test case (*case 4–7* on 1536 processes, i.e., about 120 million quadrants in the LBM grid) is with about 0.053 s well below 2% of the total runtime for grid adaption.

6   Conclusion and Future Work

We have presented a scheme to jointly partition octree grids and evaluated it for an application example, where we perform a lattice-Boltzmann flow simulation on one (adaptive) grid and short-range molecular dynamics calculations on the other (regular Linked-Cell) grid within the simulation package ESPResSo. Our partitioning is based on the finest common tree which is an intersection of two (or more) octrees. Our algorithm naturally limits interactions in sub-volumes of the domain to process-local interactions.

We have demonstrated that the combined repartitioning scheme scales in a dynamic, inhomogeneous, spatially adaptive and coupled MD-LBM setting. The runtime of the algorithm is good and so is the partitioning quality it produces. We have also reported on one of our target applications, a real-world cabin air filtration scenario. We plan to use the joint repartitioning scheme to simulate this scenario dynamically adaptive in the future as reported on in Sect. 2.

Using our joint repartitioning scheme in a way that minimizes the overall runtime is subject to future work. Weighting factors in all involved grids have to be optimized and the number of time steps between two re-balancing steps needs to be chosen in an optimal and adaptive way. This is a traditional problem in load-balancing. However, this problem gets a third dimension in our case, namely the factors for combining the weights of the different subsystems.

Acknowledgements The authors gratefully acknowledge funding provided by the German Research Foundation (DFG) as part of the Collaborative Research Center (SFB) 716 (projects D.8 and D.9), the support of our cooperation partners within SFB 716, computing resources provided by the High Performance Computing Center Stuttgart (HLRS), and the support by Carsten Burstedde concerning questions regarding the grid framework p4est.

References

1. P. Ahlrichs, B. Dünweg, Simulation of a single polymer chain in solution by combining lattice Boltzmann and molecular dynamics. J. Chem. Phys. **111**(17), 8225–8239 (1999)
2. A. Arnold et al., ESPResSo 3.1: molecular dynamics software for coarse-grained models, in *Meshfree Methods for Partial Differential Equations VI*, ed. by M. Griebel, M.A. Schweitzer. Lecture Notes in Computational Science and Engineering, vol. 89 (Springer, Berlin, Heidelberg, 2012), pp. 1–23
3. M. Brunn, Coupling of particle simulation and lattice Boltzmann background flow on adaptive grids. M.A. thesis, University of Stuttgart, Germany, 2017
4. M. Buchholz, Framework zur Parallelisierung von Molekulardynamiksimulationen in verfahrenstechnischen Anwendungen. Ph.D. thesis (2010)
5. C. Burstedde et al., p4est: scalable algorithms for parallel adaptive mesh refinement on forests of octrees. SIAM J. Sci. Comput. **33**(3), 1103–1133 (2011)
6. P.M. Campbell et al., Dynamic octree load balancing using space-filling curves. In: Williams College Department of Computer Science, Technical Report (2003), pp. 1–26
7. B. Dünweg, A.J.C. Ladd, Lattice Boltzmann simulations of soft matter systems, in *Advanced Computer Simulation Approaches for Soft Matter Sciences III* (2009), pp. 89–166
8. B. Dünweg et al., Statistical mechanics of the fluctuating lattice Boltzmann equation. Phys. Rev. E **76**(3), 036704 (2007)
9. B. Forsyth et al., Particle charge distribution measurement for commonly generated laboratory aerosols. Aerosol Sci. Technol. **28**(6), 489–501 (1998)
10. C. Helsper, W. Mölter, Determination and neutralization of the charge produced by the dispersion of powders. J. Aerosol Sci. **18**(6), 877–880 (1987)
11. S. Hirschmann et al., Towards understanding optimal load-balancing of heterogeneous short-range molecular dynamics, in *Workshop on High Performance Computing and Big Data in Molecular Engineering 2016*. Hyderabad, India (2016)
12. S. Hirschmann et al., Load balancing with p4est for short-range molecular dynamics with ESPResSo, in *Advances in Parallel Computing*, ed. by S. Bassini et al., vol. 32 (IOS Press, 2017), pp. 455–464
13. R.W. Hockney, J.W. Eastwood, *Computer Simulation Using Particles* (Taylor & Francis, Inc., Bristol, PA, USA, 1988)
14. T. Isaac et al., Recursive algorithms for distributed forests of octrees. SIAM J. Sci. Comput. **37**(5), C497–C531 (2015)
15. A.M. Johnston et al., Electrical charge characteristics of dry aerosols produced by a number of laboratory mechanical dispensers. Aerosol Sci. Technol. **6**(2), 115–127 (1987)
16. M. Lahnert et al., Minimally-invasive integration of p4est in ESPResSo for adaptive Lattice-Boltzmann, in *The 30th Computational Fluid Dynamics Symposium* (Japan Society of Fluid Mechanics, 2016)
17. M. Lahnert et al., Towards Lattice-Boltzmann on dynamically adaptive grids - minimally-invasive grid exchange in ESPResSo, in *VII European Congress on Computational Methods in Applied Sciences and Engineering, ECCOMAS*, ed. by M. Papadrakakis et al. (2016)
18. H.J. Limbach et al., ESPResSo - an extensible simulation package for research on soft matter systems. Comput. Phys. Commun. **174**(9), 704–727 (2006)
19. M. Mehl et al., Navier-Stokes and Lattice-Boltzmann on octree-like grids in the Peano framework. Int. J. Numer. Methods Fluids **65**(1–3), 67–86 (2010)
20. W.F. Mitchell, A refinement-tree based partitioning method for dynamic load balancing with adaptively refined grids. J. Parallel Distrib. Comput. **67**(4), 417–429 (2007)
21. G.M. Morton, A computer oriented geodetic data base; and a new technique in file sequencing. Technical report IBM Ltd., 1966
22. A. Pınar, C. Aykanat, Fast optimal load balancing algorithms for 1D partitioning. J. Parallel Distrib. Comput. **64**(8), 974–996 (2004)
23. M. Rohde et al., A generic, mass conservative local grid refinement technique for lattice-Boltzmann schemes. Int. J. Numer. Methods Fluids **51**(4), 439–468 (2006)

24. U.D. Schiller, Thermal fluctuations and boundary conditions in the lattice Boltzmann method. PhD thesis. Johannes Gutenberg-Universität, Mainz, 2008
25. C. Schober et al., Simulating the interaction of electrostatically charged particles in the inflow area of cabin air filters using a fully coupled system, in *Coupled Problems in Science and Engineering VII*, ed. by M. Papadrakakis et al. (2017)
26. F. Schornbaum, U. Rüde, Massively parallel algorithms for the lattice Boltzmann method on nonuniform grids. SIAM J. Sci. Comput. **38**(2), C96–C126 (2016)
27. I. Tischler, Implementing adaptive electrokinetics in ESPResSo. MA thesis. University of Stuttgart, Germany (2018)
28. C. Xiaolin, M. Zeyao, A new scalable parallel method for molecular dynamics based on cell-block data structure, in *Parallel and Distributed Processing and Applications*, ed. by J. Cao et al., vol. 3358. Lecture Notes in Computer Science (Springer, Berlin, 2005), pp. 757–764

Part VII
Miscellaneous Topics

In this chapter, another degree of interdisciplinary research is emphasized. Articles on structure mechanics, computational engineering, molecular modeling, and nuclear energy complement the research fields such as physics, solid state physics, reacting flows, computational fluid dynamics, and transport and climate which have been addressed before. The contributions generalize the field in which numerical simulations help to improve scientific knowledge. Highly sophisticated and efficient algorithms using high-performance computers yield data to develop novel scientific models. The interdisciplinary collaboration between several scientific fields drives the progress in fundamental and applied research which is optimized when numerical simulations are substantiated by experimental investigations and analytical solutions. That is, there has to be a close link between analytical, numerical, and experimental findings.

In the contribution of the Goethe-Center for Scientific Computing of the Goethe-University Frankfurt, a solver for the Biot model in poromechanics is investigated. An extrapolation scheme in time, plus a scalable monolithic multigrid method for solving the linear systems define the solver. The key ingredient for the multigrid method is an inexact Uzawa iteration, which has not been investigated with respect to its parallel properties before. Preliminary results show scalability for a 3D problem with roughly 400 million degrees of freedom. The presented model problem and solver framework serve as a prototype, which can be extended and generalized to non-linear problems easily.

The High Fidelity Monte Carlo project of the Institute for Pulsed Power and Microwave Technology of the Karlsruhe Institute of Technology aims at providing the computational resources for the development, testing, and application of advanced modeling and simulation techniques for the design and optimization of upcoming fusion reactors. Large-scale simulations are required to provide high fidelity results to assess the performance of such facilities. These include numerous time consuming Monte Carlo simulations for describing the transport of particles through the matter. The main computational tool for such application is the MCNP Monte Carlo code. The use of full-scale and detailed 3D geometry models of the different fusion facilities is mandatory to achieve a very high reliability of the final results. For the investigations of safety aspects of the fusion facilities, a rigorous 2-step method (R2S) was used to determine the decay gamma dose rates after the shutdown of the fusion neutron source. The challenging problems of radiation shielding and activation in complex geometries involving both deep penetration and radiation streaming are also tackled.

The contribution of the Chair of Thermodynamics and Energy Technology of the University of Paderborn focuses on molecular modeling and simulation which provides a broad range of possibilities to investigate various thermophysical properties. On the one hand, molecular simulation is a powerful tool for the prediction of thermodynamic data of pure fluids as well as mixtures, which are needed for process design in chemical and energy technology engineering. Especially when dealing with toxic substances or extreme conditions, elaborate experiments can be avoided with this approach. On the other hand, molecular simulation not only contributes to the availability of thermodynamic data, but also allows for a precise insight into the fluid behavior on the microscopic basis. Two simulation packages are constantly improved and extended, one for equilibrium simulations of bulk fluids, and another for the simulation of non-equilibrium systems and flow phenomena on the nanoscale. Several exemplary application cases of the two software packages are presented including model development and molecular simulations of varying system sizes and different thermodynamic properties of mixtures.

Heterogeneous materials are considered in the contribution of the Institute of Applied Materials of the Karlsruhe Institute of Technology and the Institute of Digital Materials Science of the Karlsruhe University of Applied Sciences. To be more precise, the martensitic phase transformation which is an important mechanism which strongly changes the properties of the material is discussed. On the one hand, martensite is a widely used constituent of many high-strength steels. On the other hand, martensite can also lead in some cases to an unwanted embrittlement of the material such that this transformation has to be prevented. Consequently, a better understanding of this transformation process is of high interest to control the mechanical properties of materials. For this purpose, multiphase-field simulation results of the martensitic phase transformation in a ferrite-austenite dual-phase microstructure and in an austenite-graphite microstructure are presented.

Wolfgang Schröder
Institute of Aerodynamics, RWTH Aachen University
Wüllnerstr. 5a, 52062 Aachen, Germany
e-mail: office@aia.rwth-aachen.de

Scalability of a Parallel Monolithic Multilevel Solver for Poroelasticity

Arne Nägel and Gabriel Wittum

Abstract This study investigates a solver for the quasi-static Biot model for soil consolidation. The scheme consists of an extrapolation scheme in time, complemented by a scalable monolithic multigrid method for solving the linear systems resulting after spatial discretisation. The key ingredient is a fixed-stress inexact Uzawa smoother that has been suggested and analysed using local Fourier analysis before (Gaspar and Rodrigo, Comput Methods Appl Mech Eng 326:526–540, 2017, [8]). The work at hand investigates the parallel properties of the resulting multigrid solver. For a 3D benchmark problem with roughly 400 million degrees of freedom, scalability is demonstrated in a preliminary study on Hazel Hen. The presented solver framework should be seen as a prototype, and can be extended and generalized, e.g., to non-linear problems easily.

1 Introduction

As access to high-end computational systems is nowadays widely spread and computational time is available in ever increasing amounts, this sparks the desire to solve larger and larger problems in the applications. The goal of this work is to develop an efficient and scalable solver for poroelasticity, i.e., porous media flow problems considering deformations of the medium. We focus on the linear quasi-static Biot model [2]. This system is a saddle point problem. At the core is a constraint for the stresses, coupling deformations and pressure. The second component is the flow equation, which considers the deformation of the medium in the volume terms of the time derivative.

Solvers for poroelasticity (and coupled processes in general) may be subdivided into the following classes (cf. [16]).

A. Nägel (✉) · G. Wittum
Goethe-Center for Scientific Computing, Goethe-Universität Frankfurt,
Kettenhofweg 139, 60325 Frankfurt, Germany
e-mail: naegel@gcsc.uni-frankfurt.de

© Springer Nature Switzerland AG 2019
W. E. Nagel et al. (eds.), *High Performance Computing in Science and Engineering '18*, https://doi.org/10.1007/978-3-030-13325-2_27

Fully implicit schemes (or: *monolithic* schemes) tackle the block system resulting from the spatial discretisation as single monolithic block. Efficient solvers rely on blocking strategies, thus require access to the software infrastructure, and must be implemented with proper care.

Iterative coupling is an alternative strategy. As the name indicates, convergence is achieved by iterating back and forth between the equations for the different unknowns. This approach is more modular than monolithic schemes and is less intrusive with respect to software infrastructure. For poroelasticity different algorithms have been suggested in [13, 16], with a unified presentation in [23]. In order to accelerate convergence, equations are modified in order to account for contributions from the Schur complement. The extended nonlinear case has been treated, e.g., in [14, 17].

Explicit coupling (or: *loosely, weakly coupled schemes*) suffer from severe time step restrictions, and are not considered here.

In passing we note that in the non-linear case, the linear system arising from a fully-implicit scheme can also be solved by a block-iterative preconditioner. This is similar to iterative coupling. However, since the problem at hand is linear, we do not formally distinguish these two cases here.

With respect to parallel performance, some investigations for an iteratively coupled scheme with implicit Euler time-stepping have been made, e.g., in [11, 23]. In this work, we investigate a monolithic solver with two key components: The first component is an adaptive time-stepping that is is based on (linearly implicit) extrapolation methods [7]. The building block of this method are sequences of (linearly-) implicit Euler steps, that are combined with an Aitken-Neville extrapolation. The second component is a monolithic multi-level solver. It is inspired by works on scalable solvers for the Stokes system in [9, 12]. It includes an Uzawa-type smootherwhich is constructed using the fixed stress approximation of the Schur complement [8].

The presentation is organized as follows: The mathematical model is formulated in Sect. 2, the description of the solver setup is given in Sect. 3. The work is concluded a numerical experiment in Sect. 4.

2   Governing Equations

In some domain $\Omega \subset \mathbb{R}^d$, we seek to solve the quasi-static Biot system for poroelasticity. The original formulation and derivation dates back to [2]; a recent description can be found, e.g., in the textbooks [4, 5]. The model is formulated in terms of a displacement $\mathbf{u}$ and a pressure p:

$$\partial_t(\frac{1}{M}p + \alpha \text{div}(\mathbf{u})) + \nabla \cdot [\mathbf{q}] = 0 \tag{1a}$$

$$-\nabla \cdot [\sigma] = \mathbf{f}. \tag{1b}$$

These equations describe the conservation of mass and momentum of the fluid respectively. The system includes the Biot modulus M and the Biot coefficient $0 \leq \alpha \leq 1$. We close the system by two constitutive material laws:

$$\mathbf{q} = -\kappa(\nabla p - \rho\mathbf{g}) \tag{1c}$$

$$\sigma = \lambda\operatorname{div}(\mathbf{u})\mathcal{I} + 2\mu\epsilon(\mathbf{u}) - \alpha p\mathcal{I}. \tag{1d}$$

The Darcy velocity (1c) includes the permeability κ, the acceleration due to gravitation $\mathbf{g}$, and the fluid density ρ, which is assumed to be constant. The linear elastic model (1d) for the stress tensor σ includes the infinitesimal linearized strains $\epsilon(\mathbf{u}) = \frac{1}{2}(\nabla\mathbf{u} + \nabla\mathbf{u}^T)$ and the Lamé parameters λ, μ.

Relationship to Engineering Constants

The presented model includes the material parameters α, λ, μ, M, and κ. The Lamé parameters λ, μ can also be expressed as follows::

$$\lambda = \frac{2\nu G}{1 - 2\nu} = K - \frac{2}{3}G = \frac{\nu E}{(1 + \nu)(1 - 2\nu)}$$

$$\mu = G = \frac{3K(1 - 2\nu)}{2(1 + \nu)} = \frac{E}{2(1 + \nu)}$$

The parameter μ is the shear modulus which also sometimes denoted by G. Poisson's ratio is given by $\nu = \frac{\lambda}{2(\lambda+G)}$. Moreover, we can use the elasticity modulus E, and the bulk modulus $K = \lambda + \frac{2}{3}G$.

Following [4], we introduce $\xi = \frac{1}{M}p + \alpha e$, $e := \operatorname{div}(u)$ i in (1a) as the variation of fluid content in a representative volume (relative to the original configuration). The *undrained* case corresponds to $\xi = 0$, i.e. $p = -M\alpha e$. Inserting this into (1d) yields a redefined (p independent) stress tensor with parameters $\lambda_u = \lambda + M\alpha^2$, $K_u = K + M\alpha^2$, and $\nu_u = \frac{\lambda_u}{2(\lambda_u+G)}$.

Variational Form

In the *variational form* of (1), we seek to find $(p, \mathbf{u}) \in V_p \times V_u$, $V_u \subset L^2(\Omega)$, $V_p \subset H^1(\Omega)^d$ such that

$$\int 2\mu\epsilon(\mathbf{u}) : \epsilon(\mathbf{v}) + \lambda\operatorname{div}(\mathbf{u})\operatorname{div}(\mathbf{v}) - \int \alpha\operatorname{div}(\mathbf{v})\,p + \int \mathbf{f}\mathbf{v} = 0 \tag{2a}$$

$$\int \frac{1}{M}\dot{p}\,r + \int \alpha\operatorname{div}(\dot{\mathbf{u}})r + \int \kappa\nabla p\,\nabla r - \int \kappa\rho\mathbf{g} = 0 \tag{2b}$$

holds for all $(r, \mathbf{v}) \in V_p \times V_u$. Note that boundary conditions have been neglected for the ease of presentation. In the discrete setting, we replace V_p and V_u by discrete spaces V_p^h and V_u^h. Here, we use conforming finite elements that are element-wise given by polynomials of degree 1 and 2 respectively.

Employing an implicit Euler scheme with step size τ for time discretization time discretization,

$$\mathbf{u}_{t+\tau}^h := \mathbf{u}_t^h + \triangle\mathbf{u}^h, \ p_{t+\tau}^h := p_t^h + \triangle p^h$$

results in a large *linear system* for the increments $\triangle\mathbf{u}^h$, $\triangle p^h$ in the following block form:

$$\begin{bmatrix} K_{11} & K_{12} \\ K_{21} & K_{22} \end{bmatrix} \begin{bmatrix} \triangle\mathbf{u}^h \\ \triangle p^h \end{bmatrix} = \begin{bmatrix} \mathbf{b}_u^h \\ \mathbf{b}_p^h \end{bmatrix} \tag{3a}$$

The particular form of the right hand side is not important here. In the matrix

$$\mathcal{K} := \begin{bmatrix} K_{11} & K_{12} \\ K_{21} & K_{22} \end{bmatrix} := \begin{bmatrix} A_{uu} & \alpha G_{up} \\ -\alpha G_{up}^T & (\frac{1}{M} M_{pp} + \tau\kappa A_{pp}) \end{bmatrix}. \tag{3b}$$

we identify the terms from (2) and note in passing, that the second diagonal term vanishes, if $\frac{1}{M} \to 0$ and $\tau\kappa \to 0$ simultaneously. In this case, standard smoothers are not applicable. In general (3) is a saddle point problem, as can, e.g., be seen by multiplying the second row by -1 and by using the abstract block decomposition:

$$\mathcal{K} = \begin{bmatrix} I & \\ K_{21}K_{11}^{-1} & I \end{bmatrix} \begin{bmatrix} K_{11} & \\ & S_{22} \end{bmatrix} \begin{bmatrix} I & K_{11}^{-1}K_{12} \\ 0 & I \end{bmatrix} \tag{4}$$

with $S_{22} = K_{22} - K_{21}K_{11}^{-1}K_{12}$ denoting the Schur complement. This decomposition is also important for the iterative solvers that are introduced in the next section.

3 Algorithmic Details

3.1 Iterative Schemes

Different strategies for solving (3) have been suggested in the literature. An overview on general saddle-point problems is given, e.g., in [1]. Most methods are based on iterative schemes of the following from

$$x^{(k+1)} = x^{(k)} + \widetilde{\mathcal{K}}^{-1}(f - \mathcal{K}x^{(k)}).$$

where $\widetilde{\mathcal{K}}^{-1}$ is an approximate inverse that can be computed easily. In order to devise $\widetilde{\mathcal{K}}^{-1}$, one can employ the block decomposition (4) to define different block preconditioners, e.g.,

$$\mathcal{W}_l = \begin{bmatrix} K_{11} & \\ K_{21} & K_{22} \end{bmatrix}, \ \mathcal{W}_d = \begin{bmatrix} K_{11} & \\ & S_{22} \end{bmatrix}, \ \mathcal{W}_u = \begin{bmatrix} K_{11} & K_{12} \\ & S_{22} \end{bmatrix}$$

Inexact variants can also be devised, e.g.,

$$\widetilde{\mathcal{W}}_l = \begin{bmatrix} \tilde{K}_{11} & \\ K_{21} & \tilde{S}_{22} \end{bmatrix}, \ \widetilde{\mathcal{W}}_d = \begin{bmatrix} \tilde{K}_{11} & \\ & \tilde{S}_{22} \end{bmatrix}, \ \widetilde{\mathcal{W}}_u = \begin{bmatrix} \tilde{K}_{11} & K_{12} \\ & \tilde{S}_{22} \end{bmatrix}$$

It is well know, e.g. [3, 23], that the term $K_{21} K_{11}^{-1} K_{12}$ is spectrally equivalent to a rescaled mass matrix M_{pp}. Using the approximation

$$S_{22} = K_{22} - K_{21} K_{11}^{-1} K_{12} \approx (\frac{1}{M} + \frac{\alpha^2}{K}) M_{pp} + \tau \kappa A_{pp} \tag{5}$$

yields the fixed stress splitting [13]. We do not follow this approach here, but rely on the decomposition for building a smoother [8, 15] for a monolithic multigrid method.

3.2 Monolithic Multigrid Solver

Multigrid methods, as introduced in greater detail, e.g. in the monographs [10, 20], are based on the property that all frequencies in the error are reduces equally well. This is obtained by an efficient interplay between smoothing and interpolation between grid levels. These two components need to be defined:

The grid transfer operators are given by the standard operators as defined (elementwise) by the Lagrange polynomials. The coarse grid operators are recursively defined by the Galerkin products. For smoothing we employ an inxact Uzawa method, as suggested for Biot-System in [8, 15], with the following modifications: In the forward sweep, we use $\widetilde{\mathcal{W}}_l$, where the inverse of $\tilde{K}_{11}$ is approximated by a forward Gauss-Seidel sweep for A_{uu}, and the inverse of $\tilde{S}_{22}$ is given by a damped Jacobi iteration for the right hand side of (5). The backward sweep is built using $\widetilde{\mathcal{W}}_u$ using the adjoint iterations. Note that a similar approach has successfully been used for the (related) Stokes equation before [9, 12] with excellent scalability properties.

3.3 Software

All algorithms and solvers have been implemented in the software framework UG4 [18, 22]. Details of the MPI based parallelization of UG4's multigrid solver, highly scalable on hundreds of thousands of processes, have been described in previous reports, e.g. [19].

4 Results

4.1 Benchmark Problem

As a benchmark problem, we consider the deformation of a cylinder, as suggested by de Leeuw [4, 6]: For $t > 0$ a cylinder of radius a is subject to uniform radial stress $\sigma_{rr} = -q$ at the cylinder wall, while drainage is instantaneously induced by $p = 0$ outside. This problem has an analytical solution [21]. Ibid., the initial value is given by a homogeneous pressure distribution corresponding to the undrained case for given q:

$$p(\mathbf{x}) \equiv p_0 = \frac{M\alpha}{K + \frac{1}{3}G + M\alpha^2} q$$

and the corresponding consistent displacement field $\mathbf{u}_0$. Note, that the first factor evaluates as 1, i.e., $p_0 = q$, if solid and fluid are incompressible ($\alpha = 1$, $M = \infty$). The full set of parameters is provided in Table 1.

Figure 1 shows the geometry as well as the solution after some time. Figure 2 compares the analytical and numerical solution with respect to the normalized pressure $p(\mathbf{x}_0)/p_0$ at the center $\mathbf{x}_0$ of the cylinder. As $v < 0.5$ the configuration gives rise to the well-known Mandel-Cryer effect: The pressure in the interior increases first, before it relaxes and decays to $p \to 0$. We also note that the adaptive time-stepping captures the dynamics correctly and selects the exponentially increasing time steps. The time stepping error was controlled in the L2 norm for the pressure p with a relative tolerance of 0.1%.

4.2 Load Balancing and Grid Distribution

The performance of the multigrid algorithm depends on a proper load balancing. Following [19], the grid distribution quality q_l onto P_l processes on level l of the hierarchy is computed as

Table 1 Parameters for the test case. Although computations are performed in the adimensional form, they are provided with physical units here for the sake of orientation. Gravitation is neglected ($\rho = 0$)

Parameter	Value	Unit
E	80	MPa
M	1.1×10^7	MPa
v	0.25	
α	1	
κ	10^{-6}	m^2/(Pa s)
a	1	m
q	1	MPa

Fig. 1 Benchmark problem: Illustration of pressure and deformation

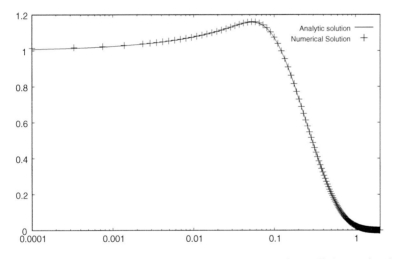

Fig. 2 Benchmark problem: normalized pressure p/p_0 at the core of the cylinder as a function of the dimensionless time. The Mandel-Cryer effect leads to an increase in pressure that is followed by a drop. Note, that time the extrapolation scheme selects an exponentially increasing time step size

$$q_l := \begin{cases} \dfrac{n_l^{total} - n_l^{max}}{n_l^{max} \cdot (P_l - 1)}, & P_l > 1 \\ 1, & P_l = 1 \end{cases}$$

Here, n_l^p is the number of elements in level l on process p, and

$$n_l^{total} := \sum_{p=1}^{P_l} n_l^p, \quad n_l^{max} := \max_{p=1,\dots,P_l} n_l^p.$$

The quality measure q_l is in the range $[0, 1]$. For $q_l = 0$ all elements of level l are contained on one process only, and $q_l = 1$ indicates that the elements are shared equally among all processes.

4.3 Numerical Experiments

The performance of the solver is investigated in a weak scaling study, using isotropic refinement. Problem size and involved number of cores increase as reported in Table 2. Using a hierarchical load-balancing and distribution scheme [18], not all processes are contributing on all levels. For each configuration, Table 3 provides a summary of number of processes used on each level of the multigrid hierarchy. Table 4 shows the levelwise distribution qualities for each run.

Aiming for representative results, performance has been evaluated in an interval of $1h$ wall clock time, which corresponds to at least 25 time steps. One time step is computed by extrapolation [7], and thus consists of three implicit Euler steps. In each of these steps, the resulting system has been solved using a BiCGStab algorithm, that was preconditioned by the monolithic multigrid solver described in Sect. 3. The

Table 2 Each line corresponds to an individual run. Recorded are the number of processes (*PE*), the number of levels (*l*), and the number of degrees of freedom (*DoFs*)

PEs	Levels *l*	DoFs
96	5	6,714,692
768	6	53,070,468
6144	7	421,991,684

Table 3 Number of processes used on each level for the individual runs

PE	lv-0	lv-1	lv-2	lv-3	lv-4	lv-5	lv-6	lv-7
96	1	1	96	96	96	96	–	–
768	1	1	96	96	768	768	768	–
6144	1	1	96	96	6144	6144	6144	6144

Table 4 Distribution qualities for each level of the multigrid hierarchy for the different runs

PE	lv-0	lv-1	lv-2	lv-3	lv-4	lv-5	lv-6	lv-7
96	1	1	1	1	0.82	0.82	–	–
768	1	1	1	1	0.84	0.84	0.99	–
6144	1	1	1	1	**0.40**	**0.40**	0.90	0.90

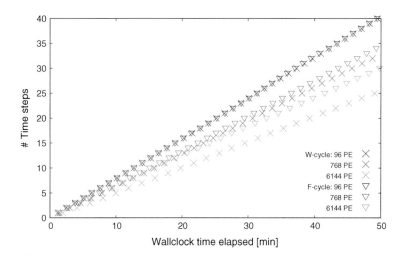

Fig. 3 Time steps per elapsed wall clock time: results of weak scaling for 96, 768, and 6144 PE for W- and F-cycle respectively

residual were reduced by 8 orders of magnitude, or, alternatively, to absolute tolerance of 10^{-14}.

Figure 3 provides the results for this setup: As a performance measure, we consider the average number of time *steps* per wall clock time unit. Results range between 0.5 *steps*/min in the worst case (W-cycle on 6144 PE) and 0.8 *steps*/min in the best case (F-cycle on 96 PE). On 96 PE, W-cycle and F-cycle show almost identical performance. However, as can be expected, the W-cycle deteriorates when increasing the level of parallelism. This issue is well known. It is due to increasing number of smoother and coarse grid correction calls on the coarse levels l, which is a accompanied by a relative reduction of the level of parallelism (as illustrated in Table 3). Using the F-cycle mitigates these issues, and results improve slightly with ≈ 0.7 steps/min on 768 PE, and ≈ 0.6 steps/min on 6144 PE respectively. Using the V-cycle is not a viable alternative, since its convergence is not independent of the mesh width.

5   Conclusion

The presented preliminary results show the potential of the combination of time-stepping and linear solver. Using an adaptive time integrator is key for improving the performance. The linear solver indicates its potential, but improving its scalability properties is still subject to further investigations and profiling. Although for the F-cycle a relative efficiency of at least 75% is obtained, it must be clarified, if this can be improved further. The imbalance shown in Table 4 (row for 6144 PE) may have a severe impact.

Acknowledgements   This work has been supported by the DFG in the *German Priority Programme 1648 - Software for Exascale Computing* in the project *Exasolvers* (WI 1037/24-2). The authors would like to thank Sebastian Reiter and Michael Lampe for support for generating the geometries and discussions on scalability aspects of *Hazel Hen*. Moreover, the technical support by HLRS staff is gratefully acknowledged.

References

1. M. Benzi, G.H. Golub, J. Liesen, Numerical solution of saddle point problems. Acta Numer. **14**, 1–137 (2005)
2. M.A. Biot, General theory of three dimensional consolidation. J. Appl. Phys. **12**(2), 155–164 (1941)
3. N. Castelletto, J. White, H. Tchelepi, Accuracy and convergence properties of the fixed-stress iterative solution of two-way coupled poromechanics. Int. J. Numer. Anal. Methods Geomech. **39**(14), 1593–1618 (2015)
4. A.H.-D. Cheng, *Poroelasticity, volume 27 of Theory and Applications of Transport in Porous Media* (Springer International Publishing Switzerland, 2016)
5. O. Coussy, *Poromechanics* (Wiley, New York, 2004)
6. E. de Leeuw, The theory of three-dimensional consolidation applied to cylindrical bodies, in *Proceedings of 6th International Conference on Soil Mechanics and Foundation Engineering, Montreal*, vol. 1 (1965), pp. 287–290
7. P. Deuflhard, One-step and extrapolation methods for differential-algebraic systems. Numer. Math. **51**, 501–516 (1987)
8. F.J. Gaspar, C. Rodrigo, On the fixed-stress split scheme as smoother in multigrid methods for coupling flow and geomechanics. Comput. Methods Appl. Mech. Eng. **326**, 526–540 (2017)
9. B. Gmeiner, M. Huber, L. John, U. Rüde, B. Wohlmuth, A quantitative performance study for stokes solvers at the extreme scale. J. Comput. Sci. **17**, 509–521 (2016)
10. W. Hackbusch, *Multi-Grid Methods and Applications* (Springer, Berlin, 1985)
11. C.J.B. Haga, Numerical methods for basin-scale poroelastic modelling. PhD thesis, University of Oslo, 2011
12. L. John, U. Rüde, B. Wohlmuth, W. Zulehner, On the analysis of block smoothers for saddle point problems (2016)
13. J. Kim, H. Tchelepi, R. Juanes, Stability, accuracy, and efficiency of sequential methods for coupled flow and geomechanics. SPE J. **16**(2), 249–262 (2011)
14. S. Lee, M.F. Wheeler, T. Wick, Iterative coupling of flow, geomechanics and adaptive phase-field fracture including level-set crack width approaches. J. Comput. Appl. Math. **314**(C), 40–60 (2017)
15. P. Luo, C. Rodrigo, F.J. Gaspar, C.W. Oosterlee, Uzawa smoother in multigrid for the coupled porous medium and stokes flow system. SIAM J. Sci. Comput. **39**(5), 633–661 (2017)

16. A. Mikelić, M.F. Wheeler, Convergence of iterative coupling for coupled flow and geomechanics. Comput. Geosci. **17**(3), 455–461 (2013)
17. A. Mikelić, B. Wang, M.F. Wheeler, Numerical convergence study of iterative coupling for coupled flow and geomechanics. Comput. Geosci. **18**(3), 325–341 (2014)
18. S. Reiter, A. Vogel, I. Heppner, M. Rupp, G. Wittum, A massively parallel geometric multigrid solver on hierarchically distributed grids. Comput. Vis. Sci. **16**(4), 151–164 (2013)
19. S. Reiter, A. Vogel, A. Nägel, G. Wittum, A massively parallel multigrid method with level dependent smoothers for problems with high anisotropies. High Perform. Comput. Sci. Eng. **16**, 667–675 (2016)
20. U. Trottenberg, C.W. Oosterlee, A. Schüller, *Multigrid* (Academic Press, San Diego, CA, 2001); contributions by A. Brandt, P. Oswald, K. Stüben
21. A. Verruijt, Theory and problems of poroelasticity. Delft University of Technology (2013)
22. A. Vogel, S. Reiter, M. Rupp, A. Nägel, G. Wittum, UG 4: a novel flexible software system for simulating PDE based models on high performance computers. Comput. Vis. Sci. **16**(4), 165–179 (2013)
23. J.A. White, N. Castelletto, H.A. Tchelepi, Block-partitioned solvers for coupled poromechanics: a unified framework. Comput. Methods Appl. Mech. Eng. **303**, 55–74 (2016)

High Fidelity Monte Carlo for Fusion Neutronics

B. Weinhorst, D. Leichtle, A. Häußler, E. Nunnenmann, P. Pereslavtsev,
Y. Qiu, P. Raj, A. Travleev and U. Fischer

Abstract The High Fidelity Monte Carlo for fusion neutronics project HIFIMC aims at providing the computational resources for the development, testing and application of advanced modeling and simulation techniques as required for the design and optimization of upcoming fusion reactors like ITER, DEMO, HELIAS and related research facilities like IFMIF or DONES. Large-scale simulations are required to provide high fidelity results to assess the performance of such facilities. These include numerous time consuming Monte Carlo simulations for describing the transport of particles (neutrons and gammas) through the complex and heterogeneous geometry. The main computational tool for such application is the MCNP Monte Carlo code, developed at LANL, US, and the coupled transport-activation tool R2Smesh, developed at KIT. Several concurrent research topics have been conducted and are reported here. Further development of R2Smesh on enhancing performance and studies on convergence issues in meshing of the activation responses are required to qualify the tools for many large nuclear performance and radiation shielding applications in fusion devices. Verification and validation of alternative codes, like GEANT 4, developed at CERN, CH, are supporting these efforts. Applications to port systems in ITER (next step fusion reactor under construction in France), breeding blankets in DEMO (Demonstration fusion reactor), full reactor of HELIAS (Helical-Axis Advanced Stellarator) and irradiation test systems of IFMIF/DONES (International Fusion Material Irradiation Facility/DEMO-oriented neutron source) are presented. It is shown that the tools on high-performance computing platforms are capable to tackle the challenging problems of radiation shielding and activation in complex geometries involving both deep penetration and radiation streaming.

B. Weinhorst · D. Leichtle (✉) · A. Häußler · E. Nunnenmann · P. Pereslavtsev ·
Y. Qiu · P. Raj · A. Travleev · U. Fischer
Institute for Neutron Physics and Reactor Technology, Karlsruhe Institute of Technology,
Hermann-von-Helmholtz-Platz 1, 76344 Eggenstein-Leopoldshafen, Germany
e-mail: dieter.leichtle@kit.edu

© Springer Nature Switzerland AG 2019
W. E. Nagel et al. (eds.), *High Performance Computing in Science
and Engineering '18*, https://doi.org/10.1007/978-3-030-13325-2_28

1 Developments and Applications of Monte-Carlo Radiation Transport Simulations

The High Fidelity Monte Carlo for fusion neutronics project HIFIMC aims at providing the computational resources for the development, testing and application of advanced modeling and simulation techniques as required for the design and optimization of upcoming fusion reactors like ITER, DEMO, HELIAS and related research facilities like IFMIF or DONES. Large-scale simulations are required to provide high fidelity results to assess the performance of such facilities. These include numerous time consuming Monte Carlo simulations for describing the transport of particles (neutrons and gammas) through the matter. The main computational tool for such application is the MCNP Monte Carlo code. The use of full-scale and detailed 3D geometry models of the different fusion facilities is mandatory to achieve a very high reliability of the final results. For the investigations of safety aspects of the fusion facilities, a Rigorous-2-Step method (R2S) was elaborated for calculations of the decay gamma dose rates after the shutdown of the fusion neutron source. The successful implementation of the R2S method with the mesh-based R2Smesh code relies on the usage of a powerful computer system that enables time consuming calculations leading to accurate and detailed results.

Several concurrent research topics have been conducted on further development of computational techniques for Monte-Carlo radiation transport simulations and their application ranging from basic calculations for optimization of self-powered neutron and photon detectors to full nuclear analyses of nuclear facilities like ITER (next step fusion reactor under construction in France), DEMO (Demonstration fusion reactor), HELIAS (Helical-Axis Advanced Stellarator) and IFMIF/DONES (International Fusion Material Irradiation Facility/DEMO-oriented neutron source). All simulations have been performed at the ForHLR II cluster at KIT.

1.1 Development of a Code System for Activation and Shut-Down Dose Rate Calculations

The neutron-induced activation of materials is an important issue for fusion facilities. Photons emitted by the activated material result in a photon field, whose spatial distribution must be taken into account when planning maintenance work in shutdown periods and for final decommissioning of the machine. One of the approaches to calculate the shut-down dose rate (SDDR) is the rigorous 2-step method (R2S) which is based on the data flow from neutron transport simulations to activation calculations to generate the photon source distribution. This source is used in subsequent photon transport simulations to calculate the dose rate distributions at different times post irradiation.

The data flow from the code for neutron flux calculations to the code for activation analysis represents a challenge for a high-performance cluster computer. One the one hand, Monte-Carlo simulation of the neutron transport is well parallelized, and activation calculations can be performed independently on a large amount of nodes. On the other hand, each activation calculation requires input data as a set of relatively small text files, whose preparation, if not performed on a local node file system, is limited by the performance of the global file system. Thus, one of the important point of the development is the design of a flexible user interface, which allows tuning the code system for a particular cluster environment without changing the source code.

The coupling code system is realized as a set of shell scripts and a fortran program. The scripts help the user to define environment variables controlling the calculations and provide some tools for intermediate data processing, based on standard linux utilities. The fortran program is a driver for activation calculations, which takes the job of preparing data, starting activation calculations and collecting their outputs. This program uses the Intel MPI library and can be started in parallel using standard cluster job schedulers.

The currently developed coupling is a new implementation of the already existing code developed at KIT [1]. Compared to the previous version, the new implementation provides a more flexible and clear user interface, assumes less simplifications and is written keeping in mind application to large problems and further development thus establishing a more error-proof and performance-optimized workflow. In particular, the new implementation allows non-uniform rectangular superimposed meshes to represent neutron flux and material distributions. It provides a more robust method to extract material compositions from the MCNP input file and utilizes FISPACT-II for the activation calculations. It allows computer-specific adjustments to FISPACT-II parallel invocation to optimize the use of global and local file systems on cluster computers. To a large extent, standard Linux command line tools are used to handle intermediate data. Dynamic array allocation is applied in the modified MCNP for sampling the decay gamma source. The code with all interfaces are written in modern Fortran in a modular way, which allows subsequent improvements and modifications.

The new implementation is being currently tested on a computational benchmark against the results obtained with the previous version. Figure 1 shows the SDDR spatial distributions. One can see visually good agreement. Also quantitatively, results are very close. This benchmark ensures that the new implementation is free of programming errors and at the same time has much better efficiency: processing of the intermediate results (formation of the decay gamma source from separate activation calculations) in the new implementation is by order of magnitude faster.

1.2 Sensitivity of R2Smesh Shutdown Dose Rates Results on Mesh Resolution

R2Smesh [1], is a Monte Carlo based computational interface developed at KIT for accurate shutdown dose rate (SDDR) analysis in fusion reactors based on the Rigorous-2-Step methodology. It combines the neutron transport using the Monte

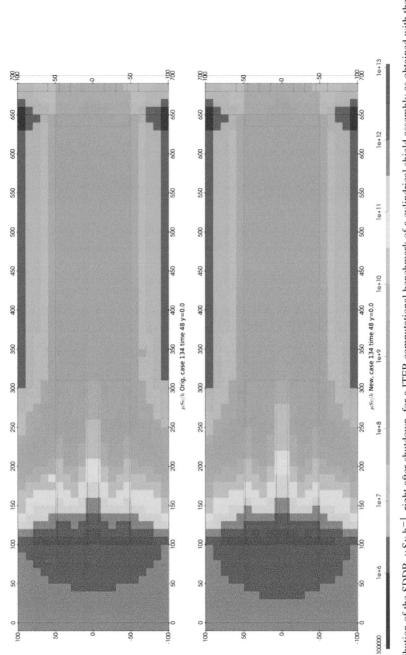

Fig. 1 Distribution of the SDDR, $\mu Sv\ h^{-1}$, right after shutdown, for a ITER computational benchmark of a cylindrical shield assembly as obtained with the previous (top) and currently developed implementation (bottom) of the coupling

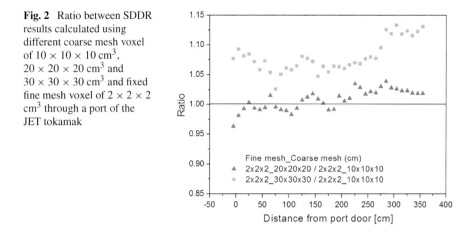

Fig. 2 Ratio between SDDR results calculated using different coarse mesh voxel of $10 \times 10 \times 10$ cm^3, $20 \times 20 \times 20$ cm^3 and $30 \times 30 \times 30$ cm^3 and fixed fine mesh voxel of $2 \times 2 \times 2$ cm^3 through a port of the JET tokamak

Carlo N-Particle Code MCNP and the activation and inventory calculation using the FISPACT code. The superimposed Cartesian mesh tally capabilities in MCNP have been introduced in its development effort for calculation of neutron intensity and multi-group energy spectra. The resulting neutron flux is averaged within each mesh element as its gradient is omitted. The accuracy and reliability of final SDDR thus depends heavily on mesh resolution and alignment. To evaluate the influence of mesh resolution on SDDR, a sensitivity analysis has been conducted on JET (Joint European Torus, Culham, UK) and EU DEMO models by using series of fine and coarse mesh. In case of rapid attenuation of neutron flux, the fine mesh is employed for accurate neutron intensity calculation. The convergence of SDDR was achieved and the optimum size of mesh voxel for JET and DEMO SDDR calculation were determined after balancing between the precision of results and computational costs. By employing the optimum mesh resolution, the R2Smesh interface was validated through comparison with JET measurements. Afterwards it was applied on DEMO for global SDDR analysis to provide support for its nuclear shielding design.

When the fine mesh spacing is fixed to 2 cm, the result is converged after setting the coarse mesh voxel to 10 and 20 cm as the ratio is not more than 1.05; the SDDR ratio between them can be seen in Fig. 2. When the coarse mesh voxel is fixed to 20 cm , the SDDR calculated with mesh voxel larger than 5 cm has high deviation up to a ratio of 1.13 compared with the use of 2 cm fine mesh voxel.

The R2Smesh code has been applied to HCPB DEMO for SDDR calculation to provide global SDDR data for the shielding design. The sensitivity study of SDDR to mesh resolution has been performed in advance to choose the optimum mesh resolution. This investigation has been carried out in part of a single blanket, i.e. inside the vacuum vessel showing particularly high SDDR, using different fine and coarse mesh voxel sizes. The fine and coarse mesh cover only part of the blanket to assess the difference due to the variance of mesh voxel in the activated volume.

Firstly, the fine mesh voxel is fixed to 3 cm and coarse mesh is set to 6, 12, 15 and 30 cm respectively. The map of SDDR can be seen in Fig. 3. The SDDR among

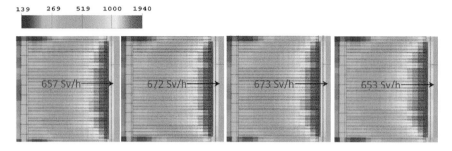

Fig. 3 Shutdown dose rate map (in Svh^{-1}) using fixed fine mesh voxel $3 \times 3 \times 3$ cm^3 and coarse mesh voxel $6 \times 6 \times 6$ cm^3, $12 \times 12 \times 12$ cm^3, $15 \times 15 \times 15$ cm^3 and $30 \times 30 \times 03$ cm^3, respectively

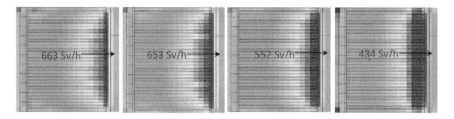

Fig. 4 Shutdown dose rate map (in Svh^{-1}) using fixed coarse mesh voxel $30 \times 30 \times 30$ cm^3 and fine mesh voxel $1 \times 1 \times 1$ cm^3, $3 \times 3 \times 3$ cm^3, $5 \times 5 \times 5$ cm^3, and $15 \times 15 \times 15$ cm^3 respectively. Scale as in Fig. 3

the four sets of SDDR data at the same positon near the front varies only slightly from 653 to 673 Svh^{-1}. Next, the coarse mesh voxel is fixed to 30 cm and the fine mesh voxel is set to 1, 3, 5 and 15 cm respectively. The map of SDDR can be seen in Fig. 4. The SDDR results calculated using fine and coarse mesh voxel of 1 and 3 cm have little deviation (663 and 653 Svh^{-1} respectively at the same position). While, when the fine mesh voxel extends to 5 and 15 cm, their results (552 and 434 Svh^{-1} respectively) have big difference compared with that calculated using finer mesh voxel. The accuracy of neutron flux actually deteriorates due to low resolution of fine mesh which is grossly averaging over steep gradients.

The sensitivity study shows the DEMO SDDR converged when setting 1 ... 3 cm of fine mesh voxel and 15 ... 30 cm of coarse mesh voxel. Based on these findings a general recommendation has been issued to work with not more than 3 cm fine mesh spacing and 30 cm coarse mesh spacing, respectively, for applications to DEMO.

This work was published in Fusion Engineering and Design [2].

1.3 V&V Analyses of the GEANT4 Monte Carlo Code with Computational and Experimental Fusion Neutronics Benchmarks

The high-energy particle physics Monte Carlo code toolkit GEANT4 has been expanded for fusion energy-range neutron transport simulations based on evaluated nuclear cross-section libraries [3]. Verification and Validation (V&V) analyses were conducted with nuclear data from the ENDF/B-VII.0 [4] and the JEFF-3.1 [5], library to show the suitability for fusion applications. Two computational benchmarks with a single fusion-relevant isotope at a time and one experimental benchmark were performed with GEANT4 in comparison with MCNP5.

The differential computational benchmark assumes a neutron beam with iso-lethargic energy distribution along the axis of a cylinder with length 2m and radius 1μm. After a single material interaction, the scattered or secondary neutrons passing through the cylinder side surface are recorded. As an example, the result for ^7Li can be seen in Fig. 5.

The integral computational benchmark is a 30 cm radius sphere with an isotropic 14.1 MeV neutron source in the centre. The neutron flux averaged over the sphere volume is recorded.

The considered experimental benchmark, performed by the Institute of Physics and Power Engineering, Obninsk, is on the transmission of 14 MeV neutrons through iron shells of different thicknesses (2.5 ... 28 cm) [6]. Geometry and source description were processed from the Shielding Integral Benchmark Archive and Database (SINBAD). In this experiment, neutron leakage spectra were measured at a distance of 679 cm from the centre of the iron shells.

This work has been presented at the Symposium on Fusion Technology (SOFT 2018) and will be published in Fusion Engineering and Design.

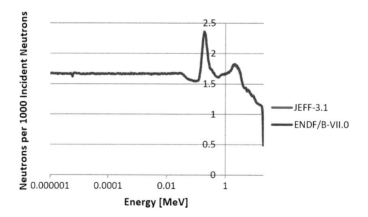

Fig. 5 MCNP5 result for the differential benchmark geometry filled with ^7Li

1.4 Development of Self-powered Neutron and Photon Detectors

Self-powered detectors (SPD) are online detectors for monitoring of neutron and photon fluxes in nuclear reactors. The simplicity of design and operation of an SPD, its compactness, low-cost, reliability, and durability make it a candidate for active detectors in European test blanket modules (TBM) of ITER. The research involved computational model development, analyses and comparison of experimental and computed results specific to the self-powered neutron and photon detectors. In [7] we analyze a chromium emitter based SPD, which is proposed as a fast neutron detector for integration in the European TBMs. A computational technique with Monte-Carlo modelling to predict the response of an SPD in the ITER environment has been developed. At first, the neutron and photon flux intensities and energy spectra, see Fig. 6, are calculated in the TBM through simulations in a MCNP model of ITER. Then, a multi-step multi-particle transport recipe is adopted for calculation of SPD sensitivity to these spectra. Starting with the information on composition of the SPD layers, main processes involved in the creation of the detector current are evaluated, corresponding to each of which a partial sensitivity is determined. Finally, a time-profile of the signal and its various components is constructed for one plasma pulse of ITER, with typical duration and fusion power. The range of electric currents achievable with SPDs is obtained, its adequacy for flux monitoring in TBMs is discussed, and finally, some design improvements are suggested.

This work has been published in Nuclear Instruments and Methods in Physics Research Section A [7].

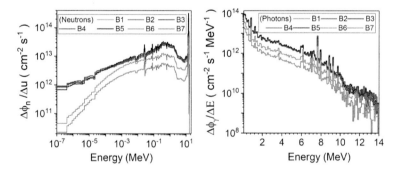

Fig. 6 MCNP-calculated energy spectra of (left) neutrons and (right) photons at selected positions in the HCPB TBM. The seven positions are chosen to show field variations along the three principal axes from the central point B5, which is near the front plate of TBM and reference for SPD response simulations

1.5   Nuclear Analysis for the ECHUL of ITER

The electron cyclotron-heating upper launcher (ECHUL) will be installed in four upper ports of the ITER tokamak thermonuclear fusion reactor. Each ECHUL is able to deposit 8 MW power into the plasma for plasma mode stabilization via microwave beam lines. These beam lines and other components needed for the guidance of the microwaves are mounted into the upper port plug. In order to protect these components and to mitigate radiation leaving the reactor, several shield blocks are also included in the port plugs design. The ECHUL components which are closest to the fusion plasma are heated by the gamma and neutron radiation and therefore will need to be actively cooled. The work investigates the influence of material homogenization by comparing results obtained in the neutronic analyses performed for the cooling system of the ECHUL blanket shield module (BSM). A small reduction in shielding performance of the BSM was found in the case of a heterogeneous description of the BSM compared to the standard approach of homogenizing the materials of the BSM by removing pipes and cooling circuits for the neutronic analyses.

This work was presented at the International Symposium of on Fusion Nuclear Technology and published in Fusion Engineering and Design [8] (Fig. 7).

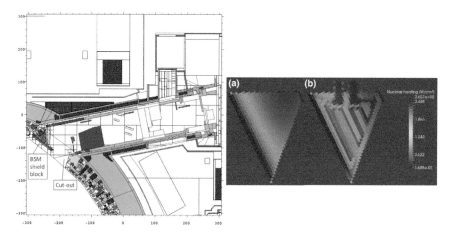

Fig. 7   The ECHUL MCNP model integrated into the ITER C-Model in the heterogeneous configuration (left). On the right a comparison of the nuclear heat load to a shield block for the homogeneous (**a**) and heterogeneous (**b**) configuration

1.6 Neutronic Analyses of the ITER TBM Port Plug with Dummy TBMs

To achieve the validation and testing of tritium breeding blanket concepts, mock-ups of breeding blankets, called Test-Blanket-Modules (TBMs), are tested in three equatorial ports of the ITER tokamak. Each TBM and its associated shield form a TBM-Set that is mechanically attached to a TBM Frame. A TBM Frame and two TBM-Sets form a TBM port plug (TBM-PP). Actually different TBM versions will be tested sequentially, each one tailored to the specific plasma operation scenario and test objective. In case a TBM-Set is not available, it can be replaced by a Dummy-TBM at any time. TBM Frame and Dummy-TBM are made of stainless steel and also need to meet ITER requirements on shielding, heat removal and safety. The maintenance operations for the TBM-PP include various activities in the Port Cell and in the Hot Cell. In order to assess the requirements, a detailed MCNP model of the TBM-PP, the Pipe-Forest (PF) inside the Port-Interspace (PI) and the Bioshield-Plug (BP) was developed and integrated into the ITER neutronics C-Model.

The performed analysis concerns the neutronic behaviour of the TBM-PP with two Dummy-TBMs by means of activation and shut-down dose rate (SDDR) calculations. The analysis includes the evaluation of the radiation fields in the rooms around the TBM-PP inside the ITER tokamak at several cooling times. Two configurations scenarios have been considered, one with a PF structure inside the PI and one without this structure. The results, obtained with R2Smesh calculations, show that the contribution to the SDDR from the TBM-PP at 12 days after reactor shut down is small compared to the contribution originating from the PF structure. Further reduction of neutron fluxes and SDDR by additional shielding structures is required.

A part of the work has been presented at the Symposium on Fusion Technology (SOFT 2018) and will be published in Fusion Engineering and Design (Fig. 8).

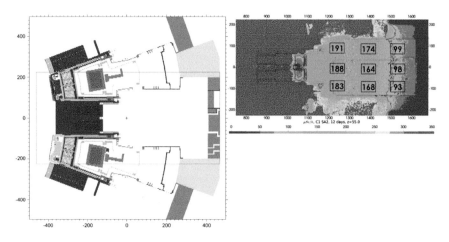

Fig. 8 The TBM PP with Dummy TBMs integrated into the ITER C-Model (left). Shutdown dose rate field, μSv h^{-1}, after 14 years of nuclear operation at 12 days of cooling (right)

1.7 Activation Analysis for the European HCPB Blanket Module in DEMO

The activity and decay heat inventories were assessed for the HCPB (Helium Cooled Pebble Bed) DEMO blankets making use of a code system that enables performing 3D activation calculations by linking the Monte Carlo transport code MCNP and the fusion inventory code FISPACT through an appropriate interface. The dedicated full scale geometry MCNP model of a 10° HCPB DEMO torus was adapted to the requirements for the coupled 3D neutron transport and activation calculations. The sensitivity analyses were performed to assess the effect of the homogenized geometry cells to the final results. The importance of the use of the impurities in the material definitions was pointed out and accounted for in the simulations. The activity of the HCPB blanket modules is dominated by the tungsten protecting layer up to one month after shutdown. After that it is dominated by the breeding zone mixture of Eurofer, breeder ceramic and beryllium pebbles with impurities. The decay heat generation in blankets is especially high in the casing that requires intensive cooling efforts for their treatment after the shutdown. The accumulation of the poisoning ^{239}Pu could make complicated the recycling of the beryllium pebbles extracted from the blankets.

The analyses of a neutron-induced activation of HCPB DEMO materials can provide detailed information required for an assessment of the safety issues related to the DEMO fusion technology. The objective of the work was to provide detailed activation and decay heat data required for quality assured safety analyses of the HCPB DEMO.

The activation and decay heat simulations in the HCPB DEMO are performed in two steps:

- A particle transport simulation to assess the neutron spectra in the different locations.
- Inventory calculations based on the neutron spectra mapped over the geometry and irradiation scenario to quantify the generation of the radioactive nuclides in different materials.

The step 1 in this procedure requires neutron spectra in all geometry cells where then the activation analyses will be performed. To this end the set of the geometry cells of interest must be specified in the MCNP input file to tally the neutron spectra. Special attention should be paid to the accuracy and convergence of the nuclear responses. This can be achieved by applying suitable variance reduction techniques.

To facilitate the activation analyses an interface code developed by KIT was applied for numerous activation calculations. This computer interface is based on MCNP5 code for particle transport simulations and FISPACT code for activation inventory analysis of the irradiated materials. FISPACT code makes use of EAF library and in the present version it is EAF-2007. The inventory results of the FISPACT outputs are collected by a versatile post-processing code.

The neutron transport calculations were carried out making use of the 3D geometry model of the HCPB DEMO and the MCNP5-1.60 code with nuclear data from the

Fig. 9 Specific activity of the different constituent materials for the detailed and the homogeneous breeder zone

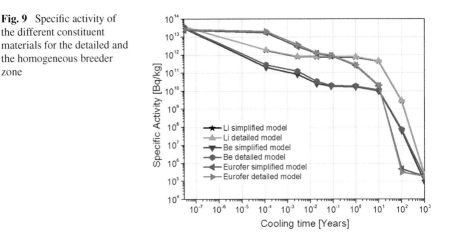

FENDL-2.1 library. The toroidal fusion neutron plasma source was simulated making use of the specially developed source subroutine linked to the MCNP executable. The results for neutron spectra were obtained with a standard VITAMIN-J 175-group structure. The statistical uncertainty in all calculations of the total neutron flux does not exceed 1%. For some low energy bins, for instance, $10^{-7} \ldots 10^{-4}$ MeV, the relative error was found to be as high as ~10%.

The results of the activation simulations depend to high extent on the size of the irradiated geometry cell. Due to the attenuation of the neutron flux the averaging of the results over the cell volume could result in overestimations of the responses. To analyse this effect the activation calculations were performed in one central IB blanket of the HCPB DEMO. To this end, three options of the blanket breeder zone were investigated: original design with the sandwich-like fully heterogeneous structure, a fully homogeneous breeder zone with the material composition discussed above and the radially segmented homogeneous breeder zone.

The neutron transport and following inventory calculations were performed both with the detailed and simplified models. Shown in Fig. 9 is the activity of the constituent breeder zone (BZ) materials as they calculated for the case with the original detailed and homogeneous BZs. The slight differences were found for Be and EURO-FER (a reduced activation ferritic-martensitic steel) for the cooling times up to 30 days, they do not exceed ~15%. The activity of the Li in detailed and homogeneous model has good agreement for all cooling times.

The activity and decay heat inventories were assessed for the HCPB DEMO blankets making use of a code system that enables performing 3D activation calculations by linking the Monte Carlo transport code MCNP and the fusion inventory code FISPACT through an appropriate interface. The dedicated full scale geometry MCNP model of a 10° HCPB DEMO torus was adapted to the requirements for the coupled 3D neutron transport and activation calculations. The analyses were done to assess the sensitivity of the results to the replacement of the real geometry with representative homogeneous material mixture. The corresponding effect was found to be not significant. The additional radial segmentation of the geometry cells does not result in the noticeable effect. Therefore the homogenized materials can be used in the DEMO activation calculations providing an acceptable accuracy of the final results. Special attention was paid to the use in the activation calculations of the commercial materials containing technological impurities. This has a severe effect on the results and the impurities must be accounted for in the calculations.

The activity and decay heat were assessed for the HCPB DEMO breeder blankets for several decay cooling times from 1 second up to 1000 years. The method used enabled the calculations of the activity and decay heat for separate constituted parts of the HCPB blankets.

This work was published in Fusion Engineering and Design [9].

1.8 Nuclear Analysis for HELIAS

A first neutronics analysis of the Helical-Axis Advanced Stellarator (HELIAS) power reactor is conducted in this work. It is based on Monte Carlo (MC) particle transport simulations with the Direct Accelerated Geometry Monte Carlo (DAGMC [10]) approach which enables particle tracking directly on the CAD geometry. A suitable geometry model of the HELIAS reactor is developed, including a rough model of a breeder blanket based on the Helium Cooled Pebble Bed (HCPB) breeder blanket concept. The CAD representation including the Last Closed Flux Surface (LCFS) is shown in Fig. 10.

The resulting model allows to perform first neutronic calculations providing a 2D map of the neutron wall loading (shown in Fig. 11, left), a 3D distribution of the neutron flux (shown in Fig. 11, right), and a rough assessment of the tritium breeding capability. It is concluded that the applied methodology, making use of MC particle transport simulations based on the DAGMC approach, is suitable for performing nuclear analyses for the HELIAS power reactor.

First calculations with application of the cell based variance reduction method were performed for a simplified stellarator model. These calculations should

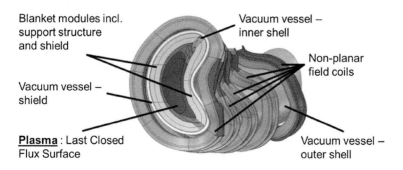

Fig. 10 HELIAS-5B CAD model including material layers and last closed flux surface

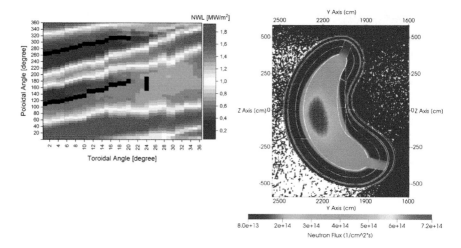

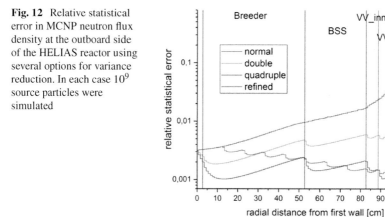

Fig. 11 Poloidal-Toroidal distribution of the Neutron Wall Loading (NWL) calculated with DAGMC (left). Neutron flux distribution with associated geometry in the beanshape side of the HELIAS reactor. Areas without any statistics are displayed in white color (right)

Fig. 12 Relative statistical error in MCNP neutron flux density at the outboard side of the HELIAS reactor using several options for variance reduction. In each case 10^9 source particles were simulated

demonstrate if this method reduces the statistical error in areas far away from the plasma chamber and if it can be applied for the complex stellarator model. Results in Fig. 12 show, that there is a significant impact by applying this method, and these results are promising to achieve the same impact for calculations with the complex HELIAS model.

The work was presented at the International Symposium on Fusion Nuclear Technology and at the Annual Meeting on Nuclear Technology and published in Fusion Engineering and Design [11].

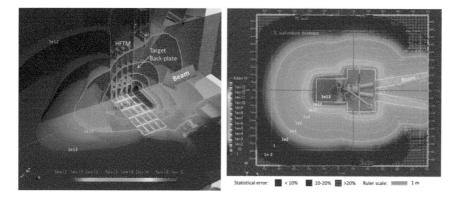

Fig. 13 Neutron flux (n/cm^2/s) distribution calculated at the centre of test cell (left) and the whole test cell (right)

1.9 Nuclear Analysis for DONES Test System

One highlight result obtained using the ForHLR II supercomputer is the modelling and nuclear analysis of DONES test system. The modelling of the DONES test system has been updated to integrate the latest target assembly, high flux test module (HFTM), the test cell (TC) updated design and the test cell surrounding rooms. Nuclear responses have been calculated on the HFTM to evaluate the damage rate, gas production, etc., Fig. 13 shows the produced neutron flux on the target assembly and HFTM region. It is concluded that the DONES can produce sufficient irradiation volume for the requirement, and also the gas production rate is very close to the condition of the DEMO reactor.

Another calculation is the global neutron flux map of the DONES test cell, as also shown in Fig. 13. Test cell is surrounded by heavy concrete walls with the thickness up to 4 m. To simulate the neutron transport with Monte Carlo method is very challenging with such a problem, and the variance reduction technique is necessary. The so-called weight-windows for variance reduction have been calculated using ADVANTG code [12], using more than 1000 CPU-hour. The neutron flux simulation uses more than 10000 CPU-hour for a single run. Without the support of the ForHLR II supercomputer, such massive calculations cannot be achieved.

This work was presented at the 18th International Conference on Fusion Reactor Materials (ICFRM-18) [13].

2 Scaling and Performance Evaluation on ForHLR II

The Monte-Carlo simulations using MCNP and R2Smesh codes have required typically a series of production runs with various demands on cores/node and total nodes. Some applications, e.g., Sect. 1.8, required about 15 GB per core, thus limiting the

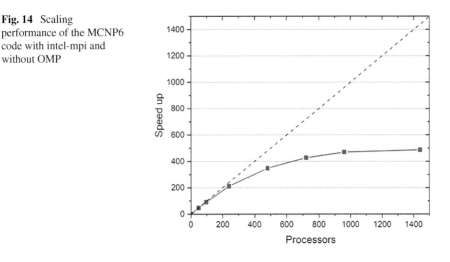

Fig. 14 Scaling performance of the MCNP6 code with intel-mpi and without OMP

node usage to 4 cores only. Meshing of the geometry requires about 10^8 facets, whereas the superimposed response meshes range from 10^5 to 10^7 regular cartesian voxels.

On average 5 nodes per run were utilized and a wall time of 1 day was specified. The minimum number of nodes used per run was 1 and the maximum number of nodes was 200. Due to very large simulation models it was often impossible to use all cores per node, in extreme cases only 4 cores/node were used during a production run over a period of several days (continued after the maximum wall time of 3 days).

The MCNP code [14], which was heavily used in most of the analysis reported in Sect. 1 is a well-established Monte Carlo transport code. Several studies on its scaling performance have been performed in the past on various super computer clusters. Figure 14 provides the scaling performance of the MCNP code up to 1400 processors compiled with intel-mpi and without OMP.

Acknowledgements This work was performed on the computational resource ForHLR II funded by the Ministry of Science, Research and the Arts Baden-Württemberg and DFG ("Deutsche Forschungsgemeinschaft"). Parts of the work has been carried out within the framework of the EUROfusion Consortium and has received funding from the Euratom research and training programme 2014–2018 under grant agreement No. 633053. Parts of the work were supported by Fusion for Energy (F4E), Barcelona, through the Specific Grant Agreements F4E-GRT-615 and F4E-FPA-395.02. The views and opinions expressed herein do not necessarily reflect those of the European Commission. Parts of the work has been funded by the ITER Organization under contract IO/17/CT/4300001445. Parts of the work was carried out using an adaption of A-lite and C-Model which were developed as a collaborative effort between AMECFW (International), UKAEA (UK), ENEA Frascati (Italy), F4E (Spain), FDS Team of INEST (PRC), IDOM (Spain), ITER Organization (France), QST (Japan), KIT (Germany), UNED (Spain), University of Wisconsin-Madison (USA).

References

1. P. Pereslavtsev, U. Fischer, D. Leichtle, R. Villari, Novel approach for efficient mesh based Monte Carlo shutdown dose rate calculations. Fusion Eng. Des. **88**(9–10), 2719 (2013)
2. L. Peng, Pereslavtsev, Pavel, Fischer, Ulrich, Wegmann, Christian: Sensitivity of R2Smesh shutdown dose rates results on the mesh resolution. Fusion Eng. Design **126**, 15–23 (2018)
3. S. Agostinelli et al., GEANT4—a simulation toolkit. Nucl. Instrum. Methods Phys. Res. A **835**, 186–225 (2016)
4. M.B. Chadwick, et al., ENDF/B-VII.0: next generation evaluated nuclear data library for nuclear science and technology. Nucl. Data Sheets **107**(12), 2931–3060 (2006)
5. R. Forrest, et al., *The JEFF-3.1 Nuclear Data Library*, ed. by A. Koning (OECD, 2006)
6. S.P. Simakov, et al., Validation of evaluated data libraries against an iron shell transmission experiment and against the Fe(n, xn) reaction cross section with 14 MeV neutron source. Report EFF-DOC-747, NEA Data Bank, Dec. 2000, Paris
7. P. Raj, M. Angelone, U. Fischer, A. Klix, Computational study of a chromium self-powered detector for neutron flux monitoring in the test blanket module of ITER. Nucl. Instrum. Methods Phys. Res. A **908**, 10–17 (2018)
8. B. Weinhorst et al., Investigation of material homogenization on the nuclear heating estimates of the electron cyclotron-heating upper launcher blanket shield module. Fus. Eng. Des. (Special Issue ISFNT, 2017). https://doi.org/10.1016/j.fusengdes.2018.03.002
9. L. Peng, Pereslavtsev, Pavel, Hernandez, Francisco, Fischer, Ulrich: Activation analyses for the HCPB blanket module in the European DEMO. Fusion Eng. Design **125**, 18–23 (2017)
10. T.J. Tautges, P.P.H. Wilson et al., *Acceleration Techniques for Direct Use of CAD-Based Geometries in Monte Carlo Radiation Transport, in International Conference on Mathematics, Computational Methods & Reactor Physics (M&C 2009)* (American Nuclear Society, Saratoga Springs, NY, May, 2009)
11. A. Haeussler, F. Warmer, U. Fischer, Neutronics analyses for a stellarator power reactor based on the HELIAS concept, Fusion Eng. Design (Special Issue ISFNT, 2017). https://doi.org/10.1016/j.fusengdes.2018.02.026
12. OAK Ridge, ADVANTG. An Automated Variance Reduction Generator, ORNL/TM-2013/416 R1 (Aug 2015)
13. Y. Qiu, F. Arbeiter, U. Fischer, F. Schwab, IFMIF-DONES HFTM neutronics modelling and nuclear response analyses, in *Proceeding of 18th International Conference on Fusion Reactor Materials (ICFRM-18)* (Aomori, Japan, Nov 5–10, 2017)
14. J.T. Goorley, et al., Initial MCNP6 Release Overview MCNP6 Version 1.0, LANL. Report LA-UR-13-22934 (2013)

Molecular Modeling and Simulation: Force Field Development, Evaporation Processes and Thermophysical Properties of Mixtures

Tatjana Janzen, Robin Fingerhut, Matthias Heinen, Andreas Köster, Y. Mauricio Muñoz-Muñoz and Jadran Vrabec

Abstract To gain physical insight into the behavior of fluids on a microscopic level as well as to broaden the data base for thermophysical properties especially for mixtures, molecular modeling and simulation is utilized in this work. Various methods and applications are discussed, including a procedure for the development of new force field models. The evaporation of liquid nitrogen into a supercritical hydrogen atmosphere is presented as an example for large scale molecular dynamics simulation. System-size dependence and scaling behavior are discussed in the context of Kirkwood-Buff integration. Further, results for thermophysical mixture properties are presented, i.e. the Henry's law constant of aqueous systems and diffusion coefficients of a ternary mixture.

1 Introduction

Molecular modeling and simulation provides a broad range of possibilities for the investigation of various thermophysical properties. On the one hand, it is a powerful tool for the prediction of thermodynamic data of pure fluids and mixtures, which are needed for process design in chemical and energy technology engineering. Especially when dealing with toxic substances or extreme conditions, challenging and expensive experiments can be avoided with this approach. On the other hand, molecular simulation not only contributes to the availability of thermodynamic data, but also allows for a precise insight into the fluid behavior on the microscopic level that

T. Janzen · R. Fingerhut · M. Heinen · J. Vrabec (✉)
Thermodynamics and Process Engineering, Technical University Berlin,
Ernst-Reuter-Platz 1, 10587 Berlin, Germany
e-mail: vrabec@tu-berlin.de

A. Köster · Y. M. Muñoz-Muñoz
Thermodynamics and Energy Technology, University of Paderborn,
Warburger Str. 100, 33098 Paderborn, Germany

© Springer Nature Switzerland AG 2019
W. E. Nagel et al. (eds.), *High Performance Computing in Science and Engineering '18*, https://doi.org/10.1007/978-3-030-13325-2_29

governs macroscopic behavior. This promotes the understanding of different phenomena and yields a connection between microscopic structure and thermodynamic properties.

To exploit the possibilities of molecular modeling and simulation, suitable computational methods combined with high-performance computing are essential. Within this project, two simulation packages are constantly improved and extended: *ms2* [1–3], for equilibrium simulations of bulk fluids, and *ls1 mardyn* [4], for the simulation of non-equilibrium systems and flow phenomena on the nanoscale. Several exemplary application cases of these two software packages are presented in this work, including model development and molecular simulations of varying system sizes and different thermodynamic properties of mixtures.

2 Force Fields for Cyclohexylamine, Aniline and Phenol

The reliable representation of thermodynamic data by molecular simulation is determined by the quality of the underlying force field, i.e. the mathematical representation of molecular configurational energy U. Numerous studies have been carried out on this matter, e.g. using sophisticated numerical methods, such as machine learning [5, 6] or by combining ab initio and classical molecular simulations [7].

The computational effort of classical molecular simulations for systems composed of small molecules may be reduced by neglecting internal molecular degrees of freedom and assuming pairwise additive interactions, i.e. taking the configurational energy as $U = \sum_{i=1}^{N-1} \sum_{j>i}^{N} U_{ij} = U_{elec} + U_{disp}$, wherein N is the number of particles of the system and U_{ij} represents the interaction energy between particles i and j. It depends on the relative position of the molecular interaction sites belonging to molecules i and j and their electrostatic interaction parameters.

According to this approach, a force field for classical molecular simulations possesses three types of parameters. The electrostatic and geometric parameters as well as the dispersion energy parameters. Molecular geometry, i.e. the location of each atom in the molecular framework, and the electrostatic parameters can be inherited from ab initio calculations, e.g. using the Møller-Plesset 2 method. The use of ab initio simulations is the backbone for developing force fields and the initial starting point of our developments as can be seen in Fig. 1.

On the other hand, the intermolecular dispersion energy U_{disp} can be described by the Lennard-Jones (LJ) potential, i.e. $U_{disp} = U_{disp}(\varepsilon, \sigma)$, where ε and σ are fitting parameters, often located at the position of each atom. Computational effort can be reduced even more by using the united-atom approach, which consists of grouping several atoms and assigning them a common center of force, represented by a single set of LJ parameters ε and σ. Indeed, the selection of the parameters to be fitted is a matter of technical art. Once the LJ parameters to be adjusted are selected, the optimization can be carried out following the workflow presented in Fig. 1. Monte-Carlo (MC) or molecular dynamics (MD) simulations are carried out for calculating the density of the pure fluid. Subsequently, vapor-liquid equilibrium

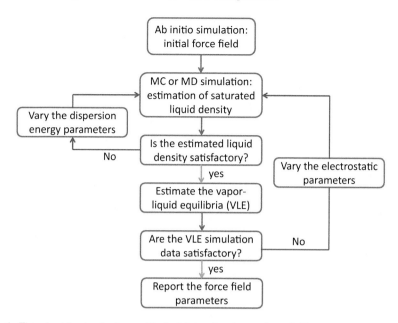

Fig. 1 Flowchart for developing multi-site LJ + point charges force fields

(VLE) simulations are carried out with the Grand Equilibrium method, which requires the chemical potential values.

The workflow given in Fig. 1 was employed for developing three new multi LJ site + point charge force fields for cyclohexylamine ($C_6H_{13}N$), aniline (C_6H_7N) and phenol (C_6H_6O). Cyclohexylamine, being technically used e.g. to delay the degradation of rubber or to prevent corrosion in steam plants, was represented by seven LJ sites and four point charges. Two of those point charges are positive and located at hydrogen atom positions, shown in white on top of Fig. 2. One negative charge is located at the center of the nitrogen atom, shown in blue on top of Fig. 2 and the last positive charge is located at the center of the united-atom methanetriyl LJ site, depicted in orange. Sites in yellow on top of Fig. 2 stand for the five identical methanediyl LJ united-atom groups of cyclohexylamine, whose parameters were selected to be fitted.

Aniline, important e.g for the pharmaceutic industry, was represented by seven LJ sites and four point charges. Similarly to cyclohexylamine, three point charges are located at the amine group. Two positive charges are at the center of the hydrogen atoms, shown in white on top of Fig. 2. One negative point charge is located at the center of the nitrogen LJ site, shown in blue on top of Fig. 2 and the last positive charge is located at the center of the carbon LJ site, depicted in black. The sites in orange represent the five methine united-atom LJ sites forming the aromatic ring of the aniline molecule. The aromatic ring of phenol, which is important to yield phenol-formaldehyde resins, is similarly represented as the aniline one. The hydroxyl group, bonded to the aromatic ring, is represented by three point charges. One positive charge is located at the center of the hydrogen atom, one negative charge is located at the

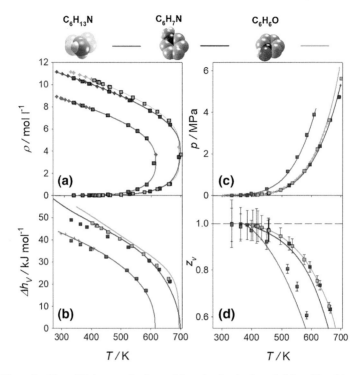

Fig. 2 Vapor-liquid equilibrium results for cyclohexylamine (red symbols), aniline (blue symbols) and phenol (green symbols). Panel **a** represents the saturated liquid and vapor density results, panels **b** and **c** stand for the enthalpy of vaporization and vapor pressure, respectively, and panel **d** shows the compressibility factor of the saturated vapor phase. Lines represent DIPPR correlations, crosses are the available experimental literature data and squares are present molecular simulation results

center of the oxygen LJ site, which is shown in red on top of Fig. 2. The LJ parameters of the methine sites in both aromatic rings were selected to be fitted to experimental VLE data.

Figure 2 gives an overview on the present molecular simulation results for these three important fluids. These data are in good agreement with experimental literature values and the DIPPR correlations. The absolute average deviations for phenol with respect to the DIPPR correlations are 3.5% for vapor pressure, 0.3% for saturated liquid density and 5.6% for enthalpy of vaporization, considering the temperature range where experimental data are available. The cyclohexylamine force field exhibits absolute average deviations of 4.7% for vapor pressure, 0.1% for saturated liquid density and 3.3% for enthalpy of vaporization. The absolute average deviations for aniline are 6.7% for vapor pressure, 0.4% for saturated liquid density and 7.2% for enthalpy of vaporization. Critical data were calculated as well and they are shown in Fig. 2. The critical temperature of phenol was slightly overestimated with respect to literature values, while critical pressure and density were underestimated. The deviations are 1.0%, −8.6% and −15.3%, respectively. Relative deviations of critical temper-

ature, pressure and density with respect to literature data for cyclohexylamine are
-1.1%, -7.1% and 27%, respectively. Analogously, for aniline these numbers are
-0.7%, -0.7% and -5.9%. Furthermore, present molecular simulation results are
consistent with a criterion recently proposed by Nezbeda [8] as can be seen in panel
(c) of Fig. 2.

3 Stationary Evaporation

The evaporation of liquid nitrogen (N_2) into a supercritical hydrogen (H_2) atmosphere
was investigated by large scale molecular dynamics (MD) simulation. This work was
motivated by the results of Dahms et al. [9], who studied that injection for two different operating pressures: 1 and 6 MPa. In case of the lower pressure, their liquid
nitrogen jet disintegrated into small droplets forming a spray, whereas in case of
the higher pressure, diffuse mixing occurred. Dahms et al. used nitrogen as a safe
substitute for oxygen in their experiments, which were aimed at the optimization of
combustion processes in rocket engines. It is well known that the interface between
liquid and vapor phase plays a key role in evaporation processes. Fundamental investigations of evaporation into vacuum based on model liquids show that the magnitude of the evaporation flux is strongly dependent on the interface temperature. With
decreasing interface temperature, the evaporation flux drops down significantly [10].
The interface properties, especially the temperature, are thus decisive for evaporation dynamics. Consequentially, it was focused in the present work on the interfacial
region between liquid nitrogen and the high pressure hydrogen atmosphere. Because
of its sound physical basis and atomistic resolution, MD simulation is an appropriate
approach to obtain a better understanding of the findings by Dahms et al. [9].

Molecular systems including an overall number of $\approx 6 \times 10^6$ molecules were set
up with a planar interface, having an extent of $A_{xy} = 85 \times 85$ nm^2, between a liquid
nitrogen film and a supercritical hydrogen atmosphere. A schematic view of how
the system was constructed is depicted in Fig. 3. A new module of the MD code *ls1
mardyn* [4], namely the *ReplicaGenerator*, was used to create comparatively large
homogeneous phases of pure nitrogen and pure hydrogen out of small ones. First, the
small systems were equilibrated to obtain physical molecular configurations. Subsequently, they were replicated and merged together, yielding larger phases. These were
equilibrated again for a short time until the replication pattern vanished. This procedure reduces the equilibration time which is particularly important for large phases.

The system was delimited on both left and right sides (z direction) by reflective boundaries, spanning a system width of $L_z = 190$ nm. These boundaries were
extended by additional molecules with fixed spatial positions to reduce their influence. To mimic an infinitely extended interface, periodic boundary conditions were
specified in x and y directions. The temperature of the liquid nitrogen $T_{N_2} = 118$ K
was maintained within a control volume on the left side, which had a width of about
ten molecule diameters only. The temperature of hydrogen was maintained within
a control volume on the right side to be $T_{H_2} = 270$ K, also leading to a constant

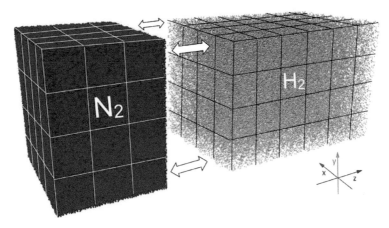

Fig. 3 Schematic view of system construction by the *ReplicaGenerator* module. The liquid nitrogen (N_2) phase and the supercritical hydrogen (H_2) phase were prepared from small fluid cubes (indicated by grid cells). These small systems were replicated to yield two larger homogeneous phases. These were equilibrated for an additional short time period and then merged together to yield the initial configuration where the two phases are in physical contact for non-equilibrium molecular dynamics simulations (NEMD)

density and composition. The latter was achieved by converting all incoming nitrogen molecules to hydrogen molecules such that a mole fraction of $y_{H_2} = 1$ mol/mol (pure hydrogen) was maintained. To establish a stationary evaporation process, a liquid reservoir added new nitrogen molecules over the left boundary of the control volume into the liquid phase. The feed rate of the reservoir was adjusted to balance the rate of evaporated nitrogen molecules entering the control volume in the gas phase. There were no interventions between the control volumes, except for the numerical solution of Newton's equations of motion of the molecular ensemble with the MD code *ls1 mardyn* [4]. The molecular models for nitrogen and hydrogen were provided by Köster et al. [11], showing a very good agreement with experimental literature data and the Peng-Robinson equation of state for VLE of pure substances and mixtures, considering a large temperature and pressure range. A series of three simulations with varying system pressure, controlled by the target density of the hydrogen gas in the right control volume, was carried out: $p = 4$, 12 and 20 MPa. During simulation, various profiles were sampled. Each profile was averaged over 10^4 time steps, with a time step length of $\Delta t = 2$ fs. At the beginning of the simulations, all profiles were step functions because the systems were set up by merging an equilibrated liquid nitrogen phase at a temperature of T_{N_2} with an equilibrated hydrogen gas phase at a temperature of T_{H_2}. Within the first nanosecond, these profiles changed rapidly, developing into a smooth transition between the liquid phase and gas phase states maintained by the control volumes. After about 3 ns, the profiles hardly changed any more such that a stationary process had been attained. Figure 4 shows the temperature, hydrogen mole fraction and density profiles of the simulation series for two pressure levels $p = 4$ and 20 MPa after an elapsed time of

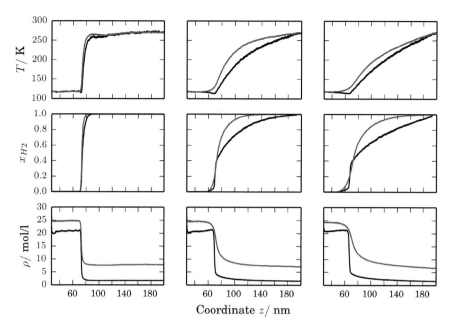

Fig. 4 Temporal evolution of the temperature T (top), mole fraction of hydrogen x_{H_2} (center) and density (bottom) profiles of the low pressure case $p = 4$ MPa (black) and the high pressure case $p = 20$ MPa (red). Columns left, center, right show profiles after 0, 1 and 3 ns

0, 1 and 3 ns. The case of $p = 12$ MPa showed qualitatively similar results as $p = 20$ MPa and the results are therefore not illustrated here. Looking at the right column of Fig. 4, several aspects can be seen that may contribute to the understanding of the macroscopic system studied by experiment [9]. The density profile of the low pressure case ($p = 4$ MPa) has a comparatively sharp interface, transitioning from the liquid density down to the gas density. The density profiles of the high pressure case ($p = 20$ MPa) exhibit a slightly higher liquid density and a significantly higher gas density due to compression. Moreover, a much broader and smoother transition between the liquid and gas density than in the low pressure case was found.

The temperature profile of the low pressure case shows a minimum within the interface region, indicating that a part of the enthalpy of evaporation is provided by the liquid phase that consequently cools down. From a molecular perspective, the liquid film cools down at the interface because molecules with higher kinetic energy preferentially leave the interface, whereas molecules with lower kinetic energy preferentially remain inside the liquid. The interface temperature decrease is only 3.5 K because heat is transported from both phases to the interface via temperature gradients. The temperature profiles of the higher pressure cases do not exhibit a temperature minimum at the interface. This can be explained by the higher pressure (and density) of the gas phase, leading to a much stronger thermal coupling with the liquid. Consequently, a sufficient amount of heat is transported from the hot gas phase to balance the amount of energy taken up by evaporating nitrogen molecules. Instead, the liquid is even heated up, indicating that more energy is transported from the gas to the liquid phase than required to drive evaporation. The mole fraction profiles show expected trends: The gas phase is rich in hydrogen and the liquid phase is rich in nitrogen. Consistent with the density profiles, the low pressure case exhibits a sharp transition, whereas the higher pressure cases have a smooth and much broader transition, indicating diffuse mixing.

4 Kirkwood-Buff Integration

Local density fluctuations within closed systems can be sampled with Kirkwood-Buff integration (KBI) [12] and lead to macroscopic thermodynamic properties. Many-body molecular simulation naturally provides detailed information on the micro-scopic structure of fluids by means of radial distribution functions (RDF) $g_{ij}(r)$, which are closely related to what is known as local composition. In case of a mixture containing components i and j, they are given by

$$g_{ij}(r) = \frac{\rho_{L,i}(r)}{\rho_j} , \tag{1}$$

where the local density of component i in a spherical shell centered around species j is

$$\rho_{L,i}(r) = \frac{\mathrm{d}N_i}{4\pi r^2 \mathrm{d}r} , \tag{2}$$

and $\mathrm{d}N_i$ denotes the number of particles of component i therein. The overall partial density is simply $\rho_j = x_j \rho$. A binary mixture of components 1 and 2 has thus three independent radial distribution functions, i.e. $g_{11}(r)$, $g_{22}(r)$ and $g_{12}(r) = g_{21}(r)$. Moreover, radial distribution functions provide access to the composition dependence of the excess Gibbs energy by means of KBI. A good convergence of their integrals

$$G_{ij} = 4\pi \int_0^\infty (g_{ij}(r) - 1) r^2 \mathrm{d}r , \tag{3}$$

can be achieved by canonic NVT ensemble simulations employing the formalism of Krüger et al. [13]. The Kirkwood-Buff integrals G_{ij} lead to several macroscopic properties, e.g. the mole fraction derivative of the activity coefficient γ_1 of component 1 is given by

$$\left(\frac{\partial \ln \gamma_1}{\partial x_1}\right)_{T,p} = -\frac{x_2 \rho (G_{11} + G_{22} - 2G_{12})}{1 + x_2 \rho x_1 (G_{11} + G_{22} - 2G_{12})} . \tag{4}$$

For studies on mass transport, the thermodynamic factor Γ is vital because it connects the Fick diffusion coefficients D_{ij} and Maxwell-Stefan diffusion coefficients $Ð_{ij}$ (for a binary mixture: $D_{11} = \Gamma_{11}Ð_{12}$) and it is given by

$$\Gamma_{11} = 1 + x_1 \left(\frac{\partial \ln \gamma_1}{\partial x_1}\right)_{T,p}. \tag{5}$$

In the field of process engineering, mixture data are of particularly interest. Properties like partial molar volumes, isothermal compressibility and, most interestingly, thermodynamic factor can be determined with KBI. To this end, KBI was implemented into the massively-parallel molecular simulation tool *ms2* [3]. However, finite system-size effects are challenging for KBI calculations by using molecular simulations of closed systems because KBI is defined for macroscopically large systems only. Further, for large distances r, the sampled RDF tend to lose statistical accuracy and may not converge to unity. Accordingly, RDF corrections are crucial [14] in the same way as an extrapolation to macroscopic size [15]. Moreover, it was shown by Galata et al. [16] that simulations with a larger number of particles N may lead to improved results. Thus, parallelization of KBI in *ms2* is essential because the computational effort is directly dependent on the number of particles N.

Finite system-size effects of KBI were assessed for a binary LJ mixture for two different system sizes in the closed NVT ensemble with $N = 4,000$ and $N = 8,000$ particles (LJ parameters $\sigma_2/\sigma_1 = 1.5$ and $\varepsilon_2/\varepsilon_1 = 0.75$). The LJ mixture was studied over the entire mole fraction range $x_1 = 0.05, 0.1, 0.2, \ldots,$ 0.95 mol/mol. The mixture density range was sampled with isobaric-isothermal NpT simulations at constant pressure $p\sigma_1^3/\varepsilon_1 = 0.03$ and constant temperature $Tk_B/\varepsilon_1 = 0.85$. MD NVT simulations for both system sizes were carried out with 800,000 equilibration steps and 15 Mio. production steps with a time step of 0.329 fs and a cutoff radius of $r_c = 5\sigma_1$. RDF were sampled for each time step with a maximum radius divided into 300 shells. Figure 5 shows the thermodynamic factor Γ_{11} over mole fraction x_1 for the two system sizes $N = 4,000$ and $N = 8,000$.

It becomes clear that the simulation results for the thermodynamic factor Γ_{11} are slightly improved for larger system sizes because of the decreased deviation to the Wilson model [17] fit. On that point, the thermodynamic factor Γ_{11} calculated by the Wilson model may serve for comparison since it is independent from the LJ model error.

The parallelization performance of *ms2* was analyzed for the MD NVT ensemble with KBI calculation turned on and off. Therefore, three different system sizes of the above-mentioned LJ mixture with $N = 4,000$; 8,000; 16,000 particles were carried out on CRAY XC40 (Hazel Hen) of HLRS and the runtime of *ms2* was measured by changing the number of nodes (each node contained 24 cores). Figure 6 shows the scaling of MD NVT simulations of *ms2* for those three systems with KBI calculation turned on and off.

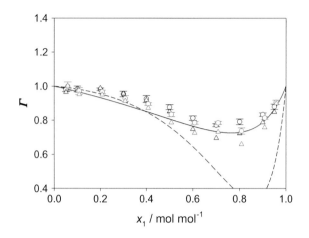

Fig. 5 Thermodynamic factor Γ_{11} over mole fraction x_1 of a binary LJ mixture calculated with MD and KBI in the NVT ensemble employing $ms2$; black symbols: $N = 4,000$; red symbols (slightly shifted to larger x_1 for visibility reasons): $N = 8,000$; circles/triangles: non-extrapolated/extrapolated data to macroscopic system size; solid line: Wilson model [17] fitted to chemical potential data calculated for the LJ mixture with Widom's test particle insertion [18]; dashed line: PC-SAFT equation of state [19]

Fig. 6 Strong scaling efficiency of $ms2$ for a binary LJ mixture in the MD NVT ensemble measured on CRAY XC40 (Hazel Hen); solid line: MD NVT (KBI turned off); dashed line: MD NVT (KBI turned on); blue lines: $N = 4,000$; black lines: $N = 8,000$; red lines: $N = 16,000$ particles

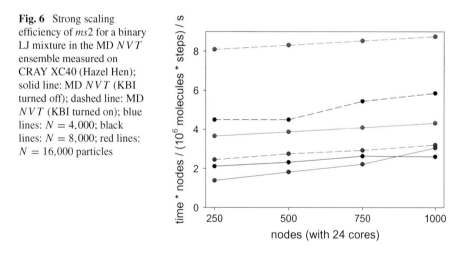

For all three systems, the cutoff radius was set to $r_c = 5\sigma_1$ and thus the NVT interaction calculation alone required the same computing effort, which only depends on the number of particles N in the cutoff sphere. However, to determine which particles are inside the cutoff sphere, the entire interaction matrix has to be traversed which allows for a computationally efficient access to the calculation of RDF shells if KBI is turned on. The computing intensity of traversing the particle matrix is proportional to N^2. As a result, the computational cost of $ms2$ should be in between N to N^2, e.g. doubling N should lead to a factor 2 to 4 in terms of computational cost. Figure 6 indicates that doubling the number of particles in *ms*2 leads to an increase of computational cost smaller than factor 4 if the number of nodes is chosen appropriately so that the overhead is small. Further, the vertical axis gives the computing power (nodes) times computing time per computing intensity (problem size) and thus, a horizontal line would show a perfect strong scaling efficiency of 100%. Consequently, Fig. 6 points out that *ms*2 is close to optimal strong scaling, but it of course does not reach 100% efficiency. Moreover, it becomes clear that for small systems with computationally cheap interaction calculations like LJ mixtures, the strong scaling efficiency decreases with increased computing power (nodes) because of communication overhead.

5 Henry's Law Constant

In an extensive study that was recently published in full [11], the mixture behavior of five substances related to the hydrogen economy, i.e. H_2, nitrogen (N_2), oxygen (O_2), argon (Ar) and water (H_2O), was examined using molecular modeling and simulation. Several thermodynamic properties were considered in this work to describe a substantial number of mixtures, including binary, ternary and quaternary systems. Apart from VLE properties, like vapor pressure, saturated densities or enthalpy of vaporization, also homogeneous density and solubility were studied. The quality of the results was assessed by comparing to available experimental literature data or sophisticated equations of state, showing an excellent agreement in many cases. Consequently, this contribution aims at an improved availability of thermodynamic data that are required for the post-carbon age.

The present work gives a short overview on the Henry's law constant that can be described particularly well with molecular simulation techniques and was used in the work of Köster and Vrabec [11] to assess aqueous systems, i.e. H_2, N_2, O_2 or Ar in liquid H_2O. Consequently, this property is used in cases where a solute is only little soluble in a solvent. It has to be noted that the purely temperature-dependent definition of the Henry's law constant was considered, wherein the solvent has to be in its saturated liquid state. Therefore, data are presented ranging from the triple point temperature to the critical temperature of H_2O.

The Henry's law constant H_i was sampled with the molecular simulation tool *ms*2 [1–3] in the NpT ensemble using 1,000 particles in total. The simulation runs typically consisted of 3×10^6 MC cycles, i.e. 2×10^5 equilibration and 2.8×10^6 production cycles. H_i is closely related to the chemical potential at infinite dilution μ_i^∞ [20]

$$H_i = \rho_s k_B T \exp(\mu_i^\infty/(k_B T)), \tag{6}$$

wherein ρ_s is the saturated liquid density of the solvent, T the temperature and k_B Boltzmann's constant. An attractive method to calculate the chemical potential at infinite dilution is Widom's test particle insertion [18], where the mole fraction of the solute can simply be set to zero such that only test particles are inserted into the pure solvent, i.e. H_2O.

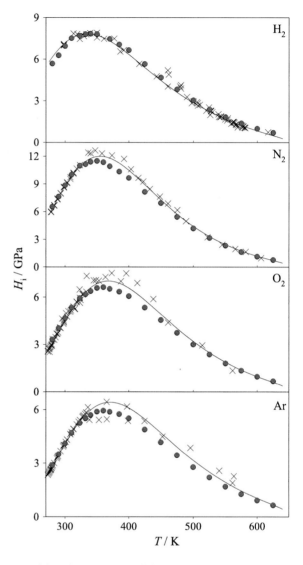

Fig. 7 Henry's law constant of H$_2$, N$_2$, O$_2$ and Ar in H$_2$O (from top to bottom): ($\bullet$) molecular simulation results, (—) official IAPWS guideline [21, 22] and (+) experimental literature data (H$_2$ [23, 24]; N$_2$ [24–27]; O$_2$ [28–31]; Ar [32–35])

All aqueous systems presented in Fig. 7 were validated against experimental data and the official IAPWS correlation [21, 22] for the Henry's law constant. They exhibit a strongly non-monotonic behavior that is qualitatively similar for the various solutes and pass through a maximum between 330 and 370 K. Due to the fact that higher values of the Henry's law constant correspond to a low solubility, the solubility decreases with increasing temperature, passes through a minimum and then increases again. On the quantitative side, however, large differences between the solutes are present. Near the triple point temperature of H$_2$O, both H$_2$ and N$_2$ are roughly three times less soluble than O$_2$ and Ar ($H_{H2} \approx H_{N2} \approx 6$ GPa compared to $H_{O2} \approx H_{Ar} \approx 2$ GPa). Moreover, the Henry's law constant values at the maximum differ significantly.

It can be seen that Henry's law constant data were predicted satisfactorily by molecular simulation, especially at temperatures close to the triple and critical point of H_2O. At intermediate temperatures, molecular simulation results for N_2, O_2 and Ar deviate slightly, whereas H_2 is reproduced almost perfectly. It has to be noted, however, that the force field of H_2 was modelled without electrostatic interactions. This was compensated by increasing the unlike LJ interaction energy, resulting in a rather unphysically large value of the binary interaction parameter $\xi = 1.52$.

6 Diffusion Coefficients of Ternary Liquid Mixtures

Numerous natural and technical processes are determined by transport diffusion of multicomponent liquids, while the underlying microscopic phenomena are still not well understood. Experimental data on multicomponent diffusion coefficients are rather scarce because strong composition dependence and cross diffusion effects aggravate experimental work. Molecular modeling and simulation is a promising route for investigating diffusion coefficients on a microscopic basis to understand the underlying phenomena and to study the relations between intradiffusion and transport diffusion coefficients. Nonetheless, not only the diffusion experiments, but also the simulations are challenging and time consuming. To gain an overall picture of the diffusion behavior of a ternary mixture, a sufficient amount of state points with varying composition must be studied, also considering the binary subsystems. Thus, sufficient parallelization of the utilized software as well as efficient data handling are essential to overcome this task.

Here, equilibrium MD simulations and the Green-Kubo formalism were used to sample intradiffusion and Maxwell-Stefan (MS) diffusion coefficients of binary and ternary liquid mixtures. Intradiffusion coefficients D_i, which describe the mobility of molecules of species i in the absence of a driving force, were obtained from the time integral of the single particle velocity auto-correlation function averaged over all N_i particles belonging to that species [36]

$$D_i = \frac{1}{3N_i} \int_0^\infty \left\langle \sum_{k=1}^{N_i} \mathbf{v}_i^k(0) \cdot \mathbf{v}_i^k(t) \right\rangle \mathrm{d}t . \tag{7}$$

MS diffusion coefficients, which describe mass transport driven by a chemical potential gradient, are directly related to kinetic coefficients which can be sampled from net velocity correlation functions of species i and j [37]

$$\Lambda_{ij} = \frac{1}{3N} \int_0^\infty \left\langle \sum_{k=1}^{N_i} \mathbf{v}_i^k(0) \cdot \sum_{l=1}^{N_j} \mathbf{v}_j^l(t) \right\rangle \mathrm{d}t . \tag{8}$$

Fick diffusion coefficients are related to gradients of composition variables and can thus be measured experimentally. For the transformation of MS to Fick diffusion coefficients, the thermodynamic factor is needed [38], which can be calculated with excess Gibbs energy (g^E) models that are typically fitted to experimental VLE data. Details on the relations between the different coefficients for ternary mixtures can be found in recent publications of our group [39, 40].

One of the ternary mixtures studied within this project was acetone + benzene + methanol. This mixture was investigated at ambient conditions of temperature and pressure, for which experimental Fick diffusion coefficient data are available in the literature [41]. Diffusion and other transport properties of the binary subsystems were already predicted by MD simulations in the present MMHBF project [42]. The utilized molecular force field models are rigid, united-atom LJ type models with superimposed point charges, dipoles and quadrupoles. These models were parametrized using experimental data of the pure components only, thus the mixture behavior was obtained in a strictly predictive manner. MD simulations were carried out with 4,000 molecules in a cubic volume over 4 to 5×10^7 time steps of ~ 1.02 fs. Further details on the simulation setup used to sample ternary diffusion coefficients can be found in the supplementary material of Ref. [40].

Figure 8 shows the intradiffusion coefficients D_i of the three components over the entire composition range of the ternary mixture acetone + benzene + methanol.

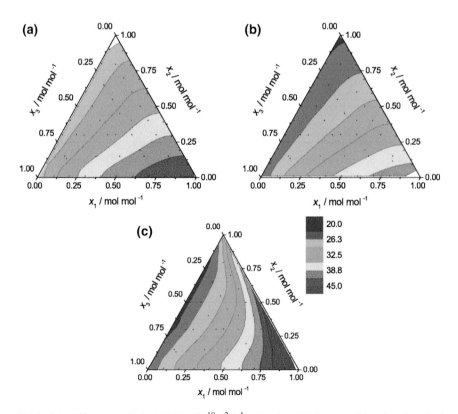

Fig. 8 Intradiffusion coefficient D_i (in $10^{-10}\,\mathrm{m^2s^{-1}}$) of acetone (**a**), benzene (**b**) and methanol (**c**) in their ternary mixture at $T = 298.15$ K and $p = 0.1$ MPa

These figures are based on simulations for 52 ternary and 40 binary state points at different compositions indicated by dots within the graphs. The coefficients of all three components have a similar composition dependence, with lowest values at small acetone content and highest values at high acetone content. Because methanol molecules form hydrogen bonds and often propagate as assemblies, the intradiffusion coefficient is similar to that of benzene. Only at low methanol concentrations, this intradiffusion coefficient is significantly higher due to the breaking of hydrogen bonds.

Figure 9 shows the three MS diffusion coefficients D_{ij} of the ternary mixture acetone + benzene + methanol. While the values of the MS coefficient between acetone and benzene D_{12} are similar to the intradiffusion coefficients, the other two MS coefficients D_{13} and D_{23} show a significantly non-ideal behavior. This is also a result of the hydrogen bonding behavior of methanol molecules, which leads to microscopic heterogeneity in mixtures with other fluids and affects their kinetics through correlated propagation. Fick diffusion coefficients predicted from the simulated MS diffusion coefficients combined with the thermodynamic factor calculated with the Wilson g^E model are in excellent agreement with experimental data [40].

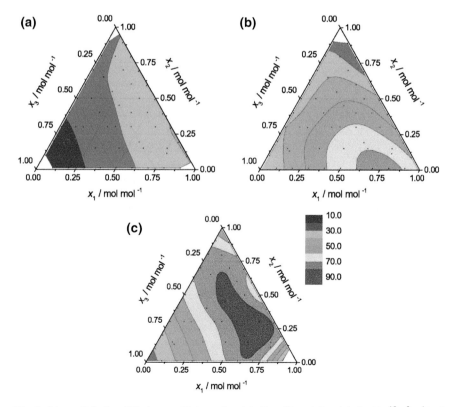

Fig. 9 Maxwell-Stefan diffusion coefficients D_{12} (**a**), D_{13} (**b**) and D_{23} (**c**) (in $10^{-10}\,\mathrm{m^2 s^{-1}}$) of acetone (1) + benzene (2) + methanol (3) at $T = 298.15$ K and $p = 0.1$ MPa

7 Conclusion

Several applications of molecular modeling and simulation were outlined in this work, considering different simulation techniques, system sizes and thermophysical properties. The workflow for the development and optimization of force field models, which are the basis of every molecular simulation, was presented. Three new force field models for cyclohexylamine, aniline and phenol were parametrized and validated against experimental VLE data.

As an application for large scale MD simulations, the evaporation of liquid nitrogen into a supercritical hydrogen atmosphere was presented. Here, special attention was paid to the behavior of the interface between the two phases. Temperature, density and concentration profiles were sampled to better understand the evaporation flux.

Further, different thermophysical properties of mixtures were predicted. Kirkwood-Buff integrals, which are related to the radial distribution functions and describe local density fluctuations, were used to sample the thermodynamic factor of binary LJ mixtures. Because the results show a significant system-size dependence, a good scaling behavior of the simulation software is necessary, which is shown to be given for the utilized simulation package $ms2$. The temperature dependent Henry's law constant of aqueous systems containing H_2, N_2, O_2 or Ar was calculated from chemical potential data sampled with Widom's test particle insertion. The results are in excellent agreement with experimental data. Finally, results for the intradiffusion and MS diffusion coefficients were presented for the ternary mixture acetone + benzene + methanol in the liquid state covering the entire composition range of the given mixture.

Acknowledgements We gratefully acknowledge support by Deutsche Forschungsgemeinschaft. This work was carried out under the auspices of the Boltzmann-Zuse Society (BZS) of Computational Molecular Engineering. The simulations were performed on the CRAY XC40 (Hazel Hen) at the High Performance Computing Center Stuttgart (HLRS) within the project MMHBF2.

References

1. S. Deublein, B. Eckl, J. Stoll, S.V. Lishchuk, G. Guevara-Carrion, C.W. Glass, T. Merker, M. Bernreuther, H. Hasse, J. Vrabec, ms2: a molecular simulation tool for thermodynamic properties. Comput. Phys. Commun. **182**, 2350–2367 (2011)
2. C.W. Glass, S. Reiser, G. Rutkai, S. Deublein, A. Köster, G. Guevara-Carrion, A. Wafai, M. Horsch, M. Bernreuther, T. Windmann, H. Hasse, J. Vrabec, ms2: a molecular simulation tool for thermodynamic properties, new version release. Comput. Phys. Commun. **185**, 3302–3306 (2014)
3. G. Rutkai, A. Köster, G. Guevara-Carrion, T. Janzen, M. Schappals, C.W. Glass, M. Bernreuther, A. Wafai, S. Stephan, M. Kohns, S. Reiser, S. Deublein, M. Horsch, H. Hasse, J. Vrabec, ms2: a molecular simulation tool for thermodynamic properties, release 3.0. Comput. Phys. Commun. **221**, 343–351 (2017)

4. C. Niethammer, S. Becker, M. Bernreuther, M. Buchholz, W. Eckhardt, A. Heinecke, S. Werth, H.-J. Bungartz, C.W. Glass, H. Hasse, J. Vrabec, M. Horsch, ls1 mardyn: the massively parallel molecular dynamics code for large systems. J. Chem. Theory Comput. **10**, 4455–4464 (2014)
5. S. Chmiela, A. Tkatchenko, H.E. Sauceda, I. Poltavsky, K.T. Schütt, K.-R. Müller, Machine learning of accurate energy-conserving molecular force fields. Sci. Adv. **3**, 1–6 (2017)
6. Y. Li, H. Li, F.C. Pickard, B. Narayanan, F.G. Sen, M.K.Y. Chan, S.K.R.S. Sankaranarayanan, B.R. Brooks, B. Roux, Machine learning force field parameters from ab initio data. J. Chem. Theory Comput. **13**, 4492–4503 (2017)
7. Y.-C. Ho, Y.-S. Wang, S.D. Chao, Molecular dynamics simulations of fluid cyclopropane with MP2/CBS-fitted intermolecular interaction potentials. J. Chem. Phys. **147**, 064507 (2017)
8. I. Nezbeda, Simulations of vaporliquid equilibria: routine versus thoroughness. J. Chem. Eng. Data **61**, 3964–3969 (2016)
9. R. Dahms, J.C. Oefelein, On the transition between two-phase and single-phase interface dynamics in multicomponent fluids at supercritical pressures. Phys. Fluids **25**, 092103 (2013)
10. M. Heinen, J. Vrabec, J. Fischer, J.: Communication: Evaporation: Influence of heat transport in the liquid on the interface temperature and the particle flux. Chem. Phys. **145**, 081101 (2016)
11. A. Köster, M. Thol, J. Vrabec, Molecular models for the hydrogen age: hydrogen, nitrogen, oxygen, argon, and water. J. Chem. Eng. Data **63**, 305–320 (2018)
12. J.G. Kirkwood, F.P. Buff, The statistical mechanical theory of solutions. I. J. Chem. Phys. **19**, 774–778 (1951)
13. P. Krüger, S.K. Schnell, D. Bedeaux, S. Kjelstrup, T.J.H. Vlugt, J.M. Simon, Kirkwood-Buff integrals for finite volumes. J. Phys. Chem. Lett. **4**(10911–10918), 235–238 (2013)
14. J. Milzetti, D. Nayar, N.F.A. van der Vegt, Convergence of Kirkwood-Buff integrals of ideal and nonideal aqueous solutions using molecular dynamics simulations. J. Phys. Chem. B **122**(21), 5515–5526 (2018)
15. S.K. Schnell, X. Liu, J.-M. Simon, A. Bardow, D. Bedeaux, T.J.H. Vlugt, S. Kjelstrup, Calculating thermodynamic properties from fluctuations at small scales. J. Phys. Chem. B **115**, 10911–10918 (2011)
16. A.A. Galata, S.D. Anogiannakis, D.N. Theodorou, Thermodynamic analysis of Lennard-Jones binary mixtures using Kirkwood-Buff theory. Fluid Phase Equilib. **470**, 25–37 (2018)
17. G.M. Wilson, Vapor-liquid equilibrium. XI. A new expression for the excess free energy of mixing. J. Am. Chem. Soc. **86**, 127 (1964)
18. B. Widom, Some topics in the theory of fluids. J. Chem. Phys. **39**, 2808–2812 (1963)
19. J. Gross, G. Sadowski, Perturbed-chain SAFT: an equation of state based on a perturbation theory for chain molecules. Ind. Eng. Chem. Res. **40**, 1244–1260 (2001)
20. K.S. Shing, K.E. Gubbins, K. Lucas, Henry constants in non-ideal fluid mixtures: computer simulation and theory. Mol. Phys. **65**, 1235–1252 (1988)
21. R. Fernández-Prini, J.L. Alvarez, A.H. Harvey, Henry's constants and vapor-liquid distribution constants for gaseous solutes in H_2O and D_2O at high temperatures. J. Phys. Chem. Ref. Data **32**, 903–916 (2003)
22. International Association for the Properties of Water and Steam. Guideline on the Henry's Constant and Vapor-Liquid Distribution Constant for Gases in H_2O and D_2O at High Temperatures (2004)
23. D.R. Morris, L. Yang, F. Giraudeau, X. Sun, F.R. Steward, Henry's law constant for hydrogen in natural water and deuterium in heavy water. Phys. Chem. Chem. Phys. **3**, 1043–1046 (2001)
24. J. Alvarez, R. Fernández-Prini, A semiempirical procedure to describe the thermodynamics of dissolution of non-polar gases in water. Fluid Phase Equilib. **66**, 309–326 (1991)
25. T.R. Rettich, R. Battino, E. Wilhelm, Solubility of gases in liquids. XVI. Henry's law coefficients for nitrogen in water at 5 to 50°C. J. Solut. Chem. **13**, 335–348 (1984)
26. B.A. Cosgrove, J. Walkley, Solubilities of gases in H_2O and ^2H_2O. J. Chromatogr. A **216**, 161–167 (1981)
27. J. Alvarez, R. Crovetto, R. Fernández-Prini, The dissolution of N_2 and of H_2 in water from room temperature to 640 K. Ber. Bunsenges. Phys. Chem. **92**, 935–940 (1988)

28. T.R. Rettich, R. Battino, E. Wilhelm, Solubility of gases in liquids. 22. High-precision determination of Henry's law constants of oxygen in liquid water from T = 274 K to T = 328 K. J. Chem. Thermodyn. **32**, 1145–1156 (2000)

29. B.B. Benson, D. Krause, M.A. Peterson, The solubility and isotopic fractionation of gases in dilute aqueous solution. I. Oxyg. J. Solut. Chem. **8**, 655–690 (1979)

30. T.R. Rettich, Y.P. Handa, R. Battino, E. Wilhelm, Solubility of gases in liquids. 13. High-precision determination of Henry's constants for methane and ethane in liquid water at 275 to 328 K. J. Phys. Chem. **85**, 3230–3237 (1981)

31. S.D. Cramer, The solubility of oxygen in brines from 0 to 300°C. Ind. Eng. Chem. Process Des. Dev. **19**, 300–305 (1980)

32. T.R. Rettich, R. Battino, E. Wilhelm, Solubility of gases in liquids. 18. High-precision determination of Henry fugacities for argon in liquid water at 2 to 40°C. J. Solut. Chem. **21**, 987–1004 (1992)

33. D. Krause, B.B. Benson, The solubility and isotopic fractionation of gases in dilute aqueous solution. IIa. Solubilities of the noble gases. J. Solut. Chem. **18**, 823–873 (1989)

34. R.W. Potter, M.A. Clynne, The solubility of the noble gases He, Ne, Ar, Kr, and Xe in water up to the critical point. J. Solut. Chem. **7**, 837–844 (1978)

35. R. Crovetto, R. Fernández-Prini, M.A. Japas, Solubilities of inert gases and methane in H$_2$O and in D$_2$O in the temperature range of 300 to 600 K. J. Chem. Phys. **76**, 1077–1086 (1982)

36. M.P. Allen, D.J. Tildesley, *Computer Simulation of Liquids* (Clarendon Press, Oxford, 1987)

37. R. Krishna, J.M. van Baten, The darken relation for multicomponent diffusion in liquid mixtures of linear alkanes: an investigation using molecular dynamics (MD) simulations. Ind. Eng. Chem. Res. **44**, 6939–6947 (2005)

38. R. Taylor, R. Krishna, *Multicomponent Mass Transfer* (Wiley, New York, 1993)

39. G. Guevara-Carrion, Y. Gaponenko, T. Janzen, J. Vrabec, V. Shevtsova, Diffusion in multicomponent liquids: from microscopic to macroscopic scales. J. Phys. Chem. B **120**, 12193–12210 (2016)

40. T. Janzen, Y. Gaponenko, A. Mialdun, G. Guevara-Carrion, J. Vrabec, V. Shevtsova, The effect of alcohols as the third component on diffusion in mixtures of aromatics and ketones. RSC Adv. **8**, 10017–10022 (2018)

41. A. Alimadadian, Phillip Colver, C., A new technique for the measurement of ternary molecular diffusion coefficients in liquid systems. Can. J. Chem. Eng. **54**(3), 208–213 (1976)

42. G. Guevara-Carrion, T. Janzen, Y.M. Muñoz-Muñoz, J. Vrabec, Molecular simulation study of transport properties for 20 binary liquid mixtures and new force fields for benzene, toluene and CCl4, in *High Performance Computing in Science and Engineering*, vol. 16, ed. by W. Nagel, D. Kröner, M. Resch (Springer, Cham, 2016)

Multiphase-Field Modeling and Simulation of Martensitic Phase Transformation in Heterogeneous Materials

E. Schoof, C. Herrmann, D. Schneider, J. Hötzer and B. Nestler

The martensitic phase transformation is an important mechanism which strongly changes the properties of the material. On the one hand, martensite is a widely used and purposefully produced constituent of many high-strength steels. On the other hand, martensite can also lead in some cases to an unwanted embrittlement of the material, so that this transformation has to be prevented. Consequently, a better understanding of this transformation process is of high interest, so as to control the mechanical properties of materials. Besides experiments, numerical tools are increasingly used to study the underlying mechanism and correlations. For this purpose, we present multiphase-field simulation results of the martensitic phase transformation, in a ferrite-austenite dual-phase microstructure and in an austenite-graphite microstructure. We discuss the influence of inhomogeneities on the resulting martensitic microstructure and on the final stress-strain state.

1 Introduction

The behavior of components in industrial applications depends on the material from which they are made. In the development of new materials, computer-aided simulation models are used more and more frequently to predict how certain process variables influence the future properties during production, and how the materials behave during the operation under load. Especially the heterogeneous appearance

E. Schoof (✉) · C. Herrmann · D. Schneider · J. Hötzer · B. Nestler
Institute of Applied Materials (IAM-CMS), Karlsruhe Institute of Technology (KIT),
Straße am Forum 7, 76131 Karlsruhe, Germany
e-mail: ephraim.schoof@kit.edu

E. Schoof · C. Herrmann · D. Schneider · J. Hötzer · B. Nestler
Institute of Digital Materials Science (IDM), Karlsruhe University of Applied Sciences,
Moltkestr. 30, 76133 Karlsruhe, Germany

© Springer Nature Switzerland AG 2019
W. E. Nagel et al. (eds.), *High Performance Computing in Science
and Engineering '18*, https://doi.org/10.1007/978-3-030-13325-2_30

of many materials on a mesoscopic length scale makes it difficult to predict the behavior on a macroscopic level. For this reason, it is advisable that the behavior is studied first by resolving heterogeneities, in order to obtain homogenized quantities that subsequently can be used in macroscopic models.

Dual-phase steel is such a heterogeneous material. It consists of 10–30 vol.% of hard martensitic islands, which are embedded in a soft ferritic matrix. Due to its combination of high strength and good formability, the use of dual-phase steel enables material savings, and thus weight reduction in industrial applications [1]. Since the morphology of the martensitic islands influences the material behavior [2], a view from the mesoscopic length scale is necessary, particularly as the local microstructure is not taken into account in homogenized approaches.

Another heterogeneous material is cast iron, which is used for brake discs, for example, in the truck sector. Since the braking system plays an important role in terms of safety, a profound knowledge of any infrastructural change in operation is necessary to predict the lifetime of brake discs with high accuracy. During braking, the brake discs are subjected to high frictional forces, which can lead to temperatures above the martensite start temperature. As a result of high cooling rates, martensitic transformations can occur [3], especially in areas close to the surface of the brake disc, where the highest temperatures are located. During a martensitic transformation, residual stresses are induced, the component becomes more brittle, and microcracks are formed. For a better understanding of the failure mechanisms in brake discs, the consideration of the martensitic transformation in cast iron plays an important role.

The phase-field approach has proven to be a powerful tool to describe phase transformations both during the production process of materials and during operation, under the influence of external fields [4–6]. In this work, we present a methodology to simulate martensitic phase transformation (MPT) in heterogeneous materials. This method is applied to study the MPT during the quenching step of dual-phase steel and during thermal loading of a brake disc. The underlying model is based on our previous work [7], and is presented in Sect. 2. In Sects. 3 and 4, we first present the simulation setup and then discuss the results of the MPT in dual-phase steel and in brake discs, respectively. In Sect. 5, we give a conclusion and show possibilities for further work.

Due to the heterogeneous character of the materials, which necessitates large domain sizes for a representative volume element, and the complexity of the involved models, both a highly parallelized software and HPC facilities are mandatory.

2 Elastoplastic Multiphase-Field Model for Martensitic Phase Transformation

In order to investigate the spatio-temporal evolution of a microstructure with the phase-field approach, an order parameter $\phi_\alpha(x, t)$ is assigned to each physically distinguishable region. These regions are called phases. Within phase α, the order parameter reaches $\phi_\alpha(x, t) = 1$, and outside, the value $\phi_\alpha(x, t) = 0$ is assumed. In contrast to sharp interface models, the interface in the phase-field approach is diffuse, which is described by values of $0 < \phi_\alpha(\boldsymbol{x}, t) < 1$. All order parameters are collected in the N-tuple $\boldsymbol{\phi}(\boldsymbol{x}, t) = (\phi_1(\boldsymbol{x}, t), ..., \phi_N(\boldsymbol{x}, t))$, where N represents the number of all existing phases. Based on the work of [8], a free energy functional is constructed

$$\mathscr{F}(\boldsymbol{\phi}, \nabla\boldsymbol{\phi}, \bar{\boldsymbol{\varepsilon}}) = \mathscr{F}_{\text{intf}}(\boldsymbol{\phi}, \nabla\boldsymbol{\phi}) + \mathscr{F}_{\text{chem}}(\boldsymbol{\phi}) + \mathscr{F}_{\text{el}}(\boldsymbol{\phi}, \bar{\boldsymbol{\varepsilon}}) \tag{1}$$

$$= \int_V \bar{W}_{\text{intf}}(\boldsymbol{\phi}, \nabla\boldsymbol{\phi}) + \bar{W}_{\text{chem}}(\boldsymbol{\phi}) + \bar{W}_{\text{el}}(\boldsymbol{\phi}, \bar{\boldsymbol{\varepsilon}})\mathrm{d}V,$$

which consists of the free energies that are required to model the MPT. The interfacial free energy $\bar{W}_{\text{intf}}(\boldsymbol{\phi}, \nabla\boldsymbol{\phi}) = \epsilon a(\boldsymbol{\phi}, \nabla\boldsymbol{\phi}) + \omega(\boldsymbol{\phi})/\epsilon$ is composed of the gradient energy density

$$\epsilon a(\boldsymbol{\phi}, \nabla\boldsymbol{\phi}) = \epsilon \sum_{\alpha < \beta} \gamma_{\alpha\beta} |\boldsymbol{q}_{\alpha\beta}|^2, \tag{2}$$

and the potential energy density

$$\frac{1}{\epsilon}\omega(\boldsymbol{\phi}) = \begin{cases} \dfrac{16}{\epsilon \pi^2} \displaystyle\sum_{\alpha < \beta}(\gamma_{\alpha\beta} + \gamma_\alpha^c)\phi_\alpha\phi_\beta + \dfrac{1}{\epsilon}\displaystyle\sum_{\alpha < \beta < \delta} \gamma_{\alpha\beta\delta}\phi_\alpha\phi_\beta\phi_\delta & \text{if } \boldsymbol{\phi} \in \mathscr{G} \\ \infty & \text{else} \end{cases}, \tag{3}$$

whereby $\omega(\boldsymbol{\phi})$ is set to infinity, if the N-tuple $\boldsymbol{\phi}$ is not on the Gibbs simplex $\mathscr{G} = \{\boldsymbol{\phi}| \sum_\alpha \phi_\alpha = 1, \text{ and } \phi_\alpha \geq 0\}$. Both parts are constructed in such a way that the interfacial energy $\gamma_{\alpha\beta}$ at the α–β interface can be reproduced. The parameter γ_α^c will be addressed hereafter. The higher-order term $\propto \phi_\alpha\phi_\beta\phi_\delta$ in Eq. (3) prevents the formation of spurious phases in binary interfaces [9]. The generalized gradient vector $\boldsymbol{q}_{\alpha\beta} = \phi_\alpha \nabla\phi_\beta - \phi_\beta \nabla\phi_\alpha$, normal to the interface, allows the separation of the individual dual contribution at an α–β interface. The width of the diffuse interface is controlled by the parameter ϵ. The chemical and elastic free energies are a sum of the phase-dependent quantities, which are indicated as follows by the superscript α:

$$\bar{W}_{\text{chem}}(\boldsymbol{\phi}) = \sum_\alpha \phi_\alpha W_{\text{chem}}^\alpha, \tag{4}$$

$$\bar{W}_{\text{el}}(\boldsymbol{\phi}, \bar{\boldsymbol{\varepsilon}}) = \sum_\alpha \phi_\alpha W_{\text{el}}^\alpha(\boldsymbol{\varepsilon}^\alpha). \tag{5}$$

Based on the work of Steinbach et al. [4, 10], the evolution equation for each phase field is given as

$$\frac{\partial \phi_\alpha}{\partial t} = -\frac{1}{\tilde{N}\epsilon} \sum_{\beta \neq \alpha}^{\tilde{N}} \left[M_{\alpha\beta} \left(\frac{\delta \mathscr{F}_{\text{intf}}}{\delta \phi_\alpha} - \frac{\delta \mathscr{F}_{\text{intf}}}{\delta \phi_\beta} + \epsilon \hat{a}(\phi_\alpha, \nabla\phi_\alpha) \right. \right. \tag{6}$$

$$\left. \left. - \epsilon \hat{a}(\phi_\beta, \nabla\phi_\beta) - \frac{8\sqrt{\phi_\alpha\phi_\beta}}{\pi} \Delta^{\alpha\beta} \right) \right] + \zeta,$$

where the bulk driving forces are calculated as $\Delta^{\alpha\beta} = \Delta^{\alpha\beta}_{\text{chem}} + \Delta^{\alpha\beta}_{\text{el}} = (\delta/\delta\phi_\beta - \delta/\delta\phi_\alpha)(\mathscr{F}_{\text{el}} + \mathscr{F}_{\text{chem}})$. Thereby, $\delta/\delta\phi_\alpha = (\partial/(\partial\phi_\alpha) - \nabla \cdot \partial/(\partial\nabla\phi_\alpha))$ is the variational derivative operator, with respect to ϕ_α. The divergence operator is given by $(\nabla\cdot)$. The movement of each α–β interface is scaled by the mobility $M_{\alpha\beta}$, and $\tilde{N}$ denotes the number of locally active phases. According to [7], the nucleation of martensitic variants is realized by the noise term ζ. As a function of the parameter γ^c_α, an additional term

$$\epsilon\hat{a}(\phi_\alpha, \nabla\phi_\alpha) = -\epsilon\gamma^c_\alpha\left(\Delta\phi_\alpha - |\nabla\phi_\alpha|\nabla \cdot \left(\frac{\nabla\phi_\alpha}{|\nabla\phi_\alpha|}\right)\right), \tag{7}$$

which does not contribute to the capillary force, is introduced into the evolution equation (6) to strengthen the diffuse interface in the context of high bulk driving forces [7].

During quenching, carbon atoms have not enough time to diffuse, and the face-centered cubic lattice of austenite transforms to a body-centered tetragonal lattice, as a result. This transformation process introduces stresses and strains into the material, which can be described as the Bain strains:

$$\tilde{\varepsilon}^0(1) = \begin{pmatrix} \varepsilon_3 & 0 & 0 \\ 0 & \varepsilon_1 & 0 \\ 0 & 0 & 0 \end{pmatrix}, \quad \tilde{\varepsilon}^0(2) = \begin{pmatrix} \varepsilon_1 & 0 & 0 \\ 0 & \varepsilon_3 & 0 \\ 0 & 0 & 0 \end{pmatrix}. \tag{8}$$

In 2D, two different variants can form, whose shear strains are directed in the opposite direction [11].

In our recent works [12–14], we proposed two different models for the calculation of the elastic driving force $\delta\mathscr{F}_{\text{el}}/\delta\phi_\alpha$ and the stresses $\bar{\sigma}(\phi)$. Both models are derived on the basis of the mechanical jump conditions and match the sharp interface solution in a binary interface, despite the introduction of a diffuse interface. In multiphase regions, a simplification is introduced to prevent a further increase of the complexity. In order to simulate the MPT in dual-phase steel with plasticity, we use the model according to [14, 15]. For the simulation of MPT in brake discs, the model according to [13] is applied, which is numerically more robust, but is not able to be coupled with a phase-dependent plasticity formulation [15]. First, the static mechanical equilibrium condition $\nabla \cdot \bar{\sigma}(\phi) = 0$ is solved implicitly for the stresses and strains, and the driving force for the phase transformation is calculated based on these results. For a more detailed description of the coupling of the used multiphase-field model with mechanical fields, the reader is referred to Refs. [12–15].

Since the chemical composition does not change during quenching, the phase-dependent chemical driving force W^α_{chem} is modeled as a constant, which can be calculated based on the chemical composition and the quenching temperature.

The set of evolution equations is implemented into the massive parallel in-house software package "Parallel Algorithms for Crystal Evolution in 3D" (PACE3D) [16]. The objective of PACE3D is to provide a package for large-scale multiphysics simulations to solve (coupled) problems, such as grain growth [17–19], grain coarsening

[20, 21], solidification [8, 22, 23], mass and heat transport, fluid flow (Lattice Boltzmann, Navier-Stokes) [24], mechanical forces (elasticity, plasticity) [7, 13, 15, 25], magnetism [26], electrochemistry [27], and wetting [28, 29]. By combining optimizations on different levels, ranging from the parameter, the model, the numerics, the setup, the application, the parallelization up to the explicit vectorization of the code, a single-core peak performance of 42.93 % is achieved for certain models [16]. On various supercomputers, the solver scales nearly ideally on up to 96 100 cores.

3   Martensitic Phase Transformation in Dual-Phase Steel

3.1   Simulation Setup

The initial microstructure, which serves as a starting condition for the simulation of MFP in a ferritic-austenitic dual-phase microstructure, is presented in Fig. 1. This microstructure, as well as the input parameters and boundary conditions, are based on our previous work [7], in which the general framework for the elastic MPT theory was presented. The domain has a physical size of $65 \times 65\ \mu$m, and was discretized with 4000×4000 voxel cells. In this work, we include plasticity, according to our recently published work [15]. A nonlinear hardening function of the form

$$\sigma_{\mathrm{h}}^{\alpha}(\varepsilon_{\mathrm{p}}^{\alpha}) = \theta_{\infty}^{\alpha}\varepsilon_{\mathrm{p}}^{\alpha} + (\sigma_{\infty}^{\alpha} - \sigma_{0}^{\alpha}) \cdot \left(1 - \exp\left(-\frac{\theta_{0}^{\alpha} - \theta_{\infty}^{\alpha}}{\sigma_{\infty}^{\alpha} - \sigma_{0}^{\alpha}}\varepsilon_{\mathrm{p}}^{\alpha}\right)\right), \tag{9}$$

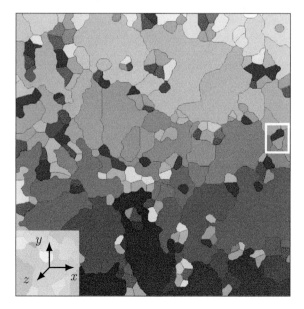

Fig. 1 Initial ferritic-austenitic microstructure. Ferritic grains are displayed in gray, and austenitic grains are highlighted in color. The microstructure was generated on the basis of an EBSD scan of a commercial DP600 finished strip, and was used in our previous work [7]

is used for all three phase classes, where $\varepsilon_{\mathrm{p}}^{\alpha}$ is the accumulated plastic strain of phase α. For the simulation, we respectively use an initial yield strength of $\sigma_0 = 0.29$, 0.65, and 1.10 GPa, for ferrite, austenite, and martensite, a yield stress of $\sigma_\infty = 0.66$, 1.10, and 1.98 GPa, at infinity, an initial hardening modulus of $\theta_0 = 70$, 15, and 190 GPa, and a hardening modulus of $\theta_\infty = 0$, 1.3 and 0 GPa, at infinity. The simulation was performed within 72h, using 1191 CPUs.

To discuss the effect of the large shear strains, we furthermore conducted a simulation without resolving the MPT explicitly. This is achieved by assuming the austenitic regions in Fig. 1 to be totally martensitic, exhibiting only isotropic eigenstrains, which accounts for the same volume increase, during the MPT, as in the simulation in which the MPT is calculated explicitly.

3.2 Results and Discussion

Martensite nucleates heterogeneously at structural defects, such as grain boundaries [30]. In Fig. 2, a section of the microstructure is shown for two different time steps. Martensite nucleates both at the existing austenite-austenite and ferrite-austenite grain boundaries, and grows into the interior of the grain. Due to the opposite shear strains, it is energetically more favorable when both variants occur alternately. This leads to the mechanism called autocatalytic nucleation, where the growth of one martensitic variant favors the nucleation of the other variant [31], which can be seen in Fig. 2a. Finally, a lath-type microstructure is formed, which minimizes the elastic free energy. Due to the large eigenstrains, the material starts to deform plastically. In Fig. 2c, the accumulated plastic strain is depicted for the time $t = 14500 \, \Delta t$. One can see that plastic deformations are present both in ferritic and martensitic regions. Due

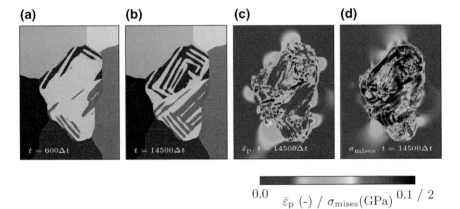

Fig. 2 Snapshots of the martensitic transformation for two austenitic grains. The selected microstructure is extracted from the section marked in Fig. 1. In **a** and **b**, the microstructure is presented for the time $t = 600 \, \Delta t$ and $t = 14500 \, \Delta t$, respectively. In **c** and **d**, the corresponding accumulated plastic strain $\bar{\varepsilon}_{\mathrm{p}}$ and von Mises stress σ_{mises} are depicted at the advanced time $t = 14500 \, \Delta t$

to the lower yield stress of ferrite, compared to martensite, the surrounding ferritic grains in the vicinity of the martensitic islands deform plastically to accommodate the stress-free transformation strain. The von Mises stress is presented in Fig. 2d. Inside the martensitic regions, high von Mises stresses prevail, whereas in ferrite, the von Mises stresses are significantly lower, which is a result of the lower yield stress.

In Fig. 3, the results are presented in the entire domain, in terms of the von Mises stress and the accumulated plastic strain. Resolving the ferritic and austenitic (respectively martensitic) microstructure by the phase-field approach, the von Mises

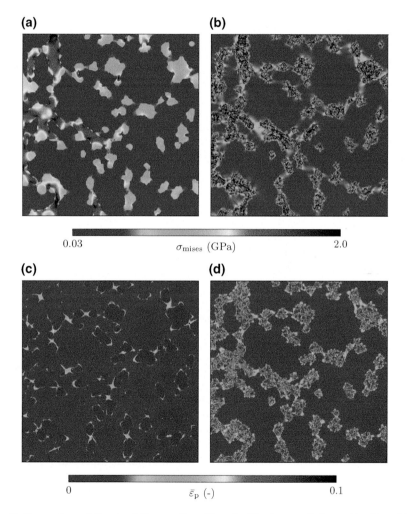

Fig. 3 Comparison of the simulation results, with and without an explicit consideration of the MPT. In **a** and **b**, the von Mises stresses are shown, and in **c** and **d**, the accumulated plastic strain is presented. The results **a** and **c** are gained by a simulation where only the volumetric parts of the eigenstrains are considered

stresses and the accumulated plastic strains are higher, compared to the isotropic assumption. This observation indicates that the shear strains of the different variants do not cancel each other out perfectly. In the isotropic case, plastic deformations occur mainly in ferritic regions close to a ferrite-martensite interface (see Fig. 3c), whereas the simulation, which resolves the lath-type martensitic structure, also predicts a noticeable amount of plastic deformation within the martensitic structures, which is consistent with experimental findings [32]. Neighboring martensitic islands interact, which leads to higher stresses and strains between them.

4 Martensitic Phase Transformation in a Brake Disc

4.1 Simulation Setup

The mesoscopic length scale shows that cast iron is a composite material, which consists of a metallic matrix, with graphite inhomogeneities as filling elements. There are different basic types of cast iron, which can be differentiated mainly by a different metallic matrix and/or graphite structure. The matrix structure can range from ferrite, pearlite, ferrite-pearlite up to all quenched and tempered steels. Furthermore, the graphite inhomogeneities differ by different shapes, sizes, and arrangements. Possible graphite forms include spheroidal, lamellar, and vermicular morphologies. In this study, cast iron with a pearlitic matrix and lamellar graphite is considered. Pearlite is a composite of ferrite and cementite, which is layered alternately in a lamellar structure.

We use an optical micrography of a cast iron specimen to extract the morphology of the lamellar graphite. Based on the assumption that pearlite transforms to austenite at a high temperature, we fill the pearlitic matrix with differently oriented austenitic grains. The initial microstructure for the simulation of MPT, as e.g. found in a brake disc base material, is shown in Fig. 4. The physical size covers $315 \times 315\,\mu$m, and was discretized with 900×900 voxel cells. The following parameters were used: The austenitic matrix was modeled as linearly elastic, with an isotropic stiffness. A Young's modulus of $E = 170$ GPa and a Poisson's ratio of $\nu = 0.3$ were employed.

At the atomic level, lamellar graphite consists of a large number of different basal planes. Covalent bonds within the basal planes connect the carbon atoms of the benzene rings, and the different basal planes are connected by much weaker Van-der-Waals forces. These different bond types result in a transverse isotropic mechanical behavior. The unrotated stiffness components of the graphite, which were modeled as linearly elastic stiffness components, were taken from [33], and are listed in Table 1.

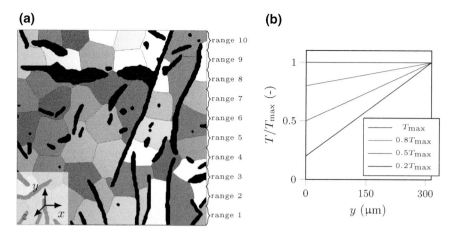

Fig. 4 **a** Initial microstructure, consisting of austenitic grains (gray) and lamellar graphite (black), for the simulation of MPT in a typical brake disc material. For the analysis, the domain is divided into ten sections (range 1–10), depending on the position in the y-direction. **b** The consequences of different temperature penetration depths are studied by using a linear line between the temperature T_{max}, at the surface of the brake disc, at $y = 315$ μm, and different lower temperatures T_{min}, at $y = 0$ μm

Table 1 Elastic parameters for lamellar graphite

	GPa
C_{11}	1126 ± 22
C_{12}	200 ± 20
C_{13}	$39,5 \pm 40$
C_{33}	$40,7 \pm 1,1$
C_{44}	$4,51 \pm 0,5$

The rotation angles of the graphite lamellae were determined using an analysis tool, based on the optical micrography.

The eigenstrains were taken to be $\varepsilon_1 = 0.11$ and $\varepsilon_3 = -0.70$, resulting in a volume increase of 3.2%, which is a typical value for the MPT. We assume a linear dependence between the temperature T and the chemical driving force Δ_{chem}, which was determined to be 3.5×10^8 J/m^3 at T_{max}, so that a temperature of $0.5T_{max}$ results in a halving of the chemical driving force. For the strain, clamped boundary conditions are used, and for the phase-fields, an extrapolation boundary condition applies.

In this work, we study the effect of different temperature penetration depths on the MPT, in an austenite-graphite microstructure, by varying the temperature T_{min} at $y = 0$ μm, as it is shown in Fig. 4b. Each simulation was performed with 55 CPUs, within 48 h.

4.2 Results and Discussion

The results for the case $T_{min} = 0.5\ T_{max}$ are depicted in Fig. 5, in terms of microstructure and von Mises stress. Since the highest chemical driving force is located at the top of the simulation domain, martensite preferentially nucleates in regions of high y-values. From the grain boundaries, the laths grow into the grain interior in a twin-related manner. When they hit a grain boundary, the martensitic nucleation in the adjacent grain is stimulated as a consequence of the autocatalytic nucleation mechanism. However, since no transformation takes place in the graphite lamellae, they act as a natural barrier to the spread of martensite. In Fig. 5b, c, it can be seen that the horizontally arranged lamellae prevent the spread of martensite towards the region of lower temperature, whereas between vertically arranged lamellae, martensite finds an energetic path to grow towards the bottom. In regions where the chemical driving force is lower, the nucleation barrier can only be overcome when a stress field, originating from surrounding variants, stimulates the nucleation.

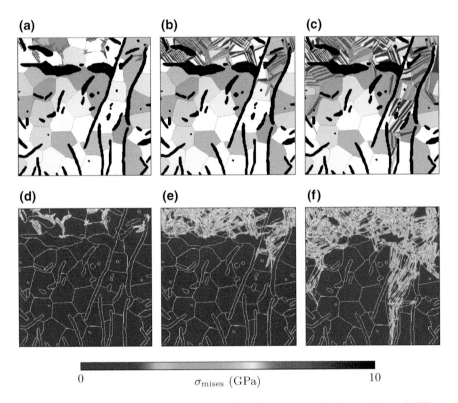

Fig. 5 Snapshots of the MPT in an austenite-graphite microstructure, for the case $T_{min} = 0.5\ T_{max}$. In **a–c**, the microstructure is shown for the time $t = 3000\ \Delta t$, $16000\ \Delta t$, and $130000\ \Delta t$, respectively. The corresponding von Mises stress is presented in **d–f**

Martensite introduces large stresses into the microstructure, which are visible in Fig. 5d, e. Due to the limitation to linear elasticity, the values are higher than in reality, but basic mechanisms can still be studied. Those stresses are unwanted, since the material becomes more brittle, and microcrack formation can be promoted. It can be observed that smaller temperatures lead to less residual strains.

In Fig. 6a, the martensitic volume fraction in initially austenitic regions is plotted over time, for different values of T_{min}. If there is no temperature gradient, i.e. $T_{min} = T_{max}$, the transformation is fastest and reaches a volume fraction of 81%, at the time $1.3 \times 10^4 \ \Delta t$. Since the volume increase leads to higher stresses, which cannot be reduced by self-accommodation, retained austenite is present. The smaller T_{min} is, the less martensite is formed. In Fig. 6b, the martensitic volume fraction is shown for $T_{min} = 0.5 \ T_{max}$, in the different ranges according to the labeling in Fig. 4, over time. It can be seen that for the ranges 8–10, MPT starts right at the beginning. The lower the y-coordinate, the later MPT starts. It is interesting to note that in range 10, less martensite is formed, compared to range 9. The vicinity of the

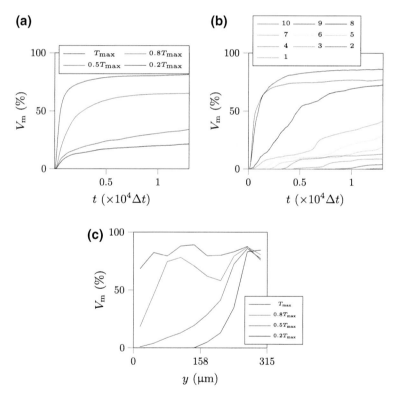

Fig. 6 Martensitic volume fraction in initially austenitic regions, as a function of time (**a** and **b**) and position (**c**). In **b**, the martensitic volume fraction is plotted for different ranges, according to the labeling in Fig. 4, for the case $T_{min} = 0.5 \ T_{max}$. In **c**, the martensitic volume fraction is presented as a function of the y-axis, for the time $t = 1.3 \times 10^4 \ \Delta t$

clamped boundaries might be the reason for this behavior. In Fig. 6c, the martensitic volume fraction is depicted as a function of the y-axis, for the time $t = 1.3 \times 10^4 \, \Delta t$. While the case $T_{\min} = T_{\max}$ shows no significant dependence of the volume fraction on the range, there is a strong correlation in the other cases. Although the temperature profile follows a linear behavior, the martensitic volume fraction shows a rather exponential form. For the case $T_{\min} = 0.2T_{\max}$, there is a threshold under which no transformation takes place, due to a chemical driving force, which is too small.

5 Conclusion and Outlook

In this work, we presented simulation results of the martensitic phase transformation in heterogeneous materials. In dual-phase steel, large-scale simulations enabled the investigation of elastoplastic effects, both within martensitic islands and ferritic regions. The heterogeneous stress-strain state after quenching is influenced by the morphology of the martensitic inclusion. The simulation results can be used as a starting condition for virtual tensile tests to extract homogenized material parameters, which then can be used to perform simulations of complex geometries on the macroscopic scale.

Additionally, we investigated the influence of a temperature gradient in an austenite-graphite microstructure, in the context of a brake disc. Simulation results indicate that a minimum chemical driving force is necessary to promote the martensitic spread. Based on these results, the model can be coupled with our recently proposed model for crack prorogation in heterogeneous material [34] to study the effect of martensitic transformation on the crack formation in brake discs. Furthermore, three-dimensional simulations are planned.

Acknowledgements The authors thank the DFG ("Deutsche Forschungsgemeinschaft") for funding our investigations within the Research Training Group 1483 and through the associated transfer project with number NE 822/16-1. Ephraim Schoof acknowledges the financial support of the Ministry of Science, Research and Arts of Baden-Wuerttemberg, under the grant 33-7533.-30-10/25/54. The authors gratefully acknowledge the editorial support by Leon Geisen. This work was performed on the computational resource ForHLR II, funded by the Ministry of Science, Research and Arts of Baden-Wuerttemberg and the DFG.

References

1. H. Hofmann, D. Matissen, T.W. Schaumann, Advanced cold-rolled steels for automotive applications. Steel Res. Int. **80**(1), 22–28 (2009)
2. K. Park, M. Nishiyama, N. Nakada, T. Tsuchiyama, S. Takaki, Effect of the martensite distribution on the strain hardening and ductile fracture behaviors in dual-phase steel. Mater. Sci. Eng.: A **604**, 135–141 (2014)
3. S. Panier, P. Dufrénoy, D. Weichert, An experimental investigation of hot spots in railway disc brakes. Wear **256**(7–8), 764–773 (2004)

4. I. Steinbach, Phase-field models in materials science. Model. Simul. Mater. Sci. Eng. **17**(7), 073001 (2009)
5. I. Steinbach, Phase-field model for microstructure evolution at the mesoscopic scale. Annu. Rev. Mater. Res. **43**, 89–107 (2013)
6. R. Spatschek, E. Brener, A. Karma, Phase field modeling of crack propagation. Philos. Mag. **91**(1), 75–95 (2011)
7. E. Schoof, D. Schneider, N. Streichhan, T. Mittnacht, M. Selzer, B. Nestler, Multiphase-field modeling of martensitic phase transformation in a dual-phase microstructure. Int. J. Solids Struct. **134**, 181–194 (2018)
8. B. Nestler, H. Garcke, B. Stinner, Multicomponent alloy solidification: phase-field modeling and simulations. Phys. Rev. E **71**(4), 041609 (2005)
9. J. Hötzer, O. Tschukin, M. Ben Said, M. Berghoff, M. Jainta, G. Barthelemy, N. Smorchkov, D. Schneider, M. Selzer, B. Nestler, Calibration of a multi-phase field model with quantitative angle measurement. J. Mater. Sci. **51**(4), 1788–1797 (2016)
10. I. Steinbach, F. Pezzolla, A generalized field method for multiphase transformations using interface fields. Phys. D: Nonlinear Phenom. **134**(4), 385–393 (1999)
11. A. Yamanaka, T. Takaki, Y. Tomita, Elastoplastic phase-field simulation of self- and plastic accommodations in cubic to tetragonal martensitic transformation. Mater. Sci. Eng. A **491**(1–2), 378–384 (2008)
12. D. Schneider, O. Tschukin, A. Choudhury, M. Selzer, T. Böhlke, B. Nestler, Phase-field elasticity model based on mechanical jump conditions. Comput. Mech. **55**(5), 887–901 (2015)
13. D. Schneider, E. Schoof, O. Tschukin, A. Reiter, C. Herrmann, F. Schwab, M. Selzer, B. Nestler, Small strain multiphase-field model accounting for configurational forces and mechanical jump conditions. Comput. Mech. **61**(3), 277–295 (2018)
14. D. Schneider, F. Schwab, E. Schoof, A. Reiter, C. Herrmann, M. Selzer, T. Böhlke, B. Nestler, On the stress calculation within phase-field approaches: a model for finite deformations. Comput. Mech. **60**(2), 203–217 (2017)
15. C. Herrmann, E. Schoof, D. Schneider, F. Schwab, A. Reiter, M. Selzer, B. Nestler, Multiphase-field model of small strain elasto-plasticity according to the mechanical jump conditions. Comput. Mech. (2018) (in print)
16. J. Hötzer, A. Reiter, H. Hierl, P. Steinmetz, M. Selzer, B. Nestler, The parallel multi-physics phase-field framework Pace3d. J. Comput. Sci. **26**, 1–12 (2018)
17. K. Ankit, B. Nestler, M. Selzer, M. Reichardt, Phase-field study of grain boundary tracking behavior in crack-seal microstructures. Contrib. Mineral. Petrol. **166**(6), 1709–1723 (2013)
18. K. Ankit, J.L. Urai, B. Nestler, Microstructural evolution in bitaxial crack-seal veins: a phase-field study. J. Geophys. Res. B: Solid Earth **120**(5), 3096–3118 (2015)
19. K. Ankit, M. Selzer, C. Hilgers, B. Nestler, Phase-field modeling of fracture cementation processes in 3-d. J. Pet. Sci. Res. **4**(2), 79–96 (2015)
20. A. Vondrous, in *Grain Growth Behavior and Efficient Large Scale Simulations of Recrystallization with the Phase-Field Method*, vol. 44 (KIT Scientific Publishing, 2014)
21. M. Selzer, in *Mechanische und Strömungsmechanische Topologieoptimierung mit der Phasenfeldmethode* (KIT Scientific Publishing, 2014)
22. A. Choudhury, M. Geeta, B. Nestler, Influence of solid-solid interface anisotropy on three-phase eutectic growth during directional solidification. EPL (Eur. Lett.) **101**(2), 26001 (2013)
23. A. Choudhury, Pattern-formation during self-organization in three-phase eutectic solidification. Trans. Indian Inst. Metals 1–7 (2015)
24. J. Ettrich, *Fluid Flow and Heat Transfer in Cellular Solids*, vol. 39 (KIT Scientific Publishing, 2014)
25. D. Schneider, S. Schmid, M. Selzer, T. Böhlke, B. Nestler, Small strain elasto-plastic multiphase-field model. Comput. Mech. **55**(1), 27–35 (2015)
26. C. Mennerich, in *Phase-Field Modeling of Multi-domain Evolution in Ferromagnetic Shape Memory Alloys and of Polycrystalline Thin Film Growth*, vol. 19 (KIT Scientific Publishing, 2013)

27. A. Mukherjee, K. Ankit, R. Mukherjee, B. Nestler, Phase-field modeling of grain-boundary grooving under electromigration. J. Electron. Mater. **45**(12), 6233–6246 (2016)
28. M. Ben Said, M. Selzer, B. Nestler, D. Braun, C. Greiner, H. Garcke, A phase-field approach for wetting phenomena of multiphase droplets on solid surfaces. Langmuir **30**(14), 4033–4039 (2014)
29. F. Weyer, M. Ben Said, J. Hötzer, M. Berghoff, L. Dreesen, B. Nestler, N. Vandewalle, Compound droplets on fibers. Langmuir **31**(28), 7799–7805 (2015) (PMID: 26090699)
30. M. Ueda, H. Yasuda, Y. Umakoshi, Effect of grain boundary on martensite transformation behaviour in Fe32 at.% Ni bicrystals. Sci. Technol. Adv. Mater. **3**(2), 171–179 (2002)
31. Q.P. Meng, Y.H. Rong, T.Y. Hsu, Effect of internal stress on autocatalytic nucleation of martensitic transformation. Metall. Mater. Trans. A **37**(5), 1405–1411 (2006)
32. S. Morito, J. Nishikawa, T. Maki, Dislocation density within lath martensite in Fe-C and Fe-Ni alloys. ISIJ Int. **43**(9), 1475–1477 (2003)
33. T.A. Pascal, N. Karasawa, W.A. Goddard III, Quantum mechanics based force field for carbon (QMFF-Cx) validated to reproduce the mechanical and thermodynamics properties of graphite. J. Chem. Phys. **133**(13), 134114 (2010)
34. D. Schneider, E. Schoof, Y. Huang, M. Selzer, B. Nestler, Phase-field modeling of crack propagation in multiphase systems. Comput. Methods Appl. Mech. Eng. **312**, 186–195 (2016)

KKRnano: Quantum Description
of Skyrmions in Chiral B20 Magnets

Marcel Bornemann, Paul F. Baumeister, Rudolf Zeller and Stefan Blügel

Abstract We present the latest version of the linear-scaling electronic structure code KKRnano, in which an enhanced Korringa-Kohn-Rostoker (KKR) scheme is utilized to perform Density Functional Theory (DFT) calculations. The code allows us to treat system sizes of up to several thousands of atoms per unit cell and to simulate a non-collinear alignment of atomic spins. This capability is used to investigate nanometer-sized magnetic textures in the germanide B20-MnGe, a material that is potentially going to play an important role in future spintronic devices. A performance analysis of KKRnano on Hazel Hen emphasizes the good scaling behaviour with increasing system size and demonstrates the extensive integration of highly optimized libraries.

1 Introduction

We have developed a unique electronic structure code, KKRnano [1–3], specifically designed for petaFLOP computing. Our method scales linearly with the number of atoms, so that we can realize system sizes of up to half a million atoms in a

M. Bornemann (✉) · S. Blügel
Peter Grünberg Institut and (PGI-1) and Institute for Advanced Simulation (IAS-1),
Forschungszentrum Jülich, 52425 Jülich, Germany
e-mail: m.bornemann@fz-juelich.de

S. Blügel
e-mail: s.bluegel@fz-juelich.de

M. Bornemann
Peter Grünberg Institut and (PGI-1) and Institute for Advanced Simulation (IAS-1),
RWTH Aachen University, 52056 Aachen, Germany

P. F. Baumeister
Jülich Supercomputing Centre, Forschungszentru Jülich, 52425 Jülich, Germany
e-mail: p.baumeister@fz-juelich.de

R. Zeller
Institute for Advanced Simulation (IAS-1), Forschungszentrum Jülich,
52425 Jülich, Germany
e-mail: ru.zeller@fz-juelich.de

© Springer Nature Switzerland AG 2019
W. E. Nagel et al. (eds.), *High Performance Computing in Science
and Engineering '18*, https://doi.org/10.1007/978-3-030-13325-2_31

unit cell if necessary. Recently, we implemented a relativistic generalization of our algorithm enabling us to calculate complex non-collinear magnetic structures, such as skyrmions, in real space. Skyrmions are two-dimensional magnetization solitons, i.e., two-dimensional magnetic structures localized in space, topologically protected by a non-trivial magnetization texture, which has particle-like properties. The focus of our work is on the germanide MnGe that is particularly interesting among the chiral magnetic B20 compounds, as it exhibits a three-dimensional magnetic structure that is not yet understood (see preliminary results [4–7]).

This report is structured as follows: In Sect. 2 we give a brief introduction on the Korringa-Kohn-Rostoker (KKR) method and explain how linear scaling is achieved. Our results on B20-MnGe are presented in Sect. 3. In Sects. 4 and 5 the focus is on code-specific issues. The parallelization scheme of KKRnano is described in detail and a performance analysis of our code on Hazel Hen is conducted.

2 Numerical Methods and Algorithms

Contrary to other Density Functional Theory (DFT) codes, which determine the Kohn-Sham orbitals by solving the Kohn-Sham eigenvalue equation, in KKRnano the Green function for the Kohn-Sham equation is obtained by solving an integral equation in real space. In this so called Korringa-Kohn-Rostoker (KKR) scheme, space is divided into non-overlapping cells around the atoms, and the calculations are separated into single-cell parts where only the potential within the individual cell enters, and a large complex linear matrix equation for the matrix elements $G_{LL'}^{nn'}(\varepsilon)$ of the Green function $G(\mathbf{r}, \mathbf{r}', \varepsilon)$:

$$G_{LL'}^{nn'}(\varepsilon) = G_{LL'}^{r,nn'}(\varepsilon) + \sum_{n''} \sum_{L''} G_{LL''}^{r,nn''}(\varepsilon) \sum_{L'''} \Delta t_{L''L'''}^{n''}(\varepsilon) G_{L'''L'}^{n''n'}(\varepsilon). \quad (1)$$

In the following we refer to $\Delta t_{L''L'''}^{n''}(\varepsilon)$ as the Δt-matrices and $G_{LL''}^{r,nn''}(\varepsilon)$ as the reference Green functions. We refer the reader to the existing literature for more details [1]. The problem given in Eq. (1) is identical to solving a complex linear matrix equation of the form

$$\mathbf{Ax} = \mathbf{b}. \quad (2)$$

As is well known, this problem scales as $O(N^3)$, where N denotes the dimension of the $N \times N$ matrix $\mathbf{A}$. Standard solving schemes therefore severely limit the system size that can be calculated. However, by choosing a reference system of repulsive potentials, $\mathbf{A}$ can be made sparse and this can be exploited by using customized iterative algorithms for solving linear sparse matrix equations. Our method uses the transpose-free quasi-minimal-residual (TFQMR) algorithm [8] which enables us to achieve $O(N^2)$ scaling. Convergence of the TFQMR iterations is achieved by working in the complex energy plane. Both concepts, the repulsive reference system and application of complex energy, were introduced by us into the KKR method [9, 10] and are used worldwide for many years. By truncating inter-atomic interactions above a certain distance, an $O(N)$ scaling behavior can be realized, and this feature should be used in any calculation involving more than 1000 atoms. In most applications the TFQMR solver accounts for the major part of the computational work.

For super-large-scale calculations (above 100,000 atoms) the electrostatics solver begins to require an amount of computing resources that is not negligible anymore [11]. The electrostatic problem is given in terms of a Poisson equation that connects electric field and potential. Solving this equation scales quadratically with system size, a drawback which will be removed in the future.

3 Complex Magnetic Textures in B20-MnGe

B20-MnGe was identified as a good candidate material to be investigated with the new version of KKRnano which now contains the feature of non-collinear magnetism and spin-orbit coupling. In a recent study [4], it was found by transmission electron microscopy that 3D magnetic objects exist in B20-MnGe. The authors of [4] came to the conclusion that their data indicates a cubic lattice of skyrmionic hedgehogs and anti-hedgehogs (see Fig. 1a) with a lattice constant of about 3–6 nm. The singularity at the center of the texture exists only in the micromagnetic description since the atomic magnetic moment of the atom, which is located at the center, remains finite [12]. Findings by Kanazawa et al. suggest that the lattice is set up by a superposition of three orthogonal helical structures, also referred to as 3Q state [13]. Here, the local magnetization is determined by the provision

$$\mathbf{M}(\mathbf{r}) = \begin{pmatrix} \sin qy + \cos qz \\ \sin qz + \cos qx \\ \sin qx + \cos qy \end{pmatrix}, \tag{3}$$

where $q = \frac{2\pi}{\lambda}$ is the wavenumber given in terms of the helical wavelength λ and x, y and z are the spatial coordinates within the unit cell. Note, that $\mathbf{M}(\mathbf{r})$ is not normalized. In contrast to other systems exhibiting a similar magnetic phase, the rather short helical wavelength in B20-MnGe allows one to perform Density Functional Theory (DFT) calculations with KKRnano. B20-MnGe is currently the subject of extensive investigation [4, 14–17]. This is mainly inspired by the discovery of skyrmions as small information-carrying particles that could potentially be used in spintronic devices [18]. However, no large-scale DFT calculation seems to have been performed, yet. At present there is a lack of a convincing explanation of what is observed in experiment. Research in the framework of micromagnetic models identified both magnetic frustration (RKKY interaction) as well as spin-orbit coupling induced Dzyaloshinskii-Moriya (DM) interaction as potentially crucial to a better understanding [19, 20]. While the 3Q state certainly constitutes the most interesting

(a)   **(b)**

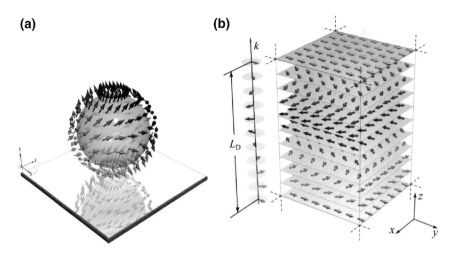

Fig. 1 Magnetic textures that are found experimentally in B20-MnGe: **a** Magnetic anti-hedgehog texture that is wrapped around a singularity at the center. Published with kind permission of ©Nikolai Kiselev 2018. All Rights Reserved. **b** Helical spin spiral that propagates in (001) direction. Reprinted from [5] and licensed under CC BY 3.0

non-trivial magnetic texture in B20-MnGe, there are also reports that the magnetic ground state in this system is actually a helical spiral (see Fig. 1b) [21] which was observed up to a temperature of 170 K [22]. These two observations are clearly contradictory and it has not yet been explained how both can coexist within the same material. The helical spin spiral in B20-MnGe forms along the (001) direction and therefore the magnetization is described by the relation

$$\mathbf{M}(\mathbf{r}) = \begin{pmatrix} \cos qz \\ -\sin qz \\ 0 \end{pmatrix}. \tag{4}$$

In the following, we refer to this as the 1Q state.

We commence our investigations by considering a $6 \times 6 \times 6$ B20-MnGe supercell (1728 atoms), where we use PBEsol as exchange-correlation functional and include only a single k-point, i.e., the Γ-point. In this initial comparison of ferromagnetic (FM), 1Q and 3Q state the respective states are imposed on the system by forcing the atomic exchange-correlation B-fields to point into specific directions. For the equilibrium lattice constant $a = 4.80$ Å, the ferromagnet is the state with the lowest total energy. As this contradicts experimental observations, we take into consideration that in experiment the crystal structure might inadvertently differ from the ideal structure. Such discrepancies can for instance be caused by strain that originates from the manufacturing process of the sample. Therefore, it is reasonable to check whether a material's magnetic properties change, when the lattice constant is varied. Such a variation is performed in the upper part of Fig. 2, where the total energy is evaluated for FM, 1Q and 3Q state. Clearly, neither the 1Q nor the 3Q state

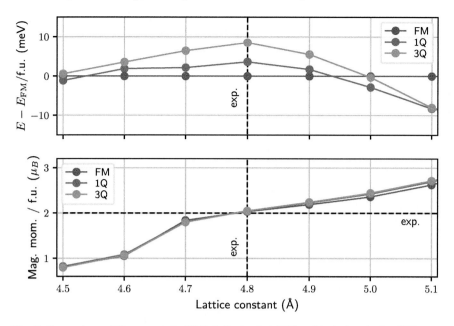

Fig. 2 Comparison of Ferromagnetic (FM), helical spiral (1Q) and hedgehog lattice (3Q) state with KKRnano. Top: Difference of total energies with the FM state as reference state for different lattice constants. The experimental lattice constant is $a = 4.80$ Å. 1Q and 3Q state are energetically preferable for $a > 5.0$ Å. Bottom: Magnetic moment per Mn atom increases with lattice constant. High-spin/Low-spin transition is clearly visible between $a = 4.60$ and $a = 4.70$ Å. Experimentally, the magnetic moment is measured to be $\approx 2\mu_B$. Published with kind permission of ©Marcel Bornemann 2018. All Rights Reserved

constitutes the ground state, when the experimental lattice constant is assumed. Yet, by increasing or decreasing the lattice constant the energetic difference can be made smaller. We focus on an increase of the lattice constant rather than a decrease since the system goes into the low-spin state below $a = 4.65$ Å [23] and according to experiment the non-trivial textures exist in the high-spin regime. A crucial transition point is found around $a = 5.0$ Å, where by imposing the 1Q or 3Q state the energy can be made smaller than for the ferromagnetic state. In general, for $a > 5.0$ Å both helical states are favored over the ferromagnetic one. Obviously, an artificial increase of the lattice constant by 0.2 Å ($\approx 4\%$) or more is fairly large. However, probes in experiment are seldom if ever perfectly clean and impurities in the sample need to be considered as a source of error in the final analysis. One potential effect of impurities is chemical pressure that causes a spatial expansion of the lattice structure. An example of the possible effects of positive chemical pressure can be found in Co-doped B20-FeGe [24]. Here, it was experimentally observed that doping can increase the melting temperature and change the magnetic properties of a B20 alloy. In the lower part of Fig. 2, the evolution of the magnetic moment with varying lattice constant is tracked. The resulting magnetic moment for the experimental lattice constant nicely falls on top of the magnetic moment of approximately $2\mu_B$/f.u. which is reported by experimentalists [21]. Furthermore, the high-spin/low-spin transition is recognizable between $a = 4.60$ and $a = 4.70$ Å. It can also be observed that the magnetic moment increases, when the lattice constant is increased. This is a common behaviour which is often observed in metallic systems. For larger lattice constants the magnetic moments of the three different magnetic textures differ more than for the smaller lattice constants. This might be connected to the observation of the differences in the total energy.

We continue our investigation of B20-MnGe by studying the influence of spin-orbit coupling (SOC) enhancement on the magnetic energy landscape. An advantage of the way SOC is implemented in KKRnano is that its contribution is added to the scalar-relativistic potential in a perturbation-like manner. This allows to scale its strength and make it artificially stronger. In this context we perform a comparison of the Bloch point (BP) state and the ferromagnetic (FM) state. The BP texture is defined by means of the four spherical parameters ϕ, θ, Φ and Θ. ϕ and θ designate the position of an **individual atom** in the unit cell which is described by the radius r and the common polar and azimuthal angle

$$\phi = \arctan(y/x) \tag{5}$$

and

$$\theta = \arccos\left(\frac{z}{\sqrt{x^2 + y^2 + z^2}}\right). \tag{6}$$

The BP texture does not depend on r and we can therefore neglect it in the following. Usually, the atomic positions are given in the Cartesian coordinates x, y and z. In the definition above, we define the origin of the coordinate system, i.e., the tuple $(x = 0, y = 0, z = 0)$, to be at the center of the unit cell. In this frame of reference, all atoms that lay in an x-y-plane which intersects with the center are described by $\theta = \pi/2$. The orientation of the **individual atomic magnetic moments** for a BP texture is then defined by the polar angle

$$\Phi = \phi + \phi_1 \tag{7}$$

and the azimuthal angle

$$\Theta = 2\arctan\left(\cot\frac{\theta}{2}\right), \tag{8}$$

where the angles designating the atomic position enter as arguments. ϕ_1 is a phase factor. An illustration of a BP is given in Fig. 3. Note, that in contrast to that illustration we conduct our investigation for a BP with $\phi_1 = \pi$, where magnetic moments are inverted, i.e., all moments point into instead of out of the center.

For our calculations we again use a $6 \times 6 \times 6$ supercell but this time with a $2 \times 2 \times 2$ k-point-mesh and LDA as exchange-correlation functional. Here, we choose LDA because it has been used extensively in all KKR codes in the past and we want to eliminate the possibility of numerical problems that could occur when

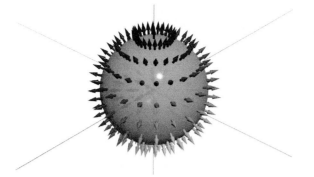

Fig. 3 Illustration of a Bloch point in real space with all magnetic moments pointing out of the center of the Bloch sphere. We use the same magnetic configuration but invert the spin direction so that all moments point into the center. Published with kind permission of ©Nikolai Kiselev 2018. All Rights Reserved

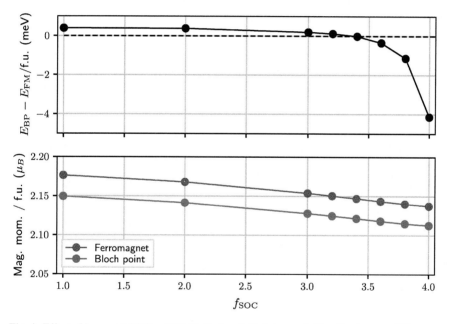

Fig. 4 Effect of increased SOC on B20-MnGe in a 6x6x6 supercell. Top: Total energy difference between (relaxed) Bloch point and (relaxed) ferromagnet. Bottom: Magnetic moment of ferromagnet and Bloch point state. Published with kind permission of ©Marcel Bornemann 2018. All Rights Reserved

SOC is artificially enhanced. The magnetic moments are allowed to relax during the convergence process. This leads to a small canting of the moments which is a known effect in B20 materials [25]. A series of calculations is conducted ranging from the physical value of the SOC to an enhancement of it by a factor of $f_{SOC} = 4.0$ (see Fig. 4). As could be expected from the investigation of 1Q and 3Q state before, the

BP state is energetically not preferred over the FM state for a small scaling of SOC. However, when SOC is scaled further up to $f_{SOC} = 3.5$, both states are energetically more or less equivalent. Above $f_{SOC} = 3.5$ the BP state is clearly preferred over the FM state with an energy difference of up to 4 meV/f.u. Within the parameter range that we checked in this study, the most beneficial scaling value for the BP is found to be $f_{SOC} = 4.0$. The effect of SOC scaling on the magnetic moment can be deemed negligible. Over the whole range it decreases by 0.07 μ_B for each of the two states (see again Fig. 4). It would be interesting to check whether a global minimum of $E_{BP} - E_{FM}$ can be found for $f_{SOC} > 4.0$. This would hint to an optimal scaling at which also other non-collinear magnetic textures, e.g., 1Q spin spiral or 3Q hedgehog lattice, can possibly be stabilized. However, this is not possible as our method becomes increasingly numerically unstable for such strong scaling factors, i.e., the total energy does not converge anymore. The reason for this is currently under investigation.

4 Parallelization Scheme of KKRnano

In order to use the full potential of modern supercomputers, KKRnano features both MPI and OpenMP parallelization. The parallelization concept is sketched in Fig. 5. There are three MPI levels to parallelize over atoms, spins and energy points. OpenMP is used in various parts of the code, mainly to parallelize important loops. The single-cell solver for calculations involving non-collinear magnetism and the TFQMR solver that solves the Dyson equation in Eq. (1) are the parts that potentially benefit the most from OpenMP.

MPI Parallelization Over Atoms

The most crucial MPI level in KKRnano is the one for atoms since application scenarios for KKRnano involve the treatment of a few thousand atoms and the KKR formalism allows us to solve the multiple-scattering problem locally for each atom, if the Green functions of the reference system and the Δt-matrices of the other scattering centers are provided (see Eq. 1). In practice, one MPI task handles 1–16 atoms. The reference Green functions are calculated by each task for the atoms it is responsible for and are then sent to other tasks that require them. The reference t-matrices for the atoms inside the respective reference cluster are not communicated but calculated by each task individually as it takes less time to re-compute them compared to communicating them. In any case, the Δt-matrices of the actual systems need to be communicated. After the necessary information is distributed, the Dyson equation can be solved independently by each task. The calculation of the local charge density can also be conducted locally since only the diagonal $n = n'$-elements are needed for this. In order to subsequently obtain the potential from the local charge moments via the Poisson equation, the moments must be shared with all other atom tasks by means of all-to-all communication. The communication pattern prevents a good

Workflow in KKRnano   **MPI levels**

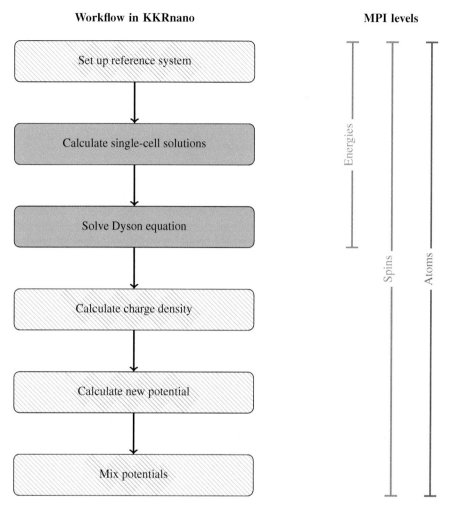

Fig. 5 Schematic representation of MPI and OpenMP parallelization in KKRnano. The most important steps in the KKR workflow are depicted on the left side and the three MPI regions over atoms, spins and energy points are indicated on the right side. Parts filled with blue comprise routines where OpenMP is used and where this can be of high importance to the overall performance while in the striped blue parts OpenMP is used but is less significant. Published with kind permission of ©Marcel Bornemann 2018. All Rights Reserved

scaling behaviour in this part of the code and we currently develop an implementation which works without all-to-all communication.

MPI Parallelization Over Spin Channels

If the system of interest is a collinear magnet, the two spin channels can be handled by two distinct MPI tasks since the magnetic Kohn-Sham equations are separable. Due to the relatively small additional MPI communication effort, this yields an almost ideal speed-up by a factor of 2 in the important multiple-scattering part of the calculation. It should be noted that in the non-collinear KKR formalism such a separation of spin channels is no longer possible because there is intermixing and all operations involving the Δt-matrices and Green functions need to be performed in full spin space, i.e., $\{\uparrow\uparrow, \uparrow\downarrow, \downarrow\uparrow, \downarrow\downarrow\}$. Thus, the spin parallelization level can only be used in connection with a collinear calculation.

MPI Parallelization Over Energy Points

The requirement to calculate the Green function at different energy points offers another possibility to introduce a parallelization level since the values of the Green function $G(\varepsilon_i)$ at energy points ε_i can be obtained in parallel. After the values are obtained, the results must be distributed among the processes within the energy parallelization level to recover the full Green function at all energy points. Assigning one MPI task to each energy point is not a promising concept because of the significantly different runtimes per energy point. Especially the points closest to the Fermi level need more TFQMR iterations. Therefore, the energy points are split into three different groups and each group is taken care of by one MPI task. In the first iteration of the self-consistency cycle the points are equally distributed to all groups. At the end of the first iteration the points are regrouped depending on how much time the iteration took for each point. The aim is to find a grouping for which all groups of energy points are converged at a similar time so that idling is avoided. This is also referred to as *dynamical load-balancing*. A correct load-balancing is of outmost importance to the effectiveness of this MPI level since all tasks need to have finished before the program can move on to the solution of the electrostatic problem, i.e., the Poisson equation. Due to the challenge that load-balancing can pose, parallelization over energy points should only be used, if plenty of processor cores ought to be utilized and the Poisson solver is well-scaling.

OpenMP Parallelization

KKRnano can be compiled either with or without support for OpenMP. If it is enabled, loops primarily in the TFQMR solver but also in the routines that calculate the single-cell solutions are executed using parallel threads. This is particularly useful on architectures that support simultaneous multithreading (SMT). However, we try to use BLAS (Basic Linear Algebra Subprograms) library routines for arithmetic operations, e.g., matrix-matrix multiplications, throughout our code. BLAS libraries usually have their own built-in SMT support. Therefore, the available SMT threads must be partitioned between the explicit OpenMP parallel regions in the code and the implicit parallelization of the BLAS library. Here, the optimal partitioning is highly architecture-dependent and a general recommendation cannot be given. On Hazel Hen the best performance is achieved, when all threads are used by the BLAS library.

Multiple Atoms Per MPI Task

Assigning multiple atoms to one MPI task is beneficial, if very large systems are supposed to be calculated on a comparatively small allocation of compute nodes. KKRnano can treat multiple atoms per atomic MPI task by using the following algebraic scheme: We rewrite the Dyson equation (see Eqs. (1) and (2)) with indices so that it reads

$$\sum_{\mu} \sum_{L'} A_{LL'}^{\nu\mu} X_{L'L''}^{\mu} = b_{LL''}^{\nu}, \tag{9}$$

where ν and μ indicate atomic indices and L, L' and L'' denote the appropriate expansion in angular momentum components. In order to describe the dimensions of the constituents of the equation above, we declare the following parameters: N_{cl} is the number of atoms in the reference cluster that defines the reference Green function. N_{tr} is the number of atoms in the truncation cluster. Only interactions with atoms that lay within the truncation cluster of an atom are considered. This is essential to achieve the linear-scaling of the multiple-scattering problem. All quantities are expanded in spherical harmonics up to l_{max}, which is usually fixed to $l_{max} = 3$. With these definitions we return to Eq. 9, where A is of dimension $N_{cl}(1_{max} + 1)^2 \times N_{tr}(1_{max} + 1)^2$, b usually has the negative local Δt as diagonal elements and X has dimension $N_{tr}(1_{max} + 1)^2 \times (1_{max} + 1)^2$. A is a matrix of blocks of size $(1_{max} + 1)^2 \times (1_{max} + 1)^2$, while X and b are vectors of such blocks. For more than a single atom per task, x and b can also take a matrix form with dimension $N_{cl}(1_{max} + 1)^2 \times N_{loc}(1_{max} + 1)^2$, where N_{loc} is the number of atoms treated by one MPI task of the atom parallelization level. The corresponding linear system of equations reads

$$\sum_{\mu} \sum_{L'} A_{LL'}^{\nu\mu} X_{LL''}^{\mu\gamma}, \tag{10}$$

where an additional index γ is introduced that runs over N_{loc}.

5 Performance of KKRnano on Hazel Hen

In this section we present performance results of KKRnano that were obtained on Hazel Hen. A runtime analysis of the parts that are related to the specific physical problems is given which is followed by an overview of how much time is spent in distinct classes of routines, e.g., library calls and MPI communication. The weak-scaling benchmarks on Hazel Hen are performed for B20-MnGe with the magnetic moments being treated as non-collinear. For this, KKRnano is compiled with the Intel Fortran compiler with optimization level O2 and linked to the Intel Math Kernel (MKL) library which handles all BLAS calls. The runs are conducted with a task distribution of 24 MPI processes per node. The OpenMP threads are reserved for the multi-threaded MKL library. We choose two atomic supercells containing 1728

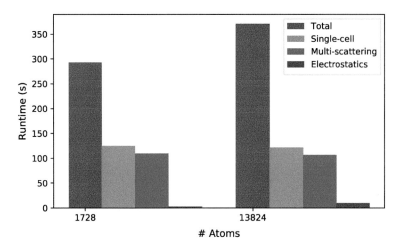

Fig. 6 Runtime of a single DFT self-consistency step for different B20-MnGe supercell sizes with KKRnano. The total runtime and the individual runtimes for the single-cell solver, the multiple-scattering solver and the electrostatics solver are given. Published with kind permission of ©Marcel Bornemann 2018. All Rights Reserved

and 13824 atoms and perform one self-consistency step for each system. We solely activate the MPI parallelization level over atoms and do not use the ones over spins and energies. Then, 72 and 576 nodes are needed for the two benchmark calculations, since each MPI task is assigned to one atom. Each MPI process reads four shared binary direct-access files before commencing the self-consistency step. Two of them are index-files of the size of a few hundred kByte while the other two can grow with system size up to several GByte. The combined size of all files that are read in is roughly $N \times 0.5$ MBytes, where N is the number of atoms. After the single self-consistency step the potential is written out. This and the read-in of the four files mentioned above accounts for most of the time spent in I/O. Figure 6 shows the total runtimes of a single self-consistency iteration and how much time is spent on solving the individual physical problems for the two supercells containing 1728 and 13824 atoms, respectively. The remaining time can be mainly attributed to Fortran direct-access I/O which does not scale well on a Lustre file system. The observation of an increase in total runtime for larger systems is a strong hint to this as the amount of file accesses is proportional to the number of MPI processes. The implementation of a more suitable I/O library (e.g., MPI I/O or SIONlib) is likely to solve this issue. The multiple-scattering solver, i.e., the TFQMR solver, and the solution of the relativistic single-cell problem are expected to account for most of the computational work in KKRnano which is why linear scaling is of particular importance in these parts of the code. Our results show that both parts scale indeed linearly, i.e., the runtime stays constant as long as the ratio of CPU cores to atoms remains constant. Furthermore, it should be noted that the walltime that is needed for a single self-consistency step of a system that includes 1728 atoms takes less than five minutes

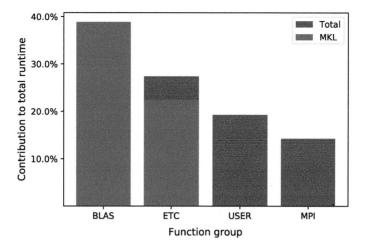

Fig. 7 Runtime results from Cray Performance Measurement Tool for a single KKRnano self-consistency step for a $6 \times 6 \times 6$ supercell of B20-MnGe (1728 atoms). 72 Hazel Hen nodes are used with 1728 MPI processes. The runtimes are divided into four subgroups which are described in the text. The fraction of the total runtime in each subgroup that is spent in routines of Intel's highly-optimized MKL library is highlighted in blue. Published with kind permission of ©Marcel Bornemann 2018. All Rights Reserved

which is a good value, especially when considering that the magnetic moments are allowed to be non-collinear. In the KKR formalism, the electron density and the potential are connected via the Poisson equation. For very large systems the Poisson solver contributes considerably to the overall runtime [11]. This is expected since the algorithm used in KKRnano scales quadratically with the number of atoms. Its impact when mid-sized systems are investigated is negligible. However, there are ideas on how to restructure the Poisson solver towards a more favorable scaling behavior. This is planned for the near future.

Next, we conduct a performance analysis on the level of individual routines. The Cray Performance Tools package provides a helpful collection of tools to gather data for this purpose. It allows to sample the time that is spent within each routine and can therefore give a good impression of which routines are performance-critical and should be considered for further optimization. Figure 7 shows the sampling results for a $6 \times 6 \times 6$ supercell calculation of B20-MnGe using 1728 MPI tasks on 72 nodes. The routines are grouped according to their origin. The *BLAS* group comprises the kernels of library-based linear algebra operations. All external routines for which no source file was given at compile time are subsumed in *ETC*. Such routines are e.g. Fortran I/O routines and MKL helper routines that are indispensable for BLAS but are not directly related to the MKL kernel. The *USER* group accounts for all routines from the KKRnano source code, i.e., our self-written routines. Finally, *MPI* contains all the routines that are needed for communication and synchronization between the MPI tasks. Performing linear algebra operations accounts for almost 40% of the total runtime. Prior tests have shown that Intel's MKL outperforms Cray's

LibSci and therefore KKRnano is linked to the former. The routines that fall into the *ETC* category consume 28% of the total runtime. The MKL routines dominate this group and it can be stated that the ratio of MKL in the total runtime is 62%. This is a remarkable value as it indicates that more than half of the walltime is spent in routines that are already highly optimized. The only group of routines that can benefit from source code tuning is the *USER* group. However, a hypothetical performance improvement by a factor of two would lead to an overall performance increase of merely 10% as the share of this group in the total runtime is only 20%. MPI routines are not crucial to performance as they amount to only 15% of the total computational effort.

6 Conclusions

In this report we introduced an extended Korringa-Kohn-Rostoker method as the foundation of our large-scale Density Functional Theory application KKRnano. Consecutively, we presented results that were obtained on Hazel Hen by applying KKRnano to the material B20-MnGe. Those results are promising and suggest further investigation of complex magnetic textures in this particular compound. The high-clocked Intel Xeon processors of Hazel Hen make a single DFT self-consistency step for a supercell of 1728 atoms feasible within less than five minutes and hence allows one to obtain results in a timely manner.

During the *Optimization of Scaling and Node-level Performance on Hazel Hen* workshop at HLRS in 2018 we were given the chance to further optimize our code and were able to identify a computational bottleneck. The performance of KKRnano on Hazel Hen was significantly improved by resolving it. The encountered problem was connected to unnecessary copying of a large complex-valued array, when the magnetic moments are allowed to be non-collinear. All results in Sect. 5 were obtained with this optimized version. Further optimization for standard calculations with a few thousand atoms is difficult since only 20% of the runtime is spent in hand-written code. The remaining 80% is spent in library calls that are already highly tuned. These routines mainly deal with MKL-BLAS and MPI communication.

Acknowledgements The authors gratefully acknowledge the Gauss Centre for Supercomputing e.V. (www.gauss-centre.eu) for funding this project by providing computing time through the project GCS-KKRN on the GCS Supercomputer Hazel Hen at Höchstleistungsrechenzentrum Stuttgart (HLRS).

References

1. R. Zeller, J. Phys.: Condens. Matter **20**, 294215 (2008)
2. A. Thiess, R. Zeller, M. Bolten, P.H. Dederichs, S. Blügel, Phys. Rev. B **85**, 235103 (2012)
3. M. Bornemann, Large-scale Investigations of Non-trivial Magnetic Textures in Chiral Magnets with Density Functional Theory. PhD thesis, RWTH Aachen University, Aachen. Submitted

4. T. Tanigaki, K. Shibata, N. Kanazawa, X. Yu, Y. Onose, H.S. Park, D. Shindo, Y. Tokura, Nano Lett. **15**, 5438 (2015)
5. F.N. Rybakov, A.B. Borisov, S. Blügel, N.S. Kiselev, New J. Phys. **18**, 045002 (2016)
6. M. Bornemann, P.F. Baumeister, R. Kovacik, B. Zimmermann, P. Mavropoulos, S. Lounis, N. Kiselev, P.H. Dederichs, R. Zeller, S. Blügel, (Frühjahrstagung der Deutschen Physikalischen Gesellschaft, Dresden (Germany), 19 Mar 2017 - 24 Mar 2017)
7. M. Bornemann, S. Grytsiuk, P.F. Baumeister, P. Mavropoulos, N. Kiselev, S. Lounis, R. Zeller, S. Blügel, (Frühjahrstagung der Deutschen Physikalischen Gesellschaft, Berlin (Germany), 11 Mar 2018 - 16 Mar 2018)
8. R.W. Freund, N.M. Nachtigal, Numer. Math. **60**, 315 (1991)
9. R. Zeller, J. Deutz, P. Dederichs, Solid State Commun. **44**, 993 (1982)
10. R. Zeller, P.H. Dederichs, B. Újfalussy, L. Szunyogh, P. Weinberger, Phys. Rev. B **52**, 8807 (1995)
11. D. Brömmel, W. Frings, B.J.N. Wylie, JUQUEEN Extreme Scaling Workshop 2017. Technical report (2017)
12. E. Feldtkeller, IEEE Trans. Magn. **53**, 1 (2017)
13. N. Kanazawa, S. Seki, Y. Tokura, Adv. Mater. **29**, 1603227 (2017)
14. N. Kanazawa, Y. Onose, T. Arima, D. Okuyama, K. Ohoyama, S. Wakimoto, K. Kakurai, S. Ishiwata, Y. Tokura, Phys. Rev. Lett. **106**, 156603 (2011)
15. N. Kanazawa, J.H. Kim, D.S. Inosov, J.S. White, N. Egetenmeyer, J.L. Gavilano, S. Ishiwata, Y. Onose, T. Arima, B. Keimer, Y. Tokura, Phys. Rev. B **86**, 134425 (2012)
16. S.V. Grigoriev, N.M. Potapova, S.A. Siegfried, V.A. Dyadkin, E.V. Moskvin, V. Dmitriev, D. Menzel, C.D. Dewhurst, D. Chernyshov, R.A. Sadykov, L.N. Fomicheva, A.V. Tsvyashchenko, Phys. Rev. Lett. **110**, 207201 (2013)
17. N. Martin, M. Deutsch, F. Bert, D. Andreica, A. Amato, P. Bonfà, R. De Renzi, U.K. Rößler, P. Bonville, L.N. Fomicheva, A.V. Tsvyashchenko, I. Mirebeau, Phys. Rev. B **93**, 174405 (2016)
18. A. Fert, N. Reyren, V. Cros, Nat. Rev. Mater. **2**, 17031 (2017)
19. E. Altynbaev, S.A. Siegfried, E. Moskvin, D. Menzel, C. Dewhurst, A. Heinemann, A. Feoktystov, L. Fomicheva, A. Tsvyashchenko, S. Grigoriev, Phys. Rev. B **94**, 174403 (2016)
20. T. Koretsune, N. Nagaosa, R. Arita, Sci. Rep. **5**, 13302 (2015)
21. A. Yaouanc, P. Dalmas de Réotier, A. Maisuradze, B. Roessli, Phys. Rev. B **95**, 174422 (2017)
22. O.L. Makarova, A.V. Tsvyashchenko, G. Andre, F. Porcher, L.N. Fomicheva, N. Rey, I. Mirebeau, Phys. Rev. B **85**, 205205 (2012)
23. U.K. Rößler, J. Phys.: Conf. Ser. **391**, 012104 (2012)
24. M.J. Stolt, X. Sigelko, N. Mathur, S. Jin, Chem. Mater. **30**, 1146 (2018)
25. V.A. Chizhikov, V.E. Dmitrienko, Phys. Rev. B **88**, 214402 (2013)

A Multi-resolution Approach to the Simulation of Protein Complexes in a Membrane Bilayer

Goutam Mukherjee, Prajwal Nandekar, Ghulam Mustafa, Stefan Richter and Rebecca C. Wade

Abstract We describe a transferable multiresolution computational approach to build and simulate complexes of two proteins—cytochrome P450 (CYP) and CYP reductase (CPR)—in a membrane bilayer using Brownian dynamics (BD) and all-atom molecular dynamics (MD) simulations. Our benchmarks showed that MD simulations of these systems could be carried out efficiently with up to 180 nodes (4320 cores) using NAMD version 2.12. Our results provide a basis for defining the ensemble of electron transfer-competent arrangements of CYP-CPR-membrane complexes and for understanding differences in the interactions with CPR of different CYPs, which have implications for CYP-mediated drug metabolism and the exploitation of CYPs as drug targets. This work was carried out in the DYNATHOR (DYNAmics of THe complex of cytOchrome P450 and cytochrome P450 Reductase in a phospholipid bilayer) project at HLRS.

1 Introduction

The complexity of complexes between proteins in membrane bilayers presents challenges for molecular simulation studies aimed at characterizing these complexes structurally, dynamically and functionally. Some of these challenges are the large size (many thousands of atoms) of many complexes, the heterogeneity (protein, lipid, cofactors, water, ions), as well as the scarcity of experimental structural information on protein-membrane interactions. Here, we describe how we have addressed these

G. Mukherjee · P. Nandekar · G. Mustafa · S. Richter · R. C. Wade
Molecular and Cellular Modeling Group, Heidelberg Institute for Theoretical Studies (HITS), Heidelberg, Germany

G. Mukherjee · R. C. Wade
Interdisciplinary Center for Scientific Computing (IWR), Heidelberg University, Heidelberg, Germany

P. Nandekar · R. C. Wade (✉)
Center for Molecular Biology (ZMBH), Heidelberg University, Heidelberg, Germany
e-mail: Rebecca.wade@h-its.org

W. E. Nagel et al. (eds.), *High Performance Computing in Science and Engineering '18*, https://doi.org/10.1007/978-3-030-13325-2_32

challenges in simulations of cytochrome P450 (CYP) enzymes and the complexes they form in a membrane with CYP reductase.

Xenobiotic metabolism can produce metabolites with physico-chemical and biological properties that differ substantially from those of the parent compound [1, 2]. Cytochrome P450 (CYP) is the cardinal xenobiotic-metabolizing superfamily of enzymes. Although the human genome contains 57 CYPs, six CYP isozymes are of particular importance to drug metabolism (CYP3A4, CYP2D6, CYP2C9, CYP2C19, CYP2E1 and CYP3A5), with about ten further CYPs playing a significant role [1–3]. Other isoforms, such as CYP1A1, play important roles in bioactivation of procarcinogens and detoxification of carcinogens [4], and CYP17 is responsible for androgen metabolism and plays a role in prostate and breast cancers [5]. CYP-mediated bio-transformation reactions are mainly oxidations. CYPs are heme-containing mono-oxygenases which activate one molecule of dioxygen and oxidize the substrate by inserting one oxygen atom into the substrate (e.g. R–H) to form the product (e.g. R–OH) and reducing the other oxygen atom to water. In the catalytic cycle, CYP requires two electrons to be transferred from a redox partner to its heme co-factor [6]. For the microsomal CYPs, the redox partner is the NADPH-dependent cytochrome P450 reductase (CPR), a flavoprotein containing three cofactor binding domains: the FMN domain, with the flavin mononucleotide (FMN) cofactor, the FAD domain with the flavin adenine dinucleotide (FAD) cofactor, and the NADP domain with the nicotinamide adenine dinucleotide phosphate (NADP) cofactor. In humans, both CYP and CPR are anchored in the endoplasmic reticulum (ER) membrane by a trans-membrane (TM) helix (Fig. 1a). Electron transfer (ET) from CPR to CYP requires association of the proteins and step-wise transfer of the two electrons from the FMN cofactor of CPR to the heme moiety of CYP. The association is mainly driven by electrostatic interactions with the positively charged proximal side of CYP interacting with the negatively charged FMN domain of CPR (Fig. 1b). However, so far, no crystal structure has been solved of a complex of the full length CYP and CPR, and little is known at the atomic level about interactions between CYPs and CPR, their behavior in the membrane, possible conformational changes of the proteins upon binding or the electron transport mechanism during the catalytic cycle.

Some modeling and simulation studies have been carried out on the basis of the crystal structure of the complex of the P450-BM3 CYP and FMN domains [7] which provides a template for a possible CYP: CPR binding arrangement [8]. MD simulation studies performed on a modelled complex of CYP2D6 and the open form of CPR in a bilayer provide an idea of the possible orientations of CPR with respect to CYP [9]. However, the complex was built manually and the time scale of the simulation (10 ns) was too short to investigate the dynamics of the complex in the membrane. Models of the CYP17-cytochrome b_5 (cyt b5) [10] and CYP2B4-cyt b5 [11] complexes built using solid-state NMR and mutation data have suggested involvement of cationic CYP residues in redox partner binding. All these models lack information on residue pair-wise interactions and TM domain interactions. It is known, however, that the TM domain of cyt b5 (another redox partner protein), plays a crucial role in formation of the protein-protein complex [12]. Solid-state NMR studies revealed the tilt of the CPR TM helix (~13°) in a bilayer but its interaction properties

(a) **(b)**

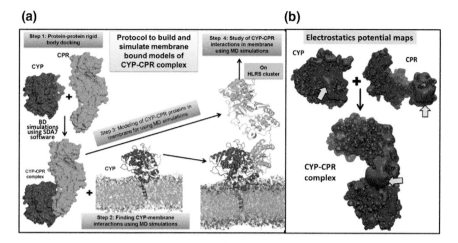

Fig. 1 **a** Schematic illustration of protocol to build and simulate a CYP-CPR complex in a membrane. **b** Molecular electrostatic potential maps of CYP and CPR proteins showing a highly positive (blue isopotential contours) patch on the proximal face of the CYP and a highly negative (red isopotential contours) patch on the FMN domain of the CPR that interact in the complex

are unknown [13]. Experimental characterization of the full length structures of complexes of CYPs and CPR is hampered by the difficulties of doing experiments with membrane-bound proteins in their native membrane-bound form [9, 14].

Here, we aim to use molecular simulation techniques to achieve a detailed understanding of the structure and dynamics of membrane associated CYP-CPR complexation, and consequently the determinants of electron transfer for the different CYPs for which CPR is the redox partner. We used multiresolution computational techniques similar to those that we developed previously to model CYPs in a phospholipid bilayers [15–18]. We have previously successfully built membrane-bound models of various CYPs, including CYP51 (human and T. brucei) [18], human CYP2C9 [15], CYP3A4 [16], CYP1A2 [16], CYP17A1 and CYP19A1 [19], and investigated the factors governing CYP insertion in the membrane. We have also developed and reported a systematic protocol to build and simulate a CYP-CPR-membrane complex using BD and all-atom (AA) MD simulations (Fig. 1a) [17]. Here, we describe the simulation of CYP1A1-CPR-membrane complexes.

2 Methods

The CYP and CPR crystal structures were retrieved from the Protein Data Bank (www.rcsb.org/pdb) [20]. The missing residues in the proteins were added using Modeller software version 9.10 (salilab.org/modeller/). Then, SDA7 [21–23] (mcm. h-its.org/sda/) was used to perform rigid-body BD simulations of the diffusional association of the globular domains of CYP and CPR in a continuum aqueous solvent. BD rigid body docking to CPR was performed for all chains of CYP in the asymmetric unit of the crystal structure. SDA is particularly useful for predicting electrostatically driven protein-protein complexes. A survey of biological ET for many redox proteins showed that the distances between redox centers typically lie within 14–15 Å [24]. Hence, during BD simulations, two constraints were imposed: (i) cofactor distance, i.e.; N_2 of FMN in CPR to Fe of Heme in CYP < 20 Å, and (ii) the distance between the centers of mass of the globular domain of CYP and the FMN domain of CPR < 60 Å. For each system, 60,000 trajectories were generated, and diffusional encounter complexes satisfying the criteria saved and clustered. ET rates for the complexes were predicted using Pathways, a plugin for VMD [25]. 6 structures representing the diversity of feasible CYP-CPR arrangements were selected and subjected to refinement by AA MD simulation in explicit solvent (with 0.15 M NaCl) using AMBER v14 (ambermd.org) for 40–100 ns. Initially, to reduce computational cost, only the globular domain of CYP and the FMN domain of CPR were used for "soluble simulations". Of the 6 CYP1A1-FMN domain complexes simulated, three maintained an arrangement suitable for ET and these were used to build membrane bound complexes of CYP-CPR by superimposing the MD-refined CYP1A1-FMN domain complexes onto a previously generated model of membrane-bound CYP1A1. The missing trans-membrane helix and linker residues of CPR were then modelled and inserted into the membrane. These three models of membrane-bound CYP1A1-CPR complexes were then subjected to AA MD simulations with the AMBER ff14SB force field for protein, Lipid14 for the membrane, GAFF for cofactors and substrate, TIP3P water molecules and 0.15 M NaCl. All MD simulations were run for at least 200 ns using the NAMDv2.12 software package on the HLRS CRAY XC40 cluster (hazelhen). Each AA CYP-CPR-membrane system contained roughly ~0.55 million atoms. The computational flow chart implemented in the present study is depicted in Fig. 2.

3 Results and Discussion

3.1 Benchmarks

The performance of the NAMD 2.12 software for atomic detail MD simulation of the protein complexes in a membrane bilayer was significantly improved over NAMD version 2.10. Performance increased approximately linearly up to 4000 cores but not beyond, see Fig. 3. This leveling off of the performance is likely due to the small number of atoms allocated to each core for this system size of ca. 550000 atoms (ca. 110 for 5000 cores), meaning considerable time for communication with neighboring cores and, due to overlapping regions, considerable duplicate work. Subsequent simulations were performed with around 180 nodes (4320 cores).

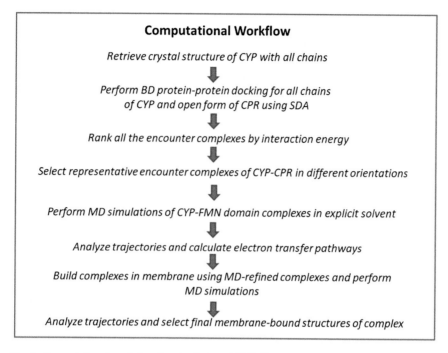

Fig. 2 Computational work flow for simulation of CYP-CPR complexes in a membrane bilayer

Fig. 3 Scaling of a NAMD 2.12 molecular dynamics simulation of a CYP-CPR complex in a membrane bilayer

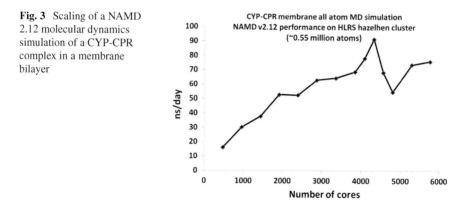

3.2 CYP1A1-CPR Encounter Complexes from BD Docking

The diffusional encounter complexes generated by BD simulations show binding driven by electrostatic interactions between basic residues of CYP and acidic residues of CPR (Fig. 1b). Three complexes had non-negligible computed ET rates ranging from about 40–21500 s^{-1}. These three complexes displayed three different orientations of the CPR FMN domain with respect to CYP1A1 although the interacting patch on CYP1A1 was similar in all cases (Fig. 4a, c, e). These complexes were used for subsequent MD simulations of full length complexes in a bilayer.

3.3 MD Simulations of Membrane-Bound CYP1A1-CPR Complexes

The MD simulations of the protein complexes in a membrane bilayer were carried out at HLRS (see Sect. 3.1). During MD simulations, the individual domains of the proteins retained their stable initial structures (Cα atom root mean square deviation (RMSD) values of ~1–2Å).

While one of the simulations (for system Ccl2; Fig. 4e) showed a very stable complex, a significant conformational rearrangement of the CYP globular domain with respect to the membrane due to interactions with the CPR was observed in the

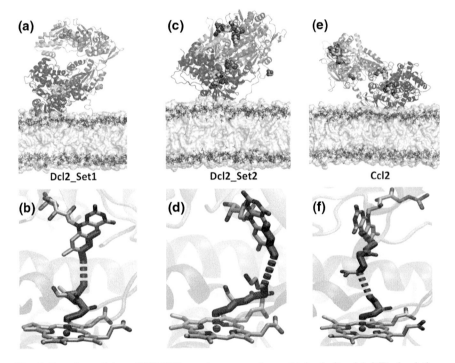

Fig. 4 Snapshots of three CYP-CPR-membrane complexes obtained after AA MD simulations (A, C, E) and predicted ET pathways (in red) in the respective complexes (B, D, F). The proteins are shown in ribbon representation: red: CYP, CPR: purple: FMN domain, orange: FAD domain, cyan: NADP domain. Cofactors are shown in yellow carbon spheres above and cyan carbon stick representation below. CPR residue Glu66 is shown above in green carbon spheres to demonstrate the different relative orientations of the two proteins

other two simulations (for system Dcl2) (Fig. 4a, c). One of the reasons for this conformational change was the weakening of CYP interactions with the membrane, facilitating reorientation of the CYP1A1 globular domain above the membrane. There was some reorientation of the FMN domain of CPR, altering the distance between the proteins and their ET reaction centers.

In the initial models, the globular domains of CPR were relatively far from the membrane. During the simulations the NADP domain came closer to CYP1A1 due to changes in the highly flexible linker loop between the FMN and FAD domain. The motion of the NADP domain shifted the CPR conformation from an open conformation towards a closed conformation but these motions had no significant impact on the interactions between CYP1A1 and the FMN domain of CPR.

3.4 Electron Transfer Pathways in Membrane-Bound CYP1A1-CPR Complexes

The FMN domain of CPR binds at the proximal face of CYP1A1 with the association governed by electrostatic interactions. The positively charged residues present on the proximal surface of CYP1A1 showed polar interactions with the negatively charged residues present near the FMN co–factor of the FMN domain. Comparison of the complexes for the three simulations showed a few common residues that may be crucial for CYP1A1-CPR interactions. The electron transfer pathways were predicted using Beratan's model for calculating the electron transfer donor-to-acceptor coupling constant (T_{DA}) [25] from which the rate of electron transfer (K_{ET}) could be computed using Eq. 1. For this purpose, the driving force, ΔG^0, for ET between the CPR FMN and the CYP heme was assigned as -0.129 eV from redox potential values of -0.140 and -0.269 eV for the high spin Fe^{III}/Fe^{II} porphyrin complex in CYP3A4 [26] and $FMN_{sq/hq}$ in human CPR [27], respectively. The reorganization energy (λ) for electron transfer was considered to be 0.70 eV [28].

$$k_{ET} = \frac{2\pi}{h} \frac{exp\left[-(\Delta G + \lambda)^2/4\lambda k_B T\right]}{\sqrt{4\pi \lambda k_B T}} |T_{DA}|^2 \tag{1}$$

It was observed that residues involved in ET pathways from the FMN cofactor of CPR to the heme of CYP changed upon refinement of the structures as the distance between reaction centers adjusted. The most favorable ET pathway for the Ccl2 system was via a glycine and a glutamic acid residue of CPR (Fig. 4e, f) and gives a computed ET rate of 62 s^{-1} that agrees well with the experimentally determined rate for other human CYPs with CPR [29–31]. For the Dcl2 system, the most favorable ET pathways were directly from the cysteine coordinating the heme and pathway gave a much high ET rate at a distance of about 15 Å (Fig. 4a, b). Note that a higher value than observed experimentally may be expected for the first electron transfer as the second electron transfer is considered rate-limiting. Together, these observations suggest that several ET competent CYP1A1-CPR complexes may exist with different orientations and interactions between the domains containing the reaction centers.

4 Conclusions and Future Perspectives

A transferable multiresolution computational approach using a combination of BD and all-atom MD simulations, both in the absence and presence of a phospholipid bilayer, was applied to modelling and simulating CYP-CPR interactions. The mechanism of electron transfer between CPR and CYP was studied using an empirical approach. An important new finding is the observation that the FMN domain of CPR may interact with a CYP in an ensemble of electron-competent binding modes with a redox center-to-center distance of ~14.5–17 Å in the membrane-bound CYP-CPR complex. The CYP-CPR interaction is governed by electrostatic interactions. Additionally, our study showed conformational rearrangements of CYP1A1 on the membrane in the presence of CPR, which might affect the substrate access and product egress tunnels. Furthermore, CPR undergoes a large degree of conformational change which resulted in transient, non-specific interactions between the NADP domain of CPR and the globular domain of CYP1A1.

We are currently extending these simulations to gain further configurational sampling of CYP1A1-CPR complexes. In further work, we will investigate whether differences in CYP-membrane interactions observed for different CYPs persist in the complexes with CPR. This will allow us to investigate whether the electron transfer necessary for CYP function is affected by specific differences in sequence that affect CYP-membrane interactions as well as to investigate the sequence dependence of the CYP-CPR interactions.

Acknowledgements We gratefully acknowledge the financial support of the Klaus-Tschira Foundation, Heidelberg University Frontiers Innovation Fund, the BIOMS Center for Modelling and Simulation in the Biosciences (Go.M.), and the German Academic Exchange Service (Gh.M., P.N.). Finally, we thank HLRS for providing computing time for the DYNATHOR project.

References

1. J. Kirchmair, A.H. Göller, D. Lang, et al., Predicting drug metabolism: experiment and/or computation? Nat. Rev. Drug Discov. **14**, 387–404 (2015)
2. J. Kirchmair, A. Howlett, J.E. Peironcely, et al., How do metabolites differ from their parent molecules and how are they excreted? J. Chem. Inf. Model. **53**, 354–367 (2013)
3. L. Olsen, C. Oostenbrink, F.S. Jørgensen, Prediction of cytochrome P450 mediated metabolism. Adv. Drug Deliv. Rev. **86**, 61–71 (2015)
4. V.P. Androutsopoulos, A.M. Tsatsakis, D.A. Spandidos, Cytochrome P450 CYP1A1: wider roles in cancer progression and prevention. BMC Cancer **9**, 187 (2009)
5. R.D. Bruno, V.C.O. Njar, Targeting cytochrome P450 enzymes: a new approach in anti-cancer drug development. Metab. Clin. Exp. **15**, 5047–5060 (2007)

6. F.P. Guengerich, Mechanisms of cytochrome P450 substrate oxidation: MiniReview. J. Biochem. Mol. Toxicol. **21**, 163–168 (2007)
7. I.F. Sevrioukova, H. Li, H. Zhang et al., Structure of a cytochrome P450-redox partner electron-transfer complex. Proc. Natl. Acad. Sci. USA **96**, 1863–1868 (1999)
8. K.K. Dubey, S. Shaik, Choreography of the reductase and P450BM3 domains toward electron transfer is instigated by the substrate. J. Am. Chem. Soc. **140**, 683–690 (2018)
9. A. Sündermann, C. Oostenbrink, Molecular dynamics simulations give insight into the conformational change, complex formation, and electron transfer pathway for cytochrome P450 reductase. Protein Sci. **22**, 1183–1195 (2013)
10. D.F. Estrada, A.L. Skinner, J.S. Laurence, E.E. Scott, Human cytochrome P450 17A1 conformational selection: modulation by ligand and cytochrome b_5. J. Biol. Chem. **289**, 14310–14320 (2014)
11. S. Ahuja, N. Jahr, S.-C. Im et al., A model of the membrane-bound cytochrome b5-cytochrome P450 complex from NMR and mutagenesis data. J. Biol. Chem. **288**, 22080–22095 (2013)
12. K. Yamamoto, U.H.N. Dürr, J. Xu et al., Dynamic interaction between membrane-bound full-length cytochrome P450 and cytochrome b5 observed by solid-state NMR spectroscopy. Sci. Rep. **3**, 2538 (2013)
13. R. Huang, K. Yamamoto, M. Zhang et al., Probing the transmembrane structure and dynamics of microsomal NADPH-cytochrome P450 oxidoreductase by solid-state NMR. Biophys. J. **106**, 2126–2133 (2014)
14. P. Jeřábek, J. Florián, V. Mart'inek, Lipid molecules can induce an opening of membrane-facing tunnels in cytochrome P450 1A2. Phys. Chem. Chem. Phys. **18**, 30344–30356 (2016)
15. V. Cojocaru, K. Balali-Mood, M.S.P. Sansom, R.C. Wade, Structure and dynamics of the membrane-bound cytochrome P450 2C9. PLoS Comput. Biol. **7**, e1002152 (2011)
16. G. Mustafa, P.P. Nandekar, X. Yu, R.C. Wade, On the application of the MARTINI coarse-grained model to immersion of a protein in a phospholipid bilayer. J. Chem. Phys. **143**, 243139 (2015)
17. X. Yu, D.B. Kokh, P. Nandekar, G. Mustafa, S. Richter, R.C. Wade, Dynathor: dynamics of the complex of cytochrome P450 and cytochrome P450 reductase in a phospholipid bilayer, in *High performance computing in science and engineering '15*, ed. by W. Nagel, D. Kröner, M. Resch. (Springer, Cham, 2016)
18. X. Yu, P. Nandekar, G. Mustafa, V. Cojocaru, G.I. Lepesheva, R.C. Wade, Ligand tunnels in T. brucei and human CYP51: insights for parasite-specific drug design. BBA-Gen. Sub. **1860**, 67–78 (2016)
19. G. Mustafa, P.P. Nandekar, T.J. Camp, N.J. Bruce, M.C. Gregory, S.G. Sligar, R.C. Wade, Influence of transmembrane helix mutations on cytochrome P450-membrane interactions and function. Biophys. J. **143**, 419–432 (2019)
20. H.M. Berman, J. Westbrook, Z. Feng, G. Gilliland, T.N. Bhat, H. Weissig, I.N. Shindyalov, P.E. Bourne, The Protein Data Bank. Nuc. Acids Res. **28**, 235–242 (2000)
21. R.R. Gabdoulline, R.C. Wade, Simulation of the diffusional association of barnase and barstar. Biophys. J. **72**, 1917–1929 (1997)
22. R.R. Gabdoulline, R.C. Wade, Brownian dynamics simulation of protein-protein diffusional encounter. Methods **3**, 329–341 (1998)
23. M. Martinez, N.J. Bruce, J. Romanowska, D.B. Kokh, M. Ozboyaci, X. Yu, M.A. Öztürk, S. Richter, R.C. Wade, SDA7: a modular and parallel implementation of the simulation of diffusional association software. J. Comput. Chem. **36**, 1631–1645 (2015)
24. A.W. Munro, et. al., P450 BM3: the very model of a modern flavocytochrome. Trends Biochem. Sci. **27**, 250–257 (2002)
25. I.A. Balabin, X. Hu, D.N. Beratan, Exploring biological electron transfer pathway dynamics with the pathways plugin for VMD. J. Comput. Chem. **33**, 906–910 (2012)
26. A. Das, Y.V. Grinkova, S.G. Sligar, Redox potential control by drug binding to cytochrome P450 3A4. J. Am. Chem. Soc. **129**, 13778–13779 (2007)
27. A. Das, S.G. Sligar, Modulation of the cytochrome P450 reductase redox potential by the phospholipid bilayer. Biochemistry **48**, 12104–12112 (2009)

28. C.C. Page, C.C. Moser, X. Chen, P.L. Dutton, Natural engineering principles of electron tunnelling in biological oxidation-reduction. Nature **402**, 47–52 (1999)
29. H. Zhang, S.-C. Im, L. Waskell, Cytochrome b5 increases the rate of product formation by cytochrome P450 2B4 and competes with cytochrome P450 reductase for a binding site on cytochrome P450 2B4. J. Biol. Chem. **282**, 29766–29776 (2007)
30. F.P. Guengerich, W.W. Johnson, Kinetics of ferric cytochrome P450 reduction by NADPH—cytochrome P450 reductase: rapid reduction in the absence of substrate and variations among cytochrome P450 systems. Biochem. **36**, 14741–14750 (1997)
31. Y. Farooq, G.C.K. Roberts, Kinetics of electron transfer between NADPH-cytochrome P450 reductase and cytochrome P450 3A4. Biochem. J. **432**, 485–494 (2010)